A CITIZEN'S DISCLOSURE ON UFOS AND ETI

VOLUME TWO

UFO DISCLOSURE AND COVERT PROGRAMS OF DECEPTION

TERENCE M. TIBANDO

A CITIZEN'S DISCLOSURE ON UFOS AND ETI

VOLUME TWO

UFO DISCLOSURE AND COVERT PROGRAMS OF DECEPTION

Copyright Page

In writing this book, I sought out the best possible evidence available on this subject whether that was from numerous UFO and ETI related books, networking with other UFO authors, researchers and first-hand witnesses to UFO sightings, from films and TV documentaries, from internet searches, or just from my personal sightings and contact experiences.

When material is quoted in this book full acknowledgement is given to the author or source of that material as indicated by the extensive bibliography, webliography and videography at the back of the book.

When photographic images are used in this book that are obtained from the internet, usually from Google Images, a full search was made to determine copyright information, the author's name, or address or email address or phone number or copyright mark in order to asked permission to use their photographs. In almost all cases where such images are posted to the internet, there was no satisfactory way to identify the owner of the image even through Google because they left no identification of themselves to be found. When an author's name does appear on an image, written permission was sought or it was not used at all; more often than not, there usually was no reply or response back from the owner.

This lack of due diligence to place a copyright mark or the owner's name is all it would take for that person to claim ownership of a picture, yet the lack of it creates major problems for many people, especially for other authors who lawfully seek their permission to use their photo images.

Because this book is one of six volumes in a series created as public educational material and is of a transitional nature, and I have quoted or referenced the websites from where I obtained the photo images and therefore, I am invoking the Fair Use Doctrine also known as Fair Usage Clause to publish these images in my book.

I will of course give full acknowledgement and credit to the author's and owners of such images in all my future book publications in recognition of their work if they come forward to be identified.

With this stated, this book and all future books in this series are copyrighted and cannot be reproduced in whole or part, in any form or by any means electronic or mechanical, including photocopying, recording, or by any information storage and retrieval system now known or hereafter invented, without the expressed permission of its author.

iv

A CITIZEN'S DISCLOSURE ON UFOS AND ETI

VOLUME TWO

UFO DISCLOSURE AND COVERT PROGRAMS OF DECEPTION

TERENCE M. TIBANDO

"Hggna"

A Cosmic Cousin
Publication

Other Publications by the Author Page

Although, this book is the author's first publication it forms a part of six smaller books or volumes that was originally written as one massive tome of UFO and ETI information entitled: **"A Citizen's Disclosure On UFOs And ETI Visiting The Earth"** which began in March 2009 and was completed in August 2016. All volumes are written but some are waiting a printing date.

Other books/volumes by the author in this series:

1. Book One (Volume One): **"Global Evidence of the UFO and ETI Presence"** **(currently available)**

2. Book Two (Volume Two): **"UFO Disclosure and Covert Programs of Deception"**

3. Book Three (Volume Three): **"Military Intelligence Industrial Complex, USAPs and Covert Black Projects"**

4. Book Four (Volume four): **"In Search of Extraterrestrial Intelligence"**

5. Book Five (Volume Five): **"Evidence of a Type Two ET Civilization in Our Solar System"**

6. Book Six (Volume Six): **"The Rosetta Stone of ETI Contact and Communications"**

Introduction

You hold in your hands the second volume of "A Citizen's Disclosure on UFOs and ETI" which was originally written as a 3500 page encyclopedic tome. It provides the latest and best evidence on this subject matter that comes from my 60 plus years of experience dating back to the early '50s when I first saw a flying saucer hover over a Canadian Air Force Base in St. Jeans, Quebec. This was followed by an ET visitation by three ghost-like beings that came into my bedroom a couple of days later.

Over my lifetime, there followed many more UFO sightings and ETI encounters and interactions many in group witness settings as a part of a CSETI field expedition to initiate human contact and communications with these diverse extraterrestrial beings visiting our planet.

This phenomenon is real, regardless of what you have read or have been told by the mainstream media or the science community. There is no transparency in their public disclosure but rather deception, lies and cover-ups designed to betray the public's trust in officialdom. Increasingly, the distrust toward the official position that dispenses their version of knowledge and truth has become suspect which has led to a public awaking to start investigating truth for themselves.

This second volume in a series of six volumes hopes to add to the public's resource for truth and honesty which will lead the reader to do further research and investigation on this subject.

There will be much in this book that will be familiar to the senior UFO researcher and much that is new, especially with some alternative and perhaps controversial perspectives that re-evaluate some long, widely-held traditional beliefs held within the UFO Community.

This and the succeeding volumes connect the dots and assemble the pieces of the puzzle to form a much larger picture of this phenomenon offering humanity potentially profound and historical implications for our future. *"The whole is always greater than the sum of its parts!"*

When writing a book of this complexity and length, it is the hope that errors in content are minimal or non-existent, although the reality is that no matter how careful the research and entry of evidence, inevitably errors will creep in because new evidence comes to light to prove an alternative reality. This author has endeavoured to minimize such errors as the UFO database is already corrupted and such errors seem to be perpetuated due to poor research or the deliberate ignoring or marginalizing of real facts. With this in mind, corrections have been made where necessary while still maintaining the integrity of the evidence as it was originally reported.

With such material gleaned from my predecessors in understanding the "Big Picture" of this phenomenon, full credit and acknowledgement is given to these Ufologists in the footnotes (author's rants), web links and bibliography of this book.

Please note the colour format used to distinguish the web links from the video links and bibliography which is noted in the Bibliography section of this book.

In essence, this book is an interactive book, perhaps the first of its kind, where web links and video links are used to enhance the research value for the student and the seasoned Ufologist and thus, it extends a 3500-page textbook into a 10,000-page tome!

The reader will be able to follow the web link references to see where I have collected the information for this book. Unfortunately, at the time of publication of this book, some of these web links may no longer be active, it is the nature of the internet and it ephemeral web links!

My apologies for any inconvenience this may cause, however, try "googling" a specific word or phrase; this may reveal a similar website!

Obviously, no one book could contain all the wealth of this particular subject which could easily fill many libraries, therefore this book should be considered a reference manual only, leading to further research and investigation for the serious UFO researcher.

As the reader goes through this second volume, they will see how the UFO community has become ghettoized with fewer people looking to be educated but instead entertained by speakers who will give presentations from careful research and investigation to pure theorizing, hype and speculation without any real proof.

UFO conventions have become a place for entrepreneurs marketing UFO and ET paraphernalia, DVDs, books and trinkets with UFO enthusiasts departing believing they have acquired a greater truth or scratching their heads even more confused with more questions than answers and who is really listening and taking them seriously.

To this arena enter the skeptics and professional debunkers to confuse the public further and the serious UFO investigators who try to lift the veil of secrecy off and shed light on this subject some people offer the wannabe believers with a new space age religion or incredible stories of their encounters with tall beautiful Venusians.

Complicating matters are the stories of cattle mutilations and alien abductions of humans for various medical experiments while behind these out-of-this-world accounts, the US Military secretly carry out their hidden agenda ensuring that false stories muddy the investigations of real UFO investigators while the real back engineering research is carried out behind closed doors away from prying public eyes.

The reader will be surprised by the reverse engineering of captured ET spacecraft, the advanced research into alien DNA and the development of programmable life forms (PLFs) that resemble alien beings, cloning and hybridization programs, mind control programs and Military abductions (MiLabs), etc.

Finally, we will look at some of the more notable Ufologists from around the world, including various government and military UFO study programs both globally and in the USA and comparing those with private UFO investigations made by citizens and scientists.

The ride into the unknown continues and hopefully this book will shed more light of understanding on the UFO and ETI subject, at the very least let it become your go-to reference manual for further investigation! **Enjoy the ride in discovery!**

Terry Tibando - September 2015

Table of Contents

CHAPTER 17

THE GHETTOIZATION OF THE UFO COMMUNITY

UFO Investigators (Ufologists)

The study of UFOs and Extraterrestrial life has attracted many people from all walks of life. It has attracted handfuls of scientists from all branches of the scientific community; it has also attracted engineers, medical doctors, policemen, soldiers and military brass from the armed forces, teachers and the everyday man off the street including your neighbour next door and the farmer out in the countryside. They all have one thing in common and that is the pursuit of truth and the discovery of new knowledge that hopefully will benefit mankind. From the early years of Ufology, people have been fascinated with the unknown and UFOs and ETI definitely represent one of the greatest mysteries of the unknown to confront investigators in the modern era.

For most private investigators, UFO research is an unpaid job, an extracurricular activity for which no one has ever gotten rich from investigating it. More often than not, UFO research is done in the spare time of one's life while trying to make a living in the real world of business. For some UFO researchers, supplemental income may come from doing the lecture tour circuit giving talks or seminars on what they have uncovered in their investigations. Depending on the quality of data gathered and its presentation or the promotion of a new book or video, some researchers are fortunate enough to receive some financial gain. Recognition from one's fellow UFO researchers and from the general public at large certainly doesn't hurt either. All of this can only enhance one's prestige, desirability as a guest speaker, and overall fame in the field of Ufology. For the entrepreneur, who interest is not in actual UFO research, but for the free enterprise of the subject, it is the selling of some UFO related trinkets at the "UFO flea market and emporium" which can be found everywhere at UFO conferences and conventions. No doubt his is perhaps the only income that may be consistent from one convention to the next.

The UFO investigator often married and depending on the passion and commitment of his research, he may be consumed to the point of obsession, so that his family and married life suffer, sometimes ending in separation from one's spouse or possibly resulting in a divorce. For most researchers, the pursuit of truth has become a balancing act between family life with a very understanding wife and an ability to focus intently all of one's drive, talents, abilities and knowledge with the amount of time allotted to UFO investigations.

Being single of course, one obviously is not prone to the problems that may be encountered from a married life and the individual can certainly be as obsessive and compulsive in his investigative work as he wants with no restrictions to time limits.

However, being single means that there is not always someone at home to share the results of the research work with, at the end of the day. Perhaps, this is a small price to pay for the pursuit of one's interests and hobbies.

A dedicated UFO investigator often spends many hours reading the literature and researching at libraries, they will spend days on the road travelling or flying from one part of the country to another in order to interview eyewitnesses and document accounts of a UFO event, and sometimes it's meeting with other researchers in a cooperative investigation. All such efforts to get to the heart of the matter is of necessity, time-consuming, but the payoff is that one potentially promising report that will reveal the elusive answer to the question of the UFO/ETI reality.

A researcher is someone who will employ the scientific method whenever possible; his notes are accurate, detailed and unbiased. His interviews with eye-witnesses are thorough, non-judgmental, or leading. His position is always neutral, he gathers whatever information is presented to him no matter how inconsequential or absurd, and his analysis and conclusions are added to the voluminous database of evidence so that he and his fellow researchers are able to obtain a clearer understanding of the phenomenon from which further assessments can be made and refined.

It is the hope of every UFO investigator to obtain enough evidence from that one major UFO event or from a compilation of documents and interviews that will eventually become the "smoking gun" evidence which will blow the whole UFO and ETI phenomenon wide open so that, no more denials and secret cover-ups will silence the subject ever again. Then, unfettered, the scientists will come forward to seriously investigate this phenomenon without fear of reprisals, threats, and intimidation or the loss of tenure. As is so often the case, when the common people lead the government leaders will follow or in this case when the UFO researchers lead the scientists will follow.

UFO Groups and Organizations

Like-minded people attract other like-minded people and this adage is also applicable to people with an interest in UFOs and Extraterrestrials. UFO groups and organizations have sprung up worldwide because many people have come forward to report their strange eyewitness accounts and because governments and scientists are non-committal in investigating this phenomenon. The military is obviously interested but prefers to work alone away from the public and they do not share information as a rule. The need to share information is critical to solving this mystery and if the one organized group with unlimited financial resources is not revealing what it knows then, it becomes paramount to organized like-minded groups of people together, regionally, nationally and globally to research and obtain the answers for themselves.

APRO

APRO (Aerial Phenomenon Research Organization – now defunct) was the oldest citizens UFO organization in the US with many affiliated branches worldwide. It was the brainchild of **Coral and Jim Lorenzen** in 1952 in Sturgeon Bay, Wisconsin as a small group to study unidentified flying objects. Jim and Coral wrote many books on the subject covering reports of eye-witness sightings from around the US and the world. The *APRO Bulletin* was the organization's news magazine which published the organization's latest UFO findings, research, and sightings which came out quarterly.

Unlike its rival, the **National Investigations Committee on Aerial Phenomena (NICAP),** which came into being a few years later after **APRO**, its main philosophy was based on hard evidence, a cooler approach to the idea of a government cover-up of UFO data, although this perception would change in later years, and its interest in sightings of humanoid-like creatures associated with the UFOs. It grew by leaps and bounds attracting many academic consultants and intellectual types, many with a string of PHDs behind their name. There were also engineers, doctors, teachers, a few military people, as well as many blue-collar people from around the world.

In 1967, APRO's membership had reached an all-time high of 1,500 members and was on the verge of becoming the number one UFO organization in the country with NICAP on the decline. Then, by 1969 APRO was broad-sided by two unexpected disasters. One came from the **University of Colorado UFO Report** on the study of unidentified flying objects, also known as the **Condon Report**, which struck Ufology (a quasi-science still in its infancy) a deliberate premeditated blow that concluded *"that nothing has come from the study of UFOs in the past 21 years that has added to scientific knowledge." "…that further extensive study of UFOs probably cannot be justified in the expectation that science will be advanced." "…the federal government should do…nothing…with them in the expectation that they are going to contribute to the advance of science." Inseparable from this conclusion is the national defense interests about UFO reports from which the history of the past 21 years has repeatedly led the Air Force to the conclusion that none of the things seen, or thought to have been seen, which passes by the name of UFO reports, constituted any hazard or threat to national security.* And on and on at ad nauseam. And as a result, the Air Force dropped its semipublic data collection effort, Project Blue Book. We will be revisiting these conclusions again in a later section. "Scientific Study of Unidentified flying Objects" Conducted by the University of Colorado under Research Contract Number F44620-67-C-0035 with the U.S. Air Force; Dr. Edward U. Condon – Project Director

The second blow APRO suffered came as a major schism when **Walt Andrus**, an APRO regional director in Illinois, decided to break away because of differences in the organization's direction and policy. He then founded the **Midwest UFO Network** (now the **Mutual UFO Network or MUFON**). Membership plummeted and APRO never recovered. In 1986, **Jim Lorenzen** died, and two years later, Coral followed. The board voted to disband the organization shortly thereafter. Clark, Jerome. The Emergence of a Phenomenon: UFOs from the Beginning through 1959
The UFO Encyclopedia. Vol. 2. Detroit: Omnigraphics, 1992.
Lorenzen, Coral E. The Great Flying Saucer Hoax: The UFO Facts and Their Interpretation. New York: William Frederick Press, 1962. Rev.:
Flying Saucers: The Startling Evidence of Invasion from Outer Space. New York: New American Library, 1966.

Lorenzen, Coral, and Jim Lorenzen. *Abducted! Confrontations with Beings from Outer Space.* New York: Berkley, 1977.
——. Encounters with UFO Occupants. New York: Berkley, 1976.
——. *UFOs: The Whole Story.* New York: New American Library, 1969.
——. *UFOs Over the Americas.* New York: New American Library, 1968.

Jim and Coral Lorenzen at APRO's headquarters in 1955
http://www.daviddarling.info/encyclopedia/A/APRO.html

Author's Rant: On a personal note, I first heard of APRO in 1963 while still attending Victoria Secondary School High when Dr. Edwards, a linguist professor from the University of Victoria came to our school one day to present a lunchtime talk on unidentified flying objects. He mentioned his involvement in APRO as a UFO investigator and I was able to get the address of the organization and became a member shortly afterward. My involvement with APRO and its Canadian counterpart: CAPRO (Canadian Aerial Phenomenon Research Organization) was like a clarion wake-up call to me to begin my own investigations into UFOs and ETI. I knew that this subject would revisit me time

after time throughout my life with ever-escalating personal experiences with this phenomenon as if I was being guided by a higher source of consciousness that wanted me to know that the ET presence was very real.

APRO Founders: Jim and Coral Lorenzen (Credit: Fortean Picture Library)
http://vintageufo.blogspot.ca/2007/11/conversation-with-coral-lorenzen.html

NICAP

NICAP (National Investigations Committee on Aerial Phenomenon) was at its time the largest private citizen's UFO organization in the States and worldwide. It was very vocal on its goals and policies of scientific investigation and research of UFOs with full disclosure of information to the public by various authorities and agencies of the government. Initially, NICAP were not supporters of aliens piloting these strange craft for as yet, there was no tangible proof to support this conclusion. Contactee cases were considered to be too irrational and boarding on the lunatic fringe of the subject to be taken seriously. No doubt, NICAP was probably the most "in your face" UFO organization that has existed to date, particularly was this no truer than in the faces of the US Air Force.

NICAP was founded October 24, 1956, not by retired Marine Corps **Maj. Donald E. Keyhoe** as some news media and some fellow UFO researchers incorrectly assumed although he did

18

eventually become its director in January 1957, but by maverick physicist, **T. Townsend Brown**, who served as its first director. Townsend Brown, a brilliant physicist demonstrated to the military in the 30's the principle of electro-gravitic propulsion using aero-form disc-shaped models and was firmly convinced that UFOs held the key to a new type of technology which used some form of antigravity propulsion system for interstellar travel. An investigation by worldwide surveillance into this phenomenon would yield an explanation he felt, which would bolster his own theories on overcoming the force of gravity through studying UFO technology. Unfortunately, he never got that chance at least in the NICAP organization due his mismanagement, financial ineptitude and his unrealistic ambitions centered on his theories of antigravity which were the cause for his resignation from the position of chairman of the board of directors.

Townsend T. Brown and Donald Keyhoe
Google Images

NICAP

NICAP attracted prominent citizens, many from the military forces and many with impeccable credentials like **Rear Admiral Delmer Fehrney** former head of the Navy's guided-missile program who was elected NICAP's board chairman. He then appointed Keyhoe to the position of director which he accepted. **NICAP Board of Governors** included besides **Donald Keyhoe, Joseph Bryan, John Fisher, Dewey Fournet, Barry Goldwater, Joseph Hartranft, Charles**

Lombard, Robert Richardson, and **Edmond Roush**. Not long afterward Fehrney left NICAP for "urgent and "personal" reasons that were never disclosed publicly.

Before long, NICAP membership increased with 5000 members and its membership became a list of who's who which included high-ranking military officers, former **CIA (Central Intelligence Agency)** officials like **Vice Adm. Roscoe Hillenkoetter**, CIA's first Director, politicians, businessmen, college professors, and clergymen. Even the great psychologist, **Carl G. Jung** was a member of NICAP who firmly believe that UFOs were extraterrestrial devices piloted by ET Intelligences. But no matter how prestigious the memberships which seem to attract a lot of military and intelligence types, NICAP was always plagued by financial mismanagement requiring regular financial injections from its membership to keep it from collapse. Keyhoe would often pay expenses out of his own pocket to keep the organization viable.

NICAP like APRO had its own newsletter magazine called the *"UFO Investigator"* and it published several documents including *"The UFO Evidence"* (1964): *"UFOs: A New Look"* (1969): and *"Strange Effects from UFOs"* (1969). These publications seem to help NICAP in its difficult financial problems.

"Soon, reflecting its director's enthusiasm, the magazine's pages were openly and more sympathetically discussing the **extraterrestrial hypothesis** of UFO origin (though a formal NICAP endorsement, by majority board vote, did not take place until 1096). It was also blasting contactees, accusing the Air Force of covering up UFO evidence, and Calling for congressional hearings." The UFO Book – Encyclopedia of the Extraterrestrial" 1998 by Jerome Clark

With the constant bombardment of written and vocal salvos from the news media supplied chiefly from its main information arms dealer, NICAP (as well as other UFO organizations) and with support of ever increasing public demand for honest and open hearings on the topic of UFOs, the Air Force had reached its breaking point with the whole matter.

With such a hotbed of controversy now firmly planted in the minds of the public and the US Air Force sufficiently agitated and angered to resolve the UFO problem once and for all, NICAP felt it was time to approach the US Congress for open hearings on the subject. But they were only greeted with a brief lukewarm interest regardless of NICAP's efforts and regardless of the fact "that the Air Force regarded it as a threat to the credibility of its repeated public assurances that there was nothing to UFO reports." The UFO Book – Encyclopedia of the Extraterrestrial" 1998 by Jerome Clark

By 1966 witnessed a dramatic increase in their membership peaking at 14,000 and a sudden turn in its financial fortunes and it certainly didn't hurt that the UFO sightings were increasing in number as well. Many UFO organizations were coming into being during this time period, though their membership credentials and numbers would never rival NICAP's, the public interest was there.

When everything was starting to go NICAP's way, The Air Force in a pretentious and in supposedly unbiased act of good will toward the public's concern for solving the enigmatic UFO mystery sponsored a scientific investigation known formerly as the **University of Colorado UFO Project** or the **Condon Report.** Believing that overdue answers to the UFO question were finally going to be forthcoming, NICAP gave its full support and assistance to the U. of C. UFO Project. Its project director **Dr. Edward U. Condon** assured the public and all interested parties that the investigation would be an objective scientific analysis of the situation. By September 1967, however, it was apparent from inside sources that the whole project was nothing than a smoke screen to placate the public and UFO organizations; that Dr. Condon had already decided that UFOs were nonsense and when the final voluminous report came out in January 1968 NICAP had already stopped its assistance to the UFO Project.

From that point on, things only got worse for Keyhoe and NICAP as everything started to spiral downward out of control. Keyhoe was eventually fired for incompetence, his autocratic behaviour, and financial ineptitude. Other board members were fired as well and membership plummeted to about 5000 followers who were probably still in a state of disbelief from all that had taken placed in the UFO field as well as within the NICAP organization. **Richard Hall** succeeded Donald Keyhoe until John Acuff took over as president in 1970; he later resigned in 1978 as membership dwindled. From when on, it was every man for himself as everyone abandoned the sinking ship "S.S. NICAP." Allan Hall assumed the steering wheel of the death bound "ship" to its "watery grave of ignominy" until in 1980 when the organization finally disbanded for good and all files were handed over to the **J. Allen Hynek Center for UFO Studies (CUFOS)** based in Chicago. The Air Force delivered the death to many UFO organizations and it shut down **Project Blue Book,** it's own small investigation team that dealt with sightings from the public with the hopes that there would be no further public interest in UFOs. **"The UFO Encyclopedia" by Margaret Sachs, 1980**

Later attempts by some Ufologists have tried to resurrect the organization but in name only. It has also, been suggested that NICAP was "scuttled" from by the CIA operatives working from within. There are many UFO researchers who feel that these are unfounded allegations against the CIA, that they are baseless charges.

Author's Rant: I am personally not convinced that the CIA, the U.S. Military, or any one of the many intelligence agencies in America is totally blameless in the eventual demise of NICAP organization. Remember, NICAP attracted a lot of military and intelligence personnel into its membership. It would not be any surprise then, that to keep tabs on an organization like NICAP, who wanted the Air Force and other military departments to reveal highly classified, secret projects and programs, would get their full undivided attention with operatives working from within it. Also keep in mind that the "S.S. NICAP" along with other major UFO organizations were "torpedoed" or "bombed" from without if we use the terminology of war. It was no surprise that the Air Force were growing tired of the verbal and written attacks against its credibility in investigating UFOs and of secret cover-ups as alleged by NICAP, so what better way than to deal a death blow to all major UFO organizations in general and the public's unwanted interest in UFOs than with a phony sponsored scientific investigation at a prestigious university from the mid-west. Sounds too conspiratorial, right? Wrong! I will show that this is not only possible but, there is freedom of information documents indicating admission by the CIA that this is a typical

ploy to destabilize any "organization of special interest" as well as their heavy involvement and control of the news media.

CUFOS

With the demise of APRO and NICAP, the somewhat dimly lit torch for continued investigations into the UFO phenomenon was handed over to another the civilian group **CFUOS (Center for UFO Studies)** founded by **J. Allen Hynek**, "a professor of astronomy at Ohio State University, and later, chairman of the astronomy department at Northwestern University. During the 1950s and 1960s, he served as the astronomical consultant to the United States Air Force's **Project Blue Book**. Essentially, his responsibility was to determine whether there was an astronomical explanation for a UFO sighting. Professor Hynek would study a UFO report and decide if its description of the UFO suggested a known astronomical object. That is, did the witness see the planet Venus or a meteor instead of a genuine UFO?"

Dr J. Allen Hynek
Google Image

Dr. Hynek was initially, skeptical of the whole UFO subject, but he changed his mind after examining hundreds of UFO reports by credible witnesses, now firmly convinced that UFOs were real and deserved serious further study. "With the closing of **Project Blue Book** in 1969, he began to seriously consider forming a private, scientific UFO organization composed of scientists and other highly trained technical experts, who would work together to solve the UFO enigma. In 1973, he started the Center for UFO Studies and served as its scientific director until his death in 1986."

CUFOS had emerged relatively unscathed from the **Condon Report** fallout mainly because Hynek decided to separate himself from not only the extremely biased conclusions reached by Edward U. Condon but also, from the US Air Force's Project Blue Book. Hynek followed the Air Force's policy as a debunker by finding plausible answers to UFO sightings, no matter how absurd or ridiculous the explanations may appear leaving many eye witnesses feeling that their experiences were not being taken seriously. It became increasingly clear to Hynek that many of the really credible and truly unknown UFO cases which came before Project Blue Book were not being given to him to research for which he had been hired by the Air Force to so, but sequestered away to the Air Force's own team of experts for internal investigation and analysis In 1972, Hynek published his classic book, *The UFO Experience: A Scientific Study,* in which he presented his categories for grouping UFO sightings and coined the phrase, "Close Encounters." Before he died, Dr. Hynek appointed **Mark Rodeghier** to succeed him as scientific director. The Center for UFO Studies continues to honour Hynek's legacy through its serious study and examination of the UFO phenomenon. CUFOS currently publishes a magazine, the Journal of UFO Studies periodically throughout the year. http://www.cufos.org/org.html

MUFON

MUFON (Mutual UFO Network), at the time of the writing of this book, is one of the oldest and largest UFO investigative organizations in the United States. MUFON is currently headquartered in Fort Collins, Colorado under the direction of James Carrion.

"MUFON was established as the *Midwest UFO Network* in Quincy, Illinois, on May 30, 1969, by **Walter H. Andrus, Allen Utke, John Schuessler**, and others. Most of MUFON's early members had earlier been associated with APRO (Aerial Phenomena Research Organization).

MUFON founders (L. to R.) Walter H. Andrus, Allen Utke, John Schuessler
https://www.mufon.com/history.html

The organization now has more than 5,000 members worldwide, with a majority of its membership base situated in the continental United States. MUFON operates a worldwide network of regional directors for field investigation of reported UFO sightings, holds an annual international symposium and publishes the monthly *MUFON UFO Journal*."
http://en.wikipedia.org/wiki/Mutual_UFO_Network

"The mission statement of MUFON is:

1. Are UFOs some form of spacecraft controlled by advanced intelligence conducting a surveillance of our Earth, or do they constitute some unknown physical or psychological manifestation associated with the planet Earth that is not understood by present-day science?
2. If UFOs are found to be extraterrestrial craft controlled by intelligent beings, what is their method of propulsion and means of achieving unbelievable maneuverability and speed?
3. Postulating that they may be controlled by an extraterrestrial intelligence, where do they originate --- our Earth, our Solar System, in our galaxy or in some distant galaxy in the universe?
4. Assuming that some of the craft are piloted by beings (humanoids), what can we learn from their apparently advanced science and civilization through study or possibly through direct communications with the occupants of these vehicles?" **"The UFO Encyclopedia" by Margaret Sachs, 1980**

"Along with **CUFOS** (the J. Allen Hynek **Center for UFO Studies**) and **FUFOR (Fund for UFO Research)**, **MUFON** is part of the **UFO Research Coalition**, a collaborative effort by the three main UFO investigative organizations in the US whose goal is to share personnel and other research resources, and to fund and promote the scientific study of the UFO phenomenon."
http://en.wikipedia.org/wiki/Mutual_UFO_Network

Major Active UFO Organizations in the USA

Apart from the four UFO organizations already mentioned above, many UFO groups and organizations are either very small consisting of a handful of members or are inactive. Some work alone and others work cooperatively with the larger UFO organizations on special UFO cases that require more professionalism or for which smaller groups lack sufficient equipment, manpower or financial resources.

Membership in most organizations attracts individuals from all fields of education and professional backgrounds, due to the organization's recognition and reputation that comes from the quality of its research work. Other factors for membership would be the collective identity and dynamics of the UFO organization i.e. professional and intellectual types will attract people of similar backgrounds: military officials will often attract other military people as in the cases of APRO and NICAP, respectively. Members will often jump from one organization to another or join more than one association at a time or start their own UFO group because of internal differences and disputes among members.

These disputes, disagreements and differences may be due to the agenda and philosophy as to what is the nature of UFOs and ETI. Perhaps, it is due to financial resource problems or not being able to participate in the investigation field work but, more often than not, the one critically serious flaw in which many UFO organization will not admit, is that it really becomes a clash of egos maneuvering for fame, recognition, power, and control. UFO organizations when they reach this sorry state become self-destructive; they are their own worst enemy and are usually marginalized to the "UFO ghettos" of unprofessionalism and poor credibility.

A worst case scenario, if this happens, would be that membership drops off dramatically and the general public, as well as any scientific establishments, will not have anything to do with it but, it will re-affirm the original notions that this whole UFO matter is nothing more than pseudo-science. This downward spiral of internal chaos and disintegration only strengthens the opposition's position to this phenomenon. The debunkers and naysayers will have won the final debate and the Air Force and other military establishments as well as the intelligence communities will be able to gloat and tell everyone who's still listening that they told them all so, that there was never anything to this whole matter whilst quietly resuming their own investigation and research in secret without any further confrontation or interruptions from some civilian UFO group. We only have to look at the now defunct NICAP as the example of this worst-case scenario. Sure the "phoenix will arise from the ashes of destruction" but, not necessarily right away and the damage may be irreparable.

Some of the currently active UFO groups and organizations in the U.S. are as follows:

- Aerial Phenomena Enquiry Network **(APEN)**
- Center for the Study of Extraterrestrial Intelligence **(CSETI)**
- Center for UFO Studies **(CUFOS)**
- Disclosure Project
- Enlightened Contact with Extraterrestrial Intelligence **(ECETI)**
- Exopolitics Institute **(ExoInst)**
- Fund for UFO Research **(FUFOR)**
- Institute for Cooperation in Space **(ICIS)**
- Mutual UFO Network **(MUFON)**
- National Investigations Committee On Aerial Phenomena **(NICAP)**
- National UFO Reporting Center **(NUFORC)**
- Project 1947
- International UFO Congress
- UFO Search Squad **(UFOSS)**

Armchair UFO Investigators

An armchair investigator is a person who has probably never gone out into the field as it were to meet with eyewitnesses who have had UFO sightings and experiences. It maybe he has no personal means of transportation, or lacks sufficient financial funds, or is socially inept or a poor education or even is misfortune with poor health. The one thing he has in common with every other UFO investigator is a passion and a drive to understand and resolve the mystery of the UFO Phenomenon.

His main means of research is reading the reports and experiences of other UFO researchers who are more fortunate than him to make contact with the eyewitnesses who have the sightings of strange aerial craft. His research comes from books and magazines that he either has purchased or borrowed from a library. Watching video documentaries or TV program specials on the subject of UFOs and ETs may also be a part of his investigation. In fact, there are very few UFO researchers today, who can state otherwise that their initial interest in the subject probably did not come from reading books and magazine or watching films on UFOs that came from someone else who had done some investigation. There is certainly nothing wrong with this approach to get one's feet wet in the subject matter and see where it leads you.

The problem arises when one considers writing a book or to go on lecture tours and orate on the subject matter and claim the source of the information as his own when in fact it is the hard researched material of someone else or from many other UFO researchers. T o build upon the shoulders of others without given full credit or acknowledgement to those whose researcher work you used is blatant plagiarism which will inevitably, not only discredit you in the eyes of your fellow researchers but, you will end up in legal prosecution, being sued for millions. If you insist on using other people's work then, give credit where credit is due, quote your sources and get out and do some research work for yourself then and only then, can you claim it for your own. This is a no brainer!

Contactees

As we've seen earlier, there are people who have close encounters with Extraterrestrial beings not just once or twice but, on a regular almost routine basis, where information, new knowledge or a warning becomes a part of the contact and communication between humans and ETI. Such people are in the UFO parlance labelled as **Contactees**.

Contactees were viewed by Hynek and Vallee when assessing these eyewitness reports with a creditability factor of very low to no creditability whatsoever. To rate these type of people with any reliability was next to impossible in light of the fact that the UFO phenomenon at the time of their earlier investigations had not as yet been identified as being Extraterrestrial in nature. Yet, here were people who claim contact with alien beings from another planet or star system, who have travelled stellar distances in spaceships commonly referred as UFOs or **"Flying Saucers."** It was still to be proven that UFOs were spacecraft, that they were occupied or even, piloted by alien beings.

During the early 50's, contactees by their statements of having contact with ETs have jumped the cue as it were, ahead of the scientists and the military of many countries. It was always assumed that if extraterrestrial contact were to occur, it would or should have occurred with our scientists, supposedly the smartest people on the planet or with some government officials or possibly with appointed diplomatic liaisons first, not with some common folk. Such monumental and life changing events happening to everyday people is unsettling to the large egos of academia and science, as well as to most government and military officialdom.

What made things truly incredible was that the reports of first contact with alien beings were the physical descriptions as being human-like in appearance and that according to the contactees, they came from either, Venus, Jupiter, or Saturn. All of these planets are according to current astronomical data hostile to human life forms, whether terrestrial or extraterrestrial in origin. These human-like beings were described as tall, with long blonde hair, having blue or gold eyes and an attractive appearance. They usually wore some type of blue jumper suit and exhibited an air of friendly but, cautious and reserve demeanour. There are so many things wrong with these accounts that logically it sounded to any rational person that these reports were pure fabrication, outright lies or hoaxes, even the communication exchange between ETI and humans were suspect.

"Wannabes"

The term "**wannabe**" applies to a person who may or may not have done any investigative UFO field work, but who fraudulently misrepresent themselves as a full-time researcher who is in the "know" regarding the subject. The wannabe will often state that they are "well acquainted" with top UFO researchers speaking to them frequently and may suggest that they are even on a first-name basis. The wannabe is at times much like **Cliff Claven** from the TV show, "Cheers" who has extensive knowledge on almost every subject under the sun, particularly on the subject of UFOs.

He desperately seeks and wants to be acknowledged by his peers. His desperation for social acceptance has more to do with being validated as a human being, that his life is worthwhile and has meaning, more so than what he happens to know and he will try to gain the acceptance of others by joining one or more clubs, associations or organizations of like-minded people with like-minded interests. Like the armchair investigator, he may be socially challenged around other people, feeling a need to overcompensate his self-importance or his very presence. Here is where the psychologist, psychiatrist, and sociologist need to apply their professional expertise to help and support the development of these individuals to live more self-confident and productive lives. Perhaps with professional help and counselling, these wannabes will be able to contribute to the subject of UFO research in a more honest and professional manner.

Your Usual Assorted Nuts, Flakes, and Granola Types

The field of **Ufology** attracts just about anyone from every walk of life and from every state of mind. Ufology is a non-exclusive arena of research where both the professional investigator and the amateur researcher can meet on the level playing field of pseudoscience, the paranormal, the mystic realm of the altered consciousness, the spiritual nexus of transcendentalism and enlightenment, where higher frequencies and higher densities of understanding and evolvement will bring about personal transformation into a higher god-like consciousness. If all of this sounds too good to be true, well the good news is, that it is too good to be true! Come on people let's get our feet firmly planted back on the ground into some solid everyday reality. This is one more example of the kind of mumble jumble new age metaphysical philosophy that creeps into the UFO database and over a period of time, slowly corrupts it from within to the point that it gains wide acceptability among UFO researchers. Metaphysical philosophy has it place in science and life in general, but as UFO researchers, we need to be on guard for this sort of thing and be careful not to embrace it as a part of Ufology. This is where the membership of "assorted nuts, flakes, and granola" types who join UFO organizations can take any serious UFO group off into another direction before the organization even realizes what has happened.

This type of acceptance of things that are unrelated to the UFO and ETI subject adulterates pure research into a mishmash of quasi-religious overtones that becomes hard to separate from real scientific research. In fact, there is nothing scientific in its protocols, its agenda or mission statement. It does, however; attract people with limited education, social skills and strong dogmatic belief systems, often commingling Christian doctrine or other religious beliefs into some new age UFO or space-oriented religion. The "**Heaven's Gate**" cult was one such self-destructive and doomsday UFO-based religion lead by **Marshall Herff Applewhite** aka. "Bo" or "Doe" and **Bonnie "*Ti*" Lu Trusdale Nettles**, a.k.a. "Beep" and both were also known as "*The Two*," together, they and 39 of their followers, all committed suicide in a belief that the Earth's civilization was about to be "recycled."

 Another example of a UFO-based religion is the **Raëlian movement**, a cult founded by a former French sports-car journalist and test driver named **Claude Vorilhon.** Followers are called **Raëlians**, who believe that Vorilhon, or "**Raël**", "received special knowledge and instruction for mankind from the creators of life on Earth, human-like extraterrestrials called whose technology enabled them to appear as "angels" or "gods" in the eyes of ancient people. Raëlians believe that previous visitation from Elohim sparked the founding of many major religions humanity knows today." Their philosophy is opposite to the Heaven Gate cult which is more akin to a hedonistic lifestyle and where an official apostasy from other religions in recognition of Elohim.
http://en.wikipedia.org/wiki/Raelism

The point here is that no matter what club, group, association, or organization people are members of, they will attract the social misfits, the socially challenged, the strange, the bizarre and the flakey and crazy people who are searching or needing acceptance by others of similar interest and the UFO subculture definitely has its share of these people.

CHAPTER 18

WHO IS REALLY LISTENING WHEN YOU FILE A UFO REPORT?

When a **UFO** sighting is reported by an eyewitness from the public sector it is because that sighting does not fit into the everyday experience of that individual and defies the conventionality of the usual aerial objects or manmade craft that people are familiar with. The individual (or group of witnesses) will at first keep the information to himself pondering the reality and meaning of the experience. The witness will try to make sense of what he has seen and perhaps, he will immediately conclude that it is one of several possibilities. The unusual sighting is nothing more than a plane or jet or some secret military aircraft that he just happens to have been in the right place, at the right time to have witnessed. He may conclude that what he saw was a meteorite or some similar astronomical body, common to all educated people who chance to look up into the sky and to the heavens above. Whatever he may think, he will try to rationalize the event in order to put it in the context of the familiar so, that he can chalk it up as a mildly interesting experience in his life.

If he feels that the UFO sighting was truly mysterious, a **Close Encounter of the First kind (CE1)** or possibly a **CE2** (evidence of landing traces and environmental disturbances) and not easily identifiable with no amount of contemplation to resolve his perplexity, he may do one of two things. Keep the mystery to himself like an itch in the middle of his brain for which he can never scratch and obtain some relief, or risk the possible ridicule of revealing what he saw with someone in hopes that that person will have a plausible answer to his dilemma. His problem now becomes, *who does he trust* to tell his story to, someone who will respect him enough to not prejudge him or make him feel like he should just keep his mouth shut because his story is too incredible, too crazy to believe. His first choice might be his wife if he's married, or some other family member or relative, or perhaps it's a good friend or work associate that he feels comfortable with and has known him for some time or even a stranger. These people who be his first choice where his "comfort zone" is probably safe and secure, where he is assured of respect and maybe only light teasing as a possible downside.

Now, let's up the complexity of his UFO sighting to a **CE3**, sighting of an Extraterrestrial being in close proximity to an ET spacecraft or possibly a **CE4**, an encounter and interaction with an alien life form (usually described as an abduction). The witness is now more troubled and emotionally upset with the whole event realizing that this is not merely a sighting of some strange, possibly extraterrestrial flying craft but, a life altering event for which he has no previous experience in his life to compare with it. His paradigm of reality as he knows it has just sent him reeling head-first into a strange new universe of alien existence and he finds himself as an unsolicited pioneer standing at the frontiers of interspecies contact and communication. Inexorably, he is compelled to share his bizarre experience in order to maintain a modicum of sanity. This is not an experience that one can keep to one's self for very long, someone else has to know what has just happened, who knows, it could be important.

He would initially have many sleepless nights until he had informed his family and friends of his experience, who no doubt, were trying to persuade him to seek more professional expert advice, to report the event to someone of authority or someone with greater knowledge and experience in such matters. The local police department may be the first to hear his report they, in turn, would do a preliminary investigation probably to assess whether the witness is not some crackpot perpetrating a hoax, for if it was a prank, he would find himself in serious trouble with the law. If his claim turned out to be legitimate with some aspects of credibility and the facts corroborated with the evidence but provided no resolvable answer to the claim then, the local police may suggest that some scientists at a university would be his next course of action to get some help, or recommend doctors or psychiatrists particularly if the witness appears to be emotionally or physically harmed by the alleged event. Assuming that this reported event appears genuine, at least the facts and evidence fit the claim of the witness then, the local Air Force or some other military officials may be contacted. They would become involved in the case, if they deem the report represented an event of national importance or as a security threat or if it appeared that an aircraft and its pilot were involved in a crash that the eye witness had reportedly mistaken. One avenue for the individual to tell his story to would be the news media who are often looking for a newsworthy event that would help sell newspapers or help fill out a thirty second or one minute news video - sound bite on television, however, the news media may in truth be the least helpful to the UFO witness's well-being.

In all these scenarios it is also possible, that not one of these people, the police or military officials, scientists or government organizations would consider the case important or serious enough to investigate and may only refer the incident to one of the many UFO organizations in the country; who would in turn send an investigator out to gather information or evidence and make an assessment of the event. Many UFO researchers will jump at the chance to investigate a UFO report where ET entities were involved in the sighting. It represents an opportunity for the UFO researcher who does his investigations carefully and methodically, to not only share his research work with his fellow researchers and with the general public, but through other investigations of similar UFO cases, to build a database of information that may reveal hidden truths and potential scientific knowledge and such investigations certainly wouldn't hurt one's reputation in the field of Ufology by becoming an expert.

Is There Anyone Else Listening?

Who is listening to this individual, who only wants to share some relevant information that he deems is important, to unload an emotional burden that he feels overwhelmed by and hopefully receive some absolution and solace to make things right again within his world? He merely wants to be taken seriously, for in his mind what he has witnessed and experienced has profoundly altered his perceptive on life but also, may have a profound impact on the rest of society. For this reason, he risks possible ridicule, censorship, and being socially stigmatized by his family, friends, and work associates and by various social agencies.

A closer examination of the type of UFO or ETI encounter will often be indicative of the type of people that are listening to what the eye witness is reporting about his experience. The details of the report may reveal something of the structure and propulsion system of the spacecraft, what type of associated technology or weapons may be on board the craft or utilized by the alien beings. The eye witness account may indicate something of the **xeno-biology** of the ETI, or what planet, star system or even galaxy the ETI are from, or what knowledge the alien beings may possess about the universe or about life in general. Perhaps, they have an undisclosed agenda, or a desire to make contact with humans and open up communications with the people of Earth. There are a thousand and one possible reasons and then some, for the truth is in the details of the UFO report and knowing as much about the details will reveal who is really listening.

In most cases, such interested parties are the same ones we are already familiar with and a few who are not so familiar but, who choose to stay in the background, off the radar of public awareness.

The interest by police law enforcement is simply to ensure that no laws are broken, that personal safety is ensured, that there are no risk to life or limb and no damage to property has occurred. Reports of UFO sightings or of "little grey or green men" are usually not followed up on unless one of the above-stated conditions have been violated or compromised.

The news media as already stated has an interest in stories of UFOs ad ET only if it sells newspapers or boosts television ratings for that particular TV station. This becomes more relevant when pictures or video are taken by the witness or by the news crew and if the story is truly sensational then, the news media can do follow up stories, perhaps an in-depth interview. Depending on how much involvement the witness had to the unfolding UFO/ETI event, a book could be written about the account which in turn might lead to a movie being produced for television or screenplay. When such cases lead to the publishing of books or movies being produced, directed and acted in, there will inevitably be many people lined up to cash-in on someone else experience. The news media should only exist to provide information to the general public and not merely for selling and obtaining ratings, profits and point shares in order to justify its existence. It should be the fifth estate or the "watch-dog" for society to ensure that the spotlight is properly aimed and focused on those items that the public needs to be duly informed about. When anything other than this policy is followed it tends to lead to manipulation of the news and "spin doctoring" of facts either to acquire more profits or in the suppression of important information from the public consciousness.

We will see later, just how the various news media manipulates information or sanitizes the news you read, hear, and see daily; occurs why it exists; and just who is behind it. Not everything that is newsworthy gets published or aired on prime time television. The "fifth estate" is now a ghetto of fluff and gonzo news reporting, shadow of its former prestigious self with shoddy journalism of inane and unimportant news events that in the scheme of things are entirely irrelevant to the informational needs of the public.

Science could make a giant leap forward in understanding and help advance humanity and civilization by light years, if the right scientists from the hallow halls of academia would only step forward out of the shadows of blind ignorance, intimidation, peer censorship and ridicule, absolved and unfettered from these fear-based mindsets that have prevented them from true scientific involvement in this phenomenon.

We know that the military of most countries will take a special interest in any UFO case where ETs have landed or interacted with humans and if the report indicates more than one witness stating the same information, the military will move quickly on the case. Such cases will have the military, like bloodhounds on the trail of a fleeing quarry sending out its own UFO investigators either in uniform or without. As much information will be collected as possible and the witness may even be shown pictures of other UFOs or ET spacecraft to see if one particular craft is seen more often than others. Other times the military may try to convince the witness that he had mistakenly misidentified what he saw as swamp gas or the "planet Venus landing in his backyard or at some other location!" The threats and intimations are many, if the military feel that the witness saw something he wasn't supposed to see, be it a new highly classified stealth fighter or an actual flying saucer. If it has anything to do with advanced technology, alien or otherwise such reports will illicit the undivided attention of the military and in some cases, it may be better to keep your mouth shut and say nothing.

From time to time, certain individuals with a great interest and with very deep pockets and fat bulging wallets will come forward to obtain information on the mysterious phenomenon of UFOs because it may represent a new technology that has financial potential to entrepreneurial billionaires. Their initial interest may be a genuine desire to find out if we are being visited by Extraterrestrials and the type of transportation mode that have enabled them to get to our planet. But, more likely it is the potential dollar signs flashing before their eyes that will motivate them into the lucrative field of alien technologies and acquisitions with all its spin-off applications.

Joe Firmage was one such enterprising internet entrepreneur having made his fortune through USWeb a dot com internet business during the hay-day of the dot com boom and is currently involved in two similar organizations. Highly successful in his businesses, he felt compelled to leave his USWeb internet company as CEO because "word began to circulate about his belief in extraterrestrial intelligences, which was the subject of a book he wrote called "The Truth." Published online at www.thewordistruth.org, it contended that extraterrestrials had appeared on Earth periodically to help spur human technological advancement. For example, important elements of modern technology were explained as deriving from an alien spacecraft supposedly recovered from the 1947 Roswell UFO incident. Firmage also related his own contact with an otherworldly figure, shortly before the USWeb IPO, with whom he discussed space travel." Joe Firmage became all fired up on the subject of UFOs and felt that his financial resources would be able to make a significant contribution toward solving the UFO mystery. But the UFO phenomenon is a tough nut to crack and does not easily give up much of its secrets. "...he has sunk more than $3 million into establishing **Project Kairos** (after an ancient Greek word meaning the right or opportune moment) *to prepare mankind for aliens."*
http://en.wikipedia.org/wiki/Joe_Firmage

The secretive and somewhat reclusive **Robert Bigelow** is another man of wealth with more than an eye on trying to solve the UFO enigma. He has like Firmage set up a website called **NIDS** to seek out and understand strange anomalies and not just to investigate UFOs but, unlike Firmage, he has developed a private business of building orbiting space habitats for commercial use as well as for NASA, et al. and all designs has proven successful in operation. Perhaps, his motive is to get up close and personal with the ETs orbiting our planet in one of his orbiting habitat platforms. If he his successful in establishing contact and communications with ETI then, acquisitions of alien technology surely can't be too far behind. He may be looking to become a "major player in the big league" of ET contact and information exchange.

David Rockefeller youngest surviving son of **John D. Rockefeller** is another wealthy individual who wanted to contribute to bringing the UFO matter to full disclosure but has found that his own staff and aids have prevented this from happening. His generous donations would never meet its intended recipients but, would be channelled to some other cause that was not designated by Rockefeller. This information was related to **Dr. Steven Greer** would he was invited to the Rockefeller ranch by David Rockefeller, himself to hear about the **Disclosure Project** that **Dr. Greer** was working on at the time.

There are other more clandestine groups and intelligence agencies who, some UFO researchers feel, know more about the UFO/ETI phenomenon than anyone else and will do whatever it takes to keep that information out of the realm of the public domain and even from some military establishments and military brass. These are the people and agencies in which very little is known about, who ensure that highly classified or above top-secret information remains secured and impenetrable. Their modus operandi is surveillance, coercion, subterfuge, bribery, intimidation, threats, violence, lethal force, assassination, and elimination, and in extreme cases the use of social chaos, disruption, and terrorism. These sinister agents of covert intelligence operate outside of government and military control, checks and balances, and are probably only answerable to a more secretive corporate group whose membership is limited to a handful of men who may be the only ones on the planet who fully understand the whole UFO/ETI picture. We will, like so many other things we touched upon, come back to this dark side of the UFO reality and try to shed the spotlight upon this deep black group.

So, who is Really Listening?

Who Is Really Listening?

When someone reports a sighting of an unidentified flying object to one of the above interested parties, it is the expectation of that individual that one of these trusted authorities will be able to take up the challenge of investigation and make some headway into shedding light on the perplexing mystery of the UFO phenomenon. They will trust and respect the opinions of these skilled and learned folk as long as they feel their attempts to solve the mystery is an honest attempt and not a display of smoke and mirrors to confuse, baffle or deceive the UFO witness with false information or with subterfuge.

When such dishonest actions take place, the eyewitness will probably start to doubt the honesty of all authoritarian figureheads when all that was asked was a reasonable explanation of the UFO event. In the mind of the eye witness, he has on a subconscious level already analyzed the event and ascertained that the only plausible conclusion to the phenomenon is one that is Extraterrestrial by nature, and with patient expectation is waiting for the same conclusion to be reached by the authorities. Even though this is not always the case for all UFO sightings, it is a reasonable assumption, particularly when the unfolding parameters of the event are truly bizarre and alien in nature.

With over 100 hundred years are excellent documentation and reporting in the modern era of UFO sightings, it must be concluded that the military officialdom, various government agencies, intelligence communities, as well as the top echelon of the private industrial sector, must know exactly what the true reality and nature of the UFO phenomenon represents. In which case any further reports of UFO sightings would not serve to shed any further light on the subject, because those who are already within the "informational loop" on the subject matter are not going to reveal anything they know to anyone who, they deem do not have a need to know, that being most of us in the general public.

The informational control on the subject of UFOs and ETI is considered to be above top secret and higher in secrecy than the development of the atomic bombs during the Second World War and is so tightly guarded that it is near impervious to any attempts to penetrate its stronghold of secrets. The potential knowledge gain from alien technology derived from these extraterrestrial spacecraft is all about power control on this planet by the first nation that is able to successfully unravel its mystery and keep it secure, away from other nations. Yes! We still live in a world of brinkmanship, where threats and sabre-rattling by old former enemies and by newly emerging industrial nations seeking an equal footing in the global political arena.

In the beginning of the modern UFO era, common folk were the first people to have encounters with strange alien craft and inevitably to Extraterrestrial intelligences and at best the reports were read in the local newspaper the next day. The military was never considered as an investigative resource to which the public could turn to seek assistance in solving matters of the paranormal or the bizarre. The reason for this attitude is that military were at the turn of the century not viewed as having the sophistication or the technology to deal with objects heavier than air that flew in the sky when perhaps, their best attempts at flight were hot air balloons and hydrogen-filled dirigibles were still years away from becoming a reality and even, the **Wright brothers** had not flown the first gas powered engine airplane, the **Kitty Hawk**. It would not be until the First World War that aircraft would be considered for aerial combat and for aerial bombing of cities on enemy territory.

The newspapers prior to and early in the twentieth century represented the voice and consciousness of millions of citizens throughout the country and would in most cases give fair and unbiased reporting of any newsworthy event. This could be the birth event of farmer Brown's new calf or colt to the politics of the day to the social gather of women demanding equal right and the vote. Nothing was considered out of the realm of the printed word.

If the local constabulary were called upon, it was to settle some dispute of ownership between two neighbours or searching and arresting some bank robber or presiding over matters where the newly evolving love affair with the automobile had resulted in an auto accident. Such were the investigative involvements of the police at the time that were limited to everyday legal matters and the upholding of societies laws to ensure peace within city borders.

The second world war more than the first world war witnessed pilot reports of fiery globes known as "**Feu**" or "**Foo**" fighters, strange craft that followed and seem to monitor both Allied and Axis aircraft. As these reports filtered in, both sides thought that these **Foo fighters** were the newly developed secret weapons of the other side but, by the end of the war, the military of various countries were pretty certain that our world was being "invaded" by alien spacecraft and not by the secret weapons of some enemy country. After the Second World War as nations began to rebuild, their countries, western nations in particular who were unscathed from the war, found themselves growing and advancing more rapidly than other nations technologically. In the US the military found itself a willing "bed-partner" with the private corporate industrial sector that was eager to build the might of the US military. Such developments particularly with nuclear missile sites, it wasn't long before the common people across America began to report strange flying objects in the sky which were not manmade aircraft, but alien in nature; the military suddenly found itself once again inundated with hundreds of UFO reports.

Because the US Military had grown much larger and stronger after the War, they appeared to the post-war public to be the right people to investigate the UFO/ Flying Saucer phenomenon. Little did the general public realize in their trust of officialdom were they going to get honest truth or open disclosure on this mysterious phenomenon.

CHAPTER 19

HOAXERS, NAYSAYERS, PROFESSIONAL SKEPTICS AND DEBUNKERS

"A skeptic is one who adheres to the conviction that true knowledge may be uncertain, who suspends judgment, and who is willing to examine new evidence." -- **Astrophysicist Bernard Haisch, Ph.D.** http://www.hyper.net/ufo/skeptics.html

In an honest and unfettered investigative search for the truth, a skeptical mind is a wonderful attribute for any seeker of truth, however, only **IF**, at the end of one's search; one acknowledges and accepts that newly discovered truth. Yet, how often it is that when truth is discovered it is not always accepted because of a former predisposition to some preconceived notion or belief or hidden agenda. Such is the mindset of the full-time skeptics, the hoaxers, the naysayers, the professional debunkers and, as Stanton Freidman would say, the "nasty negativists" toward the UFO/ETI phenomenon.

In the serious quest for truth in the field of UFO and ETI investigation and research, the professional skeptics and debunkers (if such real professionalism even really exists among this small minority naysayers), are the proverbial *"pain in the ass"* to every legitimate UFO investigator and scientist.

The professional skeptic's mission is to:

- discredit all UFO reports as unscientific in their methodology by proponents who are unfairly labelled as "crackpots" who supposedly and amateurishly "dabble at science;"

- to discredit those whose perception to correctly identify everyday conventionality and natural phenomenon in unusual and bizarre circumstances which are considered "rare undocumented natural occurrences which become the basis of establishing a fledgeling science that at best is pseudo in nature under the mantle of a scientific sounding name of Ufology;

- to hammer home their assertions through arguments of logical persuasion while leaving out the more irrefutable pieces of evidence;

- to accuse Ufology of the very same things that they, themselves are guilty of which is shouldering the *"burden of proof"* while being apologetic to the plight of the emotional state and unprofessionalism of Ufologists.

Skeptics are, for the most part, the ones that the public and UFO researchers are familiar with, tend to be highly educated and respected in their business positions and on the surface appear to express themselves with confidence, both verbally and in the written word. Skeptics, who do any kind of serious investigations but, always with a hidden motive or agenda, search for weaknesses and loopholes in the eye witness's testimony or in the evidence of a poor photograph or video, or a contradiction in another eye witness's testimony or in a lapse in memory and perception. These professional skeptics will often have a magazine or an internet site to report on their investigations and to promote their viewpoint with many gullible people buying into what they have to say. They have even been known to gate-crash UFO conventions and disrupt lectures of some guest speakers. They are given far too much credibility and airtime when it comes to the public media doing interviews with them on subjects of the paranormal and more importantly on UFOs and ETI. They will often be invited to television news interview and chat programs or in made for TV documentaries to give the appearance of a "balanced and unbiased" approach to the whole UFO question allowing the TV audience to decide for themselves as to the validity of the researcher and the reality of UFOs.

These professional skeptics are either extremely naïve and/or ignorant, or they are either being knowingly manipulated to further a well-crafted agenda for some intelligence agency. When they are interviewed on programs like CNN's **Larry King** Live, most people are becoming well aware that these skeptics are programming plants to try to win arguments or discredit credible UFO witnesses and researchers. Quite often in very short order, these people reveal themselves as incompetent and unprofessional due to their poor research of current UFO reports thus, the TV viewer no longer takes anything they say, seriously.

Quite frankly, the public doesn't give a damn about what these skeptics and debunkers have to say as their positions tend to play into the agenda of the **Military Industrial Complex (MIC),** whose only mission is to keep the lid on the secrecy and cover-up of the UFO subject matter for as long as possible!

The only time such skeptics gain credibility from the TV audiences is when they represent one of the science institutions such as astronomy with people like **Seth Shostak** of **SETI** fame or from some military official who adheres to the national security protocols and mission statements. Even, when this is the case, the viewers' knowledge and sophistication in this day and age are such that they can see through most of their arguments as based on an outmoded understanding of scientific concepts and unfounded paranoia centered on threats to national security which stem from poor foreign policies toward other nations.

By the way, there is a difference between the two, a healthy skepticism is essential in an investigation of anything new or unknown, and where the unfettered search for truth and understanding becomes the ultimate goal, whereby society or civilization advances. A skeptics search for truth is a discovery process where insight is an integral part of the discovery process and once the truth has been discovered, it is incumbent on the seeker of truth to accept that truth completely, without hesitancy or reservation, otherwise the seeker profanes all truth and dishonours himself.

In the movie "Excalibur," (*I'll paraphrase this quote*), when **Merlin** was asked by **King Arthur,** "What is the greatest of all knightly virtues? Merlin replied when pressed on the matter, *"All right then, it must be TRUTH! There must be truth in all things for when a man denies truth; he murders some part of the world and himself!"* The point here is that a skeptic searches for truth while questioning everything about it and whatever that truth may be or where it may lead him in his search for it, he accepts his findings at some point as the truth, as being conclusive with no further proof being required.

A debunker on the other hand searches for truth, perhaps not intentionally and when he finds conclusive and authenticated truth, he nevertheless, denies it and discourages self-investigation of truth by other people with the use of misinformation, disinformation, lies, and confusion. His motives are to fulfill a personal agenda or the agenda of someone else. He may be honestly deluded or blatantly deceptive to hide the truth from others. He murders some part of the world and retards its growth and advancement, while unwittingly he atrophies his own spiritual growth and development; in essence, he murders himself. It all comes down to power and control over others! A debunker is not really interested in the light of truth and the sharing of truth with others but looks at what advantage there is in the possession of such truth and how the gaining of such knowledge will bestow benefits upon him.

When we look at the hoaxers, these people, for the most part, are simply up to some mischievous and childish behaviour, perhaps, in their own eyes, it is perceived as fun but, usually it is at the expense to some hapless member of the public who may have been traumatized thinking that he had seen a UFO, when in reality he was the unwitting victim of a practical joke. In most cases, the prankster usually gets away with his candle-powered hot air laundry bag-UFO that he has unleashed onto the unsuspecting public, particularly when a legitimate UFO sighting may have occurred the night before. His prank results in the confusion of the same eyewitness as to what he had seen the night before or baffles any other people who may be star gazing that night.
In recent years the sophistication of the "UFO" device has moved away from simple ball shape balloons to saucer shape and dirigible balloons and to gas or battery powered helicopter or saucer-shaped flying toys with accompanying flashing lights and sound effects. These types of hoaxes are usually reported in the news media and given unnecessary airtime which only further discredits an otherwise legitimate phenomenon, making any serious claims as silly and unworthy of investigation. Often, many UFO organizations waste a lot of time pursuing what they hope is a genuine UFO report whenever such hoaxes are perpetrated, but fortunately, researchers have gained enough experience to quickly dismiss these cases and move on to the genuine UFO sightings.

Nay Sayers are debunkers in the UFO subculture who not only deny the reality of the UFO/ETI experience but also, have the own viewpoint that is in direct opposition to the phenomenon. They are not of the skeptical mindset but, in outright denial explaining away alien spacecraft and Extraterrestrials as mere fantasies and delusions or a deranged mind seeking public attention or just the plain mistaken perception of a conventional object or of a natural anomaly. It is as if they were afraid or intimated by the possibility of their existence not knowing how they would cope with such knowledge of their reality. Their comfortable anthropocentric viewpoint of being the centre of the universe is threatened to be pulled out like a rug from underneath them if the worst fears about ETs were to be realized.

That no scientist, government or military official, would waste their time and reputation in the pursuit of quasi- scientific knowledge because it has been bastardized by "citizen-scientists" profoundly illustrates the myopic perception of most scientists. These scientists are all too comfortable in their professions and fearful of the loss of their tenures to come out of the hallowed halls of academia to investigate what so "many intelligent, sober and trustworthy citizens with extraordinary perception" have been claiming for decades nay, centuries. That there are strange aerial machine-like devices traversing our air space with impunity, without challenge from our respective governments or military forces simply because they do not exist therefore, the onus probandi is to be played by the rules of science which clearly states: ***"extraordinary claims demand extraordinary proof"***, to which, however, Ufologist respond by stating: ***"absence of proof is not proof of absence!"***

In recent years, many scientists, military and government officials, astronomers, astronauts, etc. have stepped forward to give sworn testimony that UFOs and ETI are real. The burden of proof has been presented by these first-hand witnesses and the ball is now in the court of skeptics and scientists. It time for these lazy bastards to put up or shut! How many times throughout history has humanity witnessed that one or two individuals or a small handful of people have reported fiery rocks falling from the sky, or people could travel to and fro in horseless carriages or that two brothers could prove that heavy than air machines were capable of flight or that man would one day walk on the Moon, or that simple folk would be the first to recognize the Prophets of God for their day and age and that in this day and age extraterrestrial contact with other species and civilizations is possible and sustainable. If we choose to not acknowledge these simple common people throughout history, would we even be at the same place in our current development as a civilization or would we still be in the dark ages of medieval societies?

It is time for all courageous scientists to get off the complacent backsides and start investigating what the rest of society already knows and accepts. Your responsibility as scientists is to explore and investigates the unknown, to seek after truth, not to shift the responsibility for the burden of proof back to the ones making the claims, for you are the more learned, you have been entrusted with a duty to heed and answer the call to action by those who are perhaps not as educated as you or in a position of authority to make the wise decisions and carry out the necessary actions. Do not dismiss the testimonies of your fellow citizens with derision or contempt or as unworthy of your time. You need to prove or disprove the claims, it is your responsibility as scientists, you are the brilliant lights in society that help advance civilization forward not hinder its progress, otherwise you have no business wearing the mantle of a scientist and would be better off in some other form of employment where people might take you more seriously.

Donald Howard Menzel

Donald Howard Menzel (April 11, 1901 – December 14, 1976). He was considered as one of the leading astronomers and astrophysicists of his time and achieved notoriety by becoming one of America's most outspoken skeptics of the UFO phenomenon. His brilliant technical background in various aspects of astrophysics and radio astronomy made him the first prominent scientist to offer his opinion on the UFO matter, and doubtlessly, his stature was influential on the mainstream public and academic response to the subject of UFOs.
http://en.wikipedia.org/wiki/UFO

His explanations to the UFO phenomenon were based usually on the public's misidentification of natural atmospheric anomalies and astronomical bodies (stars and planets). Though his explanations were logical, he rarely conducted field investigations and limited himself to theoretical explanations, which he considered to be more probable than extraterrestrial visitations, even though, he accepted the probability that many technologically advanced civilizations existed throughout the galaxy, they were incapable of interstellar travel to Earth.

Dr. Donald Howard Menzel

Many researchers of the UFO phenomenon felt strongly that he harmed any serious study of the subject including fellow "atmospheric physicist and UFO researcher James E. McDonald, who judged Menzel's approach to UFOs as inadequate, dismissive and superficial." Menzel's authored or co-authored three popular books debunking UFOs: *Flying Saucers (book)* (1953), *The World Of Flying Saucers: A Scientific Examination of a Major Myth of the Space Age* (1963), and *The UFO Enigma: The Definitive Explanation of the UFO Phenomenon* (1977), all were viewed as shoddy science on UFOs when compared to his better known work in astrophysics." http://en.wikipedia.org/wiki/Donald_Howard_Menzel

The following passages are quoted in full to give the reader and student a clear picture by which his peers viewed his UFO debunking work:-

Criticism of Menzel's UFO Research

"Some observers have argued that Menzel's UFO works are lacking. Atmospheric physicist and UFO researcher James E. McDonald used the word **"Menzelian"** to describe the astronomer's approach to UFOs (which McDonald judged inadequate, dismissive and superficial).

Sociologist **Ron Westrum** writes, *"The paradox is that his UFO books represent quite shoddy science, in contrast to his better-known work in astrophysics."* (Westrum, p 34) Westrum suggests that despite Menzel's "shoddy" UFO studies, "thanks to a type of halo effect, Menzel's reputation in astronomy buttressed his loosely put together scientific arguments." (Westrum, p. 35)

Criticism also came from many scientists associated with the U.S. Air Force: **Captain Edward J. Ruppelt**, the first head of the UFO investigation **Project Blue Book**, wrote:

"The one [UFO explanation] that received the most publicity was the one offered by Dr. Donald Menzel of Harvard University. Dr. Menzel, writing in Time, Look, and later in his Flying Saucers, claimed that all UFO reports could be explained as various types of light phenomena. We studied this theory thoroughly because it did seem to have merit. Project Bear's physicists studied it. ATIC's scientific consultants studied it and discussed it with several leading European physicists whose specialty was atmospheric physics. In general, the comments that Project Blue Book received were, "[Menzel has] given the subject some thought but his explanations are not the panacea."

Menzel's critics also report that his UFO theories were literally laughable. He was an occasional consultant to the **Condon Committee** (1966-1968), a scientific study of UFOs, lead by physicist **Edward Condon** at the University of Colorado. **Jacques Vallee** recorded in his diary a story related to him by astronomer and Committee consultant **J. Allen Hynek**. While dining one evening, Hynek and some of the Committee's regular members discussed Menzel's recent trip to Boulder. Mary Lou Armstrong, the Committee's executive assistant, laughed so hard, as she recalled Menzel's speeches that she fell from her chair and landed flat on her back on the restaurant floor. Menzel's explanations for the [UFO] cases were so ridiculous that only propriety and respect for a senior colleague prevented the members of the team, including Condon, from laughing openly in his face.

Though most of Menzel's harshest critics were UFO researchers, negative criticism came from other fields: in 1959, prominent psychologist **Carl Jung** declared that Menzel *"has not succeeded, despite all his efforts, in offering a satisfying explanation of even one authentic UFO report."* (Jung, 147) Damning criticism also came from an internal U.S. Air Force analysis, which stated:

"It is easy to show that the "air lenses" and "strong inversions" postulated by Gordon and Menzel, among others, would need temperatures of several thousand kelvins in order to cause the mirages attributed to them." http://en.wikipedia.org/wiki/Donald_Howard_Menzel

Menzel even, had his own UFO sighting on May 12, 1949, which he was not able to identify initially with one of his own standard meteorological or astronomical explanations. Leaving Holloman Air Force Base, with a chauffeur bound for Alamogordo, New Mexico. Around 9.30 p.m., he spotted two bright stars in the sky which he believed them to be the stars. Realizing that the stars Castor and Pollux were in the wrong position in the sky and were motionless, he described them as "very nearly identical in diameter, nearly one-half the size of the full moon."

"A few quick tests established that the lights were not due to reflections on either Menzel's eyeglasses or the car's windows. He watched the lights for about four minutes before one of them suddenly seemed to vanish. **Menzel** wrote that he then concluded he was observing something "exceptional" and told the chauffeur to stop the car. But just as he was speaking, the second light vanished as well." **Jerome Clark, The UFO Book: Encyclopedia of the Extraterrestrial, 1998**

"In his formal report submitted to the Air Force only four days after the encounter, Menzel sketched the two round lights as he saw them, and included a few of his own calculations: if the lights had indeed been motionless, they must have been about "180 miles away" and quite large, perhaps "3/4 of a mile". (He noted that if the lights had been closer and/or in motion, their size could have been smaller.) Ultimately, he regarded the sighting as quite perplexing." **Jerome Clark, The UFO Book: Encyclopedia of the Extraterrestrial, 1998**

"Menzel revisited the sighting in his first UFO book, *Flying Saucers*: (1953), and suggested that while he could not "explain the phenomenon in every detail" it was probably "merely a reflection of the moon" which had been "distorted by a layer of haze". It might be compared to "a person riding in a fast motorboat. He might see the moon reflected in the bow wave thrown up by the boat. But the reflection would vanish when the boat stopped." **Jerome Clark, The UFO Book: Encyclopedia of the Extraterrestrial, 1998**

It would show later that even **Menzel's** UFO report demonstrated inconsistencies which added to his lack of credibility as a debunker and confirmed in the minds of serious UFO researchers his negative attitude t the phenomenon.

Menzel, the National Security Agency and MJ-12

"Physicist and UFO researcher **Stanton Friedman** reported that his own research (including examination of Harvard University archives) showed that Menzel had served as a consultant to the **National Security Agency (NSA)**.

The fact that Menzel had security clearance and worked with the U.S. government is not on its own extraordinary; many scientists participate in sensitive duties for the U.S. government. What is somewhat unusual about Menzel's case, is that he held the rarefied "Top Secret Ultra" clearance, and as Westrum notes, that Menzel's dual membership "in the academic community and in the black world of military secret projects [were] apparently unknown to many colleagues and military contacts in the [U.S. Air Force's] Air Technical Intelligence Center." (**Westrum, fn, p. 36**)

Friedman argues that Menzel's high-level clearance is evidence in favour of the existence of **MJ-12**, supposedly a secret governmental UFO study group established in 1947 which allegedly included Menzel in its membership. Friedman had earlier been baffled by Menzel's appearance on the list of purported MJ-12 members: though Menzel was a leading astronomer, he was (as of 1984 when the MJ-12 documents first surfaced), the only purported MJ-12 member who did not have extensive ties to the U.S. military or government. Only later, when the extent of Menzel's decades-long involvement with U.S. Military intelligence, did Menzel's name make sense on the list of MJ-12 members. Friedman then concluded that Menzel's shoddy "debunking" had not been "bad science", but a deliberate attempt to reduce public and professional interest in UFOs.

Given the near-consensus within the UFO research community that the MJ-12 never existed (at least by that name) and the documents supporting its reality are hoaxes or so dubious as to be unreliable, Friedman's interpretation of Menzel's security clearance is in the minority.

Also cited as possible evidence against Menzel's membership or involvement in MJ-12 is his 1949 report of a UFO encounter to the U.S. Air Force. This report was publicly unknown for nearly three decades, before being found by Sparks. Sparks has argued that if Menzel was truly privy to secret UFO information since 1947 — when MJ-12 was supposedly founded — then Menzel would have no reason to send a "confidential" UFO report to the Air Force two years later for an account he thought "exceptional". Furthermore, Menzel's 1949 report makes no mention of any such group as MJ-12. Against this, it has been argued that MJ-12 was the alleged control group, but the detailed UFO data collection was being carried out by other, lower-classified government intelligence groups, such as **Project Grudge**, then the Air Force's official public UFO study. Thus, there would be no reason for Menzel to not report his sighting for Project Grudge to incorporate into their statistics. There would also be no reason for Menzel to mention the "super secret" MJ-12 group, particularly to members of the lower-classified Grudge.

UFO debunker **Philip J. Klass**, who found discrepancies in the documents and concluded they were faked, wrote:

The addition of the name of **Dr. Donald Menzel**, a world-famous astronomer and leading UFO-debunker, is an attempt at revenge by the MJ-12 counterfeiter. Menzel was hated and maligned by the "UFO-believers" during the first two decades of the UFO era. In the eyes of a UFO-believer, there could be only one thing worse than being a UFO-debunker—and that is a debunker who *knowingly* resorted to falsehood. The counterfeiter tried to heap this final indignity on a world-famous scientist, now deceased, by listing him as a member of MJ-12. (Recently I was told by one Ufologist that he suspected that I had replaced Menzel on MJ-12 following his death.) (Klass 1988:286-*87)* http://en.wikipedia.org/wiki/Donald_Howard_Menzel

Philip J. Klass

Philip J. Klass (November 8, 1919-August 9, 2005) Klass was born in Des Moines, Iowa and died in Merritt Island, Florida. Klass was a senior avionics editor of the Washington-based Aviation Week & Space Technology, the powerful organ of the military-industrial complex and with a background as an engineer in aviation electronic, he coined the term avionics, a blending of aviation and electronics. He was also, a UFO researcher, and author of various UFO-debunking books: *UFOs — Identified, UFOs Explained, UFOs: The Public Deceived, The Real Roswell Crashed-Saucer Coverup*, and a founding member of the **Committee for the Scientific Investigation of Claims of the Paranormal (CSICOP)**, and published bimonthly *Skeptics UFO Newsletter*.

As a well known UFO skeptic, Philip Klass was a hero to fellow skeptics and debunkers but, an irksome bane to all serious UFO researchers, the proverbial unlanced carbuncle on the backside of Ufology. His detractors viewed him as a meddling and tiresome, Jerome Clark wrote that *"Klass was an obsessed crank who contributed little to the UFO debate except noise, strange rhetoric, pseudoscientific speculation, and character assassination." "A legendary truth seeker in his own mind and a pathetic liar."*

His methods varied from tactics of screaming at people who caught in a lie or a truth regarding UFOs, to gate-crashing both public and private UFO events and disrupting guest speakers. He behavior appeared to be not only foolish and immature such as his well publicized reward challenges where he bet Stanton Friedman couldn't find any "pica typeface" documents from the 1952 period to verify the MJ-12 Documents or proving he was ever involved in the CIA but, at times, his motives were coolly calculated using any opportunity for disinformation and misinformation of otherwise well researched UFO accounts. His explanation that plasma or ball lightning emulated the erratic movements reported of many of the UFOs described as bright lights as well as UFOs affecting the electrical systems of airplanes and automobiles.

This explanation of plasma used as one unverifiable phenomenon to explain away another unverifiable phenomenon (UFOs), initially split both the UFO skeptics and UFO promoters in ways that neither side was able to foresee. With UFO skeptics not fully supporting the plasma theory and UFO researchers finding some plausibility in the theory, the dust needed to settle from such an outrageous assertion to make any sense from it. The plasma theory was challenged by physicist **James E. McDonald**, a pro-supporter of UFOs as well as a "more skeptical team of plasma experts assembled by the Condon Committee" as unscientific. Some UFO skeptics like **Michael Persinger**, **Terrance Meaden**, and UFO promoters like **Paul Deveraux** and **Jenny Randles** felt strongly that plasma or earth lights represented a viable explanation for some UFO reports. To this day the original plasma theory though been discredited but adapted by others. In a game of chess, this would represent a great tactical move against the opponent which could do severe damage to the opponent's own strategy. Klass had definitely won a strategic and decisive move for the side of the debunkers more to the consternation of UFO researchers.
http://en.wikipedia.org/wiki/Philip_Klass and
http://answers.yahoo.com/question/index?qid=20070704110934AA7xodJ

Philip Klass
http://highstrangenessshow.com/the-ufologist-curse-of-philip-j-klass/

So, was the professional UFO skeptic and debunker, Phillip Klass, a seeker of truth or a disinformation provocateur? Was Klass an intelligence agent from the CIA or a misguided soul suffering from the need to over compensate from some past personal tragedy?

He certainly was by his actions and statements both oral and written, a spokesman for UFO disinformation and misinformation, lies, denials, and obfuscation; however that did not mean that as a human being that he lack morals or a consciousness for his education and business life indicates otherwise but, he obviously followed an agenda whether of his own design or someone else's. People like J. Allen Hynek and **Lt. Col. Philip Corso** were often subject to Klass's verbal attacks even though, these attacks on character or in discrediting their informational database were never done face to face, they were, however, perceived as coming from someone else and not from Klass's own origin. Lt. Col. Philip Corso knew who that someone else was: "He knows that I know he was in the **CIA**" which according to Corso is why Klass never faced him personally but vilified him from a distance. Such a proof if true, would never be revealed by the CIA until Klass's death thus, Klass's bet of $10,000 to prove he was in the CIA, unless of course, as Corso did, if you had prior dealings with the CIA.

James Edward Oberg

James Edward Oberg (born 1944) is an American space journalist, historian, and is regarded as an expert on the Russian space program. He served in the US Air Force, and in 1975 joined NASA, working at **Johnson Space Center** on the Space Shuttle program until 1997. He worked in the **Mission Control Center** for several Space Shuttle missions from STS-1 on, specializing in orbital rendezvous techniques. This culminated in planning the orbit for the STS-88 mission, the first **International Space Station** assembly flight.

In 1990s Oberg authored *Space power theory*, sponsored by United States military as a part of an official campaign in changing perceptions of space warfare, specifically deployment and use of weapons in outer space, and its political implications. "In Oberg's view, space is not an extension of air warfare but is unique in itself."

As a journalist, he writes for several regular publications, mostly online; he was previously space correspondent for **UPI**, **ABC** and currently **MSNBC**, often in an on-air role. He is a Fellow of the skeptical organization **CSICOP** and a consultant to its magazine *Skeptical Inquirer*. He was commissioned by NASA to write a rebuttal of "Apollo moon landing hoax accusations." NASA later dropped the project; however, Oberg has said that he still intends to pursue it.
http://en.wikipedia.org/wiki/James_Oberg

James Oberg appears to have stepped into the shoes of his former predecessors **Menzel** and **Klass** and attired himself with the mantle of professional skeptic and debunker of the UFO/ETI phenomenon. Whether, this is some kind of self-appointment or a coronation process from some higher power and authority, only time and further investigation will reveal this fact. However, the cape of the superhero skeptic appears well worn-out and threadbare these days but, it seems that there are always others ready to don the rags of professional skeptic and James Oberg is certainly willing to give it a go.

James Oberg

Oberg approaches the position within the confines, thus far, of his expertise which is the exploration of outer space and as it relates to astronauts, who have allegedly witnessed UFOs in the near reaches of space and on the moon. Whether it is any of the Gemini, Apollo or Space Shuttle programs, and International Space Station, astronauts have seen and continue to see alien spacecraft while in orbit around the Earth or when some of them were walking the Moon. What American, Canadian, Japanese, British, Chinese and a host of astronauts from many other countries including Russian cosmonauts have seen can only be described as intelligently controlled spacecraft not built by or from the planet Earth?

Oberg attacks the problem by stating that in space, size and distance and visual perception are factors which lose meaning when viewing objects outside the spacecraft or when looking out the portal windows. Numerous extraneous debris is either expelled intentionally from manned spacecraft (urine and waste of one kind or another) or as a periphery function of the normal operation of the spaceship e.g. ice flaking of the ship's engines or other surface structures that may have built up as the craft entered into orbit. Such objects are often mistaken as something unusual particularly when the sun reflects off its surface creating the illusion mechanical or other worldly. Booster rockets falling back toward Earth due to the pull of gravity as a normal part of flight and trajectory can also be misidentified as UFOs. Satellites already in orbit but moving in different orbits and trajectories may also become alien in appearance and movement not to mention the lost of various pieces of tools and equipment which have unusual shapes by design and function, end up taking on an orbital life of their own. With so many things hurdling around space and flying sometimes dangerously close to manned spacecraft, it's no wonder astronauts and cosmonauts are seeing strange things every time they go up in space! Then, of course, you have all kinds of micrometeorites zipping about space and other natural phenomenon that we are just discovering and learning about like "sprites" and "jets" from large thunderclouds that shoot lightning upwards into space and the list goes on.

Whoa! Put the space brakes on **Buck Rodgers, Flash Gordon, Roger Ramjet,** and **Buzz Lightyear**! If this is what our astronauts are really seeing when they're up in space or when they were walking on the Moon then, *"Huston! You have a serious problem with your astronauts and with your space program!"*

It would appear according to Oberg that their expensively well-trained astronauts have a serious defect with their vision and are incapable of recognizing common objects when it comes to their own spacecraft, equipment and orbiting satellites as well as planetary objects and near-Earth natural phenomenon. If this is truly the way it is every time they send them up into space, they are setting themselves up for failure and a potential disaster on every mission. I would bloody well hope, that every astronaut has at least 20/20 vision with an exceptional ability to identify a spent booster rocket, a sub-orbit spanner, particles of yellow urine, ice flakes, space garbage, satellites or smudges on space capsule windows or camera lenses, etc. from an unidentified flying object, if not, they sure as hell have no business being in space!

Oberg's sees these UFOs as misidentification of common space vehicle debris that has floated off from the spaceship or is merely drifting along with the spacecraft as it orbits the earth.

"Some of the less wild accounts were based on simple misunderstanding of how objects coming off a spaceship would drift along it through space. There were both small Apollo-derived fragments (insulation, ice, wiring harnesses and fired explosive bolts) and larger structures (the upper stage, and the four garage-door-sized panels that encased the lunar module for launch) that could be seen from time to time outside the windows."

Since many of these astronauts also, fly military jets and similar aircraft where keen eyesight is mandatory, it would also be expected from astronauts, in fact, more so, because your spacecraft is your home and lifeboat and with billions of dollars of equipment and spaceship hurtling through space, you cannot afford mistakes or it will cost you your life and the life of your crew. Since, excellent eyesight is a prerequisite as a part of every astronaut's training as well as the ability to identify simple things in unusual circumstances then, when an astronaut tells you Huston or Mission Control that they are seeing an alien spacecraft, ("Huston! We still have the alien spacecraft under observance!"), you sure as hell, better take notice because that statement is like gold in the bank! You don't question the authenticity of the claim, these astronauts are extremely well-trained professionals, and they have to be, because you people on the ground, on good old Terra Firma have trained them! If they're not all that they can be then, you mission control specialist are to blame for the poor training and lack of professionalism!

Oberg also states that much of what astronauts reported were taken out of context or embellished by news reporters where UFO researchers seize upon these statements as validity of an actual sighting or the news stories are outright lies and fabricated stories to illicit UFO support.

"Yet not one of these cases has any relevance to "true UFOs" (referring to Hynek's visit to NASA to get reports of astronauts accounts of seeing UFOs i9n space), as they are for the most part frauds and hoaxes conjured up by unscrupulous writers and UFO buffs (several blatant photographic forgeries have been identified in these stories), or misunderstandings by citizens concerning the meaning of ordinary space jargon, or in a few cases, reports of passing satellites which in no way appear to be extraordinary."

So, the question then, is **James Oberg** trying to mislead the public and the UFO investigators with misinformation, disinformation, lies and denials in an attempt to cover up what is routinely seen by every astronaut, who goes up into space? For who is the cover-up meant for or is he reporting factually what is really happening up there in space and in essence saving the "bacon" of every astronaut who goes up into space and who come down to a mob of news reporters looking for an "angle" for a news story? Or maybe, he, like so many other professional skeptics and debunkers is merely "talking through his hat" believing that when speaking with an air of authority in "NASA – Speak" that whatever he states will be convincing enough to satisfy the inquiries of the general public. If he really believes this is the case, then it is a delusional state of mind!

Michael Brant Shermer

Michael Brant Shermer (born September 8, 1954, in Glendale, California) is an American science writer, historian of science, founder of **The Skeptics Society**, and Editor in Chief of its magazine *Skeptic*, which is largely devoted to investigating and debunking pseudoscientific and supernatural claims and author of many books on skeptical issues.

Michael Shermer is typical of the caliber of most professional skeptics and debunkers, well educated, a confident speaker and lecturer. He has followed in the footsteps of his predecessors in taking unscientific thinking to task and to expose falsehood and fallacy in all its forms, regardless of the subject matter, even when it comes to his former religious beliefs. Often, his tirades challenge, to the annoyance of those who would wish to express or promote their own points of views or to share an unusual experience in the arena of the UFO phenomenon or the paranormal.

A once former fundamentalist Christian, he now considers himself a skeptic rather an atheist or agnostic as these terms pigeonhole what he believes in, *even though any term applied to oneself or by others is still a pigeon-holing label, whether it be viewed by others as positive or negative.* His skepticism for most things that do not fit or conform to the scientific model has become his new creed with all the religious fervour and fundamentalism as his former Christian belief; indeed, his smiling passion for science has been promoted by Shermer as the new religion for this day and age.

Michael Shermer
https://en.wikipedia.org/wiki/Michael_Shermer

However, Shermer comes off to many people who hear his talks and lectures as overly confident boarding on the arrogant with a condescending smarminess to views of those he would debunk. Perhaps, only because those that he debunks are not quick or knowledgeable in their skill sets to respond with rapier wit or irrefutable proofs of their assertions to his challenges. In comparison to his predecessors, he probably fits the category of a "wannabe" professional skeptic.

Out Debunking the Professional UFO Debunkers and Skeptics

Alright! You fed up trying to debate your point a view with professional skeptics and debunkers on your favourite topic of UFOs and ETI, only to be shot down, discredited, made to look foolish as well as being exposed as totally unprepared and unprofessional. Well, good news! Many people who are passionate about a belief in UFOs and ETI share the same boat with you on that rough sea of humiliation. The only difference is now they are disembarking from that boat of debating ineptitude and setting foot upon the firm unassailable shores of the professional debater, forearmed with the powerful skill sets of knowledge and tactics in debating those professional debunkers back from whence they come from.

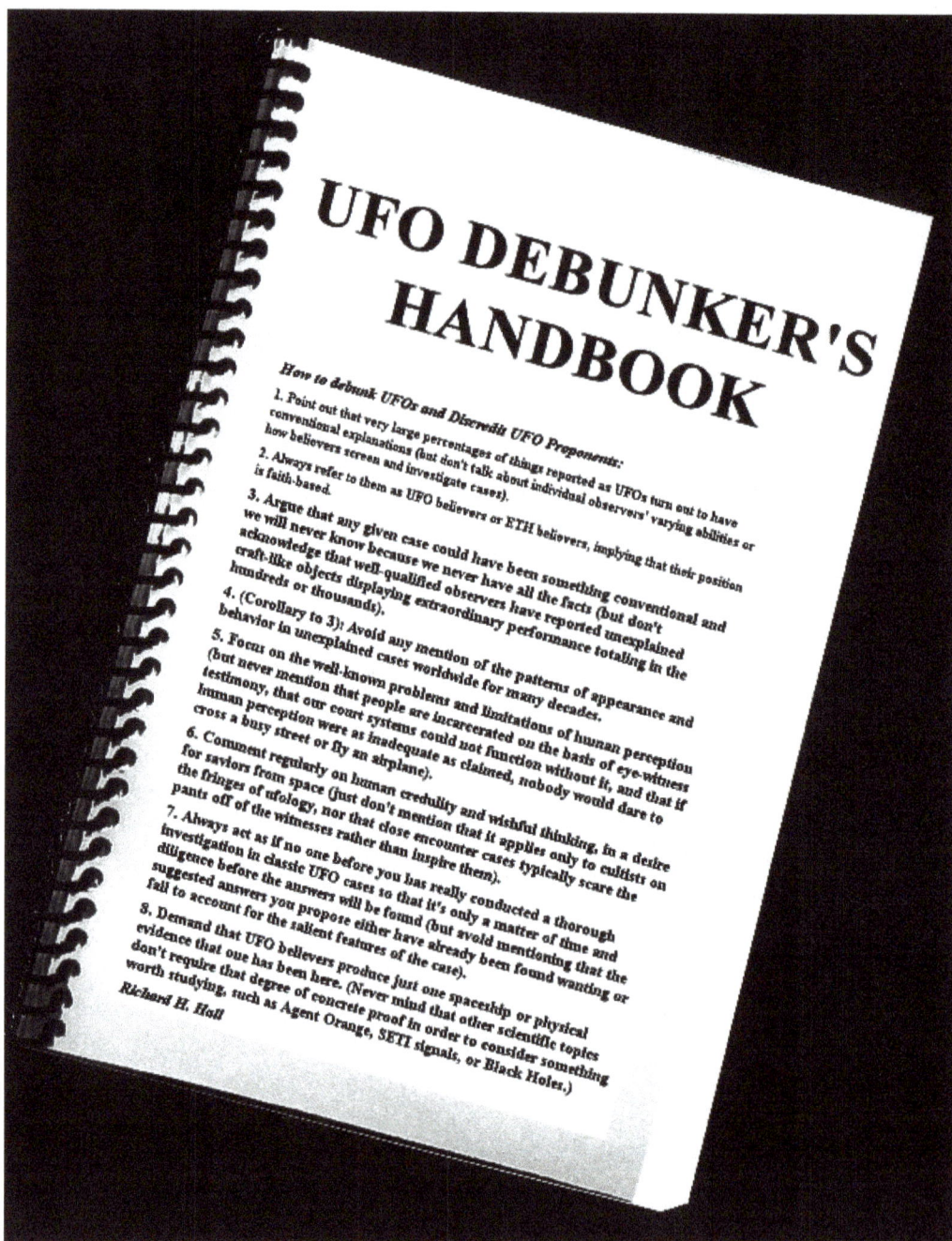

UFO DEBUNKER'S HANDBOOK

How to debunk UFOs and Discredit UFO Proponents:

1. Point out that very large percentages of things reported as UFOs turn out to have conventional explanations (but don't talk about individual observers' varying abilities or how believers screen and investigate cases).

2. Always refer to them as UFO believers or ETH believers, implying that their position is faith-based.

3. Argue that any given case could have been something conventional and we will never know because we never have all the facts (but don't acknowledge that well-qualified observers have reported extraordinary performance craft-like objects displaying extraordinary performance totaling in the hundreds or thousands).

4. (Corollary to 3): Avoid any mention of the patterns of appearance and behavior in unexplained cases worldwide for many decades.

5. Focus on the well-known problems and limitations of human perception (but never mention that people are incarcerated on the basis of eye-witness testimony, that our court systems could not function without it, and that if human perception were as inadequate as claimed, nobody would dare to cross a busy street or fly an airplane).

6. Comment regularly on human credulity and wishful thinking, in a desire for saviors from space (just don't mention that it applies only to cultists on the fringes of ufology, nor that close encounter cases typically scare the pants off of the witnesses rather than inspire them).

7. Always act as if no one before you has really conducted a thorough investigation in classic UFO cases so that it's only a matter of time and diligence before the answers will be found (but avoid mentioning that the suggested answers you propose either have already been found wanting or fail to account for the salient features of the case).

8. Demand that UFO believers produce just one spaceship or physical evidence that one has been here. (Never mind that other scientific topics don't require that degree of concrete proof in order to consider something worth studying, such as Agent Orange, SETI signals, or Black Holes.)

Richard H. Hall

How to deflate the "hot air" out of professional debunkers
http://www.theufochronicles.com/2012/03/debunkers-irrational-uninformed-and.html

So, for those wishing to up their debating skill sets, **Patrick Cooke** has put together a wonderful list of argumentative tactics that many UFO debunkers use when debating against UFO believers. By recognizing them in advance any UFO believer or researcher can win against the professional skeptic or debunker. Here is the list in point form:

"Anyone armed with the knowledge of how debunkers operate can see common threads in the way they argue their points and counter them."

- "It almost always starts with a condescending and self-assured attitude
- Dismissive terms such as ridiculous, absurd, trivial, or even pathetic are used to make the UFO believer seem ignorant and gullible.
- Arguments are as abstract and theoretical as possible, but presented in a manner that makes science superior to any actual evidence that might challenge it, making such evidence seem to be completely worthless.
- They constantly reinforce the popular misconception that anything that challenges the status quo must be inherently unscientific. They deliberately confuse the "process" of science with the "content" of science.
- The use of socially accepted authority figures, whether their expertise is in any discipline even related to the UFO field.
- The degree to which they can stretch the truth is directly proportional to the prestige of authorities they cite. This gives them the latitude of asserting that their statements are "facts", while those of the UFO believers are only "claims". They can, therefore, completely avoid examining the actual evidence and say, with impunity, that there is absolutely no evidence to support such ridiculous claims as the existence of UFOs.
- "The debunkers use the science as a weapon and accuse UFO believers of viewing science in fuzzy, subjective, or metaphysical terms and downplay the fact that free inquiry, legitimate disagreement, and respectful debate are a normal part of science.
- At every opportunity, they reinforce the notion that what is familiar is necessarily rational. The unfamiliar is, therefore, irrational and, consequently, inadmissible as evidence and, at best, an honest misinterpretation of the conventional.
- They also maintain that in investigations of unconventional phenomena, a single flaw or misstep invalidates the whole.
- They assert that if absolute proof is lacking, there is no evidence. Conversely, they claim that if sufficient evidence has been presented to warrant further investigation of an unusual phenomenon, evidence alone proves nothing.
- This will eliminate the possibility of initiating any meaningful process of investigation, particularly if no criteria of proof have yet been established for the phenomenon in question.
- And, in a seemingly logical argument, they insist that criteria of proof cannot possibly be established for phenomena that do not exist. No matter the weight of evidence proving the existence of UFOs, they simply claim that "extraordinary claims demand extraordinary evidence" taking care never to define where the "ordinary" ends and the "extraordinary" begins. This will allow them to manufacture an infinitely receding evidential horizon, which always lies just out of reach."
- Another common practice of UFO debunkers is lumping all phenomena, popularly deemed paranormal, together. In this way, they can indiscriminately drag material across disciplinary lines from one case to another to support their views, as needed.
- UFO debunkers use "ridicule, ridicule, ridicule" to hammer at the concept they are attacking. As, far and away, the single most effective weapon in the **war** against discovery and innovation, ridicule has the unique power to make people completely limp and fails to wither only those few of sufficiently independent thought.
- Trivializing the case by trivializing the entire field in question.
  http://www.bibleufo.com/debunking.htm Patrick Cooke

- They accuse investigators of unusual phenomena of believing in invisible forces and extrasensory realities.
- They also try to discredit the whole story by attempting to discredit part of the story, taking one element of a case completely out of context and finding something prosaic that hypothetically "could" explain it. With one element having been "explained" away, they can then claim that the entire case has been "explained".
- The tactic of labelling any phenomenon as occult, paranormal, metaphysical, mystical, or supernatural will turn off most mainstream scientists or people with religious or conservative leanings immediately, on purely emotional grounds.
- Asking unanswerable questions based on arbitrary criteria of proof is popular, as well. For instance, why hasn't religion or science addressed this, or if UFOs were real why aren't there clear pictures or videos? And, of course, as a last resort, why haven't they landed on the White House lawn?
- Another effective strategy used, with a long history of success, if the media reports UFO sightings, is to claim that it is for the shock or comedy value alone… or that those reporting the sightings are only looking for that elusive "15 seconds of fame" gives it an air of pure "hype".
- If an unusual or inexplicable event is reported in a sensationalized manner, they hold this as proof that the event itself must have been without substance or worth.
- When a witness states something in a manner that is scientifically imperfect, they instantly treat this statement as if it were not scientific, at all.
- If the claimant is not a credentialed scientist, they argue that his or her claims cannot possibly be scientifically correct. And, the assertion that only scientists, particularly astronomers, are "trained
- Observers" immediately dismisses police officers, pilots, air controllers, or virtually anybody else reporting a sighting as totally unqualified to verify anything they see.
- If they are unable to attack the facts of the case, they attack the participants or the journalists who reported the case.
- Ad hominem arguments, or personality attacks, are among the most powerful ways of swaying the public and avoiding the issue.
- If an investigator or chronicler of the unorthodox has profited financially from activities connected with their research, this is positive proof that they are only in it for the money.
- If their research, publishing, or speaking tours constitute their normal line of work or sole means of support, that is used as conclusive proof that they are only profiteers of sensationalism. If they have laboured to achieve public recognition of their work, they characterize them as publicity seekers.
- The tactic of "shooting the messenger" to ignore the message is common and even extends to the ridiculous practice of claiming that if someone just investigating the incident is blemished, the whole incident is questionable.
- If experts in related fields are involved, debunkers focus on the most minor details of their credentials, again pointing out the erroneous assertion that only astronomers are experts on the UFO question, with the necessary knowledge to speak on such issue.
  http://www.bibleufo.com/debunking.htm Patrick Cooke

- If all else fails, they fabricate entire research projects, by declaring that such claims have been thoroughly discredited by the "top experts in the field", whether or not such experts have ever actually studied the claims, or, for that matter, even exist.
- Finally, the tactic of choice is to debunk UFOs by debunking the concept of extraterrestrials. Debunkers declare that there is no proof that life can exist in outer space. They argue that all reports of extraterrestrials must be bogus because the evolution of life on Earth is the result of an infinite number of accidents in a genetically isolated environment.
- They completely avoid addressing the logical proposition that if interstellar visitations have occurred, Earth cannot be considered genetically isolated in the first place.
- They use nature's laws as proof that interstellar travel by extraterrestrials is impossible, because it would, obviously, violate nature's laws.
- They, of course, point out that the government-sponsored SETI program assumes, in advance, that extraterrestrial intelligence can only exist light-years away from Earth and, thus, this invalidates all terrestrial reports of ET contact.
- The important thing to consider is that debunkers are probably guiltier of practicing the very tactics they claim UFO believers are guilty of using.
- Being aware of the tactics debunkers use is important, but using logic in answering them, and not being intimidated by their self-assured and dismissive attitudes, is essential.
- They prey on weakness, thrive on ignorance, and survive, only on their ability to baffle the believer with the power of ridicule.
- While the attacked is defending against the attack, the debunker is constantly and rapidly shifting the argument in several different directions and changing the strategy of attack.
- They answer questions with questions, challenges with dismissal, and logic with unsupported facts.
- Insisting that they back their assertions, answer questions and challenges, and engage in the debate, instead of skirting every issue with a constant attack on the credibility of any issue they confront, will change the outcome.
- They do not deal in honest discourse; they only know how to use the tactics they have learned, and any diversion from those tactics will turn the tide against them. The best way to counter debunkers is to listen to their rhetoric, learn their methods, and how to counter them. This takes study and careful attention to the subtle details of the debunker's methods.

The full text of these arguments can be found at the above website."
http://www.bibleufo.com/debunking.htm Patrick Cooke

Stanton T. Freidman as a nuclear physicist, who has had a serious interest in flying saucers since 1958, reached four major conclusions which he made known on December 1997:

1. The evidence is overwhelming that Planet Earth is being visited by intelligently controlled extraterrestrial spacecraft. In other words, SOME UFOs are alien spacecraft. Most are not.

2. The subject of flying saucers represents a kind of Cosmic Watergate, meaning that some few people in major governments have known since July 1947, when two crashed saucers

and several alien bodies were recovered in New Mexico, that indeed SOME UFOs are ET. As noted in 1950, it's the most classified U.S. topic.

3. None of the arguments made against conclusions One and Two by a small group of debunkers such as Carl Sagan, my University of Chicago classmate for three years, can stand up to careful scrutiny.

4. The Flying Saucer story is the biggest story of the millennium: visits to Planet Earth by aliens and the U.S. government's cover-up of the best data (the bodies and wreckage) for over fifty years. http://www.stantonfriedman.com/index.php?ptp=ufo_challenge

When it comes to dealing with the flim-flam flamboyance of the professional debunkers, Stanton Freidman lays it on the line and exposes how they try to debate someone like himself or other researchers on the subjects of UFOs and the paranormal.

"The problem is NOT that there is not enough evidence to justify my conclusions; but that most people, especially the noisy negativists, are unaware of the real, non-tabloid evidence."

Debunkers seem to employ four major rules:

A. What the public doesn't know, we certainly won't tell them. The largest official USAF UFO study isn't even mentioned in twelve anti-UFO books, though every one of those books' authors was aware of it.

B. Don't bother me with the facts, my mind is made up.

C. If one can't attack the data, attack the people. It is easier.

D. Do one's research by proclamation rather than investigation. It is much easier, and nobody will know the difference anyway. http://www.stantonfriedman.com/index.php?ptp=ufo_challenge

In the article entitled "Skeptics, Pelicanists and Prozac Explanations" by **Dimitris Hatzopoulos**, a well respected UFO researcher from Greece states on the website Best UFO Resources:

"I have read carefully enough the "skeptical" opinions about UFOs, to be angry with the misrepresentations, fact-twisting, half-truths and even outright lies told with a straight face, in the name of "rationality" and "science", when in fact, they are unscientific and anti-scientific. I discovered that "UFO skeptics" often offer and publicize explanations that contradict the available evidence or descriptions of what happened. When I discovered that "UFO skeptics" accept explanations that are found through analysis to be invalid, I became skeptical of the "UFO skeptics". A certain ilk of skeptics (the "debunkers") will attack whenever they feel their paradigms and the status quo are questioned. Debunkers' efforts to reinvent witness testimony, to reinforce their own unexamined adherence to orthodox doctrine, have undermined understanding of a wide range of extraordinary human experiences. In several countries, after decades of anti-UFO propaganda, the "giggle factor" has become generational. Of concern is the obvious credibility gap of Science -as an institution- in the minds of people, who have

witnessed a UFO, assuming they chose to believe their lying eyes, instead of proclamations of so-called "experts". This destroys trust. Polls in the US and Canada show an overwhelming majority of people think their government is covering up information about UFOs. When the public loses trust in government experts, there is a ripple effect outwards of diminished trust in all expert scientific opinion." http://www.hyper.net/ufo/skeptics.html

The question now becomes what is the real motive and agenda of professional UFO debunkers, it certainly isn't to advance science and provide enlightenment to society, so what's really going on?

The debunkers are the Counter-UFO soldiers out to disseminate the gospel of denial. But why are they diligently debunking and covering up perhaps the most astounding phenomenon in history? Here's a list of reasons, explanations, and possibilities as to why debunkers want to debunk UFOs:

1. Intelligence agents and government operatives are on a mission to plant seeds of doubt and to distribute disinformation, to continue the ongoing cover-ups.
2. Disinformation experts are paid to debunk UFOs. Money matters most if nothing else.
3. Debunkers along with other people fear what the truth of the matter may be, so they hastily deny it. Alternative explanations will do.
4. Debunkers depend on the fact that many people believe life only exists on Earth, such as most atheists and certain fundamentalist Christians, so UFOs and ETs can't possibly exist.
5. They thrive on sheer skepticism. If it can't possibly exist, it definitely doesn't.
6. They proclaim that the concept of UFOs and ETs goes against conventional science. To stay inside the box is practical and acceptable by the norm.
7. Their DNA is obviously hard-wired in deniability - in other words, they're ornery and stubborn! Plus they're collective heads are stuck in the sand.
8. Many of them simply and honestly flat-out disbelieve in UFOs and ETs. With the exception of those getting paid - they don't care either way.
http://ezinearticles.com/?Anatomy-of-the-UFO-Debunkers---Part-Four&id=4272238

These are all reasonable explanations as to the professional debunker's motives but, the first possibility is more probable than the rest of the explanations as their modus operandi is too similar to the methods employed by government intelligence agents like those found in the CIA. If they are not CIA shills in the field of Ufology, then these people are some of the most naive people around who are the unwitting patsies of government intelligence organizations.

Either way, maintaining a position as a debunker does not exemplify one's critical thinking abilities, but instead, casts suspicion on their motives and who is really supporting and paying them to discredit any and all in the field of UFO research including the witnesses to such events. **Stanton Friedman** has already shown that astronomer **Donald Menzel** was a known NSA consultant and professional skeptics and debunkers like **Philip Klass** and kind all seem to follow in the footsteps of Menzel, perhaps maybe, more polished than Menzel.

The only way to know for sure which debunker is a government agent or shill is to do some background checks on the current crop of recognized public debunkers. Some will just be people with an opposing opinion working as the *"devil's advocate"* while others will no doubt be someone other than they claim to be. Keep in mind, as will be shown later in this book, that journalists also make some of the best government agents and CIA and NSA agents are known to work within all the major media networks so, why won't they also operate as professional UFO debunkers?

Adventures in Critical Thinking Without Ignoring the Facts

Critical thinking, depending on your understanding of the concept, may be defined as a judgment that is both purposeful and reflective, as an analytical and interpretative appraisal of what to believe, to accept, to reject, to reserve judgment for a later time or in what measure of response based on the "degree of confidence" to the claims, the observations, the experiences, the verbal or written expressions and arguments.

If we examine the concept of skepticism in the context of critical thinking, that is a positive healthy skepticism as it relates to the subject of UFOs and ETI as opposed to a negative prejudicial skepticism, we will understand the motives behind the rationale of the individual skeptic.

"Critical thinking is not always an expression of disapproval but may conclude that from a critical evaluation, an argument is positive and good. While thinking is the normal function of a rational mind, critical thinking is an intentional evaluation of distinctive thinking. The ability to think critically involves three things:

1. An attitude of being disposed (state of mind regarding something) to consider in a thoughtful way the problems and subjects that come within the range of one's experiences,
2. Knowledge of the methods of logical inquiry and reasoning,
3. Some skill in applying those methods.

Critical thinking calls for a persistent effort to examine any belief or supposed form of knowledge in the light of the evidence that supports it and the further conclusions to which it tends." http://en.wikipedia.org/wiki/Critical_thinking

We will see that many serious UFO researchers with all their foibles and fallibilities, their disposition and initial biases are nevertheless, faithful to the concepts of critical thinking and healthy skepticism in trying to understand the phenomenon before them. Their motives is to uncover potentiality of new knowledge through public disclosure, to advance science thereby, advance society and civilization in general so, that all may benefit as oppose to an elite few who wish to control and manipulate the knowledge, the acquisition, and utilization of alien technology, and essentially place an information embargo on this truth that is inaccessible to the general public.

In comparison, the professional UFO skeptics and debunkers wear the mantle of skepticism however, their motives are anything but honest and are stained with blatant obfuscation, manipulation, and omission of facts and half-truths, logic and reasoning are employed to weave an endless web of deceit and lies, fabrication, and misinformation disguised as truths. This is negative skepticism where prejudice, bias, distortion and suppression of the truth become the ultimate aim while maintaining the guise and rationale of the scientific model. Egocentrism and self-deception whether knowing or unknowingly are irrelevant when skillful arguments are used to cover up the truth by attacking the credibility of the UFO promoter or exposing the weak debating skill set instead debating intelligent fact with counter intelligent fact to discover the hidden truth of the subject matter. Many Skeptics are fully aware of and understand the truth and reality of UFOs and ETI but, ignore or refuse to acknowledge it and look for ways to convince less informed people of the UFO reality with disinformation and misinformation in close looped logic.

CHAPTER 20

THE UFO INVESTIGATOR - A HEALTHY SKEPTIC

"Skeptic – *One, who practices the method of suspended judgment, engages in rational and dispassionate reasoning as exemplified by the scientific method, shows willingness to consider alternative explanations without prejudice based on prior beliefs, and who seeks out evidence and carefully scrutinizes its validity."* http://theethicalskeptic.com/category/ethical-skepticism/what-is-ethical-skepticism/

It has been demonstrated in the above section on professional skeptics and debunkers that they do not adhere to the method of scientific investigation and impartiality, so let's look at the healthy skeptic who investigated the field of Ufology and adjust their outlook and perceptions when truth was uncovered and acknowledged.

Donald Keyhoe

Donald Keyhoe was one of the first UFO investigators, who was a strong advocate of the existence of UFOs as extraterrestrial spacecraft yet, was skeptical that they may have been piloted by extraterrestrial life forms. Relying on his military background and training as well as his association with other military personnel, he followed the trail of UFO evidence like a pit bull with a bone in its mouth regardless of the opposition which came from skeptics and from his fellow military alumni.

Though his disposition could be interpreted as being biased toward the reality of UFOs, it was based on his ability as a trained observer to understand the difference between conventional military aircraft and atmospheric phenomenon and with the unusual descriptions of craft as reported by eyewitnesses that did not fit into the normal conventionality of aerial objects. He, therefore, applied the methods of logical inquiry and reasoning, ever mindful of possible self-deception, personal prejudice, and misinformation that would arise from time to time.

Keyhoe as director of the **NICAP** organization felt throughout his investigations that UFOs were a remotely controlled type of robotic device sent from another planet either within our solar system or from another star system as intelligence gathering probes relaying information back to the home planet of origin. This particular conclusion was arrived at through the lack of evidence of ET life forms in association with the spacecraft reported by eye witness reports. Initially, his belief that UFOs were unmanned probes similar to our own remote-controlled satellites seemed to be the correct assessment of the situation given the fact that there were few eye witness reports, particularly in America, supporting the piloted spacecraft by ET life forms. Later, as such reports of alien piloted craft emerged both nationally and globally, the focus away from UFO/ET spacecraft to the alien presence took centre stage, and eventually, Keyhoe was forced to concede to the reality of the ET hypothesis before his death.

Donald Keyhoe

Dr. J. Allen Hynek

Dr. J. Allen Hynek, former astronomer from North Western University in Illinois was typical of the skeptical UFO researcher who used critical thinking and discerning judgment in his investigations. In early days of UFO research not much was known about the phenomenon and speculation surrounded the mystery so, it was normal for most serious researchers to be skeptical of what they were investigating. Hynek's skepticism was tempered with his tireless pursuit of the truth behind the UFO mystery, always questioning the evidence, always applying the theoretical constructs and the scientific method for understanding the nature of the phenomenon. His thinking was logical, with crystal clarity, with an attention to detail and accuracy, depth and breadth, and above all fairness in his appraisals and assessments.

Although highly manipulated by the protocols and hidden agenda of the US Air Force's officialdom for whom he was commissioned by to be their chief spokesman and UFO investigator. He nevertheless, through personal integrity and honesty discovered that the UFO phenomenon was indeed real and of extraterrestrial in origin. This was critical thinking employed at its very best with no disposition or preconceptions but, following the evidence to wherever it would lead him and reserving judgment until all the facts were in.

Dr. J. Allen Hynek

Jacques Vallee

Jacques Vallee Vallée was born in Pontoise, France. He received his Bachelor of Science degree in mathematics from the Sorbonne, followed by his Master of Science in astrophysics from the University of Lille. He began his professional life as an astronomer at the Paris Observatory in 1961. He was awarded the Jules Verne Prize for his first science-fiction novel in French.

He is a mathematician and computer scientist, noted in mainstream science for co-developing the first computerized mapping of Mars for NASA and for his work at SRI International on the network information center for the **ARPANET**, a precursor to the modern Internet.

Along with his mentor, astronomer **J. Allen Hynek**, Vallée carefully studied the phenomenon of UFOs for decades. His name and reputation with the study of **unidentified flying objects (UFOs)** served as the real-life model for the character portrayed by François Truffaut in **Steven Spielberg's** film *Close Encounters of the Third Kind*.

Initially, Vallée's honest and diligent style of investigation into reports of ET sightings was in defense of the scientific legitimacy of the **Extraterrestrial Hypothesis (ETH)** and later he became known for his promotion of the **Interdimensional Hypothesis (IDH).**
http://en.wikipedia.org/wiki/Jacques_Vall%C3%A9e

60

Jacque Vallee

In May 1955, Vallée first sighted an unidentified flying object over his Pontoise home. Six years later in 1961, while working on the staff of the **French Space Committee**, Vallée witnessed the ***destruction of the tracking tapes*** of an unknown object orbiting the earth. The particular object was a ***retrograde satellite*** (this satellite has become known in recent times as the "***Black Knight Satellite***" because of it coloration and is believed to have been supposedly orbiting the Earth for the last13,000 years!) – that is, a satellite orbiting the earth in the opposite direction to the earth's rotation. At the time he observed this, there were no rockets powerful enough to launch such a satellite, so the team was quite excited as they assumed that the Earth's gravity had captured a natural satellite (asteroid). A superior came and erased the tape. These events contributed to Vallée's long-standing interest in the UFO phenomenon.

His investigations soon lead him to become skeptical that the **ET Hypothesis** was the correct hypothesis to this phenomenon. Vallee was convinced by the historical and adaptive aspects of this phenomenon to mankind's evolvement on the planet that its origins had to be ***inter-dimensional*** *or* ***ultra-dimensional*** in nature. This was a brave position to take which unfortunately was not shared by the majority of his fellow UFO researchers.

Vallée began exploring the commonalities between UFOs, cults, religious movements, demons, angels, ghosts, cryptid sightings, and psychic phenomena. Speculation about these potential links was first detailed in Vallée's third UFO book, *Passport to Magonia: From Folklore to Flying Saucers*. **http://en.wikipedia.org/wiki/Jacques_Vall%C3%A9e**

As an alternative to the extraterrestrial visitation hypothesis, Vallée has suggested a **multidimensional visitation hypothesis**. This hypothesis represents an extension of the ETH where the alleged extraterrestrials could be potentially from anywhere. The entities could be multidimensional beyond space-time, and thus could coexist with humans, yet remain undetected.

Vallée's opposition to the popular ET hypothesis was not well received by prominent U.S. Ufologists, hence he was viewed as something of an outcast. Indeed, Vallée refers to himself as a *"heretic among heretics"*.

Vallée's opposition to the ETH theory is summarized in his paper, *"Five Arguments Against the Extraterrestrial Origin of Unidentified Flying Objects"*, Journal of Scientific Exploration, 1990:

Scientific opinion has generally followed public opinion in the belief that unidentified flying objects either do not exist (the "natural phenomena hypothesis") or, if they do, must represent evidence of a visitation by some advanced race of space travellers (the extraterrestrial hypothesis or "ETH"). It is the view of the author that research on UFOs need not be restricted to these two alternatives. On the contrary, the accumulated database exhibits several patterns tending to indicate that UFOs are real, represent a previously unrecognized phenomenon and that the facts do not support the common concept of "space visitors."
http://en.wikipedia.org/wiki/Jacques_Vall%C3%A9e

Five specific arguments articulated here contradict the ETH:

1. **unexplained close encounters are far more numerous than required for any physical survey of the earth;**
2. **the humanoid body structure of the alleged "aliens" is not likely to have originated on another planet and is not biologically adapted to space travel;**
3. **the reported behavior in thousands of abduction reports contradicts the hypothesis of genetic or scientific experimentation on humans by an advanced race;**
4. **the extension of the phenomenon throughout recorded human history demonstrates that UFOs are not a contemporary phenomenon; and**
5. **the apparent ability of UFOs to manipulate space and time suggests radically different and richer alternatives.**

Vallée has contributed to the investigation of the **Miracle at Fatima** and **Marian apparitions**. His work has been used to support the **Fatima UFO Hypothesis**. Vallée is one of the first people to speculate publicly about the possibility that the "solar dance" at Fatima was a UFO. The idea of UFOs was not unknown in 1917, but most of the people in attendance at the Fatima apparitions would not have attributed the claimed phenomena there to UFOs, let alone to extraterrestrials. Vallée has also speculated about the possibility that other religious apparitions may have been the result of UFO activity including **Our Lady of Lourdes** and the revelations to **Joseph Smith.Jacques Vallée** and other researchers have advocated further study of unusual phenomena in the academic community. They don't believe that this should be handled solely by theologians. http://en.wikipedia.org/wiki/Jacques_Vall%C3%A9e

Vallée proposes that there is a genuine UFO phenomenon, partly associated with a form of non-human consciousness that manipulates space and time. The phenomenon has been active throughout human history and seems to masquerade in various forms to different cultures. In his opinion, the intelligence behind the phenomenon attempts social manipulation by using deception on the humans with whom they interact.

Vallée also proposes that a secondary aspect of the UFO phenomenon involves human manipulation by humans. Witnesses of UFO phenomena undergo a manipulative and staged spectacle, meant to alter their belief system, and eventually, influence human society by suggesting alien intervention from outer space. The ultimate motivation for this deception is probably a projected major change of human society, the breaking down of old belief systems and the implementation of new ones. Vallée states that the evidence if carefully analyzed, suggests an underlying plan for the deception of mankind by means of unknown, highly advanced methods. Vallée states that it is highly unlikely that governments actually conceal alien evidence, as the popular myth suggests. Rather, it is much more likely that that is exactly what the manipulators want us to believe. Vallée feels the entire subject of UFOs is mystified by charlatans and science fiction. He advocates a stronger and more serious involvement of science in the UFO research and debate.[11] Only this can reveal the true nature of the UFO phenomenon.

Vallée is often highly critical of UFO investigators overall, both believers and skeptics, asserting that what often passes for an acceptable level of investigation in a UFO context would be considered sloppy and seriously inadequate investigation in other fields. He has pointed out logical flaws and methodological flaws common in such research. Unlike many critics of UFO investigative efforts, his critiques are not condescending and dismissive and he indicates that he is simply interested in good science. http://en.wikipedia.org/wiki/Jacques_Vall%C3%A9e

His research has taken him to countries all over the world. Considered one of the leading experts in UFO phenomena, Vallée has written several scientific books on the subject which of particular interest to other Ufologists are listed here:

- *Anatomy of a phenomenon: unidentified objects in space – a scientific* appraisal *(1st hardcover ed.). NTC/Contemporary Publishing. January 1965. ISBN 0-8092-9888-0.*

  Reissue: UFO's In Space: Anatomy of A Phenomenon (paperback reissue ed.). Ballantine Books. April 1987. p. 284. ISBN 0-345-34437-5.

- *Challenge to Science: The UFO Enigma – with Janine Vallée (1966)*
- *Passport to Magonia: From Folklore to Flying Saucers. Chicago, IL, U.S.A.: Publ. Henry Regnery Co. 1969.*
- *The Invisible College : What a Group of Scientists Has Discovered About UFO Influences on the Human Race (1st ed.). 1975.*
- *The Edge of Reality – Jacques Vallée and Dr. J. Allen Hynek (1975)*
- *Messengers of Deception: UFO Contacts and Cults (paperback ed.). Ronin Publ. June 1979. p. 243. ISBN 0-915904-38-1.*
- *Dimensions: A Casebook of Alien Contact (1st ed.). Contemporary Books. April 1988. p. 304. ISBN 0-8092-4586-8.*

- *Confrontations – A Scientist's Search for Alien Contact* (1st ed.). Ballantine Books. March 1990. p. 263 hardcover. ISBN 0-345-36453-8.
- *Revelations: Alien Contact and Human Deception* (1st ed.). Ballantine Books. September 1991. p. 273 hardcover. ISBN 0-345-37172-0.
- *UFO Chronicles of the Soviet Union : A Cosmic Samizdat* (1992)
- *Forbidden Science: Journals, 1957-1969* (1992)
- *Wonders in the Sky: Unexplained Aerial Objects from Antiquity to Modern Times* (1st ed.). Tarcher. 2010. p. 528 paperback. ISBN 1-58542-820-5.

Vallee's research papers on UFO include:

- Five Arguments Against the Extraterrestrial Origin of Unidentified Flying Objects – Jacques Vallée, Ph.D.
- Six Cases of Unexplained Aerial Objects with Defined Luminosity Characteristics – Jacques Vallée Ph.D.
- Physical Analyses in Ten Cases of Unexplained Aerial Objects with Material Samples – Jacques Vallée, Ph.D.
- Report from the Field: Scientific Issues in the UFO Phenomenon – Jacques Vallée, Ph.D.
- Crop Circles: "Signs" From Above or Human Artifacts? – Jacques Vallée, Ph.D.
- Are UFO Events related to Sidereal Time – Arguments against a proposed correlation – Jacques Vallée, Ph.D.

http://en.wikipedia.org/wiki/Jacques_Vall%C3%A9e

Dr. Steven Greer

Dr. Steven Greer is arguably, the world's foremost proponent of the **Extraterrestrial Hypothesis** having considerable firsthand experience with UFO and ETI encounters, but also, with an astounding knowledge and understanding of the cosmology of astral and spirit entities as well as ET life forms. Dr. Greer has pioneered the successful contact protocols for human initiated encounters with Extraterrestrial Intelligences as well as bringing many top military officials, scientist, astronauts, intelligent operatives, and whistle- blowers to the forefront in a public disclosure event at the **National Press Club** in Washington, D.C. in 2001. These men and women, all with impeccable credentials, who are considered heroes of the country, have sworn affidavits of first-hand involvement with UFO crash retrievals, simultaneous radar and visual sightings, alien captures, interrogations and examinations and other damning evidential material. But, **Dr. Greer** is also, a healthy skeptic particularly, when it comes to abduction scenarios and cattle mutilations as allegedly caused by malevolent ETI. With irrefutable evidence which proves that quasi para-military intelligences are behind most, if not all human abductions and cattle mutilations, Steven Greer has done more to resolve the mystery that surrounds the UFO/ETI phenomenon that anyone else before him.

These men stand out as healthy skeptics in pursuit of the truth and who were not easily dissuaded by the opposition to block the truth from becoming public knowledge. In stark contrast to these seekers of truth and understanding, are the professional skeptics, hoaxers, naysayers, and debunkers, who are the suppressors of truth. Their sole purpose appears to be in clouding the reality of the existence of Extraterrestrial Intelligences visiting this planet thus, creating confusion through outright denial and falsehoods which many UFO researchers feel borders on the deliberate cover up and suppression of the scientific knowledge to be gained from the UFO reality.

Dr. Steven M. Greer
http://www.jerrypippin.com/UFO_Files_disclosure_project.htm

It should be stated here, that not all skeptics are agents of some intelligence agency or a member of a covert cover-up group; many may simply have an honest difference of opinion to the existence of UFOs and ETI and desire further scientific inquiry into the matter. Perhaps, they want to make a name for themselves by disproving the evidence of UFOs/ETI and "stealing the thunder "away from the people who had actually have seen UFOs.

Debunkers are in many ways similar in motive and agenda to their counterpart, the UFO promoters; they seek the attention of the news media in order to put themselves in the limelight whenever a UFO story, picture, or video grabs the media headlines. However, they may also, be the unsuspecting players of the covert intelligence groups unwittingly aiding them in a never ending game of "UFO hide and seek," the suppression and elimination of the truth.

"Critical thinking gives due consideration to the evidence, the context of judgment, the relevant criteria for making the judgment well, the applicable methods or techniques for forming the judgment, and the applicable theoretical constructs for understanding the nature of the problem and the question at hand. Critical thinking employs not only logic but broad intellectual criteria such as clarity, credibility, accuracy, precision, relevance, depth, breadth, significance and fairness." http://en.wikipedia.org/wiki/Critical_thinking

"Within the framework of scientific skepticism, the process of critical thinking involves the careful acquisition and interpretation of information and use of it to reach a well-justified conclusion. Critical thinking is important because it enables one to analyze, evaluate, explain, and restructure our thinking, decreasing thereby the risk of adopting, acting on, or thinking with, a false belief. However, even with knowledge of the methods of logical inquiry and reasoning, mistakes can happen due to a thinker's inability to apply the methods or because of character traits such as egocentrism. Critical thinking includes identification of prejudice, bias, propaganda, self-deception, distortion, misinformation, etc."
http://en.wikipedia.org/wiki/Critical_thinking

Within the parameters of critical thinking, one's dispositional perspective is uniquely individualistic or character-logical. As a truth seeker, a reality check of one's personal beliefs and values is always necessary in order to stay true to a logical and conclusive outcome. "Its focus is in developing the habitual intention to be truth-seeking, open-minded, systematic, analytical, inquisitive, confident in reasoning, and prudent in making judgments." Those who have an opposite disposition and are denigrators of truth" are intellectually arrogant, biased, intolerant, disorganized, lazy, heedless of consequences, indifferent toward new information, mistrustful of reasoning, imprudent; are more likely to encounter problems in using their critical thinking skills." If one fails to recognize one's own disposition, various forms of self-deception and closed-mindedness, both individually and collectively, will result.
http://en.wikipedia.org/wiki/Critical_thinking

Author's Rant: I suspect that that the motives and agendas for most professional UFO skeptics and debunkers, is not the pursuit of truth as the end game result whereby, new knowledge is discovered but, its suppression and cover-up in unbridled arrogance and closed-mindedness, heedless of the consequences to others with a total disregard to any new knowledge that may be gained. These are harsh words no doubt towards most skeptics and debunkers but, over the decades, I have not seen any change by these people in their thinking or attitude or in their being more open-minded and inquiring toward this subject by which I can be dissuaded from this conclusion. I am left concluding that certain professional skeptics and debunkers are in full knowledge of the reality of the UFO phenomenon and are duplicitous in the cover-up process or are the unwitting pawns in a well-orchestrated agenda by the military industrial complex for reasons that only they know.

We will see, as we go deeper still into this subject, just how far its Machiavellian design and agenda really extend. An agenda that is slowly and methodically being imposed upon an unsuspecting public, that may lead to an irreversible outcome for most of humanity, the

likes of which make World War Two Nazi Germany's imperialistic machinations look like childish school playground bullying!

"When persons possess intellectual skills alone, without the intellectual traits of mind, *weak sense critical thinking* results. Fair-minded or *strong sense critical thinking* requires intellectual humility, empathy, integrity, perseverance, courage, autonomy, confidence in reason, and other intellectual traits." From the above four acknowledged leaders and promoters of the UFO and ETI Hypothesis; It becomes readily apparent that a fair-mindedness, personal integrity, intellectual humility, perseverance, etc. becomes the hallmark of their character and reputation.

"Thus, critical thinking without essential intellectual traits often results in clever, but manipulative and often unethical, thought. In short, the sophist, the con artist, the manipulator often uses intellectually defective but effective forms of thought." Here, we see the modus operandi of the UFO skeptics and debunkers. They are clever, perhaps an intellectual with degrees beside their names but, they are unethical and highly manipulative, they are con artists "par excellence," they do not use open-mindedness and intellectual humility with fair play, their debating tactics utilizes "intellectually defective but effective forms of thought," often deflecting any facts away as untrue, unproven and unscientific. In short, there is no sense of integrity about their character. They give an air of being objective but, they uncritically fail to discern their own "subjectivity" and one-sidedness."

Can one paint or "tar and feather" all skeptics and debunkers with the same brush? Of course not, many skeptics do have a legitimate opposing point of view based on personal knowledge and experience which makes their viewpoint just as valid as the UFO promoter. In fact, there are many UFO promoters who fit the same profile as the professional UFO skeptics and debunkers, poor character traits such as egocentrism and weak intellectual traits of mind, who force their agenda of UFO reality upon others like some fanatical religious proselytizer. There are extremes on both sides of this subject and if at all possible, they should be avoided at all costs as they add no understanding to the phenomenon other than the type of mindset they display. These are the assorted nuts, flakes, and granola types within the field of Ufology and professional skepticism! Once again, this is where the psychiatrists, psychologists, and sociologists are needed in this aspect of the UFO phenomenon.

"To develop one's critical thinking traits, one should learn the art of suspending judgment developing one's intellectual empathy and intellectual humility. The second requires extensive experience in identifying the extent of one's own ignorance in a wide variety of subjects (ignorance whose admission leads one to say, "I thought I *knew*, but I merely *believed*"). One becomes less biased and more broad-minded when one becomes more intellectually empathic and intellectually humble, and that involves time, deliberate practice and commitment."
http://en.wikipedia.org/wiki/Critical_thinking

CHAPTER 21

WHY IS THE SEARCH FOR TRUTH ALWAYS LANDMINED WITH MISINFORMATION, DISINFORMATION, EXCORIATION, DENIALS, LIES, AND OBFUSCATION?

In our search for understanding the nature and reality of UFOs and ETI, many questions jump to mind such as *"Are we alone in the universe?"* and *"Are we being visiting by Extraterrestrial Intelligences?"* However, in uncovering the answers to these questions, another very important question arises which ranks equally as high in importance as these first two: *"Why is there misinformation, disinformation, excoriation, denials, lies, and obfuscation in the search for truth?"*

The answer is extremely complex, steeped in subterfuge and completely off the radar scopes of most UFO researchers and what is currently known in the public sector is mostly misinformation, disinformation, and obfuscation, etc. like the title of this subsection states.

The question that needs to be answered first is "If we are being visited by Extraterrestrial beings then, is there a benefit or advantage or conversely, is there a danger or disadvantage to mankind by the ETI visitation? Now, a corollary to this question is that, if they are here, where and what is the benefit or danger, thus far for mankind?

The superficial answer is they are not here because we would have already been benefitted by their presence assuming they are benevolent. Or the question may be understood that they are here but, the ETs have decided not to interact with us at this time, therefore, there has been no benefit to humanity as yet. Or the question could be understood that ETs are here, however, any benefit and advantage from them toward the humanity has been hijacked by the military and government powers and the private industrial sector for selfish reasons. From this, one may still conclude that they are not here at all because there is no evidence of their benefit to us.

In all of this critical thinking and skepticism to hunt down the truth of the matter, the biases and agendas of both the UFO supporter, promoters and denigrators often cloud the real issue, that is, a real someone, an eyewitness to a real unusual event for which he alone he is the only one who can offer an interpretation as to what he has seen. All the professional investigators and naysayers are merely bit players in his experience and unless, called upon by him to assist in solving the mystifying experience, should objectively and impartially record the event, examine the evidence, assess the report in comparison to the tens of thousands of reports worldwide, draw up some conclusions and setup plans and actions based upon assessments and conclusions to resolve the matter to the satisfaction of the general public. Nothing short of this is acceptable!

The second part of the question, if we have been visited then, they must be malevolent to our existence in which case, we need to arm ourselves to the teeth and get ready to duke it out with them to the bitter end! This second part of the question also, has a corollary, "What damage have they done to us? Where is the occupational landing forces and the global carnage and death, the internment camps, the massive human slavery environments, etc., etc? We observed that there is no destruction of ourselves or our planet, therefore, they are not here on this planet or they have decided not to attack us just yet, waiting instead to bide their time until; they can find sufficient weaknesses in our defenses to be able to conquer us. Again, it may be concluded that they are simply not visiting us at all!

Keep in mind that these questions we ask are based on the expectation that there be some physical interaction as proof of their existence and so far, it appears that all we've done is prove what the skeptics and debunkers have been saying all along, that there are no Extraterrestrials visiting the planet! This is typical of the argument, the quasi-logic and deductive reasoning of most skeptics. There is, of course, another possibility which involves an alien agenda that most ufologists, scientists, skeptics and debunkers have not seriously considered which we will explore later on in another part of this book.

We've already proven with hardcore evidence, the reality of this phenomenon and no further proof or evidence is required. We are at this point in our journey to understand the big picture of the UFO/ETI question searching for answers to the secrecy and cover-up for a reality that we already know exists. The clue has been given in the question: "Is there any benefit or advantage to humanity?"

The benefits and advantages have been understood and seized by a small group of well-organized people who do not want to share the benefits with the rest of humanity. The benefits and advantages come in the form of alien technology, alien DNA, alien intelligence and knowledge and the potential for alien communications, exo-sociopolitical interstellar trade, and commerce. The downside or disadvantage is the weaponization of space and it may also, bring to realization our worse fears of an interstellar war! However, the real question is who really is benefitting from all this alien hardware, technology, and knowledge? Well, it hasn't been you and me, or we would have seen the advantages by now. Our world would have transitioned into a new global civilization, **(by the way folks, it is coming, we're almost there, however, we still have a little way to go!)** where we would have been invited into an interstellar community and would have become, as **Dr. Michio Kaku**, the famous Theoretical Physicist, Professor, and bestselling Author has stated so often, a **Type I Civilization**!

So, who stands to benefit from the misinformation, the disinformation, the excoriation, the denials, the lies, and the obfuscation of the truth?" As it turns out, there is more than one group or organization that has broken away as a separate society or civilization becoming the Type I Civilization as spoken about by Kaku.

Most people would say that it is the military industrial complex and for the most part, they would be correct but, the reality is that US military is merely the beneficiary of the ill-gotten alien technology which they obtained by force. Their benefactors, the ones who control the purse strings; who wield the actual power in the country and influence the other power brokers around the world, can be characterized as a shadowy multi-transnational group with rogue subgroups within its ranks, all intent on fulfilling their own personal agendas to the detriment of the rest of humanity.

The influence wielded from this powerful elitist group reaches across all national boundaries, they coerce most other countries politicians and governments by bribery or threats, they control and manipulate most of the world's economies, they command most of the Earth's resources regardless of environmental damage, they "spin doctor" and "sanitized" the news content of most of the world's news media, they eavesdrop and spy from on high in secret orbiting surveillance satellites upon anyone who they feel is a potential threat to their well-crafted agenda.

Their ultimate goal is nothing short of seizing absolute power and control over all of humanity in servitude to the whims of a few wealthy corporate elite. They do not care, whether the Earth and her citizens have a future, they care not about the rape and pillage of mother Earth resources to serve their own greedy and ruthless ends. Their evil and corrupt machinations rival former Nazi Germany, but at this current time it is on a planetary scale and as **Dr. Steven Greer** has coined so succinctly, we are killing our planet; we witnessing the **"planeticide"** of the Earth.

The rape and pillage of planet Earth by its own inhabitants is truly alien and bizarre even, to any visiting extraterrestrial life forms to our planet, who in absolute amazement and horror must view us a suicidal race bent on its own self-destruction!

The Power Players of Mistruth - Behind the Curtain of OZ

A little history is needed at this point to understand the concept of social power and control, the need to wield it and its influences on society at large. In former bygone ages, absolute power and control of the people and a country's natural resources, its government, its economy, it physical, intellectual and spiritual growth and development fell upon two orders in society, namely the *"rulers"* and the *"ecclesiastics,"* or the monarchies and the clergy. Then, in the middle of the 1800's something unique, wondrous and prophetic occurred, something that had never been witnessed before in history. Power was seized from these two powerful and ancient institutions and was given to the common people. Corruption and despotism from within, like a cancer spreading to all the healthy organs of the body, was attacking the very foundations of these social orders. *"The world's equilibrium hath been upset through the vibrating influence of this most great, this new **World Order**. Mankind's ordered life hath been revolutionized through the agency of this unique, this wondrous System – the like of which mortal eyes have never witnessed."* **Gleanings, LXX, p.136 and The Power of the Covenant – Part One, pp.18**

Many kingdoms, empires, and monarchies around the world were toppled or overthrown into extinction. Power from many ecclesiastic orders, whether they were of the **Buddhist, Jewish. Christian or Islamic religions**, all became impotent in their control over their adherents. Though these same religions have continued to grow either, through blind religious observance or through intimation and threats, the potency that once made them unique and vibrant has long since gone leaving a mere husk of resemblance to their past former glory. Many people around the world but, particularly from the undeveloped industrialized nations who were once oppressed, under educated, in extreme poverty from the burdens of excessive taxation and from military armament build-up (Yes! It all began in the mid 1800's) and who were spiritually atrophied were given a "rebirth" of their basic human rights and along with these rights, they were given the powers that were once wielded by the rulers and the clergy.

Though the general populace of the world is largely unaware of the implications of so profound a spiritual and world embracing concept that began from such a small nascent social movement in 1850's Persia (now known as Iran), its effects would cascade forward through time to our present age and beyond for thousands of centuries to come. Its greatest influence and effect would be felt in the West, the power base of great wealth and materialism unlike any other country at the time. The Middle East and Persia, in particular, was the centre of spiritual atrophy and moral decadence. The West, personified by the United States became the centre for ultimate political and material decadence. Both halves of the same apple; both represent a dichotomy of extremes, opposite ends of the spectrum in the social order of civilization on this planet. Yet, each intrinsically and intimately intertwined, an inseparable part of each other destined to bring about a new world order that will lead to the golden age of civilization on this planet.

However, at this current time, the future for mankind looks anything but hopeful, golden, peaceful, or just. It would seem that we are all in a hand basket to hell, whistling, happy, and gleeful as we make our way along a path to a destination, we know nothing about and fill with potential dangers for us all. Our future appears to have been hi-jacked, albeit, temporarily by a few powerful and unscrupulous people within secret covert organizations. As yet, the world as a collective commonwealth has not recognized it newly acquired God-given powers and has not seized the reins of control for its destiny. It is still asleep on its bed of heedless and ignorance unaware of the new day and age in which we live. While the world is in this morose stupor, there have been people who have recognized the profound implications of this unique period of time and they have craftily seized upon the "reins of destiny" to fulfill their own personal agendas at the expense and detriment of everyone else on the planet. Enter the **Wealth Corporate Elite**!

It was back in the late 50's when **President Eisenhower** became aware of the UFO/ETI phenomenon and its implications which were to be used in a hidden agenda for global domination. Realizing that he had lost control over the UFO phenomenon to a "rogue group" of military officials, scientists, and private industrialists, in one of his last public addresses, he warns the public to be aware of the "military-industrial complex."

"In the councils of government, we must guard against the acquisition of unwarranted influence, whether sought or unsought by the military-industrial complex. The potential for the disastrous rise of misplaced powers exists and will persist. We must never let the weight of this combination endanger our liberties or democratic processes."
President Dwight Eisenhower

"The very word secrecy is repugnant in a free and open society. And we are as a people inherently and historically opposed to secret societies, to secret oaths and to secret proceeding. We decided long ago, that the dangers of excessive and unwarranted concealment, of pertinent facts, far out-weigh the dangers which are sighted to justify it." **President John F. Kennedy**

This clandestine *"rogue group"* (aka. The wealthy corporate elite) has orchestrated the development of a world order that is so profane as to be secular and eschatological in its orientation, so perverse in its disposition as to be corrupt on all levels of government and law, and so highly manipulative and materialistic as to polarize society into extremes of an oligarchic wealthy elitists governing over a global populace of poverty stricken indentured servants. This false aberration of a world order that would be forced upon us through a gradual removal of our basic rights, freedoms, and privacy, would enslave most of humanity in a way not known nor witnessed before in human history and would benefit only a small populace of the planet's wealthy corporate elite. Lacking in any world embracing vision or global support, such a world order is doomed to failure at it very inception or would become so, intolerable as to cause a massive global revolution to overthrow such an oppressive regime relegating it upon the dung heap of miserable social failures of history

.

So, diametrically opposed is this concept of a world order from the original spiritual concepts and teachings set down by that 19ᵗʰ century Prophet, **Baha'u'llah** for a world order based on spiritual laws and precepts which is not only our spiritual birthright and but as citizens of this Earth, it is our inheritance from an ever-evolving social order destined to culminate in *a world civilization lasting 500,000 years*!

The question that needs to be asked is who specifically stands to benefit from a secular and corrupt system of world order? If we accept what we read in the newspapers and what we watch on television, we see that regardless of the political turmoil in the world; no matter who holds the top seat in a political office or what party is in power, (although, one party may be more acceptable than another); whether there are wars (which are always financially lucrative for arm dealers); whether the price of oil, gold and blue chip stocks move up or down; whenever, there are catastrophic disasters on a scale affecting tens to hundreds of thousands of people (whether natural or manmade); there are always, opportunities for profiteering and seizing more control over the same hapless populace who are suffering in the extreme. Behind the scenes of chaos and turmoil, wars and catastrophes, there are private industrial corporations operated by wealthy families that are constantly scheming with dark self-vested interests, who are able to profit in almost any financial climate and to control more of the world's resources and thus, its populations.

Private corporate businesses like the oil, gas and coal companies, as well as the power utility companies who are heavily invested in hydroelectric and thermal nuclear plants, are those who stand to benefit most. By controlling the supply and demand of natural resources used in power generation thus, keeping the general public tied into the power grid system, the power base and status quo of a greedy and power-hungry corporate elite is thereby, maintained. Control is also maintained by suppressing and sequestering other forms of energy power generation.

The fossil fuel burning combustion engine depletes its fuel supply quickly and thus required endless refilling to keep it running. This particular fact ensured the longevity of fossil fuel as a source of power to the automobile industry and ensured that the public's love affair with the automobile made them captive to a closed loop grid system. The fuel source began more important that the performance factor of the automobile, even after a hundred plus years of existence, there has been no significant improvement in the combustion engine now than when it was first was invented. Better carburetors have been developed and improved upon since the early days but, they never see the light of day because the automobile industry is like a whore in bed with the oil companies. The auto industry builds powerful combustion engines with higher volumes of fuel consumption per distance travelled, with poor carburetion, this fact meant that more fuel was required in a larger fuel tank in order for it to travel a comparable distance by a more fuel efficient smaller vehicle with a smaller gas tank.

72

With the suppression and black shelving of alternative power plant technology such as battery power or hydrogen and water powered engines, people have little choice but to continually drive the inefficient fossil fuel engine automobiles. Even though, consumers and politicians are demanding these alternative energy sources for their cars, trucks, and buses, the public will still

However, we still haven't given a name or a face to this "rogue group" or even identified any of its members. We understand that their quasi-religious belief has less to do with a Supreme Deity and more with a materialistic belief system, where their temple of unholy worship is the "church" of deceit, lies, and manipulation, their "religious chant" or credo is power and control, their "sacrament" is the acquisition of all earthly resources, and their "wine of life" is the misery and enslavement of the human race. What organization or people fit his description?

It appears that this reclusive and secretive group has finally poked it head from out of that dark cave of seclusion to reveal itself in a supreme air of arrogance and confidence. The **Council on Foreign Relations (CFR), the Trilateral Commission** and the **Bilderberg Group** are just a few of the corporate entities that have identified themselves with a new world order ruled by a one world government. Wealthy families like the **Bushes**, the **Rothschilds**, the **Rockefellers**, and the **Bildebergs, Gettys** and **Buffets** as well as numerous other wealthy families from Europe, the Middle East and from Asia are all interconnected and know each other intimately. All have the same goals and agendas for a one world order. Into this mix can be found royalty and various international politicians but for this dissertation, we will restrict ourselves to the U.S. as they represent the economic and resource power base in the world at this time.

David Rockefeller is an American banker, world statesman, and globalist and in any other time period these positions of status and station would be highly laudable but, these are not ordinary times. **David Rockefeller** is also, a multi-billionaire (and there is nothing wrong with being rich), he is also, CFR Director and Founder of the Trilateral Commission.

Giving credence to his involvement in a world order system, in a letter to the Trilateral Commission, David Rockefeller has been quoted as saying:

"We are grateful to the Washington Post and the New York Times, Time Magazine and to the other publications, whose directors have attended our meetings and respected their promises of discretion for almost forty years. It would be impossible for us to develop our plan for the world, if we are subject to the bright lights of publicity."

What follows next are quotes from various politicians and UN officials and their perception and reaction to the one world order concept as envisioned by **Rockefeller** and company.

"The drive of the Rockefellers and their allies to create a one world government combining supercapitalism and Communism under the same tent, all under their control. I am convinced there is such a plot, international in its scope, generations in the planning, and incredibly evil in intent." - **Congressman Larry P. McDonald, 1976.**

Corporate entities identified with a new world order rule:
Council on foreign Relations, Trilateral commission and Bilderberg Group
Google Images

"To achieve world government, it is necessary to remove from the minds their individualism, loyalty to family traditions, national patriotism, and religious dogmas." - **Brock Adams, Director UN Health Organization**

"No one will enter the New World Order unless he or she will make a pledge to worship Lucifer. No one will enter the New Age unless he will take a Luciferian Initiation." – **David Spangler, Director of Planetary Initiative United Nations**

"In the next century, nations as we know it will be obsolete; all states will recognize a single, global authority. National sovereignty wasn't such a great idea after all." – **Strobe Talbot, President Clinton's Deputy Secretary of State**

"We shall have world government whether or not you like it, by conquest or consent." – **James Warburg, Council on Foreign Relations (CFR) member**

"The CFR uses individuals and groups to bring press from below, to justify the high-level decisions for converting the U.S. From a sovereign Constitutional Republic into a servile member state of a one-world dictatorship." – **Former Congressman John Rarick**

*"The case for government by elites is **irrefutable.***" – **Senator William Fulbright, Former chairman of the US Senate Foreign Relations Committee**

"Fifty men have run America and that's a high figure." – **Joseph Kennedy, father of JFK**

"The real rulers in Washington are invisible, and exercise power from behind the scenes." – **Felix Frankfurter, Supreme Court Justice**

"In my view, the Trilateral Commission represents a skillful, coordinated effort to seize control and consolidate the four centers of power: political, monetary, intellectual, and ecclesiastical. All this is to be done in the interest of creating a more peaceful, more productive world community. What the Trilateralists truly intend is the creation of a worldwide economic power superior to the political governments of the nation-states involved. They believe the abundant materialism they propose to create will overwhelm existing differences. As managers and creators of the system, they will rule the future." - **Senator Barry Goldwater from his book, "With No Apologies"**

What is really interesting about this is that the topic of a one world government is constantly talked about by many powerful people while the various news media is silent on the subject or refuses to comment on its reality until you realize the very people behind this movement are the ones who control the news media. You are not told what you are supposed to know!
https://www.youtube.com/watch?v=7AAJ34_NMcI

The Search for Truth is the Foundation of Enlightenment and the Advancement of a Global Civilization

We have seen that some of the players of mistruth are in the "game" of world domination and power control on this planet. They are considered as the "movers and shakers" who determine the political and economic direction for all of humanity. This would certainly be an unalterable reality if we merely sat back, observed and did nothing to participate in the direction of our destiny; foolishly leaving it to the misguided false imaginings and the corrupt hands of the corporate wealthy elite.

Though these people are knowledgeable in the ways of the world, their knowledge is not based upon divine knowledge: though they have faith and confidence in what they orchestrating upon an unsuspecting world, they lack true faith in mankind's hopeful future; though they are steadfast in their agenda of a one world order and government, their steadfastness is founded on the ephemeral materialistic realities of life; though they speak in the guise of truths about the potential benefits of a new world order under the control of governing economic bodies, their truths are distortions of reality that benefit only themselves: though their actions may demonstrate an upright, honest and just character where peer recognition awards and heaps upon them acclamations and honours yet, in secret, their actions are contradictions in stark contrast to the image of an upstanding pillar of society, whose hidden true agenda is nothing less than the downfall of society; their fidelity or loyalty to their country and to the world at large in which they promise so much is a facade of their true intent, their true loyalties lie with their secret societies and associations; finally their humility in their speech and in their actions is self-serving as they serve not humanity, they do not promote truth in all things and are not self-sacrificing of their time, of their financial resources or their stations in life. These people and their subversive organizations are neither enlightened, nor seekers of truth but, operate in a delusional state of mind wholly inseparable from the materialism of this earth which all the religions of past and present have continuously cautioned us is ephemeral and is the basis of corruption. As Jesus the **Christ** has stated in the New Testament, "It is easier for a camel to pass through the eye of a needle than for a rich man to enter into heaven."

We must remember **Baha'u'llah's** declaration that power had been seized from the *"rulers and ecclesiastics"* and given to the people. This was not an emotional outburst to some tyrannical 1800's time period in Middle Eastern history, it was a declaration vouchsafed to all mankind and not to a selected few of its populace. In this new day and age, we are experiencing the process of disintegration and integration, the death spasms of an old world order being rolled up and the birth pangs of a new world order being rolled out, destined to culminate in a world civilization that will last for the next 500,000 years!

We are living in an age that has been prophesied by all the world's religions, an age unlike no previous age before it. It is a time in which everyone has the responsibility for their own spiritual growth and not dependent on the perceptions and understandings of the clergy or the layman. The independent and unfettered search and investigation of the truth of all things is paramount to one's spiritual growth and to the advancement of society. No longer will people tolerate suppression and cover-up of truth and knowledge that benefits only a few people, all of Earth's citizens must reap the benefits of this age of fulfillment and enlightenment. We will witness a world embracing commonwealth where mankind is finally living in peace and harmony with itself. It is a time in which the golden age of human civilization will be crowned with unimaginable achievements and advancements in all areas of the human condition. In light of our subject at hand, the UFO/ETI phenomenon is an integral part of that process of truth, knowledge, and enlightenment which will form a major cornerstone in the foundation of human civilization on this planet.

Author's Rant: There is a very strong spiritual aspect to this UFO/ETI subject matter which many Ufologists, scientists, military and government officials often seem to overlook or fail to grasp the significance of this relationship and its implications. I will illustrate later in this book series, the universal concepts of spirituality and its interconnectivity with all sentient intelligences, regardless of whether the life forms are human or extraterrestrial. In fact, with a little deductive reasoning on the part of the reader, he will discover one of the reasons why Extraterrestrial Intelligences are here visiting our planet! Can you give an educated guess as to what that may be?

How about the birth or the emergence of a global civilization?!

CHAPTER 22

UFO GHETTOS, UFO SUBCULTURES AND UFO RELIGIONS

An assessment of the field of UFO and ETI research over the past sixty years finds that it is still a nascent science experiencing the usual growth problems of development, trying to survive its adolescent years. Eventually, with experience and with adherence to rigid scientific methods it will mature into a fully recognized and legitimate field of science. Until that moment in time unfolds, they are nevertheless, confronted with the task of constant course correction steering away from potentially self-destructive tendencies inherent in those awkward adolescent years as well as those unforeseen threats from without. This may be an over-simplistic evaluation but, the reality of the situation, if we are, to be honest with our assessment, Ufology at this critical period in time is seriously in a grave predicament.

To those UFO researchers involved in one or more UFO organizations, they will no doubt be in a state of denial of the above assessment, courageously defending their positions and all of Ufology in general. They will state that their UFO organization is growing and surviving, scientific and professional attracting highly qualified and highly educated professionals every day to the ranks of membership. Quite frankly, one should expect this as in any organization. New blood infused into any organization should see it thriving, growing, and becoming a viable entity contributing to the overall growth of society. But, we have seen how some of the largest UFO organizations in the U.S. have gone into extinction for various reasons, already discussed.

Currently, the field of Ufology is a quagmire of competitive egos where any serious UFO investigator must tread precariously through a perpetual haze of unsolved ufological mysteries upon a field that is land-mined with continuous internal wrangling among its researchers, side-stepping the half truths from poorly investigated eyewitness accounts, all the while, dodging the bullets and missiles of faulty assessments and conclusions derived from inconclusive proofs and leaps of logic. The database of UFO research is so seriously flawed and corrupted that it finds itself severely wounded and hemorrhaging, in desperate need of triage, before it becomes a rotting corpse upon the battlefield of scientific inquiry.

It's no wonder that the professional skeptics and UFO debunkers take Ufology and its researchers to task and rake them over the coals of pseudo-science. It's no wonder that the news media has a field day in playing up the "giggle factor" to any UFO story while giving only the slimmest margin for any credibility. It's no wonder that scientists run the other way from becoming involved in an unprofessional organization that is at constant schism with itself, any involvement would be professional suicide. It's no wonder that the military industrial complex feels no threat whatsoever, from the various UFO organizations or from individual investigators as they are generally perceived as chaotic and rarely cohesive in their efforts. When someone or some organization gets too close to discovering the truth, the military industrial complex merely releases disinformation or denials to distract the public's attention away from the claims of the UFO researcher thus, continuing to cloud the issue and to throw doubt on the existence of UFOs and ETI. It's no wonder the intelligence communities finds very little that is intelligent with these organizations, other than to keep an eye on its senior officers.

This then, is the **UFO Ghetto** which all serious researchers at one time or another have come across, and have unwittingly become mired in its membership or have tried assiduously to avoid any association with it. Does this mean that all UFO researchers and all UFO organization are incompetent or that they are incapable of any great investigative results? No, but many are systemically afflicted with these problems arising from within and/or from without. This tends to keep them off balance, unstable and in an endless pursuit of reports that perpetuate an unsolvable UFO mystery.

An Extraterrestrial Smorgasbord of Evidence in a Waste Basket of Paranormal Confusion

Some of the major problems with Ufology are its tendency to lump more than one mysterious phenomenon together with the UFO and ETI phenomenon using faulty logic and circumstantial evidence for which there is no relationship. We find that most UFO groups and organizations will research many **cryptozoological** and **metaphysical phenomena** such as **Bigfoot** or **Sasquatch,** the **Chupacabra,** the **Mothman,** and the **Loch Ness Monster,** the **Earth spirit, spirit guides,** and **ghosts** other monsters. Because some reports of ET spacecraft are seen in areas where a Bigfoot creature was sighted or where the infamous Chupacabra of Puerto Rico or Mexico was seen, the conclusion is to draw an immediate association to the UFO phenomenon. They are viewed as possible extraterrestrial life forms set loose upon our planet to see how we interact with them. This is mere supposition, a blatant leap of illogic!

Does this mean that a real event did not occur? No, of course not and it isn't meant to be quickly swept under the rug of ridicule or pseudo-science and forgotten about. But to draw some irrational conclusion about the event without further investigation is a disservice to the eye witness of the event and to any knowledge that could be gained through proper scientific inquiry. There may be two or more unrelated events occurring at the same location but at different times. Each deserves its own investigation and if there is a relationship between the two occurrences this will come out from proper scientific methodology and research, not from some speculative and un-researched conclusion.

Another major problem with Ufology is its ghettoization by mainstream science because it suffers from an identity crisis of scientific credibility due in part from manipulation from without and from within. This identity crisis is characterized by behaviour as perceived by the general public, the scientific community and other levels of officialdom as almost carnival-like in nature. A closer inspection into the UFO Ghetto reveals a subculture that is akin to a Star Trek or science fiction convention with all the ballyhoos, costumes, books, videos, toys, trinkets, t-shirts and even with the entertainingly colourful parade of aliens in full regalia. The UFO cottage industry is alive and well and needless to say, enterprisingly profitable.

The UFO Ghetto is populated with a denizenry from diverse economic backgrounds and social standings who can best be described as the closest representation of an alien culture on earth who frequently appear at every UFO convention, lecture tour and seminar that pops up. There are the professional UFO researchers making the rounds on the UFO lecture circuit or the budding amateur investigator looking to cross paths with the professionals of Ufology and to share in any news-breaking reports or to listen to a lecture on the research work by a fellow investigator.

Author's Rant: Admittedly, I must include myself as one of its denizens having frequented the many aisle ways and lecture rooms of the "Whole Life Expo" in Seattle back in the 90's, perhaps, though, more as an observer than a participant.

In the continuous stream of the bumping and jostling crowd of people, one cannot help but notice the curiosity seekers looking for their next "fix" of alien knowledge to be imparted to them and the odd assortment of the disenfranchised and socially challenged, who stare into every face they meet, desperately seeking for some sign of peer recognition among the throng of the lost, bemused, and somewhat tortured souls in attendance.

In the **UFO subculture** one can find acceptance from many like-minded individuals who are truly supportive of your hobby and interest, and of course, everyone is an expert and has an opinion or a pet theory on the relevant UFO topic in question, no matter how misguided or incorrect it may be. On occasion, one does come across that rare gem of authenticity in the guise of a researcher or a book with the tidbit of knowledge which may lead the serious investigator into a new avenue of research but, these are rare circumstances. If your intent is to cover the social circles of Ufology for the 6 O'clock newscast, then you will not be disappointed as the UFO subculture has an abundance of Ufologists and UFO stories to keep the TV viewers informed and entertained.

Within this eclectic membership particularly among the senior members and those who hold an office position, other problems arise. Group dynamics, the clash of egos, the direction, goals, mission statements and policies of the organization, as well as to what associated phenomenon should be investigated and studied, and the on-going treasury needs to stay viable, are just some of the problems that will pull an organization apart causing many to jump ship to some other UFO group or to go it alone. UFO researchers, it appears are their own worst enemy, they certainly don't need any help from the outside!

Compounding these ever present problems that plague most UFO organizations is the unsuspected amount of disinformation and misinformation skilfully planted within various UFO accounts usually in the form of staged or hoaxed UFO events that are made to appear on the surface as legitimate sightings. The perpetrators are the secret deep black tactical units of the military industrial complex who continually maintain the misperceptions of the UFO phenomenon are kept front and centre in the public's minds. These well-orchestrated, hoaxed UFO events are designed to misdirect the public's attention, instilling a sense of fear and foreboding regarding the subject, and perpetuating the evil alien scenario, particularly when abductions become a part of an eye witness account. Thus, ultimately these investigative reports end up corrupting the database of most UFO investigators and their organizations. As if this wasn't enough, now, the problem becomes further acerbated by not knowing which UFO reports are legitimate and which are stage crafted in a continued campaign of disinformation.

There are few, if any, UFO organizations that are perceived as credible and methodical in scientific research of the phenomenon. **MUFON** comes close but, they too suffer from some of the above-stated problems in Ufology.

UFO Magazines, Books, Videos and Blog Talk Radio

There have appeared many UFO magazines and journals throughout the history of the UFO phenomenon relating the experiences of eyewitnesses, the reports of strange flying craft, and the accounts of the sometimes elusive alien creatures and ETI. Whether it came from the historic accounts of an ancient Japanese emperor, millennia ago, or from the reports of flying shields in the skies above the battlefields of **Alexander the Great,** or the strange glowing craft emerging from the ocean waters of the Atlantic as **Christopher Columbus** set out to discover a western trade route to the Indies, they all had one thing in common, and that was to inform others who came after them that we were not alone and that our world held strange secrets yet to be discovered.

Today's UFO magazines continue the same important role of informing and educating it readership as its forerunner, the ancient and historic manuscripts of ancient cultures and societies. Their content is now more technical and sophisticated reflecting a more technical and sophisticated society which currently seems to be experiencing an ever increase surge in the UFO/ETI phenomenon. Editorial reviews, opinions, and thoughts; guest interviews; in-depth reports of the most current and intriguing eyewitness accounts, pictures and pictures and more pictures of UFOs fill the pages and there are the mandatory advertisements to pay for the publishing of the magazine. UFO magazines are compelling literary sound bites to entice further reading and investigation with the promise of greater insights into a mystery that seems to know no rational answers.

If magazines don't provide you with enough information on the UFO subject then a 500 page book may do the trick and if one book isn't enough then there are literally tens of thousands of books covering every aspect of the UFO and ETI phenomenon to educate and turn you into a walking encyclopedia of UFO knowledge.

For those requiring more than just pictures, there are thousands of videos showing almost every conceivable brilliant light, glowing orb, flying saucer or triangular craft in a day or night time setting in just about every location around the planet.

Blog talk radio programs over the internet have taken off like a blazing saucer out of hell streaking into the sky-blue airwaves of the worldwide web with increasingly new blog talk sites coming online yearly enabling a worldwide audience to tune in and download their podcasts. This is great for those who know nothing about the UFO/ETI phenomenon, however, most blog radio programs seem to be only regurgitating the same old UFO mythology and their guest speakers, now are able to do a lecture tour over the internet via their phones without leaving their own homes. New reports and relevant information become global news within minutes and then just as easily it becomes yesterday's news, almost as quickly.

YouTube, StumbleUpon, even Facebook and similar websites are another source of current UFO news and the latest uploaded videos on the UFO subject, accessible to all who have a computer. But once again, what should be a trustworthy and legitimate outlet for information sharing and networking becomes a visual forum for CGI hacks wanting to flaunt their expertise in hoaxed UFO videos in order to see how many people will believe their forgeries. Bragging rights of their abilities and talents are their main motivation, much to the chagrin and annoyance of all serious UFO researchers. These people unwittingly play right into the hands of the professional UFO skeptics and debunkers, as well as the disinformation and misinformation campaign programs of the military industrial complex. Perhaps, if they but realize that they are only muddying the waters of real serious UFO research and unwittingly aiding those who wish to keep such matters covered up, a more honest legitimate approach for their talents could be found. But, here again, these CGI hoaxers could be paid shills of the military intelligence complex whose task is the same as the professional debunkers which is to create mayhem in for Ufologists to keep them side tracked from the real UFO investigation. These hoaxed UFO videos become, unfortunately, part and parcel of the thousands of reported UFO case files that only add to an already corrupt database in Ufology.

Regardless of your particular sources of information on everything UFO and ETI, there is a vendor or a publishing house that can satisfy your heart's desire to fulfill that need to know. To that end, however, there is no guarantee that what you read or watch will not be a complete and unadulterated source of truth or a well-fabricated set of lies, denials and falsehoods, to which the odd adage still applies: *"Let the customer beware!"* When you enter the UFO arena of discovery, not everything is as it appears and so, vigilance and a discerning eye, as well as healthy skeptical mind of inquiry, become the prerequisites in your search for the truth.

The UFO Cottage Industry is Alive and Well

Like every other business out there in the white and blue-collar world, it's all about making money in sufficient quantities to be profitable and self-sustaining and the UFO Cottage industry is no different. The only difference is that the research of UFO/ETI accounts is not a lucrative business and in fact, it rarely breaks even, and the old saying is very true that no one makes any money in UFO research. So how do investigators and organizations support themselves and stay operational from tear to year?

Researchers are either financially independent with income from another career or through family wealth or inheritance or from working for an organization with sufficient venture capital to do research work with a potential future payoff from the knowledge gained or the acquisition of foreign or alien technology. This type of organization is military, an intelligence agency or a private corporate organization like the **Bigalow organization**. The typical UFO organization like MUFON is entirely dependent on its membership fees to subsidize its research endeavours and the promotion of conventions or conferences but it usually is a break even proposition.

A UFO researcher who comes across that one big case which carries with it significant insight into the UFO mystery may decide to write a book on that case and if he also does the lecture tour circuit promoting his book then, the cash registers will start to ring a pleasant tune to his ears. Getting radio interviews as a guest speaker on such talk shows as **"Coast to Coast"** show or as a UFO expert on one of the major TV networks UFO "specials" will only add to his popularity and reputation making him a sought out guest speaker.

But for most UFO researchers, this scenario rarely happens to this degree of success, if at all and so, for the most of us, who have spent many years reading other people's books or watching their videotapes and DVDs, and occasionally have the opportunity to actually talk to some eyewitnesses about their experiences or to attend a UFO conference or seminar, this will probably be the height of our research attainment.

This means that we can, not only learn a lot from the investigations of other researchers who have the ability and financial resources to travel from one case to another but, be presented with the opportunity as a free enterprising entrepreneur to ride the coattails of these investigators at UFO conventions with the selling and promotion of UFO souvenirs and memorabilia. This is where the money is and where one can make a lively and sustainable income while pursuing their interest in the UFO phenomenon. Shirts ball caps, books, tapes, CDs, DVDs, toys, trinkets, cups and mugs and anything related to UFOs and ETs is a saleable commodity to the convention's UFO enthusiasts. In fact, the UFO lecturer and the UFO commodity promoters work hand in hand even though both will never admit their mutual cooperation or their appreciation for each other. If both parties succeed in their own agendas through unacknowledged cooperation, then their mission is accomplished.

The UFO Conventions have become a breeding ground for everything new age, paranormal and just plain wacky besides the venue for the serious UFO researcher trying to enlighten a public on the latest developments in the UFO field. Not all conventions are assembled in this type of forum as there are many which have restrictions placed on the content and manner of operation so that a reasonable semblance of scientific methodology and presentation is maintained. Where this forum deviates is when the convention is open to everyone and everything under the sun to generate the most profitable returns. The degree to which you, as the serious UFO researcher wish to come across to your fellow researchers, the public and the science community (if they are interested and paying attention) will depend on the venue, its location, and its preparation and organization.

If what you want is to attract scientific inquiry and legitimacy, with a desire to have to get more people with responsible power and authority on board then, you tailor fit the venue to fit that forum and the financial returns become secondary. If you open the venue to attract the most number of guest speakers and sales vendors then, you will attract a large segment of the population interested in the paranormal, etc. and the scientific community will probably stay away. You may even attract one or two intelligence agents skulking around just to see who shows up to these types of events.

Dr. Steven Greer and his **Disclosure witnesses** is a prime example of having chosen the correct venue location, along with the most compelling witness testimonies that may be video documented by the largest gathering possible of the nation's top new media thus, providing the greatest news impact both nationally and globally.

The UFO conventions, on the other hand, are held every year in Las Vegas, Nevada or in Laughlin, Texas or in Roswell, New Mexico which attracts a lot of UFO enthusiasts and vendors. They don't however, attract the national news media because of the circus-like atmosphere of the venues. However, some local or state media may show up to do a 6 O'clock news piece and but this type of venue certainly won't attract the scientific community. Everyone who attends these typical UFO conventions learns a little bit more than they did, usually first-hand from those who do the research work, and everyone goes back home feeling pretty good about their time spent at the convention site.

So, which venue do you think science is most advanced by and perhaps, sparks further inquiry into the subject matter?

CHAPTER 23

MISINFORMATION, MISINTERPRETATION AND MISPERCEPTIONS OF UFOS AND ETI

In the previous subsections, we've seen some of the flaws afflicting most UFO organizations, from investigating unrelated paranormal events and cryptozoological phenomenon believing that such accounts have less than six degrees of separation from the UFO and ET phenomenon. We've seen also, how the cottage industry mentality is alive and well and has once again caused science not to take the UFO/ETI subject seriously, though it has not stopped some scientists from venturing into this elusive mystery in hopes of making some sense out of it. With such flaws in Ufology, that on the surface may appear minor and perhaps, manageable and repairable, it is, unfortunately, a lot worse than what we discussed so far!

We are going to delve deeper into this corrupt database from which Ufology has built a faulty foundation upon and from which it propagates these flaws as truths unwittingly not knowing the difference and thus, making any attempt to correcting the mistakes.

The UFO experience is often viewed as being closely related to a religious experience as it has all the hallmarks and affiliations with everyday life as does a religion. When asked quite frequently by the news media or family members or friends and strangers what do you believe? Do you believe in UFOs and aliens? Right off, we see that the question becomes a matter of belief. Belief is an act of faith found in religion and religion is often viewed as steeped in tradition, superstition and paranormal events following a set of laws and practices that are considered as outdated and not having kept up with the current times. This is remarkably a good description of Ufology!

Author's Rant: When asked, "Do I believe in UFOS and aliens?" My answer like so many others before me would have been, Yes! But upon reflection based on over 60 years of personal experience with the phenomenon, my answer is now a resounding, No! I don't believe in UFOs, I am a "witness" to the Extraterrestrial Intelligence phenomenon! This means that I have had experiences with UFOs and ETI. Not everyone may be as fortunate enough to have such experiences in which case their perception would be grounded in a belief that such phenomenon does exist without the proof of a sighting or an experience.

This tendency to believe is a personal matter and should not invalidate the whole UFO subject but, in the untrained and unbalanced minds of a few people, as we shall see, there is an associated religious aspect with the UFO phenomenon. Because of this religious aspect, it has not stopped some opportunists from developing a new age UFO religion based upon Extraterrestrial visitations to the planet mixed with some aspects of traditional religious belief, like Christianity or based upon the writings of ancient "**Ascended Masters**".

Besides the obvious religious connotations associated with the UFO/ETI phenomenon, we will examine the mindsets that come from the ***"Deifying and Demonizing"*** of the Extraterrestrial Intelligences and as well as the nomenclature of such terminology as ***"Abductions"*** and ***"Cattle Mutilations."***

**Will Jesus Christ in a Flying Saucer Save Us or Will Darth Vader
and His Little Gray Demons Eat Us For lunch?**

Historically, religions whether they be **Hindu, Buddhist, Judaic, Christian, Islamic**, etc. have
personified and made "flesh" the spiritual concepts of evil, golems, Satan, the Devil, Jinns or
Djinns, unclean spirits, fallen angels and/or any creatures that are either unusual, bizarre or most
recently, not native to this planet. Whenever, they are mentioned in any biblical or scriptural
texts, they are always perceived and thought of as the literal physical incarnations of all that is
evil and ungodly.

Such perceptions have carried over to the current times when some people of religious
persuasion have had an interaction with Extraterrestrial biological life forms. Whether they have
acknowledged them as alien to this planet or something unusually rare and native to the Earth, if
they have not encountered it before then, their religious training and belief system will override
any rational common sense. Such encounters with ETs will then be labelled as a demonic or as
minions of Satan or the Devil.

These perceptions based on religious teachings are for the most part due to interpretations by
clergymen or doctors of religious learning, and if these men have misunderstood their own
religious writings because some biblical passage was meant to be understood in a spiritual
context rather than a literal physical meaning then, the error of understanding is passed on to the
community's congregation. The misinterpretation of religious writings can affect the whole
population of its believers, both nationally and globally, and not just in one time frame of history
but, down through the centuries even, for millennia. All this present day confusion that we've
experienced is due largely in measure to the misinterpretation of religious writings by a few men
either, knowingly or unknowingly and has in fact occurred **repeatedly** in many religions, down
through the ages. You would think that humanity would get it right with each successive renewal
of religion. This is a historic fact which no major religion to date has come forward and openly
admitted to its adherents or to the world at large and has only further compounded the matter.

The lack of understanding of the holy writings and teachings has even, caused the leaders of the
current religions of the day to reject outright, the Manifestations of God when they appear in a
particular age because they did not fulfill prophesy in the manner that was expected. This due to
a lack of understanding of their own holy writings. It has even lead to the imprisonment, torture,
and exile or to the death of each successive **Messenger of God**! And you wonder why this planet
is so screwed up in it moral outlook on life!

Humanity goes through this exercise approximately every thousand years or so, in the hopes of
getting it right. Yes! Religion is progressive and renews itself periodically according to a divine
plan to ensure that mankind never forgets its Creator and to help civilization to advance along a
spiritual as well as a material path.

If religious leaders demonize ET intelligence because it does not fit in with everyday religious
orthodoxy because of some personal bias or interpretation, then the adherents of that religious
order, who receive spiritual guidance from their Pastor or Mullah, also, develop the same biases
and misunderstandings as their religious leader. People need to investigate matters of truth

unfettered for themselves, instead of relying on spiritual guidance from someone else. They need to question things and come to understand things for themselves and not from the minds and mouths of other people.

In this day and age, it has become expected among fundamentalist Christians that **Christ** may even appear in a flying saucer at the end of days to save us from ourselves and from the little evil invading ETs who plan to wreck **Armageddon** havoc on the Earth. If you are Muslim, a similar prophecy of the **Twelfth Imam** would appear to save the faithful from the evil hands of the infidels or to save us from the likes of a "***Darth Vader** and his hungry little gray demons who have plans to eat humanity for lunch*". Such superstitious beliefs hold sway over much of the common population as to keep them fear-bound and unable to progress rationally or spiritually beyond their own worn-out belief system.

Fortunately, there are some rational minds within the various sects of Christianity like Catholicism that through church council meetings in Rome with its own Church astronomers, people like the late **Father Baldacci**, have come out and stated publicly that Extraterrestrial beings are also the children of God and should be accepted as such and should not be feared. This is almost a public declaration by the **Church of Rome** saying that ETs are real and visiting the planet, even if it wasn't stated by the Pope, himself.

At this present time, when the public acceptance of the reality of Extraterrestrial life visiting our planet is between 75% to 80% plus with yet, no official acknowledgment from government leaders, it appears that some formal institutions in society are already jockeying for position to provide salvation to the visitors from the stars! It would seem to depend on who you talk to from within the various sects of Christianity, that Christ is deemed to be the ET's saviour as well. Thank heavens, that God in His infinite wisdom has decidedly appointed an individual saviour for every intelligent species of being in the universe!

This brings us to the opposite side of the same coin which is the deifying of Extraterrestrial intelligence as god-like in their omnipotence based on their superior technology. Historically, this has been how humans have viewed and acknowledged ETs whenever an advanced civilization meets a primitive or undeveloped technological civilization. We've already seen in the beginning of this book, that almost every pre-bronze age society has had visitations from space-faring ET civilizations. Where this has occurred, human cultures have been dutiful in their worship of these star people perhaps, because we either marvel at their strangeness or covet their technology and their ability of flight. We record their visits and make images to their likeness and try to emulate something of their abilities. If first contact is fear- based or hostile in the initial encounter then, we demonize them as evil.

We have perceived them as angels with or without wings (although, these avian protrusions may, in fact, have a basis in reality), who have imparted knowledge, culture and the beginnings of a more advanced civilization. It would seem that worshipful appreciation was a small way of saying thanks for giving your culture a kick-start out of the stone age and into the bronze or iron age.

The reality is that alien civilizations are merely different from us and may appear superior to our own because their circumstances may be due to early planetary formation and evolutionary processes, greater consciousness and spiritual development and perhaps, a history that has never known a time of war amongst its own kind. All these higher qualities of civilization make it attractive to be more closely associated with it in the hopes that our own society and ourselves will be greatly improved by that relationship. In the next section, we will see how these star beings have inspired some humans to leave behind the mainstream existence of everyday earthly life by developing a religion based on the existence of ET beings.

Contactees Revisited

As we have already seen earlier in this section, "Contactees" are individuals who claim to have experienced contact with Extraterrestrials and who were given messages of advanced knowledge or profound wisdom by these Extraterrestrial beings. The contactees felt compelled that it was their mission to share these messages. These suspect claims of alien encounters are often described as frequent, while some contactees claim only a single encounter. As a cultural phenomenon, contactees perhaps had their greatest notoriety from the late 1940s to the late 1950s, but individuals continue to make similar claims in the present, such as Swiss cult leader **Billy Meier**. The forum for most contactees who wish to share their alien messages has usually been with newsletters and on the lecture tour circuit at UFO conventions.

Some of these contactees include **George Adamski** (1891–1965), probably the best known contactee, followed by **Truman Bethurum** (1898–1969), **George Van Tassel** (1910–1978), **Daniel W. Fry** (1908–1992), **Orfeo Angelucci** (1912–1993), **George King** (1919–1997)**, Buck Nelson** (1894–1982), and many lesser known reputation. With **Betty** and **Barney Hill** (1920–2004 and 1922–1969, respectively), the era of contactee-ism abruptly ended, the age of the '**Space Brothers**" was replaced by the Abduction scenario and the little "Grays"!

An unfortunate fact for most contactee claims is that their stories of contact contain much material that has not stood the test of time. Claims of unknown planets within this solar system, or that the planets of our solar system are inhabited by so-called "Space Brothers", beings who are physically similar to humans but more spiritually evolved. *Although recent ETI research of ET exo-biology and xeno-morphology types indicate that a species of ETI does appear to be very human looking and could very well be considered as our "Cosmic Cousin"* [my italics].

Contactee accounts usually describe beneficial experiences involving human-like aliens; these are generally different from those who allege alien abduction, where abductees describe their experiences rather negatively.

George Adamski

George Adamski, one of the first "contactees" in the 1950s, was (April 17, 1891 – April 23, 1965) a Polish-born American citizen who claimed to have photographed ships from other planets, met with friendly **Nordic alien**, "Space Brothers", and to have taken flights with them through the solar system.

At the age of 22, from 1913 to 1916, he was a soldier in the 13th U.S. Cavalry Regiment K-Troop fighting at the Mexican border during the Pancho Villa Expedition. He then moved out West settling at Laguna Beach in California where he founded the "Royal Order of Tibet," which held its meetings in the "Temple of Scientific Philosophy." By 1940, Adamski and some close friends of his moved to a ranch near California's Palomar Mountain where they built a new home called Palomar Garden and a new restaurant called Palomar Gardens Café.

George Adamski
https://web2.ph.utexas.edu/~coker2/index.files/sbrothers.shtml

It was here on Mount Palomar campground on October 9, 1946, during a meteor shower, thatAdamski and some of his friends claimed they witnessed a large cigar-shaped "mother ship. This event lead to his interest in astronomy with the purchase of a telescope and in 1947, Adamski took a photograph of what he claimed was the 1946 cigar-shaped "mother ship" crossing in front of the moon over Palomar Gardens.

On May 29, 1950, Adamski took a photograph of what he alleged to be six unidentified objects in the sky, which appeared to be flying in formation. Adam ski's May 29, 1950, UFO photograph was depicted in an August 1978 commemorative stamp issued by the island nation of Grenada in order to mark the "Year of UFOs". http://en.wikipedia.org/wiki/George_Adamski

On November 20, 1952, Adamski and several friends had decided to take a car ride into the Colorado Desert near the town of Desert Center, California when they are said to have seen a large submarine-shaped object hovering in the sky. Believing that the ship was looking for him, Adamski asked his friends to stay behind, by the car and then walked away from the main road into the desert. Shortly afterward, according to Adamski's accounts, a scout ship made of a type of translucent metal landed close to him, and its pilot, a **Venusian** called **Orthon** disembarked and sought him out.

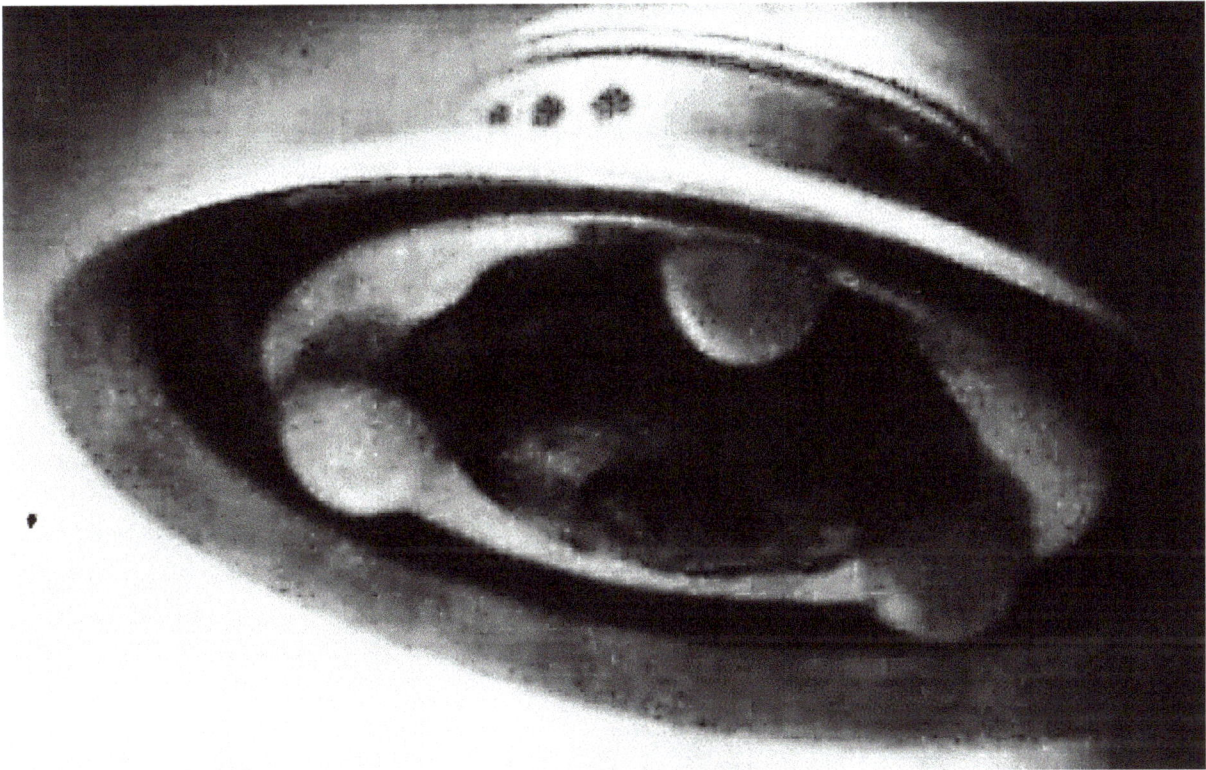

Adamski's photograph a UFO, taken on December 13, 1952, which
some Ufologists and skeptics have labelled as a "chicken feeder"!
http://forgetomori.com/2007/ufos/ufo-photos-adamski-scout-ships/

Orthon, as Adamski described him was a humanoid of medium height, with long blond hair, and tanned skin, and as wearing reddish-brown shoes with trousers that were not like Adamski's (perhaps, they were tightly cuffed as seen in various drawings of Orthon).

Orthon communicated with Adamski via telepathy and through hand signals warning him of the dangers of nuclear war and later arranging for Adamski to be taken on a trip through the solar system including the planet Venus, the location where late Mrs. Adamski had been reincarnated. Oddly, Orthon had refused Adamski to photograph him but, instead asked Adamski to provide him with a blank photographic plate, which Adamski says that he gave him. When Orthon left, Adamski said that he and **George Hunt Williamson** were able to take plaster casts of Orthon's footprints and that the prints contained mysterious symbols. Williamson, George Hunt, *Other Tongues—Other Flesh.* Amherst, Wisconsin: Amherst Press, 1953.

Orthon is said to have returned the plate to Adamski on December 13, 1952, at which point it was found to contain new strange symbols. It was during this meeting that Adamski is said to have taken a now famous UFO photograph using his 6-inch (150 mm) telescope. UFO researchers and some scientists have regarded this picture as a fabricated hoax similar to a chicken brooder or a streetlight.

While Adamski's highly dubious **Flying Saucer** photographs and false claims of contact with ETs were reaching the public's attention in the U.S., the news hadn't yet reached most of Europe's news media. In May 1959, Adamski received a letter from the head of the **Dutch Unidentified Flying Objects Society** informing him that **Queen Juliana** of the Netherlands desired a meeting to hear of his of experiences. After the meeting, Dutch Aeronautical Association President Cornelis Kolff said, "The Queen showed an extraordinary interest in the whole subject."

There were rumours of other arranged meetings with **Queen Elizabeth II** of England which never took placed and an alleged meeting with Pope John XXIII whereupon, Adamski claimed he had received from the Pope, a "Golden Medal of Honour" for his visit.

Adamski's outspoken assertion that the photographs of the far side of the Moon taken by the Soviet lunar probe Luna 3 in 1959 were, in fact, fake and claimed instead, that there were cities, trees, and snow-capped mountains there. Skeptics had a field day with Adamski's claims is that the planet Venus was inhabited with human type beings. The skeptics argued that Venus is unable to sustain intelligent life due to its environmental conditions. These conditions include an atmospheric pressure at the planet's surface which is 92 times that of the Earth, clouds composed of sulfuric acid and an average surface temperature of 461.85 °C. Of course, no one could live under the surface of the planet, and as a result, most consider Adamski's claims to be a scientific impossibility. http://en.wikipedia.org/wiki/George_Adamski

By 1962, his announcement that he was going to attend a conference on the planet Saturn saw his reputation plummet. It became obvious to those who knew him or the news media who reported on his activities that the public was rapidly losing interest in his "far out" stories.

On April 23, 1965, at the age of 74, Adamski died of a heart attack in Maryland.

For most UFO researchers Adamski is viewed as a pioneer in the contactee filed basing much of his claims on the former works of **Madame Helena Petrovna Blavatsky's** Venusian Ascended masters with some minor variations on this theme. But what is extraordinary about Adamski's claims is that there have been some well-known Ufologists notably **Timothy Good** has come out in defense of some of the photographs taken by Adamski. Good compares Adamski's infamous photo as seen above to those taken later by young **Stephen Darbishire** at Coniston, Lancashire in February 1954 and to the Polaroid photo taken by architect **Hugo Vega** near Lima, Peru in October 1973. Even the description and alleged photos of "cigar-shaped" craft have been since reported on numerous occasions by witnesses around the world.

Author's Rant: I also, have seen this particular type of craft by Golden Ears Mountains near Maple Ridge, B.C. along with three other witnesses, my daughter Annika, her girlfriend Chloe and my aunt-in-law, Marie T.

One most also compare the striking similarity of the Adamski Venusian saucer with the German Nazi Haunebu ll flying saucer which the US and its WW ll allies claimed does not exist or was ever built.

Compare these German Nazi photos (top two and bottom left) with the Adamski Venusian flying Saucer; note the similarities
Bottom right photo courtesy of Dr Greer from his movie "Unacknowledged
http://discaircraft.greyfalcon.us/HAUNEBU.htm **and** https://www.youtube.com/watch?v=ehmnGolJZq0

Finally, there have also been many ET sightings of beings described as very human-like in appearance commonly referred as the "blond Nordics" as reported in the now famous **Travis Walton** case, the **Billy Meier** Pleiadian encounters, the **Howard Menger** contact reports, and the case reports from the **CE-4** Conference at M.I.T. This is one contactee case that needs to be revisited as there is even a theory proposed by **Leon Davidson** that Adamski was duped by one of the intelligence agencies (CIA) for reasons known only to them. Jacque Vallee goes even further to state that this covert group is composed of fascist-oriented intelligence operatives who set up Adamski. "Let us note in passing that the Adamski "Venusian"… and many other similar extraterrestrials were all tall Aryan types with long blond hair."

This pre-supposes that there had been a breakthrough in reverse engineering of either German WWII designed flying saucers or the captured saucers of Roswell, New Mexico in 1947. There has been rumours going around in Ufology that such breakthroughs in reverse engineered alien technology, especially saucer technology had been made in the early '50s and that such craft were test flown and maybe humans who were of German extract or who had German accents as claimed by some contactees role-played the part of Nordic type ETs who gave warnings to early contactees like Adamski, et al.

Travis Walton, The Walton Experience, 1978; and Travis Walton, *The Fire in the Sky*, 1996; also Timothy Good, Alien Base – Earth's Encounters with Extraterrestrials 1998; Jerome Clark, *The UFOBook: Encyclopedia of the Extraterrestrial*, 1998; C.D.B. Bryan, Close Encounters of the Fourth Kind 1995

Truman Bethurum

Truman Bethurum was born in Gavalin, California, (August 21, 1898 – May 21, 1969) and was one of the "graduates" of the 1950s class of contactees following in the footsteps of his predecessors, George Adamski, George Van Tassel, Daniel Fry, George King who were already leaders of their own new age religious movements. Like his predecessors, but mostly inspired by the exploits of George Adamski, he too, claimed to have spoken with humanoid aliens and ridden on their flying saucers.

He worked as a mechanic as a part of a road-building crew in the early 1950s and unlike most of his road crew members, he moonlighted as a fortune teller and spiritual advisor. In 1953, as a novice magazine and newspaper publisher, Bethurum published the Redondo Beach *Daily Breeze*, and in his newspaper of September 25, 1953, recounts of being contacted numerous times by a humanoid landing crew from a flying saucer. In repeated encounters, he conversed with their beautiful female captain, **Aura Rhanes**, who had a "slender Latin-type face" and wore a radiant red skirt, black velvet short sleeve blouse and a black beret with red trim.

Bethurum noticed that the **Clarions** all dressed like Greyhound bus drivers, not only did they speak perfect English, but they also spoke in rhyme, and they enjoyed polkas and square dances. The ship itself was 300 feet in diameter, 6 yards deep and made of burnished stainless steel which could hover silently inches above the ground. They didn't refer to their ship as a saucer but, called it a "scow." They came from the unknown planet Clarion, which from the earth, and in contradiction to astronomical laws of planetary motion, always remains out of sight behind our moon. According to Bethurum, the **Clarionites** were smaller than humans, lived for 1000 years, and were all good Christians who attended church every Sunday. His book, *Aboard a Flying Saucer*, gave few details of Clarion and its people, but ample details of his suffering at the hands of skeptics.

Some of Bethurum's later books include *The Voice of the Planet Clarion* (1957), *Facing Reality* (1958), and *The People of the Planet Clarion* (1970), published after his death. The first 44 pages of the final book are an autobiography of Bethurum covering his life up to 1953. He mentions in this last book that astronomers had told him that Clarion couldn't possibly orbit either Sun or Earth in a manner in which it always remains behind the Moon as seen from Earth. As if caught in an ever evolving lie, he states that **Captain Rhanes** must have meant to say Clarion is in the same orbit as our Earth, but always behind the Sun from our viewpoint. Since the 1960s, it has been known that no other planet exists in either spot. Earlier, in 1954, Bethurum had told audiences during his lectures that Rhanes probably meant Clarion was in another solar system. http://en.wikipedia.org/wiki/Truman_Bethurum and http://ufocasebook.com/saucerstory.html

Truman Bethurum
https://en.wikipedia.org/wiki/Truman_Bethurum

Howard Menger

Howard Menger (February 17, 1922 – February 25, 2009) was another American contactee who claimed to have met extraterrestrials throughout the course of his life, meetings which were the subject of books he wrote, such as *From Outer Space To You* and *The High Bridge Incident*. Menger. His first alleged contact at the age of ten was with a person from another planet, a beautiful young woman sitting on a rock in the woods near his hometown High Bridge. He felt hormonally attracted to this woman who sensing his feelings for her, told him that it was not to be but, that he would have a lifetime relationship with her younger sister whom he would recognize at first sight at a later date. Shortly after leaving high school, he entered the Army and was attached to the 17th Tank Battalion. In later life, he was often employed as a sign painter.

Howard Menger

He rose to prominence as a handsome and charismatic contactee regaling those who listen with accounts of conversations with friendly Adamski-style Venusian "space brothers" in the late 1950s. He was widely dismissed as a charlatan, who simply jumped on the bandwagon in the wake of publicity following publication of George Adamski's wild stories of chit-chatting with Nordic-looking spacemen, and during at least one live TV appearance he admitted as much. Nonetheless, his various stories, photographs, and films have been accepted by some UFO believers. Other contacts followed with other humanoid beings. Then in 1946, the woman disembarked from a spaceship and announced that a wave of contacts was in humanity's immediate future as many space people were coming to Earth to assist in solving its problems.

In 1956, in the wake of the publicity given contactee George Adamski, Menger took some photos of flying saucers and claimed he took a ride in a Venusian ship. An examination of his pictures led to denouncements that they were a hoax, and they caught Menger lying about his having read (and drawing material from) Adamski's books. Amid the controversy, a young blonde woman came to a gathering at the Menger home. He recognized her as the sister of the space person who had originally contacted him as a child. They began an affair and were

94

eventually married. The woman, Connie Weber, wrote her story, which was published in a book under the pseudonym Karla Baxter. It actually appeared in 1958, a year prior to Menger's first book. The title, *My Saturnian Lover,* continued Menger's claim that he was actually an extraterrestrial who had reincarnated on Earth.

He rose to prominence as a handsome and charismatic contactee regaling those who listen with accounts of conversations with friendly Adamski-style Venusian "space brothers" in the late 1950s. He was widely dismissed as a charlatan, who simply jumped on the bandwagon in the wake of publicity following publication of George Adamski's wild stories of chit-chatting with Nordic-looking spacemen, and during at least one live TV appearance he admitted as much. Nonetheless, his various stories, photographs, and films have been accepted by some UFO believers. Other contacts followed with other humanoid beings. Then in 1946, the woman disembarked from a spaceship and announced that a wave of contacts was in humanity's immediate future as many space people were coming to Earth to assist in solving its problems.

In 1956, in the wake of the publicity given contactee George Adamski, Menger took some photos of flying saucers and claimed he took a ride in a Venusian ship. An examination of his pictures led to denouncements that they were a hoax, and they caught Menger lying about his having read (and drawing material from) Adamski's books. Amid the controversy, a young blonde woman came to a gathering at the Menger home. He recognized her as the sister of the space person who had originally contacted him as a child. They began an affair and were eventually married. The woman, Connie Weber, wrote her story, which was published in a book under the pseudonym Karla Baxter. It actually appeared in 1958, a year prior to Menger's first book. The title, *My Saturnian Lover,* continued Menger's claim that he was actually an extraterrestrial who had reincarnated on Earth.

While most contactees have religious revelations to impart after their "experiences," Menger came back from his saucer-rides with a far more practical message: a new outer-space-approved diet for losing excess weight.

The authors believe that the underlying pattern of the UFO phenomenon is controlled by a central network (CIA) from which contrived information, such as sightings, holograms, photographs, specimens and documented accounts is fed to the public. This central network consists of various top secret agencies in our government working with specialized personnel of the Army, Navy, and Air Force in secret locations. Other UFO researchers have confirmed this aspect as well, stating that government intelligence agents had approached him to make up a story to see how the public would react in 1953.

We have seen this happen before, where some intelligence agency has tested the "public waters" to see the extent that the general public would ***"buy into"*** alien contact and the messages announced by the contactees of the 1950s. It would appear that a deep dark agenda was afoot to affect the American consciousness which as we will see later would escalate to more Machiavellian proportions.

Howard Menger died recently on February 25, 2009, at the age of 87.

Daniel W. Fry

Daniel William Fry was born in Verdon Township, Minnesota, July 19, 1908, and passed away on 20, 1992. He was an American contactee who claimed he had multiple contacts with an alien and took a ride in a remotely piloted alien spacecraft on July 4, 1949. Fry worked at the **White Sands Proving Ground** in New Mexico, where one evening on July 4, 1949, in nearby Las Cruces, (although, some researchers have stated that the original date Fry gave was on July 4, 1950) Fry had planned to join the town's holiday festivities but missed the last bus to the event. Finding the Bachelor Officers Quarters (BOQ) where he stayed, too hot, he decided to explore a path in the desert he had never been down. There, Fry claimed a 30 foot (10M) diameter, 16 foot (5M) high "oblate spheroid" landed in front of him, where he talked remotely with the pilot who operated the craft from a "mother ship" 900 miles (1400 km) above Earth.

Daniel Fry
http://ufologi.net/kontaktpersoner/kontakt.htm

Fry claimed he was invited aboard and flown to New York City and back in 30 minutes. During the flight and subsequent meetings, Fry asserted that he talked with the pilot named **Alan**, **(pronounced "a-lawn")** who gave Fry information on physics, the pre-history of earth including **Atlantis** and **Lemuria** and the foundations of civilization.

96

Shortly after Fry went public with his story in 1954, he failed a lie detector examination about his claims. Fry also took photos and 16 mm film of supposed UFOs, but subsequent analysis of the original footage has provided strong evidence the UFOs were faked.

Later, Fry received a doctorate; however the "degree" was from a mail-order outfit in London, England called Saint Andrew College and was a ***Doctorate of Cosmism.*** In 1954, Fry published his first book called **"The White Sands Incident"** and a year later started an organization called **"Understanding"** which published a monthly newsletter by the same name.

Understanding Inc. peaked in the early sixties with about 1,500 paid members. In 1974, Enid Smith donated 55 acres of land including eight buildings near Tonopah, Arizona to the Understanding group. The buildings, first intended as a religious college, had the ironic feature of being round and saucer shaped. Understanding Inc. had fully taken the property over by 1976 but, given Daniel's tight finances during his retirement and the declining Understanding membership, the property fell into disrepair. In early October 1978, the kitchen and the library were burned to the ground by an arsonist and never rebuilt.

From 1954 onward, with little reimbursement, Fry gave thousands of lectures to organizations such as service clubs, radio and television stations. He also published other books such as "Atoms, Galaxies, and Understanding", "To Men of Earth", "Steps to the Stars", "Curve of Development", "Can God Fill Teeth?" and "Verse and Worse". He, along with other contactees would attend the yearly **Spacecraft Convention** at Giant Rock in Yucca Valley for the next twenty years, hosted by a friend and fellow contactee, George Van Tassel.
http://en.wikipedia.org/wiki/Daniel_Fry

Author's Rant: I have personally met Daniel Fry back in 1968 when he came to Victoria, British Columbia on a lecture tour circuit through the country. I had already had my own recent sighting (one of many in my lifetime) that year and as a teenager, this was the first UFO event I attended to listen to another person talk about his own personal UFO experience. He lectured with a southern drawl and I found him to be sincere about his experiences but, something inside told me that not everything he said was above board. Maybe, it was the slightly blurry photographs and books that he was selling after his lecture presentation that were somewhat, high priced or the air of a business transaction after the sales pitch to convince the audience that what he was selling was the genuine article. I left the auditorium perplexed by what I had heard as to whether or not to accept Daniel Fry's story of riding on board a remotely piloted spacecraft. Certainly within the realm of possibility but I needed more evidence. At that point, I considered my journey as a novice researcher into UFO investigations as open minded with a healthy skepticism.

George Van Tassel

Van Tassel (March 12, 1910 - February 9, 1978) was born in Jefferson, Ohio and grew up in a fairly prosperous middle-class family and was a classic 1950s contactee in the mold of **George Adamski, Truman Bethurum, Orfeo Angelucci** and many others. He dropped out of high school in the 10th grade and got a job at a Cleveland airport; he also got a pilot's license.

At 20, he headed for California, where at first he worked for a garage owned by an uncle and where he met **Frank Critzer**, a German immigrant, who was an eccentric loner, and prospector who claimed to be working a mine somewhere near **Giant Rock**, a 7-story boulder near Landers, California. During World War II, Critzer was suspected of being a spy for Germany and later, died during a police siege at the Rock in 1942. Van Tassel upon hearing news of Critzer's death applied and successfully obtained a lease of the abandoned airport near Giant Rock from the Bureau of Land Management with a renewable Federal Government contract to develop the airstrip.

George Van Tassel

Van Tassel became an aircraft mechanic and flight inspector who at various times between 1930 and 1947 worked for **Douglas Aircraft**, **Hughes Aircraft**, and **Lockheed**. While at Hughes Aircraft he was reportedly a Top Flight Inspector. He finally left Southern California's booming aerospace industry for the desert in 1947. He and his family at first lived a simple existence in the rooms **Frank Critzer** had dug out under Giant Rock. Van Tassel eventually built a home, a cafe, a small airstrip, and a dude ranch beside the Rock.

Meditating beside Giant Rock in 1951, Van Tassel claimed to have been transported astrally to a huge alien spaceship orbiting the earth, where he met the all-wise "Council of Seven Lights." The following year, Van Tassel reported he had been visited physically by human-appearing, friendly space aliens from Venus, who suggested that he attempt to build a structure aimed at extending human life, to help people take advantage of the wisdom acquired through age. The structure became known as the **"Integratron"** and it became his obsession for the next 25 years.

The structure actually, was completed 1959 but seemed non-operational and Van Tassel spent most of his life in a fruitless endeavour to get it working

The building was a domed time/energy machine built utilizing the theories of **Nikola Tesla** and following telepathic communications from an advanced ET being, in order to recharge and rejuvenate people's cells. The wood structure lacks a rotating metal apparatus on the outside which was to be the functioning part. Now it is simply an empty all wood dome, lacking even metal screws or nails. In recent times New Agers have declared the structure a power spot and claim to be rejuvenated by staying there, and experiencing sound baths inside.

The Integratron

He hosted **The Giant Rock Spacecraft Convention** annually beside the Rock, from 1953 to 1978, attracting at its peak in 1959 as many as 10,000 attendees. Guests trekked to the desert by car or landed airplanes on Van Tassel's small airstrip, grandly called *Giant Rock Airport*.

Every famous contactee appeared personally at these conventions over the years, and many more not-so-famous ones. References often state that the first and most famous contactee, **George Adamski**, pointedly boycotted these conventions. In fact, however, Adamski attended the third convention, held in 1955, where he gave a 35-minute lecture and was interviewed by **Edward J. Ruppelt**, once head of the Air Force *Project Blue Book*. It was the only such convention Adamski attended.

Like most 1950s contactees, he founded a paranormal research organization called *The Ministry of Universal Wisdom*, and *The College of Universal Wisdom* to codify the spiritual revelations he was receiving via "psychic resonance" from the "Space Brothers".

Van Tassel's now-rare book, *I Rode a Flying Saucer* (1952, 1955), recounts some of the cosmic wisdom he received from "Solgonda" and a large number of other god-like Space Brothers. Among his other works are *The Council of Seven Lights* (1958), *Into This World and Out Again*, *Religion and Science Merged*, and *When Stars Look Down*.
http://en.wikipedia.org/wiki/George_Van_Tassel

Most notably, Van Tassel will be remembered by his followers for not only the Integratron and the **Giant Rock Spacecraft Conventions** but for also, channeling and receiving messages from "**Ashtar**," commandant of sector, patrol station Schare, not the pluralized form known as the "**Ashtar Command**" started by Robert Short (or also known as Bill Rose), an Editor of a 1950s UFO magazine - "Interplanetary News" and at one time friend of George Van Tassel.

Orfeo Angelucci

Orfeo Matthew Angelucci (*Orville Angelucci*) (June 25, 1912 – July 24, 1993) was one of the most unusual of the mid-1950s contactees who claimed to be in contact with extraterrestrials.

Angelucci claimed that he suffered from poor health and extreme nervousness for most of his life, and eventually moved for health-related reasons from Trenton, New Jersey to California in 1948, where he got a job on the assembly line at the Lockheed aircraft plant in Burbank. Fellow contactee George Van Tassel was also employed for a time at this plant.

In his books, Angelucci says he was particularly terrified of thunderstorms and was attracted to California because he heard thunderstorms were very rare there. Angelucci wrote the first version of his pseudoscientific account of matter, energy and life, *The Nature of Infinite Entities* in 1952, based on "research" done earlier in Trenton, including the launching of a giant cluster of weather balloons.

Beginning in the summer of 1952, according to Angelucci in his book *The Secret of the Saucers* (1955), he began to encounter flying saucers and their friendly human-appearing pilots during his drives home from the aircraft plant. These superhuman space people were handsome, often transparent and highly spiritual. Eventually, Angelucci was taken in an unmanned saucer to earth orbit, where he saw a giant "mothership" drift past a porthole. He also described having experienced a "missing time" episode and eventually remembered living for a week in the body of "space brother" Neptune, in a more evolved society on "the largest asteroid," the remains of a destroyed planet, while his usual body wandered around the aircraft plant in a daze.

Orfeo Matthew Angelucci
https://alchetron.com/Orfeo-Angelucci-1370576-W

In his later book, *The Son of the Sun*, Angelucci related an account that he claimed had been told him by a medical doctor calling himself, Adam whose experiences were similar to Angelucci's. He also published several pamphlets on space-brotherly themes, such as "Million Year Prophecy" (1959), "Concrete Evidence" (1959) and "Again We Exist" (1960).

Eduard (Billy) Meier

"Billy" Eduard Albert Meier born February 3, 1937, in Bulach, the Swiss Lowlands of Switzerland is one of the most remarkable contactees to step upon the UFO stage to claim ET contact. His many UFO photographs remain controversial which he states are evidence of his encounters. Meier's contacts are described as human in appearance known as the **Plejaren** from the Pleiades star system, who are visiting the Earth to impart spiritual and philosophical wisdom. Eduard "Billy" Meier's, a farmer claimed his first extraterrestrial contacts occurred in 1942 at the age of five with an elderly extraterrestrial human man named **Sfath**. Contacts with Sfath lasted until 1953. From 1953 to 1964 Meier's contacts continued with an extraterrestrial human woman named **Aske**. Meier says that after an eleven year break, contacts resumed again (beginning on January 28, 1975) with an extraterrestrial human woman named **Semjase** the granddaughter of Sfath.

As a teenager he once enlisted in the French Foreign Legion for a short period of time before returning home. He traveled extensively around the world pursuing spiritual exploration, covering some forty-two countries over twelve years. In 1965 he lost his left arm in a bus accident in Turkey. In 1966 he met and married a Greek woman, Kalliope with whom he has three children. The nickname "Billy" was given to him by an American friend who thought Meier's cowboy style of dress reminded her of "Billy the Kid".

Eduard "Billy" Meier
http://www.thelivingmoon.com/47john_lear/02files/Billy_Meier_001.html

Meier's large collection of controversial photographs showing alleged spaceships known as beamships as well as alleged humanoid extraterrestrials, the Plejaren has divided the UFO community as to their authenticity as well as the verbal attacks of skeptics. Many of Meier's photographs are some of the most crystal clear photos ever taken of the ET spacecraft to date which he said that the Plejaren gave him permission to photograph as well as film their beamships as evidence of their extraterrestrial visitation.

Michael Horn who has appeared on popular late-night paranormal programs such as Coast to Coast AM has become Meier's main supporter in America reporting that the Billy Meier beamships have appeared in the skies over Florida. Retired U.S. Colonel Wendell Stevens has documented extensively the Billy Meier case, his photos and films and feels strongly that Meier is a credible witness.

Meier's first alleged contact with extraterrestrials began on January 28, 1975 communicating both directly in person and by telepathy with a core group of the Pleiadians/Plejaren, or Errans as he also refers to them (Erra being their home planet), who gave their names as **Ptaah**, **Semjase**, **Quetzal**, and **Pleja**, among numerous others.

"Semjase the Plejaren"

These visitors reportedly hail from the **Plejares star system** which is beyond the **Pleiades** and in a dimension that is a fraction of a second in the future from our own (an alternate timeline). These Plejaren have allegedly afforded Meier a more interesting sampling of evidence than that derived from most such encounters, including highly detailed photography, videos, multi-toned sound recordings, the temporary use of a weapon which he employed for trial on a nearby tree, and metal alloy samples.

Meier claimed the visitors charged him with certain informational and consciousness-raising tasks. As he undertook this mission, he met with a great deal of scorn and derision in addition to (according to his center) twenty-one assassination attempts. Some of these were allegedly initiated by hostile extraterrestrial entities and subsequently defeated largely through the intervention of his Plejaren friends. Meier was uncomfortable with the megalomaniacal associations some would attach to his role as a representative (such as the use of the term "prophet", e.g.) but he undertook the effort nonetheless.

In 1975 Meier established the "**Free Community of Interests for the Fringe and Spiritual Sciences and UFOlogical Studies" (FIGU)**, a non-profit organization for the benefit of researchers into this field, and headquartered it at the Semjase Silver Star Center.

Meier is a true enigma in that it is hard to imagine that with only one arm he was able to achieve such quality photos and film evidence unless, as some UFO researchers have strongly suggested he used elaborate models and was aided by accomplices, yet he has demonstrated repeatedly that it is possible for him to have taken these photographs. When it come to the evidence of what was communicated to him by the Plejaren, the information falls apart as too incredible and unscientific in it premise leaving the UFO researcher with too many contradictions to overcome in order to accept the claims of Meier. http://en.wikipedia.org/wiki/George_Van_Tassel

CHAPTER 24

GIVE ME THAT NEW TIME SPACE RELIGION BECAUSE THESE EARTH RELIGIONS ARE JUST NOT WACKY ENOUGH!

The development of new age **UFO religions** finds its origins and influence in the mid 1800's with **Theosophy** and the **Great I AM** founded by **Helena Petrovna Blavatsky** who was an occult writer. She "proposed a complex supernatural order with a hierarchy of "ascended masters"' who originated from Venus. Later, in the 1930's Guy and Edna Ballard Americanized this theme of "Venusian ascended masters" by declaring that he had met with them inside the Grand Teton Mountains. From the frequent meetings with the masters, he received messages which he delivered to the worldwide contactee movement. These contact sessions and messages from the masters help Ballard develop a religion that was not based on an Earth-centered religious belief but on contact with Extraterrestrials with the I AM hierarchy of ascended masters now having been replaced with the space command hierarchy. Jerome Clark, The UFO Book: Encyclopedia of the Extraterrestrial, 1998)

The study and belief in UFOs have all the hallmarks and similarities to any of the world's major religious belief systems. UFOs are at this period in time, beyond the ken of man's understanding and knowledge as is the supreme being of any religion, God. UFOs make themselves manifest, sometimes in physical or holographic form and perform mysterious and amazing aerial maneuvers, display colourful brilliance, they can travel at astounding speeds, come to a dead stop and merely hover and on occasion land and disembark their ETI occupants. Religion manifests angelic beings and prophets who display miracles and cause astounding events to occur in their name of a higher divine source. Extraterrestrial visitors may appear enigmatic and elusive or personable dispensing warnings and admonitions or providing new knowledge to their human contacts. Religion educates its faithful followers through warnings, admonitions, and proscriptions and through prescribed sacred laws, teachings and rituals. Both UFOs and religion require adherence and obedience to the legitimacy of a higher source of power, authority, and control. Failure to do so results in consequences that are punishable, which are built into the very act of non-compliance; reward for faithfulness is enlightenment, knowledge, spiritual and physical evolvement. The list of comparisons is almost limitless. It is no wonder that the transition from traditional Earth-based religions to an Extraterrestrial space age religion has very few bumps along the way to acceptance particularly if it offers something that traditional religion doesn't offer.

We see in the above example the beginnings of the contactee movement or at least the inspiration and impetus for the development of a new order of spiritual belief. With the mystery of unidentified flying objects traversing the skies of many countries, people turn to their religious leaders or to the scientific community for answers to these aerial enigmas. UFOs and ETI make us question the very core of our spiritual beliefs and force us to re-evaluate our understanding of the physical universe in which we live. We are in need for the sake of our sanity and reassurance that the world we woke up to this morning still operates under the spiritual and physical laws that we are familiar with from yesterday. Yet, the contactee forces us to perceive the world that is less familiar, to challenge the concepts of spirituality and science which have now, become inseparably intertwined in a new paradigm of reality; to explain the unexplainable in a rational world that leaves no room for the irrational.

The contactee lives in a world, where personal religious beliefs are often restricted and channeled into neat and tidy, tightly packaged Sunday morning sermons dispensed by an entrusted spiritual authority for the salvation of those lost and wayward souls. This is also, a world where science has gradually over the last 500 years or more become arrogantly dismissive toward the paranormal, the supranatural and the metaphysical having not learned the spiritual lessons of man's history but, instead finds itself cocooned in its own technology and sheltered by the techno-babble of its scientific creed and methodology to which the lay person is excluded. The contactee in his search for earth-bound answers from either, **Christianity**, **Judaism** or **Islam**, etc. leaves him wanting for more than what is being offered by these religious institutions. Science with its supporters from the government and the military fares no better as a substitute for rational answers believing itself to be at the apex of understanding where anything that does not fit into its paradigm of knowledge of the universe is irrelevant or non-existent. However, for most people this type of world is enough and whether by apathy, complacency or timidity, they seem contented to go about their everyday lives without the need to question the inexplicable quirks and oddities of life, relying instead on faith or science to provide guidance and answers to the mysteries of life.

When such lost and wayward souls by happenchance receive an epiphany of wonderment, of enlightenment from a chance encounter with a UFO or possibly from an ET contact, suddenly, the universe opens itself up for discovery and provide some answers for questions which were formerly unanswerable. The Irrational now appears rational and the unexplainable now becomes explainable, the formerly lost are now found and saved. This, then, is the world of the "contactee" and the "new-ager" who seeks salvation from a higher source than is found on Earth.

In such a delusional state of spiritual perception, the contactee believes he has received a divine enlightened message that may advance him and those of his fellow men to a higher level spiritually. When the contactee is viewed as being emotionally agitated, or confused on some unaddressed social aspect of life, or suffering from depression stemming from a difficult episodic period in his life, he seeks answers and possible salvation from the stars. It is possible that he may be looking for a position of leadership among his peers or simply peer recognition to validate how he feels and perceives life.

Studies over the years into UFO research, particularly in the socio-psychological aspects of the UFO contactee, find that they are neither emotionally agitated, confused or depressed in any sense but are in fact, stable and rational contributing members of society. We have seen elsewhere in this book that the message revealed to the contactees by their extraterrestrial contact is usually a message of warning and admonition as well as of hope for mankind. This is for the most part, the extent of the peaceful and rational side of the movement.

There are, however, as in most things in life, a duality of nature to the contactee movement that certainly lives up to it infamous reputation of kooks and crazies running around propagating a religion from the stars. As Jacques Vallee warned in his book: "Messengers of Deception" there are those contactees who will try to turn these ET messages into a UFO religious movement, due

106

in part to direct communications with ET beings who appear with supernatural powers that seem angelic-like, (the opposite to this is that ETs are demonic based on similar elements of contact that are considered diabolic in nature), and to the spiritually bankrupt and moribund society in which we live that has failed to address the irrationality found in the modern era.

The UFO contactees offer us an invitation to step into a "dream world far off," into a "pleasant mirage," to "reprogram our consciousness." Such views would need to be challenged as too radical, as a possible threat to orthodox religions, to the government, science and the military industrial sectors of society, particularly when the public's perception of the offices of authority are already viewed with cynicism and suspicion. The need to maintain the status quo becomes paramount in ensuring perceived societal stability, to maintain the illusion that everything is under control and that society's best interests are being looked after by those who know best. *"Oh! What a tangled web we weave, when at first we practice to deceive!"*

As Vallee points out, he doubts that censorship cannot be effective for very long, given the continued UFO sightings by the public. He feels that the time to overhaul "existing disciplines" and "promoting pioneer research" in the sciences of "physics and biology" is now, as well as initiating a new kind of "research that will take UFO data as empirical observations and try to use them to initiate an interaction with the phenomenon itself." (This is a prophetic statement because this is exactly what Dr. Steven Greer has accomplished with his CE5 Initiative and Contact Protocols! This CE5 Initiative will be covered in a later section of this book). Further, Vallee states, *"And we should do this now, before the new myth is created, before the myth of extraterrestrial revelation replaces belief in the rational acquisition of knowledge."* **Jacques Vallee; Messengers of Deception; 1979**

Vallee warns that there is an emotional powder keg ready to blow in the UFO phenomenon that may create sociological changes in such ways as new church beliefs based on UFOs or a new political movement. We could even foresee (again prophetic) scientists tempted to respond to this emotional outburst by building multi-billion dollar projects like radio telescopes to eavesdrop on galactic communications between Extraterrestrial civilizations. The effect will be to indirectly divert public attention away from UFO sightings; however, they will do nothing to explain how these alien spacecraft have already appeared at our door step. Such multi-billion dollar projects as Vallee goes on to say, may be a ploy by politicians to not only condition the public psychologically to the ET presence but, as a way to ensure their own political position and power.

I personally believe that the reality of the situation is that very few politicians, whether in the U.S., Britain, Canada or elsewhere, know what is really going on within the UFO phenomenon. That this knowledge is not only, so far beyond the reach of the public domain but, it is even beyond the oversight and control of most governments.

We live in a world of comfortable conventionality that is resistant and slow to any change to our paradigm of reality; UFOs are irrational forces that challenge that paradigm of reality and the rational nature of man. Contactees in accepting the UFO reality are in essence, pioneers in the forefront of irrationality and their power for societal change comes not from the UFO phenomenon itself but, from society's failure at recognizing the UFO phenomenon. As Vallee points out this failure forces each individual to deal with the challenge based on his own spiritual development.

Vallee also comes to some startling insights and conclusions where he states that there is "*a machinery of mass manipulation behind the UFO phenomenon,*" that is intent on social and political changes, that contactees are a part of that machinery who are "*helping to create a new form of belief,*" resulting in the hope and realization of that an age-old dream: salvation from above by far wiser celestial navigators from the stars. This manipulation creates deceptions and conditioning which diverts the attention of the public away from the social changes that are reshaping and perhaps, re-inventing the old worn-out institutions of society by keeping our focus on the stars.

Are the manipulators of societal change alien in nature or merely coming from a covert group of humans who have access to very advanced technology and power? Finally, Vallee concludes that UFOs are "psychotronic technology that are physical devices used to affect human consciousness." The purpose of which is to bring about "social changes on the Earth using methods of deception; systematic manipulation of witnesses and contactees; covert use of various sects and cults; control of the channels through which the alleged space messages can make an impact on the public." **Jacques Vallee, Messengers of Deception; 1979**

It is not my intention here, to do an in-depth critical analysis and dissection of Vallee' whole book: "Messengers of deception." Vallee touches on many of the basic concepts and currently accepted viewpoints of the UFO phenomenon. He has also, pioneered new perspectives of thought which still remain highly controversial to this day, particularly his view that ET intelligence is not a physical intergalactic reality or as he states "*the spacecraft theory"* or *"spacecraft hypothesis"* but rather, the phenomenon demonstrate an inter-dimensional or ultra-dimensional nature. It would appear that Vallee's focus and concern are that UFOs by their interaction over the centuries with our physical reality is to manipulate either the weak-minded or the easily influenced and coerced part of society into developing a space age religion of subservience and adoration toward a higher intelligence from the stars. This is a real sociological concern that should not be taken lightly and we will look at a few examples of how this has manifested but, before we get into this, a quick this step is required to clear up a few pointers that Jacques Vallee brings up.

 One has to question the depth of his knowledge and understanding of the concepts on inter-dimensionality as he treats it as a rather unique occurrence in nature as it pertains to sentient intelligence. Vallee views it as a mathematical reality that imposes itself onto our three dimensional universe in ways that are sublime and yet unique and even, acknowledges that other life forms exist in these other dimensions. The problem is he views them as distinct and separate from our physical universe and not a part of the whole cosmology or spectrum of life in which all dimensions co-exist and interface together, simultaneously. The aspect of inter-dimensionality is, in fact, a commonplace reality of nature that permeates throughout the universe whereby, all living creatures, substances, and elements are imbued and are interconnected, particularly is this true of all sentient intelligence. Humans are, therefore, inter-dimensional by nature, as is a glass of water or a rock or plant! All religions for thousands of years have as a part of their written teachings been stating this reality of mankind, that we are spiritual or inter-dimensional beings.

The other interesting point is that Vallee does not explore far enough into the covert intelligence programs that operate within the military industrial sector, which has been for years perpetrating a worldwide deception of truly Machiavellian and totalitarian proportions with the suppression and manipulation of the truth behind the UFO phenomenon. At the time he wrote "Messengers of Deception," a greater understanding and evolvement of the UFO phenomenon has taken place and so, I will assume, so has Vallee's knowledge and understanding on this subject. What he posited then was that the covert intelligence groups within the military -industrial sector was either not thought of or capable of targeting and using advance psychotronic weapons to influence stealthfully, the minds of anybody off the streets or anywhere for that matter. Only such technology could be wielded by extraterrestrials or ultra-dimensional beings for purposes and agendas that only they were aware of or understood.

Now, let's take a look at a few of the space age religious cults that were or still are actively operating in American society and elsewhere.

Heaven's Gate

In the last decade, no other cult received more publicity than did the cult known as **Heaven's Gate** only because most of its followers back in 1997 under the psychotic leadership of **Marshall Applewhite** committed mass suicide. California Police from San Diego made the disturbing discovery of 39 bodies of the members of the Heaven's Gate cult in a rented mansion in the upscale San Diego-area community of Rancho Santa Fe, California, on March 26, 1997. All had died from an apparent suicide which coincided with the appearance of the **Hale-Bopp Comet** of 1997 believing that an alien spacecraft followed closely behind the comet that was to pick up the souls of the cult followers to take them to a better place to live.

Back in the early 1970's Marshall Applewhite had suffered a major heart attack requiring his hospitalization during which he claimed a near-death experience. He felt that he and his nurse at the time, **Bonnie Nettles** were kindred spirits and were "the Two", that is, the two witnesses spoken of in Book of Revelation 11:3 in the Holy Bible. It would seem that these two delusional minds supported each other's delusions. A psychiatrist named Marc Galanter posits that Applewhite and Nettles may have suffered from "the psychiatric syndrome of *folie à deux,* in which one partner draws the other into a shared system of delusion."

For a brief and unsuccessful period of time, they tried to manage an inspirational bookstore, they then began traveling around the U.S. promoting by lectures their belief system. Like so many New Age faiths, they combined the Christian doctrines of salvation and apocalypse with the concept of evolutionary advancement and a belief in the existence of **Extraterrestrial Intelligence**. They believed that they had come from the Next Level (i.e., heaven) to find individuals who would dedicate themselves to preparing for the spaceship that would take them there Their belief that they were from the Next Level is evidenced by both their assertions that they were the two witnesses referred to in Revelation 11 who had risen from the dead after being killed for spreading the word of God, and their belief that Applewhite was the Second Coming of Jesus Christ incarnate, and Nettles was the Heavenly Father. Balch, Robert W. *"Waiting for the Ships: Disillusionment and Revitalization of Faith in Bo and Peep's UFO cult."* and James R. Lewis, ed. *The Gods have Landed: New Religions from Other Worlds.* Albany: SUNY. 1995.

What Christianity could not fulfill or explain, answers obtained from Extraterrestrial messages would fill the void and what was not explicitly revealed by ET messages, then Christianity was the mainstay and fall-back support system to round out the new age belief. Each new and old belief system feeding off of and providing the necessary support to the other, when required.

Applewhite and **Nettles** were known as **"Bo and Peep"** and **"Do and Ti"** with other aliases being adopted over the years. The group also had a variety of names. At the time Vallée studied the group, it was called HIM (Human Individual Metamorphosis), finally arriving at the group's last name that of Heaven's Gate.
Jacques Vallee, "Messengers of Deception" 1979

Their belief system is essentially based on detachment from all things earthly, the giving up of all personal possessions, the severing of all earthly ties to family and friends, to establishment and authority, and to other religions, even to the extent of castration among male members and the avoidance of the sex act by all members. Hating all things earthly meant assurance to the "Next Level" beyond human.

"Heaven's Gate members believed that the planet Earth was about to be recycled (wiped clean, renewed, refurbished and rejuvenated), and that the only chance to survive was to leave it immediately. While the group was formally against suicide, they defined "suicide" in their own context to mean "to turn against the Next Level when it is being offered", and believed that their "human" bodies were only vessels meant to help them on their journey."
http://en.wikipedia.org/wiki/Heaven%27s_Gate_%28religious_group%29

Identified *as "The Two," Marshall H. Applewhite and Bonnie Lu Trusdale (sitting at table)*
hold a meeting in Waldport, Oregon, September 14, 1975, to recruit followers.
Over twenty years later, thirty-nine Heaven's Gate members committed suicide at
the cult's mansion in Rancho Sante Fe, California. **BETTMAN/CORBIS**"
http://www.deathreference.com/Gi-Ho/Heaven-s-Gate.html

The group believed in several paths for a person to leave the Earth and survive before the "recycling", one of which was hating this world strongly enough: *"It is also possible that part of our test of faith is our hating this world, even our flesh body, to the extent to be willing to leave it without any proof of the Next Level's existence'"*.
http://en.wikipedia.org/wiki/Heaven%27s_Gate_%28religious_group%29

Initially, recruitment by the group was disorganized with members coming and leaving routinely, until socialization techniques were refined and the restructuring of the group halted the decline in defection. Social influence processes on new recruit members was a form of brainwashing where a tightly controlled daily regimen of physical tasks and mental exercises, and deprivation of various kinds eliminated all independent thinking and brought conviction and loyalty to the group. With compliant members, they were now deployable to travel around through the western states to spread the message of Bo and Beep. But commitment issues were an on-going source of problems for the group where members' freedoms were restricted to an intolerable state and thus, new recruitments were needed for a rapidly dwindling membership.

Studying the group by various sociologists reported that the life in the group could be equated to a "medieval monastic order." Re-socialization was reviewed and improved upon by Bo and Beep which required members not to be alone or to have all conversations and occasional phone calls monitored and to have members write about their "spirit list of influences they needed to overcome. Bonnie Nettles died in January 1985 of cancer leaving Applewhite despondent for years which may have lead to the need to commit suicide to be with her in that next spiritual realm.

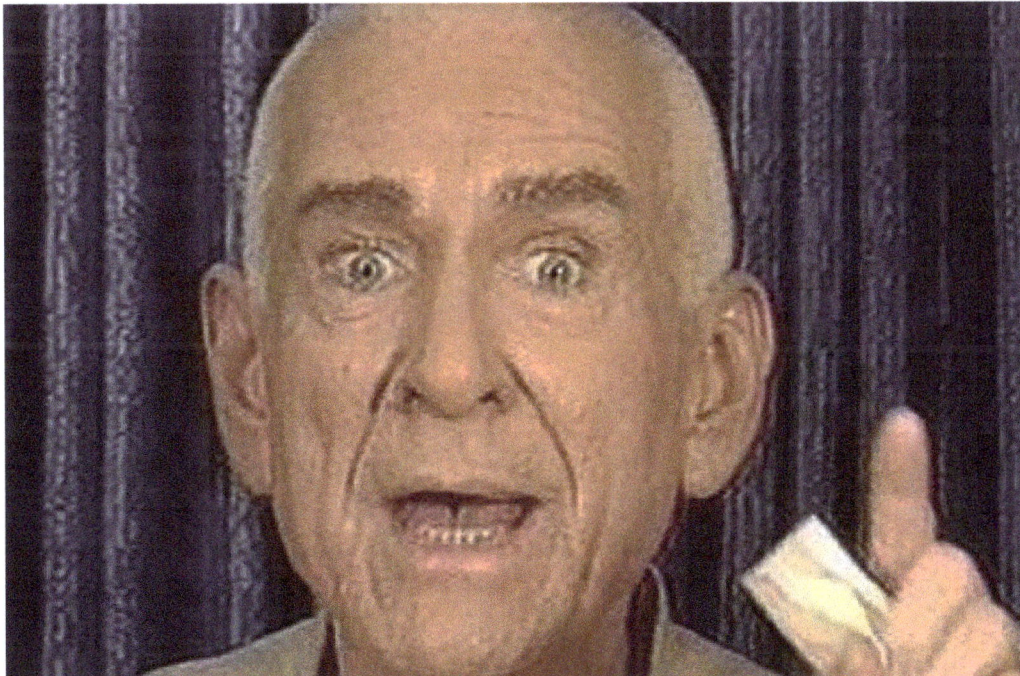

**Marshall Applewhite aka. "Peep" or "Ti" appears on
a videotape before his eventual suicide**
http://dldeprez.blogspot.ca/2011/07/seekers-of-exotic-escape-my-classmates.html

Feeling to evacuate the Earth as quickly as possible because Earth's citizens were refusing to evolve, Applewhite convinced the remaining 38 followers to commit suicide. He claimed that "a spacecraft was trailing the comet Hale-Bopp so that their souls could board the supposed craft. This belief was not exclusive to the Heaven's Gate cult as people like **Dr. Courtney Brown**, a remote viewer who was interviewed on the Art Bell radio talk show, Coast to Coast claimed the same UFO/comet Hale-Bopp connection. **Balch, Robert W. The Gods have Landed: New Religions from Other Worlds. Albany: SUNY. 1995.**

"Applewhite believed that after their deaths, a UFO would take their souls to another "level of existence above human", which Applewhite described as being both physical and spiritual. This and other UFO-related beliefs held by the group have led some observers to characterize the group as a type of UFO religion."
http://en.wikipedia.org/wiki/Heaven%27s_Gate_%28religious_group%29

Recent information has come to light that indicates that the Heaven's Gate cult was deliberately targeted and carefully manipulated by a covert intelligence group to test their ability to coerce the minds and actions of members of the general public, i.e. Applewhite and Nettles. Hallucinogenic drugs and psychotronic weapons were used over a prolong period of time against these two people, who were picked as prime candidates because they were suffering from sociological and psychological challenges in their lives.

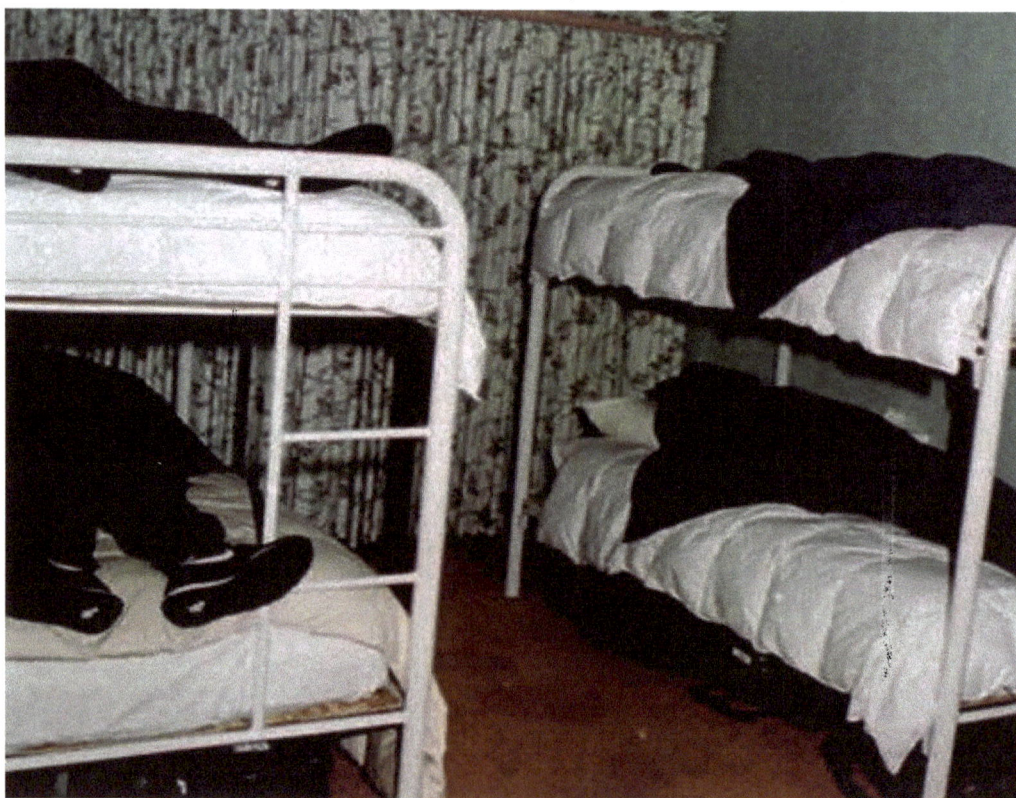

**The death scene of the Heaven's gate cult on March 26, 1997
in Rancho Santa Fe, California**
http://www.findadeath.com/Deceased/h/Heavens_gate/Heavens_gate_mass_suicide_cult.htm

"The **Heaven's Gate** cult committed mass suicide just days after the Arizona ("**Phoenix Lights**" – added by author) sighting. **Marshall Applewhite** received a "sign" that it was time for him and 38 of his followers to shed their human containers and join the Evolutionary Level Above Human in a spaceship hiding behind the comet **Hale-Bopp**. The acronym for **Evolutionary Level Above Human** is **ELAH**, the name for God, and, when inverted, spells **HALE**. Was this part of a **psy-op** trigger to elicit a response?

Coincidentally, three members of the **Heaven's Gate** "away team" worked for **Advanced Development Group, Inc. (ADG)**, a company that *developed computer-based instruction for the U.S. Army*. ADG later became **ManTech Advanced Development Group**; these organizations have connections to the First Earth Battalion, a psy-op group formed within the U.S. military to allegedly handle extraterrestrial affairs such as abductions or contact through telepathy or remote viewing, and to hold simulated extraterrestrial "drills" using holographic technology. First Earth Battalion eventually became the Jedi Project. All of these secretive psy-op groups are, or were, well skilled in spreading information and disinformation about the occult, the **Heaven's Gate** cult, the **attacks of 9/11**, remote viewing and UFO's.

Lee Johnson was a member of a Utah rock group called Dharma Combat. He was a **Heaven's Gate** away team member and committed suicide with the cult members. After his death, the surviving members of the band released an EP that featured a song called MESLIM written by Johnson. The song featured some odd lyrics about a possible attempt by the U.S. government to stage an alien threat. What did he know, and why did he and the 38 followers of **Heaven's Gate** believe that there was an alien spaceship following the **Hale-Bopp comet**? Was all of this a test run for a mass exercise?

Jacques Vallée wrote a book called "Messengers of Deception" where he revealed that groups like **Heaven's Gate** and the **Raelians** and events like the "alien autopsy" seemed to be linked to military intelligence.

Allegedly, branches of the military are engaging in psychological warfare involving false flag exercises. These include simulated extraterrestrial attacks. The frightening part is that the military psy-ops groups are conducting these alien-based psychological experiments to test the collective acceptance of a real alien landing. They are allegedly faking occasional UFO reports and YouTube videos to gauge public reaction. They are also allegedly taking real events shown on YouTube and creating a sense of confusion by creating very sophisticated fakes that appear to be eyewitness documentation.

These fakes are used to discredit any and all anomalous events as hoaxes. The mainstream media then takes the story and jokingly ridicules those who believe that the event was out of the ordinary. This has been lost on many armchair skeptics who believe that they know CGI when they see it. Of course, there is CGI; your taxes paid for the intelligence blackouts and disinformation. The psy-op begins and soon everything is a hoax and evidence is overlooked in the race to debunk and not to decipher."
http://synchromysticismforum.com/viewtopic.php?f=4&t=2257

This has become a common practice by covert intelligence and black op military groups who want to target specific people for hoaxed scenarios or long term "staged" disinformation campaigns and agendas. As **Vallee** points out that *"exposing unsuspecting people to staged scenes in order to further a certain belief is an old trick"* routinely used during the Second World War by both the Allied and Axis powers to deceive each other with disinformation and propaganda. Why would such an effective intelligence tool ever go out of style or practice? In fact, such covert tactics have only become more sophisticated and more finely polished in this day and age. **Jacques Vallee; Messengers of Deception; 1979**

The purpose of staging an event like **Heaven's Gate** or the abduction scenario of **Betty and Barney Hill** in September, 1961 may have been to serve no other purpose than to reinforce the belief in alien visitors or to create a distraction or diversion away from some other covert agenda or perhaps, to convince other people that a space invasion was imminent of malevolent in design.

Author's Rant: As a side note to this real psychological threat by covert military psy-op groups upon an unsuspecting populace, Vallee leads us to a promising door of inquiry, the real probability of a covert military intelligence campaign with known practices of deception and disinformation utilized since World War II and perhaps, even earlier. Yet, he knocks on this door of inquiry, peeks through and briefly glimpses another reality of what's on the other side but, his lack of curiosity or perhaps, his fears stop him from entering through that door to fully explore what is beyond it. We are left tantalized by some real intriguing possibilities in Vallee's pioneering research, but instead, we are disappointed by his desire to pursue an interdimensional hypothesis and rationale characterized by an alleged hidden presence of faerie folk, demons, and leprechauns, et al as somehow manipulating the minds of humanity and always evolves three steps ahead of us as we evolve socially.

Followers of Vallee's work may be left wondering if the ET involvement with humans is more multi-dimensional in nature and perhaps, less extraterrestrial. It appears his own personal biases and perceptions get the best of him as he pursues an interdimensional hypothesis leading us to search elsewhere, and which overrules the possibility of another answer to this enigma, one that is perhaps more down to earth in its explanation with very real extraterrestrial connections.

Raelian Movement

One has to wonder, what would possess someone to create and institute a new religious movement in a world full of existing religious belief systems that are pre-ordained by the Almighty Creator, Himself? Yet, the very understanding of what is the nature and the identity of a supreme, universal deity and mankind's relationship to it was in the mind of **Claude Vorilhon**, a young reporter and racing car enthusiast from Clermont-Ferrand, France open to interpretation. Back in December 13, 1973, while hiking through a volcanic parkland area near his home, he suddenly found himself in the company of a brightly lit conical spacecraft descending close to him. He was greeted by a child-like ET being wearing a green spacesuit which bore a **Star of David** emblem with a swastika in the middle which was also reproduced on the side of the occupant's spacecraft. This symbol came to mean "As above, so below, and everything runs in cycles."

The ET being asked Vorilhon to come back the next morning and again for a half dozen times to be educated in the sacred passages of the Bible which he should bring with him to these educational learning sessions. Vorilhon was given the name, **"Rael"** by the little alien and was told that supreme extraterrestrial scientists known as the **Elohim**, as mentioned in the Bible were the true creators of mankind. He was told that humanity was created by a cloning procedure from DNA and by hybridization with the Elohim and "implanted" on the Earth. Vorilhon-Rael was to be the last of forty great prophets which included Jesus Christ, Buddha, Mohammed, Joseph Smith, etc., whose mission is to unify and "demythologized Christians, Jewish and Muslims Scriptures as a true history of space colonization." Followers were to deny the existence of God and the rational soul and that immortality could be achieved by regeneration through the science of DNA cloning. Jacques Vallee; Messengers of Deception; 1979

Claude "Rael" Vorilhon, then and now
http://www.merseyskeptics.org.uk/tag/what-is-it/ and http://www.merseyskeptics.org.uk/tag/what-is-it/

Democracy was also, to be wiped out and the military in all countries was to be discontinued as these institutions inhibited the natural evolution of mankind. Man's aggressive ways were to be curbed and subdued, only then would the Elohim return to share their vast scientific knowledge and technology with humanity. Failure to be a peaceful, non-hostile civilization would result in a **Sodom** and **Gomorrah** ending of mankind.

There many tenants and practices in this new age religion from space that is similar and positive in many ways to those beliefs of the world's major religious like achieving success in their careers and life's pursuits, to eschew good health by the avoidance of harmful drugs and stimulants, to foster their physical pleasure which it claims will improve one's immune system, intelligence and telepathic or psychic abilities. On the opposite end of the spectrum of practices and beliefs that are far removed from everyday earth-based religions, is that among adults only, nudity is accepted and practiced, the full liberty for individual sexual expression is permitted in all its forms, between the opposite sex, the same sex and with ETI; the ultimate goal being the "*Big C.O.*" or *"Cosmic Orgasm!"*. Robert W. Balch; 1995; *"Waiting for the ships: disillusionment and revitalization of faith in Bo and Peep's UFO cult."* **In James R. Lewis, ed. The Gods have Landed: New Religions from Other Worlds; Albany: SUNY.** and http://en.wikipedia.org/wiki/Raelian_movement

Marriage is not condoned and long-term relationships between couples are not encouraged and termination of any relationship is respected and considered positive to the society as a whole. Children should only be brought into existence when desired and if they become too much to deal with like some sort of annoying nuisance, then society is entrusted to raise them with no guilt or pressure placed on the woman. To many people attracted to this purely physical and hedonistic lifestyle, this simplistic religious belief system offers only simplistic solutions, without all the worries and cares of the world.

Raëlians at Love Hug Festival in Seoul, South Korea. According to Raelian teachings, the practice of nudity among Raelian women members is encouraged.
http://jobpagoda-blog.tumblr.com/post/35475319258/ra%C3%ABlians-at-the-love-hug-festival-in-seoul

Raëlians have founded Clonaid, a company that envisions that someday human beings can be scientifically recreated through a process of human cloning. In recent years this cloning company has, like so many aspects of the **Raelian Movement** come under attack because of its search for a country in which the cloning of humans is not considered illegal or morally improper. Quebec

116

was even considered as a possible home for the world's first human cloning, as the sympathy from French-speaking people would perhaps make its construction a reality, but alas for this cult, this has not yet occurred or even, in any other country.

In keeping with other controversies, Raelians have also founded Clitoraid, an organization whose mission is to repair genitally mutilated clitorises.
http://en.wikipedia.org/wiki/Ra%C3%ABlism.

Author's Rant: My dear friend Lloyd B. a former CSETI member of Vancouver, now departed from this world was at one time a practicing Raelian having met Vorilhon-Rael in Quebec during one of his many lecture trips to this country. Currently, the Raelian Movement is about 20,000 strong worldwide, its strongest centres being France, Japan, and Canada, notably in Quebec. Lloyd related to me that Rael was given a trip to the Elohim's home planet in1975 which is populated with about ninety thousand free-loving quasi-immortal citizens. They are not allowed to have children and have been sterilized to ensure unregulated population growth.

He stated that Rael was shown how they created people from cloning DNA. One of the Aloha's robots demonstrated the process by making a beautiful woman based on Real's specifications appear in a glass cube device. The young gorgeous lady materialized almost instantly, and then with little or no emotion they proceeded to destroy the beautiful specimen by disintegration. They told Rael not too feel bad that they had destroyed the woman as they could easily re-create her again, if he so wish, whereupon the beautiful woman reappeared, along with a brunette, a blonde, and red haired women as well as a beautiful black woman and a Chinese woman, all voluptuous, all very beautiful. He was asked which one he desired to be with. Rael having difficulty deciding, the Elohim made it easy for him by allowing him to be with all six women to spend the night in wild abandoned sexual revelry.

I asked Lloyd why he joined this cult group and he told me that his Catholic Italian upbringing was so rigid and restrictive in his childhood right through to adult years that he needed to break away from these religious mores and find a new direction, one that allowed a lot more personal freedom without the guilt or recriminations. He found that the Catholic Church could not provide satisfactory answers to questions that came from an ever-changing, ever-evolving world. Some of the queries he had about science were answered or confirmed by the philosophy of the Rael Movement. Here, we find the typical reasons why someone would find this cult religion attractive because, it allows an individual to break from the traditional values of society, of religion and even the scientific paradigms of a mechanistic universe. It allows one to explore their sexual and perhaps, carnal natures without prohibitions or restrictions. The sexual instinct is a powerful driving force in human nature which can be reduced to its basic animalistic side or elevated to a creative almost ethereal reality. The Raelian movement offers a strong inducement that is heavy on hedonism for the spiritually and morally lax or challenged individual.

Here then, is the downfall of this controversial movement which stands in stark contrast and contradiction to all the major world religions. It truly alienates (pardon the pun) every religion based upon a monotheistic belief in an omnipotent, omnipresent, omniscient supreme deity while accepting certain beliefs and tenants from these earth-based religions and at the same time denying the majority of their scriptures, laws, tenants, practices, services and history, in essence the very spiritual nature that comprises every religious belief system.

The logo of the **Star of David** with its **Swastika** in the centre represents a frontal assault on the memory of the tragic loss of life during WWII, to the people of the Israel as a result of the holocaust against the Jewish people of Europe by Nazi Germany under Hitler's flag waving swastika. It denigrates a sacred symbol venerated by a religious community with the placement of a symbol of blasphemy within its centre. Given these reasons and the desire by Rael's cult to build a temple in Jerusalem, it's easy to see a nation's unbearable intolerance toward their cause. The nation of Israel has had more than its fill of prejudice and anti-Semitism in its history and desires not invite further strife into its borders.

Two symbols of Raelism
https://en.wikipedia.org/wiki/Ra%C3%ABlism

Interestingly, as a point of fact, the swastika symbol or its reverse image is an ancient sacred symbol that was used almost globally during the Neolithic period of India in the *Indus Valley Civilization* and by ancient cultures such as Greece and by native American Indians. It is sometimes used as a geometrical motif and sometimes as a religious symbol that remains widely used in Eastern and *Dharma religions* such as *Hinduism*, *Buddhism*, and *Jainism*. In Japan, the left-facing version is used to point out temples on maps. The symbol was revered as a good luck or victory symbol. It is only in Western cultures and societies since the Second World War who opposed Nazism that the Swastika has fallen into disgrace as a religious symbol.

The **"Swastika of David"** on the left and the **"Wormhole of David"** on the right, used from 1991 to 2007 The symbol on the right has now become the standard logo for the Raelians in the

118

hopes it will allow Israel to permit the cult to build a third temple in their country. No permission has been given as yet. http://en.wikipedia.org/wiki/File:Raelian_symbols.svg

A scale model of the Raelian Extraterrestrial Embassy. The intended building to be constructed where the Elohim visitors will meet with earth political leaders and scientists to share their technology and wisdom
https://raellaserie.wordpress.com/2013/01/01/ramuel-le-fils-de-rael/

The Urantia Book

Here is a strange case in UFO space-based religions that of which came first, the UFO space-based religion or the book about religion from outer space? In this case, the book appears to have been revealed through channeling by an unknown author, probably a stockbroker while in a deep state of sleep. The book is **The Urantia Book.** It is a hefty 2178 pages in length including 51 pages of content, 1814 pages of text and an amazing 312 pages of index. It could be considered as five books or volumes in one (see below the five sections of the book). The book was revealed over a period of years from 1924 to 1935 with its eventual first publishing in 1955.

Dr. William Sadler and his wife, **Dr. Lena Sadler** were first approached by a neighbor who was upset and concerned for her husband who seem to be talking in his sleep about many unusual things which she said came from spirit beings. She felt the good doctors needed to hear what he was muttering in his deep trance-like sleep in case he was suffering symptoms of psychotic dissociation or worse. http://en.wikipedia.org/wiki/The_Urantia_Book

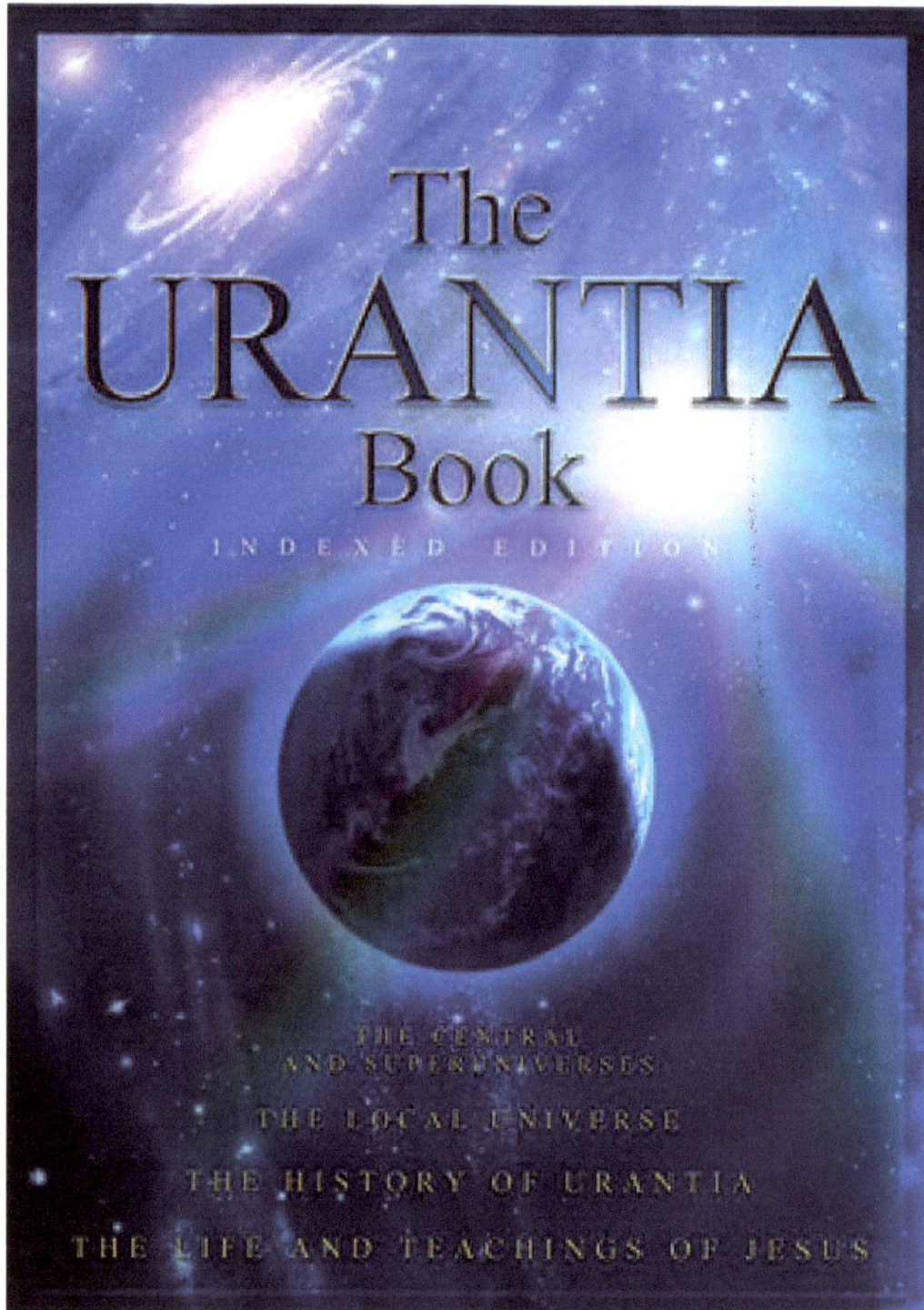

The channelled, voluminous 2178 page Urantia Book containing the history of the universe and the Earth. It is exceeded only by this author's book, "A Citizen's Disclosure on UFOs and ETI" with 3457 pages!

http://www.abzu2.com/2012/03/24/the-urantia-book/

120

Dr. William Sadler was a professor Physiologic Therapy at the Post-Graduate Medical School of Chicago, consulting psychiatrist at Columbus Hospital, Director of the Chicago Institute of Physiologic therapeutics, and for over twenty-five years, a professor and chairman of the department of pastoral psychology at McCormick Theological Seminary. Among the many books and papers to his credit as a talent writer, was his book, published in Chicago in 1920. As a very religious man and a re-known skeptic and debunker of various forms of paranormal fraud and psychic charlatanism, he wrote in his book *The Physiology of Faith and Fear*:

"It is not uncommon for persons in a cataleptic trance to imagine themselves taking a trip to other worlds. In fact, the wonderful accounts of their experiences, which they write out after these cataleptic attacks are over, are so unique and marvelous as to serve as the basis for founding new sects, cults, and religions."

Sadler explained that these phenomena were the emanations of "marginal consciousness without the awareness of the central consciousness." Yet, here before him was a mystery that seems to contradict his psychiatric understanding of most of the things he had written about. What started off with automatic speaking from a neighbour's spouse while in a heavy, unarousable sleep transitioned to automatic handwriting of such coherence and informational content that a stenographer was required to write everything this man was saying while in his cataleptic trance. So voluminous was the channeled information that it filled 2097 pages and when all was said and done, Sadler after eighteen years of study and careful investigation had failed to reveal the psychic origin of these messages. "I find myself at the present time just where I was when I started."

The Urantia Book, though not known by that name at the time was not easily understood. Sadler would share passages from it to his family, friends and former patients at their regular Sunday discussion meetings. Because most of the book's meaning was elusive, his Sunday group formed a "Forum" to ask questions of the "channeler" of which five individuals and Sadler himself formed a smaller group called the Contact Commission responsible for gathering the questions from the Forum, acting as the custodians of the handwritten manuscripts that were presented as answers, and arranging for proofreading and typing of the material. From 1935 to 1955, the papers of the book were studied, further clarifications and revisions were made, and finally collation and a table of content were completed for publication.

The Urantia foundation was formed to promote the book and it was later determined "that a man named **Wilfred Kellogg** was the sleeping subject and authored the work from his subconscious mind, with William Sadler subsequently editing and authoring parts." Through writing analysis and methods of stylometry at least nine authors were believed to be involved in the authorship of the book. http://en.wikipedia.org/wiki/The_Urantia_Book

In the book, the word **Urantia** is the name of the planet Earth and sometimes **The Urantia Book** (called the **Urantia Papers** or *The Fifth Epochal Revelation*) is a spiritual and philosophical book that discusses God, Jesus, science, cosmology, religion, history, and destiny. The book consists of an introductory statement followed by 196 "papers" divided into four parts:

- **Foreword** - a guide to the terminology - explanations for concepts and words that are "in designation of Deity and certain associated concepts of the things, meanings, and values of universal reality."
- **Part I**: **The Central and Super Universes** consists of 31 papers that address what are considered the highest levels of creation, beginning with the "First Source and Center of all things and beings", the eternal and infinite "Universal Father".
- **Part II**: **The Local Universe** is composed of 25 papers pertaining to the origin, administration and personalities of "local universes." It presents narratives on the inhabitants of local universes and their work as it is coordinated with a scheme of spiritual ascension and progression of different orders of beings, including humans.
- **Part III**: **The History of Urantia** includes 48 papers that compile a broad history of the Earth, presenting a purported explanation of the origin, purpose, and destiny of the world and its inhabitants. An additional 15 papers cover various topics such as "Deity and Reality", the concept of the Thought Adjuster, "Personality Survival", and "The Bestowals of Christ Michael".
- **Part IV**: **The Life and Teachings of Jesus** is presented in 77 papers and narrates "The Life and Teachings of Jesus", about his childhood, teenage years, family life, various employment experiences before the commencement of his public ministry, an exhaustive recital of his public ministry and the events that led to his crucifixion, death, and resurrection. It continues with papers about appearances after he rose, Pentecost, and finally, "The Faith of Jesus."

There are many facets of other world religions incorporated in the book, including from Islam, Taoism, Judaism, Hinduism, Shinto, and Confucianism as well as Buddhism and the many obvious Christian concepts that run throughout the book. There is no church or clergy, therefore, worship and spiritual development if one accepts the book as divine scripture becomes an act of self-immersion in the writings. With no central organization or religious administration, there is no way to do a head count of the number of practicing believers, The Urantia Brotherhood, founded in 1955 claims approximately twelve hundred members, with the highest concentrations in the western part of United States.

The most common contentious issues with this loosely based organization is its claim to a written revelation from celestial beings, its conflict with modern scientific theory of the universe, the alleged plagiarized material of its content, sexist, racist and discriminatory passages in The Urantia Book are viewed as not in harmony with modern societal perceptions, Christians who believe that the **Bible** is the only true **Word of God**, deny The Urantia Book because it denies some fundamental Christian doctrine.

Personally, this last critique sets up Christians as being fundamentally and fanatically dogmatic, narrow-minded and highly competitive and defensive about their own religious beliefs. This unattractive attitude has been maintained for over 2000 years and has been a source of constant conflict, intolerance and religious wars throughout its history. I should be pointed that this meant not to bash Christian belief and it adherents but not everything under the sun is the approval of Christianity to be acknowledged as being from God or as a source of truth and wisdom. Admittedly, The Urantia Book is however, a distortion on some of the Bible's sacred text which I attribute to unhappy and disgruntled Christians who find that their own religion does not live up to the exigencies of the time or who feel restricted and confined to certain practices and belief structure, and thus, seek new avenues of religious freedom without the condemnation, censorship or the restrictions on self-expression in all its forms.

Unarius

Another example of a UFO space-based religion is the flying saucer group, Unarius or the **"Unarius Academy of Science"** which is one of America's most enduring contactee movements for the last fifty plus years. Unarius is an acronym which stands for **"UN**iversal **AR**ticulate **I**nterdimensional **U**nderstanding of **S**cience". It was founded in February 1954 in Los Angeles, California by Ernest Norman and his wife **Ruth Norman** until his death in 1971 and then Ruth became "the mission" (this was the name of the movement, initially) leader until her death in 1993. The organization was founded on the concept of bringing in of the interdimensional science of life through the books channeled by **Ernest Norman** and it was never thought of as a religion by Ernest, even though spiritual aspects were evident from the teachings of Christianity and various other ancient belief systems. As with most channelers, the cosmology of spiritual belief has its source in the channeled transmissions from Extraterrestrial beings, the "Space Brothers" in this case from Venus and with one particular Venusian by the name of **Mal-Var.**

From such channeled communications from Venus and later from Mars, Ernest wrote many books, his first being *The Voice of Venus*. "In the work, Venusians are described as having "energy bodies" and living in a higher vibratory plane that would be invisible to a human, were he to stand in the middle of the capital city known as *Azure*. The planet Venus and its culture are said to be more spiritual than that of the Earth and that more advanced Earth-dwellers visit and study on Venus when they sleep. Healing wards for human suicides, alcoholics, the mentally impaired and similar human wreckage exist in Azure and these souls are treated with positive energy and light to help them reincarnate with greater integration."

In his book, *The Truth About Mars*, "the Chinese evolved from ancient interstellar migrants who began colonizing Mars a million years ago. They are reported to have returned to Mars, where they live in underground cities, after being attacked by natives of the Earth. A group which had become separated did not return with them and this group branched off and formed the various Asian racial genotypes." ***In this day and age, this is considered an obvious racially prejudiced perspective.*** [my bold italics] http://en.wikipedia.org/wiki/Unarius

Through other channelings, Ernest wrote a series of books known as the Pulse of Creation Series describing seven spiritual planets: Venus, Eros, Orion, Muse, Elysium, Unarius, and Hermes. Each book in the series detailing one of the **seven planes of Shamballa** in the spiritual or astral realm of existence where each contained "advanced teaching centres and enlightened universities that were homes of great teachers and ascended masters" such as Jesus Christ.

Throughout the mid 60's, the Normans delved into reincarnation and past life regression with Ernest and Ruth seeing themselves as formerly Jesus and **Mary Magdalene** with some of their followers as admitting complicity in the persecution of Christ*! Once again, it doesn't take much to piss off the Christians with such outrageous statements as in the above two paragraphs.* [My bold italics added].

These are the primary principles as explained by Norman in *The Infinite Concept of Cosmic Creation*: Everything is energy which is never created or destroyed but merely changes form. People are indestructible metamorphic energy forms possessing souls with reincarnation ability. Personal development is karmic requiring positive action instead of negative action. Spiritual realms or levels exist beyond the physical worlds or universe inhabited by beings of both celestial and of lower nature.

Ruth Norman aka. "Queen Uriel, Queen of the Archangels"
http://beforeitsnews.com/alternative/2015/05/meet-unforgettable-pioneer-lightworker-ruth-norman-aka-queen-uriel-et-ambassador-prophecy-messenger-more-flamboyant-than-cher-or-liberace-3147688.html

Upon the death of Ernest in 1971, Ruth followed in her late husband's footsteps and became a channeler par excellence with the aid of two student sub-channelers who were given the spiritual names of Antares and Cosmon. Together, they spent several days channeling with Ruth where she recalled a past life on ancient Atlantis as Ioshanna which she adopted as her new name. She then changed her name again after having visions of a planetary paradise on the planet of Eros where she married the archangel Michael and where she was crowned by her deceased husband in celestial form as "Queen Uriel, Queen of the Archangels." She and her followers re-enacted the vision at a local hotel ballroom with Uriel regaled in a wedding dress and her followers dressed in ancient Grecian-style garments.

Uriel as the "Cosmic Generator" and wearing the costume by the same name which is made of black velvet, embellished with planets and stars that are electrically illuminated, she provides life-force to the Earth-like worlds
http://www.laweekly.com/calendar/welcome-space-brothers-the-films-of-the-unarius-academy-of-science-4760457

Many messages through channeling sessions indicated that many great scientists, ancient philosophers, former American Presidents and even deceased astronauts, as well as other UFO contactees, were supportive of Uriel and the Unarius movement. Prophecy became important as the space-age religion continually evolved with a channeled message in 1972 from previously unknown spiritual planets, the "33 worlds of an interplanetary confederation" that would arrive in a stunning landing event, each with a 500,000 square foot spacecraft landing on top of each other in a multi-stacked fashion. They were coming to earth to share their technology and help mankind evolve to a higher level of existence. They were supposed to have arrived in 1975 but, because of a former past life karma transgression by one of the group's sub-channeler, **Louis Spiegel**, his prophecy was transmitted incorrectly. This lead to a schism within the ranks of the

Unarius cult and it was up to Uriel to set the situation right by stating that the intended ET landing would take place in the new millennium in 2001.

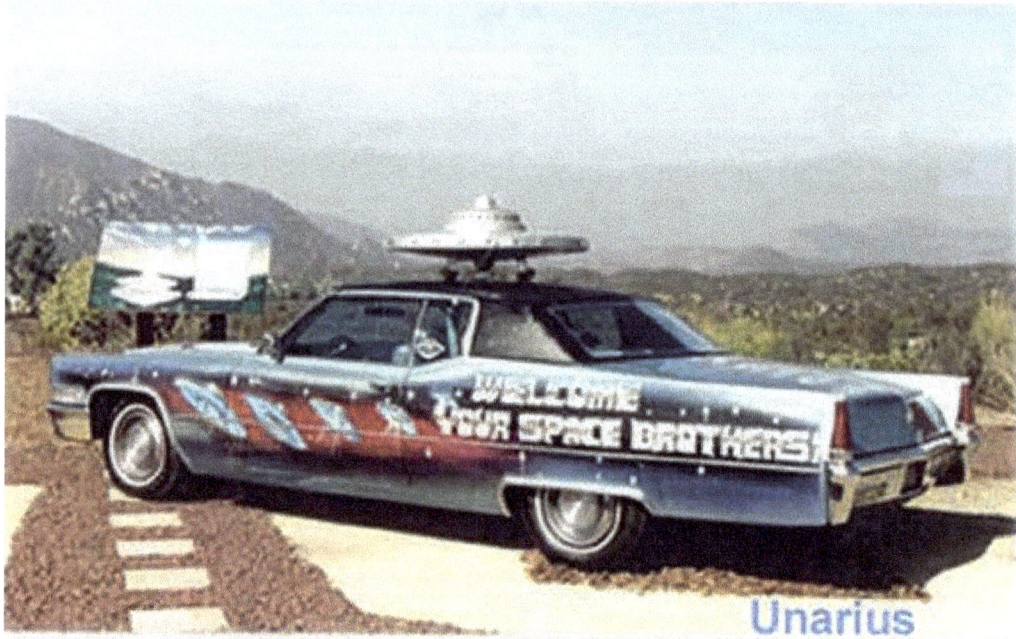

The Unarius Space Cad taken in East San Diego County, California Copyright© Unarius
http://www.unarius.org/spacecad/

However, the damage and internal differences could not be undone and Uriel found herself less involved with the movement mostly due to advancing age and poor health. **Diana Tumminia and R. George Kirkpatrick "Unarius: Emergent Aspects of an American Flying Saucer Group" In James R. Lewis, ed.** *The Gods have Landed: New Religions from Other Worlds.* **Albany: SUNY. 1995**

Louis Spiegel assumed leadership in the organization when Ruth Norman died in 1993, which lead to further schism in the membership because not everyone agreed with Spiegel's channeled messages and many people fell away from the group.

Finally, to add insult to injury, 2001 has since come and gone with many other expected unfulfilled events not the least was the failure of no space fleet landing having occurred and the passing of Louis Spiegel in 1999, the organization has returned to its roots with a small council providing administration duties to the cult group.

Besides offending the easily offended Christians, there is criticism even from scientists from many disciplines that find numerous inaccuracies with their scientific viewpoints as well as some racial insinuations towards the Chinese people is totally speculative.

Uriel towards the later years of her life was colourfully eccentric particularly in her unusual costumes and her Cadillac with the flying saucer mounted on the roof top which she drove around to UFO conventions and Unarius meetings. The best comment albeit, the least flattering

126

was made by Alex Heard, in *Apocalypse Pretty Soon: Travels In End-Time America*, where **Ruth Norman** has been criticized for her flamboyance & costumes worn in later years, he quipped of her that she was, **"a true American original who combined the couture sensibilities of a drag queen with the *joie de vivre* of a Frisbee-chasing Irish Setter."**

Aetherius Society

One of the longest surviving UFO religions to date originated outside of the U.S. that being the **Aetherius Society** founded by **George King** in London, England in 1955. Dr. George King (he received his doctorate is in Divinity in 1960 in America) had been a taxi driver at one time but, has had a life of notable distinction having been awarded university degrees and civil recognitions (National Fire Service during the Second World War), he is a metropolitan archbishop, a prince, and Knight of Malta.

King claimed that he was contacted by an extraterrestrial intelligence known as **Aetherius** as well as other advanced intelligences from within this solar system. These ETI in turn claim, through communications referred to as "**Cosmic Transmissions**", to represent a Cosmic and Solar Hierarchy referred as the "Interplanetary Parliament". John A. Saliba, "Religious Dimensions of UFO phenomena", In James R. Lewis, ed. *The Gods have Landed: New Religions from Other Worlds*. Albany: SUNY. 1995

George King Founder, Cosmic Master, Primary Terrestrial Mental Channel and Archbishop, of the Aetherius Society, 1955
http://www.theninefreedoms.org/dr-george-king.html

The main guiding principle of the organization is service to humanity, through the manipulation of subtle energies, through prayer, healing and other, technology-based, means. Its teachings combine yoga spiritual teachings with other teachings from advanced extraterrestrial beings channeled through the yogic mediumship of **King**. The society's beliefs are similar to **Theosophy** and the **Ascended Master Teachings** blending traditions and teachings from both the Eastern and Western occult philosophies. The society's "Transmissions" seem to be mainly concerned with the dangers of atomic experimentation and of humanity's inability to monitor or to control its effects.

King claimed that he received prophetic warnings in the form of telepathic communications conveyed to him by his extraterrestrial hierarchy that an atomic accident in the Urals in 1958 would occur as recorded in its journal, the *Cosmic Voice*, and later, yet another nuclear accident would happen, this would become the **Chernobyl disaster of 1986** that befell the Soviet Republic.

The following is a partial list of the society's core beliefs drawn from the society's literature:

- Service to mankind is probably the most important belief of the society. "That action speaks louder than words" to use a borrowed and often used adage. It is what a person does for others that counts, not what they profess to believe. The society's motto, "Service is the Jewel in the Rock of Attainment" reflects this importance.
- There is only one God. In this sense, all major religions are similar in nature and all religious paths are leading towards the same ultimate end.
- Spiritual people should cooperate with each other since they have a greater power when working together.
- There is advanced, intelligent life on other planets.
- Unidentified flying objects really are intelligently controlled extraterrestrial spacecraft visiting this Earth. Unlike some UFO groups, their belief is that extraterrestrials are friendly and are here to help mankind in its development.
- A belief that man, and indeed all life, is a divine spark of the Creator, our God, and that Earth is a classroom on the evolutionary ladder of life back to the source from which all came.
- Mother Earth as a living breathing entity which is thousands of lives more evolved than we are.
- Jesus, Buddha, Krishna and certain other great religious leaders were of extraterrestrial origin and came to Earth to help mankind.
- They consider yoga and meditation as very important and George King to be a "Master of Yoga". In common with traditional yogic teaching, Aetherians deem meditation or **Samadhi** to be an experiential state of Adeptship "when the soul is bathed in the Light of pure Spirit and one becomes a knower of truth." Mantras are also frequently repeated during services of the group.

Karma and reincarnation are considered two natural, all pervasive laws of God.
http://en.wikipedia.org/wiki/Aetherius_Society

The Aetherius Society mission is to spread the teachings of the Master Aetherius, the Master Jesus, and other **Cosmic Masters** and to prepare the way for the next Master. The use of spiritual healing and prayer by tuning in and radiating the "Power" transmitted during a "Holy Time", or "Spiritual Push", regardless of one's religion. To promote a brotherhood and the Society's growth based on the Teachings and Knowledge of the Cosmic Masters. Other mission objectives include Operation Prayer Power, Operation Sunbeam, Operation Starlight, and Operation Space Power utilizing technology invented by King.

Such technology includes the "Spiritual Energy Battery", designed to hold a charge of spiritual energy for an indefinite period. This battery when connected to a "Spiritual Energy Radiator" can be discharged much like an electrical capacitor thus radiating spiritual energy in a concentrated stream which is directed to help in world situations of war, disaster, etc. Effectively this is healing prayer energy on a global scale that alleviates suffering and hardship in stricken areas. The Aetherius Society believes that Operation Prayer Power and other such "missions" have averted, and helped to bring relief to, many disasters.
http://en.wikipedia.org/wiki/Aetherius_Society

Operation Prayer Power, a global healing mission lead by George King
https://wrldrels.org/profiles/AetheriusSociety.htm

Once again, skeptics and fundamentalist conservative Christians are their chief critics. Many skeptics perceive the society's views as strange and even more incredible than the beliefs of many other new religions. Scientists see their belief in UFOs as irrational and without scientific basis, for example, the group views most of the planets in this solar system, and the Sun itself, are inhabited by life forms that are much more advanced than life on Earth. Historically, this has led the society to petition governments to release information they feel they are hiding on UFOs. Add to this, the group claims to use physical equipment for the manipulation of spiritual energy.

Ever alert and ever competitive to new age religions which appear to threaten and dethrone the sacred position of Jesus Christ and Christianity in general, has conservative and fundamentalist Christians offended by the society's view of Jesus. The Aetherians say that Jesus is living on Venus. King first announced this in the British press in the late 1950s. This caused the largest controversy in their history as Britain was still a Christian nation at that time. The group was accused of blasphemy or being in service to Communists. Most Christians would have found the idea offensive at that time, especially as King claimed to channel the voice of Jesus.

Added to this, some of the Aetherians' statements on life on Mars or Venus are suspiciously like that found in **C. S. Lewis's** "Out of the Silent Planet" trilogy: for example, Jesus being on another planet and the beings on other worlds being invisible due to "living on higher vibratory planes" (the same belief is held by **Theosophists**—not that the Master Jesus lives on Venus, but that some beings on other planets, including Venus, are normally invisible, but they can appear to us if they so desire. Another issue is that the group deems "orthodox Christianity" to be unsatisfactory.

The society's stated position since the death of George king in 1997 is that it regards him as an **Avatar** – in other words, the incarnation of a great soul, with a great purpose to fulfill. Unlike many new age religious cults, membership growth for this long-lived cult group has been extremely slow with only 650 followers over the last five decades focusing. Its focus instead has been on a committed membership that sustains its spiritual activity with the development of values and the mission policy of the society, rather than in membership numbers.
http://en.wikipedia.org/wiki/Aetherius_Society

Church of Scientology

When asked, what is the first thing you think of when the word, **Scientology** is mentioned, almost immediately most people's response will be **Tom Cruise** or **John Travolta** as celebrities whose lives are often in the limelight of controversy by their beliefs in Scientology? It may be that people will think of the interrelationship of extraterrestrial life with some aspects of the world's major religions or the strange practice of "auditing," the removal of all stress and negativity from one's painful past life incarnations and experiences. But most people will connect Scientology with **L. Ron Hubbard** as its founder and author of the self-help system, "Dianetics" from which its members derive their teachings and spiritual beliefs. Most of what follows is a mere scratch on the surface of such a complex cult religion that the few pages offered here, simply do not do it justice. There are many books written about Scientology that cover the subject matter extensively and in depth from every angle imaginable. The reader is encouraged to investigate these books for himself.

Lafayette Ronald Hubbard (March 13, 1911 – January 24, 1986) was an American science fiction author of many science fiction books and stories as well as the book Dianetics. Over the next three decades, Hubbard developed his self-help ideas into a wide-ranging set of doctrines and rituals as part of a new religion he called Scientology. Hubbard's writings became the guiding texts for the **Church of Scientology** and a number of affiliated organizations that address such diverse topics as business administration, literacy and drug rehabilitation.
http://en.wikipedia.org/wiki/Scientology

130

Hubbard was a controversial public figure, and many details of his life are still disputed, whether, it was his travels to far Eastern countries at an early age in life or his less than lustrous military career that was full of leadership problems or his dabbling in the black arts of magic and the occult. He felt himself to be psychologically troubled, and was financially destitute after his military career for a period of time, yet, his writings enabled him to eventually get ahead of the game and start living the American dream of success. Official Scientology biographies present him as a "larger-than-life" figure whose career is studded with admirable accomplishments in an astonishing array of fields. Many of these claims are disputed by former Scientologists and researchers not connected with Scientology.

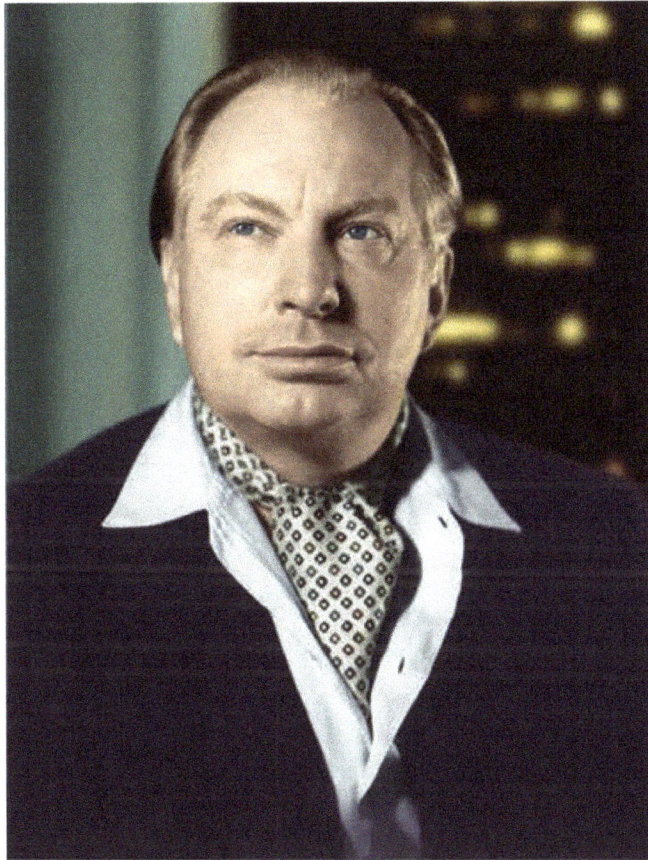

Lafayette Ronald Hubbard
http://www.centrumbasic.sk/studijne-centrum/nasa-metoda/

Scientology has superseded most of the traditional new age space-based religions of the 50's to become America's fast-growing cult religion of the twentieth century. The word *Scientology* is a pairing of the Latin word *scientia* ("knowledge", "skill"), which comes from the verb *scīre* ("to know"), and the Greek *lógos* ("word" or "account [of]")
http://en.wikipedia.org/wiki/Scientology

Scientology has been surrounded by controversies since its inception. It has often been described as a cult that financially defrauds and abuses its members, charging exorbitant fees for its spiritual services. The Church of Scientology has consistently used litigation against such critics, and its aggressiveness in pursuing its foes has been condemned as harassment. Further controversy has focused on Scientology's belief that souls ("thetans") reincarnate and have lived on other planets before living on Earth. Former members say that some of Hubbard's writings on this remote extraterrestrial past, included in confidential Upper Levels, are not revealed to practitioners until they have paid thousands of dollars to the Church of Scientology. Another controversial belief held by Scientologists is that the practice of psychiatry is destructive and abusive and must be abolished.

Dianetics posits the existence of a mind with three parts: the conscious "analytical mind," the subconscious "reactive mind," and the somatic mind. The goal of **Dianetics** is to remove the so-called "reactive mind" that Scientologists believe prevents people from becoming more ethical, more aware, happier and saner. The Dianetics procedure to achieve this is called "auditing". Auditing is a process whereby a series of questions are asked by the Scientology auditor, in an attempt to rid the auditee of the painful experiences of the past which Scientologists believe to be the cause of the "reactive mind". http://en.wikipedia.org/wiki/Scientology

Dianetics has achieved no general acceptance as a *bona fide* scientific theory. Scientists have described Dianetics as an example of pseudoscience. Objective experimental verification of Hubbard's physiological and psychological doctrines is lacking.

To date, no regular scientific agency has established the validity of his theories of prenatal perception and engrams, or cellular memory, or Dianetic reverie, or the effects of Scientology auditing routines. Existing knowledge contradicts Hubbard's theory of recording of perceptions during periods of unconsciousness.

J.A. Winter, M.D., originally an associate of Hubbard and an early adopter of Dianetics, had by the end of 1950 cut his ties with Hubbard and written an account of his personal experiences with Dianetics. He described Hubbard as "absolutistic and authoritarian", and criticized the **Hubbard Dianetic Research Foundation** for failing to undertake "precise scientific research into the functioning of the mind". He also recommended that auditing be done by experts only and that it was dangerous for laymen to audit each other. Hubbard writes: "Again, Dianetics is not being released to a profession, for no profession could encompass it."

Commentators from a variety of backgrounds have described Dianetics as an example of pseudoscience, that is, information which claims to be scientific but which fails to meet the basic criteria for science. For example, philosophy professor Robert Carroll points to Dianetics' lack of empirical evidence:

"What Hubbard touts as a science of mind lacks one key element that is expected of a science: empirical testing of claims. The key elements of Hubbard's so-called science don't seem testable, yet he repeatedly claims that he is asserting only scientific facts and data from many experiments. It isn't even clear what such "data" would look like. Most of his data is in the form of anecdotes and speculations ... Such speculation is appropriate in fiction, but not in science." http://en.wikipedia.org/wiki/Scientology

W. Sumner Davis similarly comments that:

"Dianetics is nothing more than an example of pseudoscience trying to legitimize itself ... Hubbard, had he indeed been a scientist, would have known that truth is not built on axioms, and facts cannot be found from some a priori knowledge. A true science is constructed on hypotheses, which are arrived at by the virtue of observed phenomena. Scientific knowledge is gained by observation and testing, not believing from some subconscious stipulation, as Hubbard would have us believe."

The procedure of Dianetics therapy (known as *auditing*) is a two-person activity. One person, the "auditor", a **Scientology** counsellor guides the other person, the "preclear". The preclear's job is to look at the mind and talk to the auditor. The auditor acknowledges what the preclear says and controls the process so the preclear may put his full attention on his work. The auditor's task is to help the person discover and understand engrams, and their limiting effects, for themselves. It bears a superficial similarity to confession or pastoral counselling, but the auditor does not dispense forgiveness or advice the way a pastor or priest might do. Other critics and medical experts have suggested that Dianetic auditing is a form of hypnosis.

Also in 1951, **Hubbard** introduced the **electropsychometer (E-meter)** as an auditing aid. Based on a design by Hubbard, the device held by Scientologists is a useful tool in detecting changes in a person's state of mind. http://en.wikipedia.org/wiki/Scientology

Most auditing requires an E-meter, a device that measures minute changes in electrical resistance through the body when a person holds electrodes (metal "cans"), and a small current is passed through them. **Scientology** asserts that watching for changes in the E-meter's display helps locate engrams. Once an area of concern has been identified, the auditor asks the individual specific questions about it, in order to help them eliminate the engram, and uses the E-meter to confirm that the engram's "charge" has been dissipated and the engram has in fact been cleared. As the individual progresses, the focus of auditing moves from simple engrams to engrams of increasing complexity. At the more advanced OT auditing levels, Scientologists perform solo auditing sessions, acting as their own auditors. An article in Newsweek stated that "the Dianetics concept is unscientific and unworthy of discussion or review." Some practitioners of **Dianetics** reported experiences which they believed had occurred in past lives or previous incarnations.

In early 1951, reincarnation became a subject of intense debate within Dianetics. Campbell and Winter, who was still hopeful of winning support for Dianetics from the medical community, championed a resolution to ban the topic. But Hubbard decided to take the reports of past life events seriously and postulated the existence of the *thetan*, a concept similar to the soul. This was an important factor in the transition from secular Dianetics to the religion of Scientology.

The United States **Food and Drug Administration (FDA)** began an investigation concerning the claims the **Church of Scientology** made in connection with its E-meters. On January 4, 1963, they raided offices of the Church of Scientology and seized hundreds of E-meters as illegal medical devices. The devices have since been required to carry a disclaimer saying that they are a purely religious artifact. http://en.wikipedia.org/wiki/Scientology

In 1979, as a result of FBI raids during Operation Snow White, eleven senior people in the church's Guardian's Office were convicted of obstructing justice, burglary of government offices, and theft of documents and government property.

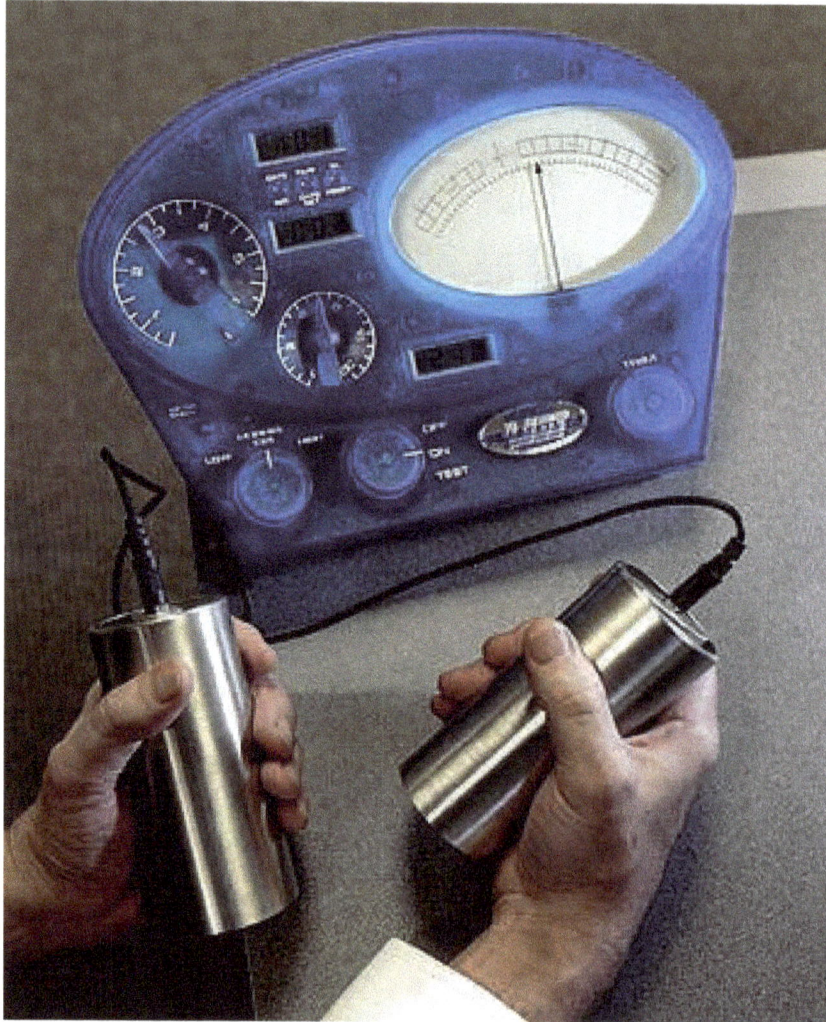

The Infamous E-Meter (*Electropsychometer*) used for "Auditing"
https://luckyottershaven.com/2017/01/07/my-love-affair-with-scientology/

On January 24, 1986, **L. Ron Hubbard** died at his ranch near San Luis Obispo, California and David Miscavige became the head of the organization.

In a 1993 U.S. lawsuit brought by the Church of Scientology against Steven Fishman, a former member of the Church, Fishman made a court declaration which included several dozen pages of formerly secret esoterica detailing aspects of Scientologist cosmogony.

As a result of the litigation, this material, normally strictly safeguarded and only used in Scientology's more advanced "OT levels", found its way onto the Internet.

This resulted in a battle between the Church of Scientology and its online critics over the right to disclose this material, or safeguard its confidentiality. The Church of Scientology was forced to issue a press release acknowledging the existence of this cosmogony, rather than allow its critics "to distort and misuse this information for their own purposes." Even so, the material, notably the story of Xenu, has since been widely disseminated and used to caricature Scientology, despite the Church's vigorous program of copyright litigation. http://en.wikipedia.org/wiki/Scientology

Xenu, (**Xemu**) (pronounced /ˈziːnuː) was, according to writer **L. Ron Hubbard**, the dictator of the "Galactic Confederacy" who, 75 million years ago, brought billions of his people to Earth in a DC-8-like spacecraft, stacked them around volcanoes and killed them using hydrogen bombs. Official Scientology dogma holds that the essences of these many people remained and that they form around people in modern times, causing them spiritual harm. Members of the Church of Scientology widely deny or try to hide the Xenu story.

These events are known within Scientology as "*Incident II*", and the traumatic memories associated with them as *The Wall of Fire* or the *R6 implant*. The story of Xenu is part of **Scientologist** teachings about extraterrestrial civilizations and alien interventions in Earthly events, collectively described as *space opera* by **Hubbard**. Hubbard detailed the story in *Operating Thetan level III (OT III)* in 1967, warning that the R6 "implant" (past trauma) was "calculated to kill (by pneumonia) anyone who attempts to solve it."

The **Xenu** story is part of the Church of Scientology's secret *"Advanced Technology,"* normally only revealed to members who have already contributed large amounts of money. The church avoids mention of Xenu in public statements and has gone to considerable effort to maintain the story's confidentiality, including legal action on the grounds of both copyright and trade secrecy. Despite this, much material on Xenu has leaked to the public via court documents, copies of Hubbard's notes, and the Internet. http://en.wikipedia.org/wiki/Scientology

Other former **Scientologists** such as **Arnie Lerma**, who this author met in Rio Rico, Arizona in October 2009 at a three day **CSETI** conference informed me that he had also divulged secrets of the **Church of Scientology** on the internet back in 1996. He was also, subsequently sued for millions of dollars by the Scientologist lawyers on behalf of the Church for his revealing publicly what he knew about their practices and beliefs on the internet. Fortunately, Arnie won his case against Scientology proving their coercive brainwashing techniques and their racketeering methods and since has gone on to help de-program other members of **Scientology**.

In December 1993, the Church of Scientology experienced a major breakthrough in its ongoing legal battles when the IRS granted full tax exemption to all Scientology Churches, missions and organizations. Based on the IRS exemptions, the U.S. State Department formally criticized Germany for discriminating against Scientologists and began to note Scientologists' complaints of harassment in its annual human rights reports.

In 2000, the Italian Supreme Court ruled that Scientology is a religion for legal purposes. In recent years, religious recognition has also been obtained in a number of other European countries, including Sweden, Spain, Portugal, Slovenia, Croatia, and Hungary, as well as Kyrgyzstan and Taiwan. Other countries, notably Canada, France, Germany, Greece, Belgium and the United Kingdom, refuse to grant Scientology religious recognition.

In 2005, the Church of Scientology stated its worldwide membership to be 8 million, although that number included people who took only the introductory course and did not continue on. In 2007 a Church official claimed 3.5 million members in the United States, but according to a 2001 survey published by the City University of New York, 55,000 people in the United States would, if asked to identify their religion, have stated Scientology. In 2008, the same survey team estimated there to be only 25,000 Americans identifying as Scientologists.

The general orientation of **Hubbard's** philosophy owes much to Will Durant, author of the popular 1926 classic *The Story of Philosophy*; *Dianetics* is dedicated to Durant. Sigmund Freud's psychology, popularized in the 1930s and 1940s, was a key contributor to the Dianetics therapy model and was acknowledged unreservedly as such by Hubbard. Another major influence was Alfred Korzybski author of General Semantics. Hubbard was friends with fellow science fiction writer A. E. van Vogt, who explored the implications of Korzybski's non-Aristotelian logic in works such as *The World of Null-A*, and Hubbard's view of the *reactive mind* has clear and acknowledged parallels with Korzybski's thought; in fact, Korzybski's "anthropometer" may have been what inspired Hubbard's invention of the E-meter.

Beyond that, Hubbard himself named a great many other these include philosophers from Anaxagoras and Aristotle to Herbert Spencer and Voltaire, physicists, and mathematicians like Euclid and Isaac Newton, as well as founders of religions such as Buddha, Confucius, Jesus and Mohammed – but there is little evidence in Hubbard's writings that he studied these figures to any great depth.

As noted, there are elements of Eastern religions evident in Scientology, in particular, the concepts of karma, as present in Hinduism and in Jainism, and dharma. In addition to the links to Hindu texts, Hubbard tried to connect Scientology with **Taoism** and **Buddhism**. Scientology has been said to share features with **Gnosticism** as well.

Of the many new religious movements to appear during the 20th century, the Church of Scientology has, from its inception, been one of the most controversial, coming into conflict with the governments and police forces of several countries (including the United States, the United Kingdom, Canada and Germany). It has been one of the most litigious religious movements in modern history, filing countless lawsuits against governments, organizations, and individuals.

Reports and allegations have been made, by journalists, courts, and governmental bodies of several countries, that the Church of Scientology is an unscrupulous commercial enterprise that harasses its critics and brutally exploits its members. *Time* magazine published an article in 1991 which described Scientology as "a hugely profitable global racket that survives by intimidating members and critics in a Mafia-like manner. http://en.wikipedia.org/wiki/Scientology

**Tom Cruise, one of Scientology's more active
and prominent celebrity members**

Hubbard envisaged that celebrities would have a key role to play in the dissemination of Scientology, and in 1955 launched *Project Celebrity*, creating a list of 63 famous people that he asked his followers to target for conversion to Scientology. Former silent-screen star **Gloria Swanson** and jazz pianist **Dave Brubeck** were among the earliest celebrities attracted to Hubbard's teachings.

Today, Scientology operates eight churches that are designated *Celebrity Centers*, the largest of these being the one in Hollywood. Celebrity Centers are open to the general public, but are primarily designed to minister to celebrity Scientologists. Entertainers such as **John Travolta,**

Kirstie Alley, Lisa Marie Presley, Nancy Cartwright, Jason Lee, Isaac Hayes, Tom Cruise, Chick Corea and **Katie Holmes** have generated considerable publicity for Scientology. http://en.wikipedia.org/wiki/Scientology and https://www.youtube.com/watch?v=JKORjwXqh70

CHAPTER 25

CATTLE MUTILATIONS AND ABDUCTIONS

In the late summer of September 8, 1967, a strange event occurred on the King ranch situated at the base of the western slope of **Mount Blanca Massif** and the southeast of the **Great Sand Dunes**, part of the **San Luis Valley** in Colorado. It has become the first known case of an **"Unexplainable Animal Death" (UAD)** from unusual circumstances that has now become legendary not only in newspaper journalism and other news media but also, among most UFO organizations. The case is, of course, that of the three-year-old mare, **Lady** also, referred to by the press as **"Snippy"** the horse owned by the **King family**, Agnes, her two sons, Harry and Ben (Harry being the ranch boss) and one of her daughter's, Nellie and her husband, **Berle Lewis.**

The San Luis Valley has been an area historically for unusual events and aerial sightings dating from current times back to at least 1898 with a sighting witnessed by Agnes's deceased husband C. M. King. The Valley's native Indians have reported strange occurrences of animals and creatures for dating back centuries. This exceptionally large valley according to traditional native belief has never known war or seen conflict and is considered a valley of peace where problems of conflict are resolved amicably. It resonates with a spiritual energy and presence as witnessed by the number of diverse religions and their temples dotting the valley floor and slopes.

On a gusty morning, Harry noticed only two of their three horses were outside waiting to be fed and watered, and not seeing Nellie's filly, Lady among them, waited for her return. By the next morning, realizing the horse had still not shown up and sensing something was wrong, Harry set out to find what had become of Lady. After searching for about an hour, Harry spotted something curious which caught his attention right away, lying in the meadow about a quarter mile from the main ranch house. The hair on the back of his neck rose as he gazed upon what remained of Lady's carcass. The horse was missing all the tissue from the shoulders to the tip of its nose exposing the bleached white bones of the animal as if it had been in the sun for years. There was a strong smell of medicine that Nellie thought could have been embalming fluid around the dead horse.

It had rained for several days previously and the ground around the meadow was soft and muddy but no other animal tracks were found other than Lady's hoof marks. King determined from the hoof prints that Lady appeared to have been cut from the other horses that were in a full gallop and Lady's tracks indicated that she had come to a sudden full stop.

Conjecture arose among different researchers from **APRO (Aerial Phenomenon Research Organization)** and newspaper and magazines articles which seem to confirm King and Lewis's account that the carcass was found one hundred feet further along the meadow and while the second version supports the investigation that found the tracks were in tight circles and the corpse was found some twenty feet away.

Several sets of burnt marks about 15 in number were found nearby Lady's corpse almost a week later by the **Kings** and **Lewises** further searching of the area as well as what appeared to be "giant horse tracks in around nearby Chico bushes.

On September 8, 1967, the corpse of Lady, erroneously referred to and mistakenly reported as "Snippy" the horse by most of the world's news media was discovered by the King and Lewis Families near Mount Blanca, Colorado
http://www.santafeghostandhistorytours.com/UFOMUTILATIONS.html

It wasn't long before word got out to the local newspapers in Alamosa which eventually reached the national and international press about the bizarre death of the horse, now known as Snippy, the name that has since stuck in the social memories of almost everyone who has ever heard the strange case, due in part to poor and inaccurate journalism from the news media. As it turns out Snippy was the horse that Berle Lewis rode and Lady was her colt that died under unusual circumstances as reported to **Christopher O'Brien** in his book, *The Mysterious Valley*.

Speculations and erroneous conclusions abounded in the news media and by all those who investigated like the police, the U.S. Forest Service, veterinarians, animal pathologists, university sociologists, and even some military and government officials. UFO investigators from **APRO** and **NICAP**, even **Robert Low** from the **University of Colorado's UFO Study** headed by **Dr. Edward U. Condon** was sent to investigate the case, there was also, the news press, soil specialists, plant and wildlife experts, curiosity seekers, and anyone who had an interest, a theory or an opinion.

Theories as to what was the cause of Snippy's death ranged from natural causes like wild animal predators such as coyotes, vultures, however, there were no other animal tracks other than the horses tracks described earlier and the fact there was never any blood found on or around the corpse; or maggots feeding on the carcass; to witchcraft or pseudo-religious cult worshippers who live in the Valley area or from another district area; to a lightning strike as the cause of

140

death, but no burnt marks or scorching were found on the carcass; some of the Kings and Lewises felt that there was a UFO connection to the horse's death as did some of the UFO organizations based on prior and post UFO sightings in the immediate area (this was the first suggested connection of UFOs associated with an animal's death); and the U.S. military has also been strongly implicated as carrying out unauthorized experiments on livestock in the area initiating their program with Snippy, the horse and then, later with other ranchers cattle, sheep and even family pets. Support for this comes from ranchers and from the general public seeing military helicopters in the vicinity of cattle mutilations. **Saunders David R, Harkins R Roger (1969),** *UFO's? Yes! Where the Condon Committee Went Wrong,* **World Publishing, ASIN B00005X1J1**

Because there wasn't any blood found on or by the carcass and the wound area above the shoulder area appeared clean and sharp made perhaps by a scalpel as opposed to being jagged from a knife (ruling out hunters) or animal bites. Later, they also discovered that the brain and vital organs were missing and all this evidence has been assumed incorrectly to be beyond human know-how or technology at the time, leaving many people to jump upon the UFO hypothesis by association due to the frequent sightings in the San Luis Valley.

"A curious element of the modern Ufological mythos was created (or mis-created) as the facts twisted subtly in the wind of dissemination. The Snippy case and the San Luis Valley's apparent corresponding UFO activity provided "guilt by association scenario which led to the inexorable linking of UADs with UFOs that October 1967. As a result of this questionable process of dissemination, this link has endured ever since." **Christopher O'Brien (1996),** *The Mysterious Valley,* **St. Martin's Press, ISBN D-312-95883-8**

Once again, we see the same leap of illogic being used as the method to establish some conclusive proof of connecting two separate events into one common reality, even if the coincidence of the events fit a similar time frame or share one of two elements of synchronicity. The need to find patterns to establish trends and thus confirm theories becomes paramount to most scientists and UFO investigators but, like the UFO phenomenon itself seems to evolve from one stage to the next always staying two or three steps ahead of the investigation.
If the mutilation of Snippy, the horse was but a single incidence perpetrated by the cruel actions of hoaxers or a cult group of Satan worshippers, it would have soon become a distant and forgotten memory in the life of the **King** and **Lewis** families, the whole thing being explainable. However, everything points to some hidden agenda at work here and with all the news media attention focused on this bizarre event playing up all the angles to squeeze out the last ounce of sensational journalism, the time was right for the players of covert black operations to up the ante. The news media had created the right mindset and fear base amongst the mid and southwestern state ranchers and even up into Alberta, Canada that the stage was set for the next level in the mutilation saga. This time, cattle were the intended livestock victims and not merely in the hundreds but according to **Linda Moulton Howe's** research in her books, "*A Strange Harvest*" and "*An Alien Harvest,*" the numbers were in the thousands, in fact, cattle mutilations numbered an incredible ten thousand by
1979https://www.youtube.com/watch?v=xfkbv_a8yc4

With such a staggering number of cattle mutilations and a lot of extremely upset cattle ranchers losing their livestock on an almost weekly and monthly basis and extending from one decade into the next, the local and neighbouring police across district counties were called in. when they came up short with conclusive answers as to the cause of the cattle mutilation phenomenon, the FBI got involved. The phenomenon took off in full earnest in the fall of 1973 however, some put the date close to the time around the **Snippy,** the horse mutilation event of 1967.

In a nutshell, the cattle mutilation once a North American phenomenon had become a worldwide problem wherever, cattle ranching in any country is a major industry but currently, the phenomenon appears to be concentrated in the States. Cattle are usually found mutilated under abnormal circumstances where the cause of death is undetermined and the carcass of the cow was exsanguinated (has had all its blood and bodily fluids drained).

Most major organs appear to have been surgically removed, the reproduction or sex organs and the rectum have been cored out in one smooth operative procedure, usually the skin and flesh around the jaw is removed along with one or both eyes, as well as the whole tongue right down to the base of the throat or in most cases the whole head is skeletonized leaving only clean white skull bone. Sometimes, the legs are amputated and left beside the carcass or the skin and flesh are also removed. Sometimes, it appears that the animal had been dumped or dropped from a great height as there may be broken vertebrae, ribs or broken legs. Cattle that were victims of the mutilation process were four and five years old. There has been evidence of chemicals sometimes found in the flesh of the corpses upon autopsy as well as in pieces of flesh found nearby that were in unusual chemical combinations that reacted to human skin upon contact with a burning sensation like an acid burn. Good Timothy, (1993), Alien contact: top-secret UFO files revealed, William Morrow & Co., ISBN 0-688-12223-X; Linda Moulton Howe (1980 1989), *A Strange Harvest* (Documentary)

There is usually no blood found on, in or around the corpse, and scavengers will not touch the carcass. Upon close inspection, there are not footprints found around the dead cows even where the ground is soft or muddy. There is evidence of clamp marks found on the animal's legs suggesting that the cattle had been transported away from their grazing or pen areas and mutilated in some other location. Investigators have also found abnormally high radiation levels detected near the dead animals where they are found.

Close examination of the wound areas or excisions indicated cauterizing using extremely high heat with what has been described as laser-like precision giving credence to the scenario that UFOs sometimes sighted in the same or neighbouring areas and piloted by Extraterrestrials with laser scalpels are responsible for the cattle mutilation. There have been reports about UFO's lifting the animal off the ground and here again, though this is not beyond the realm of possibility, these reports are usually not considered airtight. To date no one has video footage of an animal being carried away by an alien spacecraft - but there are dozens of witnesses who report seeing this phenomenon.

Jaw area of cow has been cleanly excised of flesh and the whole tongue removed
(Photo by Judy Messoline, Owner/Operator/Author of UFO Watchtower)

While researchers, police, and scientists suggest, upon superficial investigations, that predators such as coyotes and vultures are to blame, yet there are no bites, chewed or tearing marks on the open areas of the wound. Some say, that the animal died of natural causes and fly larvae and maggots are responsible for the vital organs that are missing and the skeletonizing of the corpse as insects travel into and through the body's natural openings and cavities when decay takes place.

Satanists have also been blamed for causing numerous cattle mutilations as a part of some sacrificial ritual. The problem with this theory is there are no signs of vehicle tracks or human footprints in the area of the mutilated corpse or even people when known cattle mutilations have taken place within a short time span of an hour or less after the cow, bull or calf was in the presence of the rancher.

Strange black unmarked helicopters have been linked with these mutilations since they have been sighted at the same time or later in the same area where the mutilated bovines have been found. This last theory is probably the most likely cause for what is really going on, strongly implying secret experiments being carried out by the military or by some covert black sector of military intelligence. There have been reports where witnesses to cattle mutilations have seen humans who appear to come from what looked like UFO or flying saucer but when cockpit lights were turned on, they revealed helicopter rotors above the craft. It is alleged that these events were done by humans arriving in helicopters which were disguised as flying saucers, as disinformation to distract US public attention from real secret aircraft development. There is a theory that the systematic removal of the same parts from all the abducted cattle was a covert test for chemical or radioactive contamination. We may never know the truth about that, but as we

143

have been shown over and over again, the government has been covering up information for decades. http://www.crystalinks.com/animal_mutilation.html; and Fawcett Lawrence, Greenwood Barry (1993), *UFO Coverup*, Fireside, ISBN 0671765558

Mutilated cows (top and below) are often found with their udders and rectum cored out and internal organs removed.
http://www.qsl.net/w5www/mutilation.html

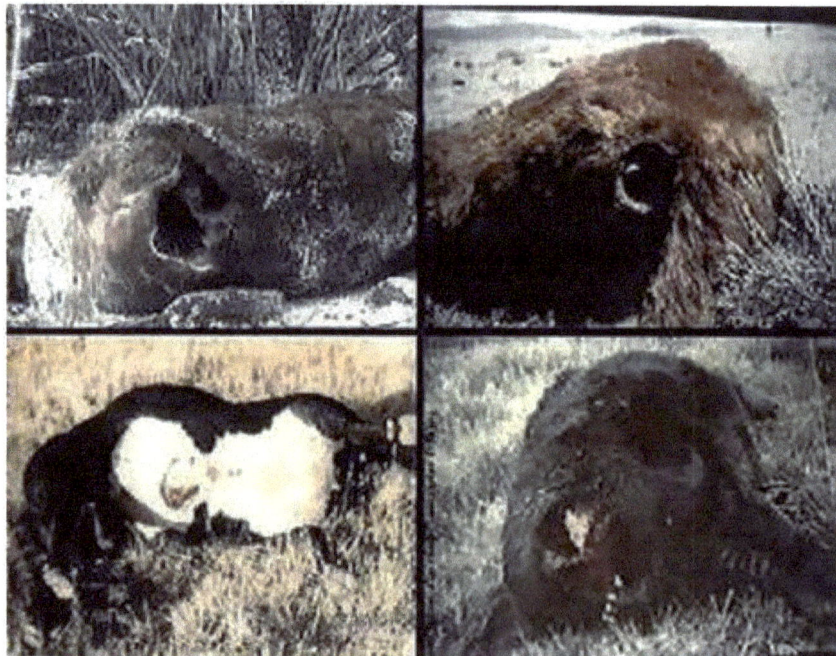

http://www.qsl.net/w5www/mutilation.html

In April 2001, **NIDS - National Institute for Discovery Science** (founded in 1995 by real-estate developer Robert Bigelow, who set it up to research and advance serious study of various fringe

144

science, and paranormal topics, most notably Ufology) interviewed a rancher (a former Vietnam War veteran) then living in Cache County, Utah who had reported his strange cattle deaths and injuries over a two decade period. What makes this case different from the other reported cattle mutilations had less to do with the mutilations but, more intriguingly, with a run-in the rancher had with mysterious military personnel in 1976.

The account was corroborated via a separate interview with the witness's former superior officer, the retired sheriff of Cache County. NIDS notes that they verified the interviewees' identities and past employment claims, but the report withholds their names due to promises of privacy. For some time prior to 1976, several ranchers in and near Cache County had witnessed unmarked helicopters which they associated with the ongoing mutilations.

As the retired sheriff reported, "every time we'd see those black helicopters, a day or so later we'd have a farmer or rancher call in with two or three dead cows." The helicopters were not only witnessed flying in the skies but also landing at the small Cache-Logan Community Airport. The ranchers organized armed patrols and determined that if the chance presented itself, they would fire upon the helicopters.

The rancher insisted that late one evening, three unmarked and unusually quiet helicopters were sighted. He and several others rushed to the Cache-Logan airport where they saw a small airplane on the runway alongside one of the three helicopters. After trying to block the airplane's departure route with their cars, they observed a man wearing coveralls who left a Huey-type helicopter "with a shiny suitcase in hand." (*Keep this "shiny suitcase" fact in mind, as it is very important!*) [My italics]

The man jogged to the airplane and gave the briefcase to the airplane's pilot. Much to the unnamed witness's surprise, the airplane began to take off, narrowly missing the impromptu police vehicle barricade and forcing one fellow deputy to crash his police cruiser to avoid a collision. The man left the helicopter's door open, and the former sheriff noted "on the bottom of the door on the helicopter" the words "Property of U.S. Army" were inscribed; "that little insignia they have on all their equipment."

The witness then turned his attention to the man, a tall figure perhaps 35 years old, wearing green military-style fatigues which lacked, however, the characteristic identification of name, rank or military unit. The witnessed asked the tall man for identification; the man replied: "I have none." The witness said he touched the man's chest, partly to see if the man wore dog tags under his clothing. He did not.

When the witnesses touched the man's chest, the airborne helicopters both made moves that the witness "Interpreted as a beginning gun pass". The witness speculated that "if I attempted to physically arrest that man, they'd a killed us both, right there, on the spot, with the same mental attitude that the pilot had ... that ran my patrol car off a runway."

Due to this perceived threat (and the lack of any firm evidence of criminal wrongdoing), the witness decided to not attempt an arrest. He did, however, tell the man that the police and ranchers in the area were frustrated with the inexplicable deaths of their cattle and that they suspected the unmarked helicopters were involved. He added that eventually, some armed rancher or farmer with a rifle would shoot at the helicopters, "and we intend to bring you down, sir."

After a pause, the man said, "May I go?" The witness said yes, and the man returned to his helicopter, which flew off to the west. The witness (and several other individuals) asserted that after this showdown that the unmarked helicopters were no longer seen, and the cattle mutilations in northern Utah and southern Idaho ceased for some five years.

Later in 1976, a National Sheriff's Association colleague in Texas was investigating a number of cattle mutilations, and, inspired by the Utah police's actions, soon organized armed patrols, and began closely monitoring the unmarked helicopters. Law-enforcement officers in Texas had no face-to-face encounter with the pilots, but they reported that the mutilations in the area ceased.

Another belief recently put forward concerns a large amount of mutilations taking place in close proximity to former nuclear test sites. It has been speculated that the U.S. Military has been dissecting these cows to examine the amount of exposure they receive from radiation fallout. Since these cows would account for a section of the meat and dairy industry, as well as represent an estimate of the human population's exposure to the fallout. Random samples would need to be taken since varying times of exposure and other factors must be accounted for. Obviously, the need for a "cover-up" would be mandatory since a population finding out that they've been poisoned would not be beneficial. http://www.crystalinks.com/animal_mutilation.html

Some plausible and probable explanations are required to make some sense from these bizarre animal deaths which have associated itself with the UFO/ETI phenomenon. It should now be obvious to all who study this UFO phenomenon that the public mindset has adapted and become conditioned to the world of the mysterious but, along with cattle ranchers and UFO researchers, they are divided in opposing camps as to what is causing the animal mutilations.

Natural Causes

It is safe to say that animal predators and insect life account for a very small percentage of the unusual wounds found on the carcasses. One reason cited by ranchers is that the animals were healthy and showed no sign of disease prior to death, and were large and strong enough not to be a likely target for a predator. In some cases, ranchers have reported that the mutilated cattle were among the healthiest and strongest animals in their herd. Natural death by accidents, diseases and life cycles also, account for only a small percentage of mutilation cattle corpses. Lightning strikes are so remote a cause for animal death that the surgical-like wounds found on the animal simply doesn't justify lightning as a valid explanation; lightning tends to leave burn wounds and scorching on animal flesh.

Human Intervention

The FBI and the **ATF (Alcohol, Tobacco and Firearms)** launched their own investigation of the phenomenon. followed by a state-level investigation carried out by enforcement officials in New Mexico It also contended that alleged surgical techniques performed during mutilations had become 'more professional' over time. Deviant attacks against animals are a recognized phenomenon. There have been many recorded cases around the world, and many convictions. Typically the victims of such attacks are cats, dogs, and other family pets, and the actions of deviants are usually limited to acts of cruelty such as striking, burning or beating animals. However, attacks have also been recorded in larger animals, including sheep, cows, and horses. Deviants, particularly those with sociopathic disorders, have been found to have mutilated animals in elaborate ways using knives or surgical instruments. Certainly, cult groups and Satanists could be behind the mutilations and they too, have their methods that give credence to their involvement but to date, no cult group has been singled out by the police or FBI.

Still, with law enforcement from various regional, federal and international agencies and many scientists from diverse professional backgrounds, all investigations point to as yet, an undiscovered intelligence hiding behind some elusive agenda known only to themselves. This leaves us with a covert government or military intelligence and the alien intelligence hypothesis.

Government Involvement

It has been speculated that cattle mutilations are the result of covert research into emerging cattle diseases which could be transmitted to humans. Other federally funded bodies are believed to be involved, and they are supported by the US military. This is based on allegations that human pharmaceuticals have been found in mutilated cattle, and on the necropsies that show cattle mutilations commonly involve areas of the animal that relate to oral, (alimentary canal) anal, and reproduction systems. It has also been suggested that there is a clandestine U.S. Government effort to track the spread of **Bovine spongiform encephalopathy ("mad cow disease")** and related diseases, like scrapie that affect sheep and goats.

Further investigation into this and other related theories lead to and concluded that another, more sinister clandestine program was in operation in and around the Dulce, New Mexico area as news reporter Leigh Black Irvin for "The Daily Times" out of Farmington, New Mexico, discovered in his interview with Greg Valdez.

For decades, strange lights in the night sky and mysterious cattle mutilations have sparked rumors of a secret underground alien base near the small northern New Mexico town of Dulce, which is tribal headquarters of the Jicarilla Apache Nation.

A new book, "Dulce Base: the Truth and Evidence from the Case Files of Gabe Valdez," purports to solve the mystery. It claims that humans, not aliens, are behind the strange happenings. The book's author is **Greg Valdez**, son of former New Mexico State Police Officer **Gabe Valdez**. http://www.daily-times.com/four_corners-news/ci_24351581/new-book-says-government-not-aliens-were-behind

In 1976, ranchers found many mutilated cows, and Gabe Valdez became one of the lead investigators in the case, his son said.

Greg Valdez says his father began a decades-long investigation into the large number of mutilations in northern New Mexico. Prior to his death in 2011, the former police officer determined that the mutilations and strange aircraft were, in fact, ***human-caused.***

After pouring over recently declassified documents, Gabe Valdez concluded that ***the federal government was using the Jicarilla Apache Nation to test environmental contamination caused by nuclear testing in the late 1960s.***

Greg Valdez says this contamination was caused by an experiment known as **"Project Gasbuggy"** that took place 21 miles southwest of Dulce on Dec. 10, 1967. The project's goal was to identify peaceful uses for nuclear explosions, and it involved the detonation of a 29-kiloton device located 4,227 feet underground. The intent was to release pockets of natural gas that could be used commercially.

Author's Rant: This sounds like another bogus explanation used by the government to cover up its true intentions which may have been to see what leaked radiation would do to an already marginalized racial minority over a long term period.

Gasbuggy was carried out by the U.S. **Atomic Energy Commission (AEC),** the **Lawrence Radiation Laboratory** and the **El Paso Natural Gas Company**, according to the book.

While the device was successfully detonated and gas wells were drilled at the site, the gas was too radioactive for commercial use. ***Like, as if the AEC hadn't already determined what the after-effects of nuclear radiation would be, ever since the first atomic bombs were tested and dropped upon Japan at the end of WWII.*** (Bold italics added by author for emphasis).

Greg Valdez said his father found out that the federal government conducted the cattle mutilations to determine the effects of radiation from Gasbuggy.

Author's Rant: This may be true but, other researchers believe that a far more nefarious program that involved the engineering and manipulation of alien genetics were the real reasons behind cattle mutilations.

"They were testing the cattle to avoid panicking the public," he said. *"They were also testing advanced aircraft from a nearby off-site air base. The aircraft was invisible and silent and used optic camouflage. The technology has since been declassified."*

Greg Valdez alleges that several government agencies and military entities, such as the U.S. Air Force, were heavily involved in the cover-up. He says that the CIA and National Security Agency also became involved when Albuquerque businessman Paul Bennewitz discovered evidence of secret military projects on the Kirtland Air Force Base, which also had ties to Dulce. http://www.daily-times.com/four_corners-news/ci_24351581/new-book-says-government-not-aliens-were-behind

To protect the secrecy of their operations, Gabe Valdez learned that the government started a disinformation campaign and encouraged rumors about UFOs and aliens, said Greg Valdez.

By 1979, when the government realized Gabe Valdez had discovered the truth about who was behind the strange happenings, they began monitoring him, his son said. He said the family found hidden listening devices in their home.

Aztec resident Brooks Marshall is a UFO enthusiast and local paranormal expert. Marshall has followed reports of the supposed Dulce underground base for years and has attended many symposiums and lectures on the issue.

Marshall believes there is truth to a secret military presence centered around Dulce. But he doesn't think government involvement comes close to explaining the cattle mutilations.

"From early on, the mutilations indicated a technology that we just didn't have," he said. *"There was a complete absence of blood, and the incisions looked like microsurgery and laser technology had been used -- technology humans didn't have at the time. Plus, there were no tracks, and other animals -- even predators -- would avoid the carcasses, which also looked like they had been dropped from a very high distance. Many of the bones would be broken."*

Marshall said specific organs, like the tongue and reproductive organs, were removed from the cattle. The mutilations still periodically occur, he said.

From his research, Marshall concludes there are two pieces to the Dulce mystery: the cattle mutilations, which he says remain unexplainable, and the government activity.

"It's my opinion that there was an extremely large underground military base," he said. *"There's evidence that there was strong military activity there. One theory is that this base housed aliens."*

Marshall, who once worked for El Paso Natural Gas, is familiar with the Gasbuggy Project and said he worked with employees who were directly involved with it. While he believes there have been disinformation campaigns conducted by the military to deflect attention from a possible underground base, he doesn't find credence in the theory that the cattle mutilations were the result of government tests for radioactive effects.

"It doesn't make sense that the government would test cattle in that way," he said. *"All they'd have to do to test radiation levels is to walk up to the cow with a Geiger counter."*
http://www.daily-times.com/four_corners-news/ci_24351581/new-book-says-government-not-aliens-were-behind

This conclusion by Valdez hit the proverbial nail on the head and raises many other questions as to the true motives behind such government secret programs where the public who are affected are told absolutely nothing. By the time the public start probing and asking questions of their government representatives as to what's really going on, a significant amount of time has passed and it usually it too late to do anything or to bring those responsible to justice.

Covert Military Agenda

There has been a high probable reason to suspect the military as the primary culprit in ranchers finding mutilated cattle on their ranches. When we speak about military involvement, let us be clear, they represent a branch of the government and thus the government is duplicitous when the military is implicated. It is believed the military are engaged in a genetic R&D program which uses certain body parts of cattle in creating a new hybrid species. The covert military black ops teams fly around over cattle ranches usually at night day but, military daytime exercises would also act as a cover for covert programs where certain animals are targeted from out of a herd. These select animals are tagged with phosphorescent marker dyes that are invisible to the naked eye but can be seen with high powered ultraviolet light equipment.

Theories of government involvement in cattle mutilation have been further fueled by "**black helicopter** " sightings near mutilation sites. On April 8, 1979, three police officers in Dulce, New Mexico reported a mysterious aircraft which resembled a U.S. military helicopter hovering around a site following a wave of mutilation which claimed 16 cows. On July 15, 1974, two unregistered helicopters, a white helicopter and a black twin-engine aircraft, opened fire on **Robert Smith Jr.** while he was driving his tractor on his farm in Honey Creek, Iowa. This attack followed a rash of mutilations in the area and across the nearby border in Nebraska. The reports of "helicopter" involvement have been used to explain why some cattle appear to have been "dropped" from considerable heights.

Alleged photograph of a flying saucer in the process of abducting a cow (the picture is probably a clever hoax) Google images

It's even believed that this cattle mutilation program is so sophisticated that some stealth helicopters use holographic imagery of a flying saucer to "cloak" their aircraft so that the public will think that aliens are indeed responsible for the mutilations. But it goes even further than this. It now appears that the **MIC (Military Industrial Complex)** has black programs and projects in which they utilize reverse engineered alien spacecraft along with genetically engineered aliens or

programmable life forms (PLFs) that resemble the ET species known as the "**Greys**" to carry out cattle mutilations and human abductions!

Human Mutilations

A horrifying fact came to light on September 29, 1988, when the mutilated body of a man (supposed UFO-related **human mutilation**) was discovered nude at the Guarapiranga reservoir in Brazil (actual location is unknown at this time). Photos which are extremely graphic of the victim's cadaver indicated that he suffered the same fate as a typical UFO-related animal mutilation as described throughout most western States of America. There were no signs of struggle from the victim's body indicating that he may have been drugged first before the indignity was perpetrated upon him. The actual graphic details will not be discussed here and anyone requiring such detailed information or photographic evidence, merely has to search the internet or read the magazine article: "UFO-Related Homicide in Brazil: The Complete Story" [*from International UFO Magazine*] and no doubt there are other reading sources of information on this case.

Immediate association of malevolent alien intervention was the conclusion that the Brazilian, U.S. and British UFO researchers had reached after investigating this first human mutilation case. Brazilian police have hinted that there have been a dozen or so similar cases which would seem to build the conclusive strength of this case. There is too much that is wrong with the conclusions reached by UFO investigators on this case and on the cattle mutilations that are reported in many countries around the world. I will try to go into some of the false presumptions and conclusions to give a more logical analysis that will make sense and put everything into perspective. It will be less to do with alien malevolence and more to do with human malevolence against other humans.

First and foremost, the term "mutilation" conjures up a negative mindset that instills an image of a cruel, intentional, malicious act against people or animals in the minds of most people. The image is so strong, that the thought of the action is hard to remove from our consciousness even, when we are able to put such negative thoughts to the back of our conscious minds, the thought is still in our subconscious, ready to surface very quickly whenever, a similar circumstance or action arises to re-affirms the original thought and imagery.

As Dr. Steven Greer has stated, *"Words are powerful encapsulations of thought and we need to choose our words carefully!"*

Alien Agenda

Finally, the Alien/UFO theory needs to be examined as this is a popular theory when combined with the human abductions scenario. Various hypotheses suggest cattle mutilations have been committed by aliens gathering genetic material for unknown purposes. Most of these hypotheses are based on the premise that humans are not capable of performing such clean dissections in such a short space of time without being seen or leaving evidence behind at the mutilation site, and that we simply do not possess such technology or advanced cutting tools. Because cows are the main part of the global human diet, ETs, it is speculated are carrying out a study on this aspect of the human food chain. Other speculative theories center on possible specific nutrient requisites, hormone procurement, species propagation (reproduction), and rote experimentation on mammalian populations. Some UFO researchers have theorized that because humans and cattle share a substantial number of chromosomes this makes them the logical choice for large scale biological incubation and experimentation on subjects for human pathogens, in the same way that horses have been used to produce Tetanus treatments. http://www.daily-times.com/four_corners-news/ci_24351581/new-book-says-government-not-aliens-were-behind

Black Unmarked Helicopters and That Bright "Shiny Suitcase"

Here again, the military whether it be the US or the UK or even friendly countries like Canada, Australia, and Brazil, etc. are linked to the subversive, covert agenda of the US **M.I.C. black programs** such as cattle mutilations. There is enough evidence from ranchers all over the States and Canada and this story is now becoming all too familiar, indicating that "black unmarked military helicopters (these are recognized as being the property of the U.S. military and this fact is not disputed) are seen in the vicinity before, during and after reports of cattle mutilations. These helicopters have been shot at by ranchers and in some instances, there are reports of helicopters returning or instigating gun fire on ranchers to get them to move out of an area. When such black helicopters are fired upon or ordered by some official authority to leave an area where cattle mutilations have been reported, the incidence of mutilations drops off almost immediately, for a few years at least, before the problem arises again in the same or neighbouring ranch area. These helicopters have even been reported as being initially perceived as UFOs or flying saucers without sound and then as almost, as if some partial malfunction had occurred the rancher or witnesses were able to see the rotor blades of a helicopter above what would be the cupola of the disc revealing the saucer's true identity and nature as being manmade in origin. This malfunction "is very telling as it indicates the use of stealth and holographic technology which are of an obvious human design and not extraterrestrial in origin.

Remember that *"shiny suitcase"* that was seen and reported by the rancher and the county sheriff at the Cash-Logan airport carried by one of the men from a Huey-type helicopter? While it may be just speculation as to what was in the shiny suitcase, it is conceivable that it contained a portable field surgical laser use to *perform surgical operations and possible amputations under combat conditions.* It would not be a hard stretch of the imagination to see how such a medical device like a **laser-scalpel** would be used for other purposes, than in combat conditions, such as in excising body parts from a sedated cow or bull that was tranquilized and then moved from the ranching area to a location of seclusion and privacy. **Lasers** were spoken about by **Einstein** in 1917 and their existence was realized by the late 50's and certainly lasers were conceptualized into rifle and cannon models by the early 60's. No doubt further research and development would produce second, third and fourth generation, etc. even though the military speak as if they are still trying to develop and build the first practical combat handheld laser weapon. In the military industrial world, whatever, you can conceive has already been researched, developed, and built and is at least three to five or more generations ahead of what you can imagine!

When you link this with the stories of reverse engineering of alien technology then, it is not hard to imagine that major breakthroughs in aeronautics and space technology, associated ground-based weapons technology, electronic micro-circuitry, and advancements in medicine, cloning and exo-biology occurred in the early fifties. Much of this technology was written about by the late **Colonel Philip J. Corso**, in his book *"The Day After Roswell"* including the serious proposal by the United States Army to build lunar bases on the Moon!

Below is an ad extolling the specifications and commercial viability of the portable field-ready surgical laser built by Phillips Laboratory that is battery powered, built to military standards, and certainly smaller than a "shiny suitcase" in size. There's even, a U.S.A.F. contact address for interested buyers.

"The **Laser Medical Pac"** was developed in-house by the USAF's **Phillips Laboratory** and has been operational since 1969 by the Air Force, (for far longer than was originally imagined by UFO researchers who investigate cattle mutilations), is a very compact device that provides the field paramedic or physician a unique, portable, and battery-operated laser capability. The laser is able to cut like a scalpel, as well as coagulate bleeding, and close wounds. The laser component is now commercially available. The commercial variety, however, requires an electrical power hookup.

The **Phillips Laboratory** system consists of a completely self-contained laser package that fits inside a beltpack. Laser energy is delivered to the instrument by a fiber-optic cable, the fiber providing very intense power density at the tip of the instrument. The output wavelength, which ranges from visible red to the mid-infrared, can be designed to provide different tissue interactions."

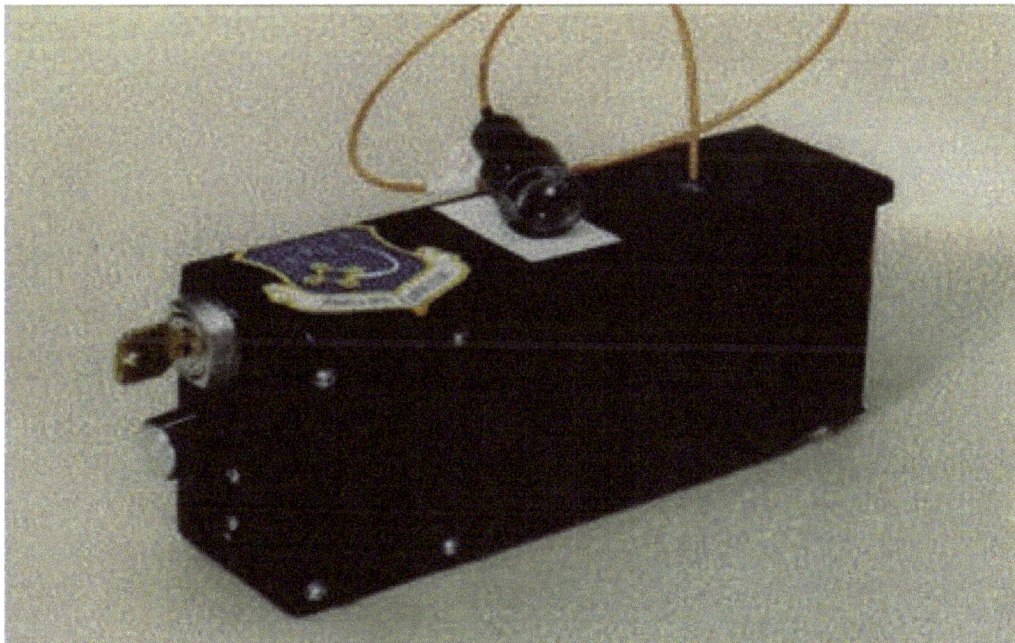

The belt pack Laser Medical Pac used by the USAF since 1969

"The Pac is powered by two 2-volt batteries to operate the laser and one 9-volt battery to power the electronics. It features a unique phase change heat sink that allows 20 minutes of continuous operation. (Under normal usage the heat capacity should allow "***unlimited thermal capacity.***") [Italics added] The laser is protected against over-temperature by a thermal switch. A battery recharger port is also provided, as is a key lock for safety and security reasons. The fiber-optic is pig-tailed into the laser array and "pipes" the laser light to the variable focus lens. The light at the tip of the fiber is very intense (one kilowatt per square centimeter).

Different disposable tip designs that can be directly applied to the wound are being evaluated. Seeding the infrared beam with a single red diode for visual operating cues is also being evaluated. The actual dimensions are currently 7" x 3" x 2.5", which will easily fit into a small beltpack, only the fiber and lens will extend from the beltpack. An alternate package is being developed to operate with common camcorder batteries. These 12-volt batteries provide two amp hours and are quickly removable.

This device is being evaluated as a means of ***"stopping bleeding."*** [Italics added] Currently, it takes a 2-watt Argon laser about l0 seconds to stop a bleeding artery. ***The Argon laser is roughly half the size of an office desk and requires a wall plug.*** In comparison, ***the Medical Pac laser delivers l0 watts and fits into a beltpack."***

http://www.zyn.com/FLCFW/fwtproj/medpac.html (this hyperlink is no longer active and the website has been removed from the internet. Perhaps, the general public is starting to figure things out which means for the USAF, this is hitting too close to home!)

Potential Commercial Uses
- Military Portable Laser Applications
- Medical and Military Rescue Applications

Benefits of Technology
- Lightweight laser allows portability
- Increased power capabilities for remote medical and military laser equipment

Options for Commercialization
Phillips Laboratory is currently working with the Air Force's Armstrong Laboratory at Brooks Air Force Base, Texas, to expand the engineering, manufacturing, and development of this new semiconductor laser for medical purposes. This team is currently seeking industry partners for continuing research and development of lightweight, semiconductor laser technology that will benefit Phillips and Armstrong laboratories, the partners involved and end users of any products produced. Contact for partnership comes from (not surprisingly):

United States Air Force
AIR FORCE MATERIEL COMMAND
Office of Public Affairs, Phillips Laboratory,
3550 Aberdeen Ave SE, Kirtland AFB, NM 87117-5776

United States Air Force
AIR FORCE MATERIEL COMMAND
Office of Public Affairs, Phillips Laboratory
3550 Aberdeen Ave SE, Kirtland AFB, NM 87117-5776
(505) 846-1911

LASER MEDICAL PAC

The Laser Medical Pac, being developed in-house by the USAF's Phillips Laboratory, is a very compact device that provides the field paramedic or physician a unique, portable, and battery-operated laser capability. The laser is able to cut like a scalpel, as well as coagulate bleeding, and close wounds.

The Laser Medical Pac has military applications for advanced trauma life support on the battlefield. It can be used by special operations personnel, pararescue jumpers, squadron medical elements, and flight surgeons. Civilian uses for the Pac are in stabilizing highway accident victims before they are transported to a hospital.

The laser component is now commercially available. The commercial variety, however, requires an electrical power hookup.

The Phillips Laboratory system consists of a completely self-contained laser package that fits inside a beltpack. Laser energy is delivered to the instrument by a fiber-optic cable, the fiber providing very intense power density at the tip of the instrument. The output wavelength, which ranges from visible red to the mid-infrared, can be designed to provide different tissue interactions.

The Pac is powered by two 2-volt batteries to operate the laser and one 9-volt battery to power the electronics. It features a unique phase change heat sink that allows 20 minutes of continuous operation. (Under normal usage the heat capacity should allow unlimited thermal capacity.) The laser is protected against over-temperature by a thermal switch. A battery recharger port is also provided, as is a key lock for safety and security reasons. The fiber-optic is pig-tailed into the laser array and "pipes" the laser light to the variable focus lens. The light at the tip of the fiber is very intense (one kilowatt per square centimeter).

-MORE-

ILLUSTRATION 49 - This medical laser is portable enough to be worn on a belt pack around the waist, and can be used to either make cuts or close wounds. According to the Air Force, "It can be used by special operations personnel ..." *Reprinted with permission from Phillips Laboratory, Office of Public Affairs, Kirtland Air Force Base, NM.*

From the book: "Underground Bases and Tunnels: What Is the Government Trying to Hide?" by Richard Sauder

Factor into this, the handful human mutilations in Brazil that were carried out by either a paramilitary organization of Brazil or by some covert US military group either intruding into the country or with Brazilian government's cooperation to stage a carefully orchestrated hoaxed alien scenario to essentially send a clear message to not only many of the world's intelligence agencies, but also, to the many Ufologists throughout the world, that ETs are hostile. In other words, someone or some powerful covert black operative groups are trying *"to scare the living shit out us"* into believing that we are being invaded by malevolent aliens from outer space for reasons that are not as yet, clear.

Author's Rant: Now, I would be the first to say this sounds like outright paranoia and conspiratorial claptrap if it was not for the overwhelming evidence which we will explore in the next section of this book. The proverbial "rabbit hole" goes ever deeper and it's a matter of how deep do you want to go down that "rabbit hole." In the next subsection, we will look at some of the cryptozoology that is associated in the waste basket of Ufology.

On June 16, 1994, Jon Denton published on the website NewOK an article on this very same laser that is being studied by Ken Bartels, a veterinary research professor at OSU. He is investigating a company like **Phillips Laboratory** to make what the military calls a **Laser Medical Pac (LMP)** for animal and commercial hospital use.

Instead of a 600-pound laser, like those now used in hospitals, the palm-sized laser weighs about 4 pounds and is powered by a belt-mounted battery, Bartels said. The device is equally useful in battlefield operations or at the scene of an auto collision.

Surgical lasers have been around for 25 years, Bartels said! http://newsok.com/medical-tool-of-trade-researcher-helps-develop-portable-laser/article/2469190

The important point here is that such technology as used by the military or in the **"Black World of Science"** is at least a generation (20 + years) or more (possibly 100 years) ahead of the **"White World of Science"** (the everyday, common world of mainstream science and technology). In terms of the military mindset, this is *outdated technology* which can be easily passed down into the public domain without the worries of breaching **National Security** protocols!

CHAPTER - 26

CHUPACABRAS, SASQUATCHES, AND THE MOTHMAN, OH, MY!
EXPERIMENTS IN DNA AND GENETIC ENGINEERING

As was mentioned earlier in this section, the field of Ufology is a mishmash of belief systems and wild theories that include in its files and on-going research, a waste basket of cryptozoological phenomenon. Among these cryptoids are most notably the **Chupacabra**, the legendary **Sasquatch or Bigfoot** and the **Mothman**. There are other mythological creatures but, these stand out as the ones that are often associated with the UFO phenomenon. In some accounts, the UFO phenomenon and one of the above mentioned creatures have been coincidently sighted in the same vicinity, but never simultaneously together at the same time. UFO researchers have historically tied the two phenomena together by similar case reports, either in the same country or in other countries.

An eye witness may sight a **Sasquatch** for example and then, later that week or within a month, someone else has a UFO sighting in the same location as the Sasquatch sighting. When you get a number of these types of case reports even when they are spread out through a country, the tendency is to look for patterns of behaviour and draw conclusions even when the phenomenon are totally unrelated. Thus, the elusive and enigmatic Sasquatches that are seen by people hiking through the wilderness are in fact, actually alien creatures dropped off on terra firma by UFO piloting Extraterrestrials to see how humans will interact with them! If the extraterrestrial explanation doesn't suit your liking, then they can be interdimensional creatures, either way, they are intelligent and appear non-aggressive for the most part. **Chupacabras, on the other hand,** appear predatory especially toward a farmer's livestock and pets. Fortunately, children and adults don't seem to be on the preferred diet of these elusive and intelligent creatures who are also, known by the Spanish euphemism from the locals of Puerto Rico as the *"goat-sucker."*

The Chupacabra is a creature reportedly sighted in most of the Americas, from as far north as the Carolinas, and as far south as Chile. Initial sightings began in Puerto Rico in the early 1990s with reports of strange deaths among farm animals. The Chupacabra or goat-sucker derived its name from the reported habit of attacking and drinking the blood of livestock other animal deaths. Reports of the strange animal have come from countries like the Dominican Republic, Argentina, Bolivia, Chile, Colombia, El Salvador, Panama, Peru, Brazil, the United States and, most notably, Mexico; there have even been recent reports in Russia.

"The Chupacabra is described as having different appearances but, with several common traits. They are typically described as being 3 ft. (1 m) or taller, and roughly humanoid in shape. Usually, Chupacabras are said to appear in three specific forms:

- The first and most common form is a lizard-like being, appearing to have leathery or scaly greenish-gray skin and sharp spines or quills running down its back. This form stands approximately 3 to 4 feet (1 to 1.2 m) high and stands and hops in a similar fashion to a kangaroo. In at least one sighting, the creature hopped 20 feet (6 m). This variety is said to have a dog or panther-like nose and face, a forked tongue protruding

from it, large fangs, and to hiss and screech when alarmed, as well as leave a sulfuric stench behind.
- The second variety bears a resemblance to a wallaby or dog standing on its hind legs. It stands and hops as a kangaroo, and it has coarse fur with grayish facial hair. The head is similar to a dog's, and its mouth has large teeth.
- The third form is described as a strange breed of wild dog. This form is mostly hairless, has a pronounced spinal ridge, unusually pronounced eye sockets, teeth, and claws. This animal is said to be the result of interbreeding between several populations of wild dogs, though enthusiasts claim that it might be an example of a dog-like reptile. It is, in fact, an animal with the mange which can cause almost any fur covered animal to lose its hair.

Mange is a class of skin diseases caused by parasitic mites which can also infect plants, birds, and reptiles causing the poor condition of the hairy coat due to the infection. Thus, mange includes mite-associated skin disease in domestic animals (cats and dogs), in livestock (such as sheep scab), and in wild animals (for example, coyotes, cougars, and bears). Parasitic mites that cause mange in mammals embed themselves either in skin or hair follicles in the animal, depending upon their genus.

The account during the year 2001 in Nicaragua of a **Chupacabras** corpse being found supports the conclusion that it is simply a strange breed of wild dog. The alleged corpse of the animal was found in Tolapa, Nicaragua, and forensically analyzed at UNAN-Leon. Pathologists at the University found that it was just an unusual-looking dog. There are very striking morphological differences between different breeds of dog, which can easily account for the strange characteristics.

Some reports claim the Chupacabra's red eyes have the ability to hypnotize and paralyze their prey—leaving the prey animal mentally stunned, allowing the Chupacabra to suck the animal's blood at its leisure. The effect is similar to the bite of the vampire bat, or of certain snakes or spiders that stun their prey with venom. Unlike conventional predators, the Chupacabras sucks all the animal's blood (and sometimes organs) through a single hole or two holes."
http://www.monstropedia.org/index.php?title=Chupacabra

The Chupacabra legend began in about 1992, when Puerto Rican newspapers began reporting the killings of many different types of animals, such as birds, horses, and as its name implies… goats. At first, it was suspected that a satanic cult was behind the random killings but later, these killings spread around the island, and many farms reported a loss of animal life. The killings had one pattern in common: each of the animals found dead had two punctured holes around their necks.

In March 1995 in Puerto Rico, eight sheep were discovered dead, each with three puncture wounds in the chest area and completely drained of blood.

In July of 2004, a rancher near San Antonio, Texas, killed a hairless, dog-like creature which was attacking his livestock. This creature was later determined to be a canine of some sort, most likely a coyote, with demodectic mange. This particular case of hairless coyotes with demodectic mange would become more frequent as reports came out of the Texas-Mexico border region.

In April 2006, MosNews reported that the **Chupacabra** was spotted in Russia for the first time. Reports from Central Russia beginning in March 2005 tell of a beast that kills the animals and sucks out their blood. Thirty-two turkeys were killed and drained overnight. Reports later came from neighboring villages when 30 sheep were killed and had their blood drained.

There are the usual theories whenever there are strange things that cannot be readily explained. Chupacabras have been likened to the gargoyle statues of Medieval Europe. This may explain their appearance down through the ages in Europe and as such were recorded in stone sculptures on churches and cathedrals to keep the public afraid of any place with gargoyles or in this case Chupacabras.

Certain South American rain forest natives believe in the **"mosquito-man"**, a mythical creature of their folklore that pre-dates modern Chupacabras sightings. The mosquito-man sucks the blood from animals through his long nose, like a big mosquito. Some say mosquito-man and Chupacabras are one and the same.

**An artist interpretation of a genetically engineered Chupacabra (left) and
a misidentified hairless (mange) Coyote as the "Chupacabra" (right)**
Google Images and http://mythicalcreatures666.tripod.com/id1.html

Some cryptozoologists speculate that Chupacabras are alien creatures. Chupacabras are widely described as otherworldly, and, according to one witness report, NASA may be involved with this particular alien's residency on earth. The witness reported that NASA passed through an area in Latin America, with a trailer that was thought to contain an incarcerated creature. Some people in the island of Puerto Rico believe that the ***Chupacabras are a genetic experiment*** from some United States' government agency, which escaped from a secret laboratory in El Yunque, a mountain in the east part of the island.

Examining these theories for a sense of rationality and credibility, we find that they all have some basis of merit but, the last theory stands out as being more credible than the others and it may actually be tied into the NASA agency theory as well.

In order to establish the legitimacy of the last theory, we will need to look at the current science of DNA research, the human genome project and the various cloning programs taking place throughout the world. All these branches of research make a very strong case that there may very well be secret genetic experiments into creating hybrid creatures that have never existed before on this planet by covert agencies of the government or the military or both.

It is a well-known fact that there are worldwide programs into genetic research and DNA manipulation by many countries. The **Human Genome Project (HGP)** that began in October 1990 was an international effort coordinated by the U.S. Department of Energy and the National Institutes of Health with contributions from the U.K. Japan, France, Germany, China, and others. The human genome program has been successfully mapped by 2003 ahead of its 15 year schedule and is now offering up its secrets for the further advancement of medical research and development.

The Project goals were to determine the complete sequence of the 3 billion DNA subunits (bases), identify all human genes, and make them accessible for further biological study. As part of the HGP, parallel sequencing was done for selected model organisms such as the bacterium *E. coli* to help develop the technology and interpret human gene function. An important feature of the HGP project was the federal government's long-standing dedication to the **transfer of technology to the private sector.** By licensing technologies to private companies and awarding grants for innovative research, the project catalyzed the multibillion-dollar U.S. biotechnology industry and fostered the development of new medical applications.
http://www.ornl.gov/sci/techresources/Human_Genome/home.shtml

Vacanti "Earmouse"

Through such transfer of technology to the private sector, many interesting and controversial projects came from the HGP. One such genetic breakthrough was reported in 1996 on a BBC TV broadcast of a mouse and what appeared to be a "human ear" growing on its back! Pictures of the mouse with the human ear soon appeared after the British newscast. The Vacanti mouse as it was known named after its creator, **Dr. Charles Vacanti**, an anesthesiologist at the University of Massachusetts and **Dr. Linda Griffith-Cima**, an assistant professor of chemical engineering at M.I.T. in 1995. It was created to demonstrate a method of fabricating cartilage structures for transplantation into human patients; a resorbable polyester fabric was infiltrated with bovine cartilage cells and implanted under the skin of a hairless mouse. The mouse itself was specifically bred with a genetic mutation which, apart from causing baldness, inhibited the mouse's immune system, preventing a transplant rejection.

160

It later it became known as the "earmouse and was actually an ear-shaped cartilage structure grown by seeding cow cartilage cells (there was never any human tissue used) into a biodegradable ear-shaped mold. The pictures were circulated worldwide, appearing on many front pages and on TV programs such as the **Jay Leno** show. It provoked horror among the animal rights and advocacy groups and anti-genetics groups, who protested the use of the mouse for the experiment. http://en.wikipedia.org/wiki/Vacanti_mouse and BBC News, Sci/Tech, Monday, April 13, 1998, "Girl may be first to grow artificial ear"

DNA and genetic manipulation just didn't end there, as more brilliant minds demonstrated their ingenuity and logic by producing a "spider-goat" with the ability to milk the goat and extract the silk web material through a filtering process from the milk! The end result was an abundance of silk thread to manufacture police and military **Kevlar bullet-proof vests**, a definite breakthrough in the medical and agricultural arenas of science! No longer will scientists be bothered with the tedious task of milking spiders for their silk thread when goats could produce the same thing in their milk but, in liters.

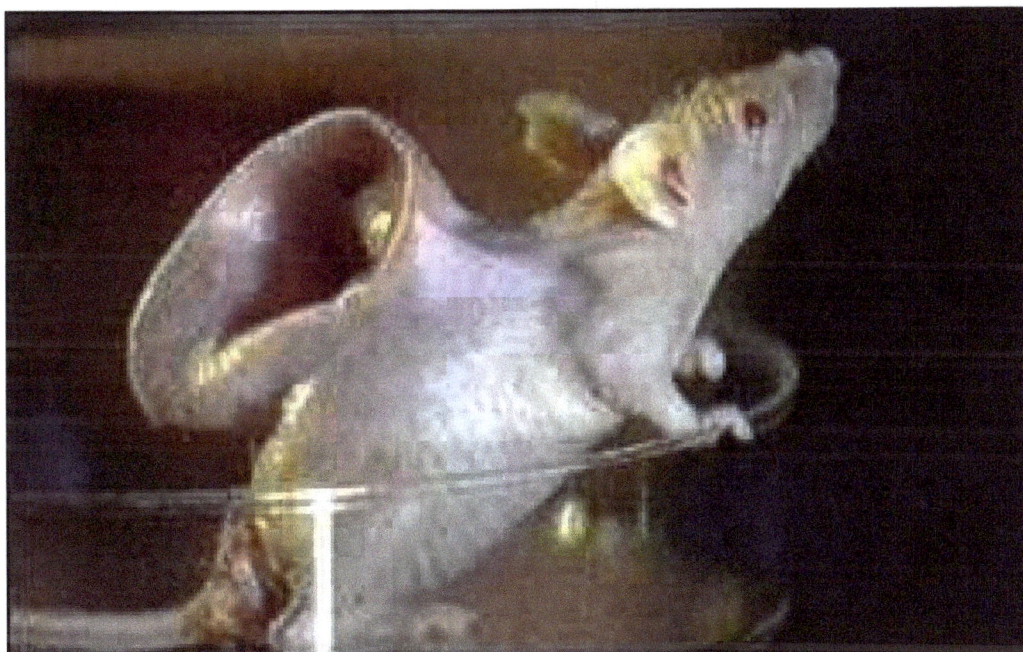

Hello, Then! What's this ear? The world famous genetically engineered Vacanti mouse or the "Earmouse" of 1995
http://news.bbc.co.uk/2/hi/health/1949073.stm

Silk Web Producing "Spider-Goats"

Nexia Biotechnologies in Quebec were approached by scientists at the U.S. Army's **Soldier Biological Chemical Command (SBCCOM)** in Natick, Mass. and together with "Canadian know how" and "American ingenuity" have taken the specialized silk producing gene from a spider and inserted it into a goat embryo. About five percent of the time, a spider silk gene becomes part of one of the goat chromosomes, thus the **goat-spider** was born!

"The result is that you have 70,000 goat genes and one spider gene," he says. "That's the reason why the goat doesn't start to look like a spider. It's very diluted so the change is only in the goat's mammary gland and the silk is eventually exported."

After the egg is transplanted back into a foster mother, the result is a goat that looks like a goat, acts like a goat, BUT produces milk which contains proteins which, when treated, produce a very close imitation of the valuable spider silk. A single goat only produces small amounts of the desired material, so an extremely large herd is required to acquire useful quantities.
http://www.sciencechannel.com/video-topics/sci-fi-supernatural/kapow-superhero-science-spider-silk-gene-goats/

With silk web producing spider-goats a reality, could a genetically engineered "web-spinning, web-slinging superhero " possibly be on the horizon in the very near future with all the abilities of the comic book superhero that we come to know?

"Web-spinning" Spider Goats. Spider silk is produced from the goat's milk.
A spider's web silk is lighter and stronger than steel of the same size
http://www.jesus-is-savior.com/End%20of%20the%20World/Genetics%20Nightmare/spider_goats.htm
and http://www.dpughphoto.com/animals_spots.html

"Glow in the Dark Pigs"

Genetic engineering it seems was on a roll when the world once again was astounded by news coming out of the Far East, Taiwan to be precise. Scientists in Taiwan say they have bred three pigs that "glow in the dark." Though there had been previous partially phosphorescence pigs, they claimed that their pigs are the only ones in the world which are green through and through.

Phosphorescence is a specific type of photoluminescence related to fluorescence. Unlike fluorescence, a phosphorescent material does not immediately re-emit the radiation it absorbs.

The slower time scales of the re-emission are associated with "forbidden" energy state transitions in quantum mechanics. As these transitions occur very slowly in certain materials, absorbed radiation may be re-emitted at a lower intensity for up to several hours after the original excitation.

Commonly seen examples of phosphorescent materials are the glow-in-the-dark toys, paint, and clock dials that glow for some time after being charged with a bright light such as in any normal reading or room light. Typically the glowing then slowly fades out within minutes (or up to a few hours) in a dark room. http://en.wikipedia.org/wiki/Phosphorescence

Fluorescent or "Glow in the dark" pigs
https://www.buzzfeed.com/alisonvingiano/scientists-in-china-made-glow-in-the-dark-pigs?utm_term=.xmArwqAbx#.mkOBkjaD2

In the case of the **glow in the dark pig**, the pigs are transgenic, created by adding genetic material from jellyfish into a normal pig embryo. The scientists, from National Taiwan University's Department of Animal Science and Technology, say that although the pigs glow, they are otherwise no different from any others. To create them, DNA from jellyfish was added to about 265 pig embryos which were implanted in eight different sows.

They are the only ones that are green from the inside out. Even their heart and internal organs are green, the researchers say. In daylight, the researchers say the pigs' eyes, teeth, and trotters look green. Their skin has a greenish tinge; this is particularly noticeable when an ultraviolet light shone on them will make the pigs glow in the dark. The researchers hope the pigs will boost the island's stem cell research, as well as helping with the study of human disease.

163

The scientists will use the transgenic pigs to study human disease. Because the pig's genetic material encodes a protein that shows up as green, it is easy to spot. BBC News, Thursday, 12 January 2006, "Taiwan breeds green-glowing pigs"

"Glow in the Dark Kitties" or How Science Created the "Cheshire Cat"

At the time of the writing of this textbook, things in the genetic engineering world of science were going from the amazing to the ridiculous as the news media and the World Wide Web reported the creation of "**glow in the dark kittens**"! Were geneticists trying to help the pet owners help find their lost kittens and cats that mysteriously wandered away from the homes at night or just trying to prove they could create fluorescent in any animal, perhaps even, to create the **C.S. Lewis** fictional "**Cheshire Cat**" as a pet? However, the **Mayo Clinic** scientists who created this glowing cat had a bigger goal in mind: fighting AIDS and other human diseases.

In 2007, South Korean scientists altered a cat's DNA to make it glow in the dark and then took that DNA and cloned other cats from it — creating a set of fluffy, fluorescent felines. Here's how they did it: The researchers took skin cells from Turkish Angora female cats and used a virus to insert genetic instructions for making red fluorescent protein. Then they put the gene-altered nuclei into the eggs for *cloning,* and the *cloned embryos* were implanted back into the donor cats — *making the cats the surrogate mothers for their own clones.*

What's the point of creating a pet that doubles as a nightlight? Scientists say the ability to engineer animals with fluorescent proteins will enable them to artificially create animals with human genetic diseases. http://www.mnn.com/green-tech/research-innovations/photos/12-bizarre-examples-of-genetic-engineering/glow-in-the-dark

In the mainstream world of science, the motivation to find cures for human diseases is a driving force in the frontiers of medical advancement which ultimately will advance the civilization of humanity, but one has to wonder to what lengths does this research into gene splicing and manipulation extend? Is it all pure science and what can we expect next…designer glow in the dark pets and other fluorescent animals?

The substance that makes the cat glow is a version of the **green fluorescent protein (GFP)** that lights up the crystal jelly, a type of jellyfish that lives off the West Coast of the United States. Years ago scientists realized that the gene for GFP is a perfect marker when they insert another new gene into an organism. By inserting a version of GFP along with their gene of choice, they could easily see if they were successful because the organism would glow. Since the technique was first developed, researchers have made many glowing animals, including pigs, mice, dogs, mice, rats, even fish, salamanders and frogs you can buy in the pet store. http://www.smithsonianmag.com/science-nature/the-glow-in-the-dark-kitty-77372763/?no-ist
Breakthroughs in science come by leaps and bounds and genetic engineering is a big part of the new frontier of science. As everyone was still trying to catch their breath from the news of "spider goats" and "green pigs", science once again announced another breakthrough, this time in cloning, the stuff that was once considered science fiction is a now reality in our lifetime.

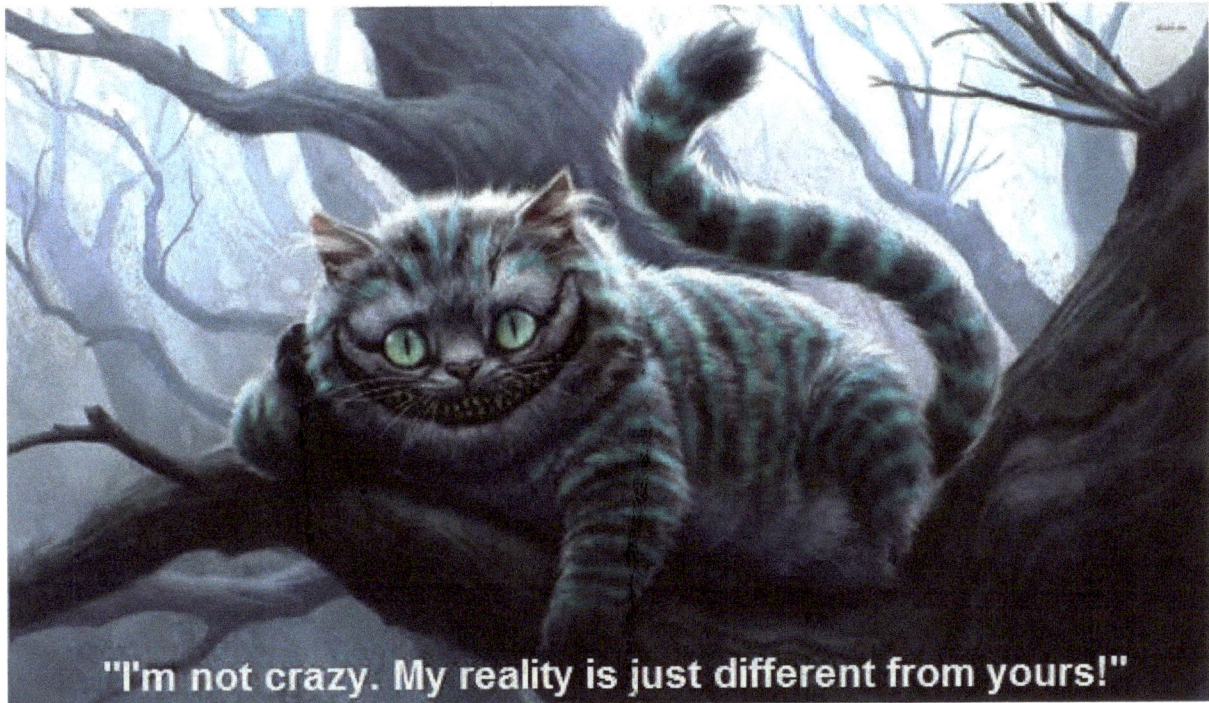

"I'm not crazy. My reality is just different from yours!"

The Cheshire Cat gives Alice a foreboding check on reality. Perhaps, the same can be said of the Black World of Science's warning to mainstream science and society!
https://www.walldevil.com/748653-cheshire-cat-wallpaper.html

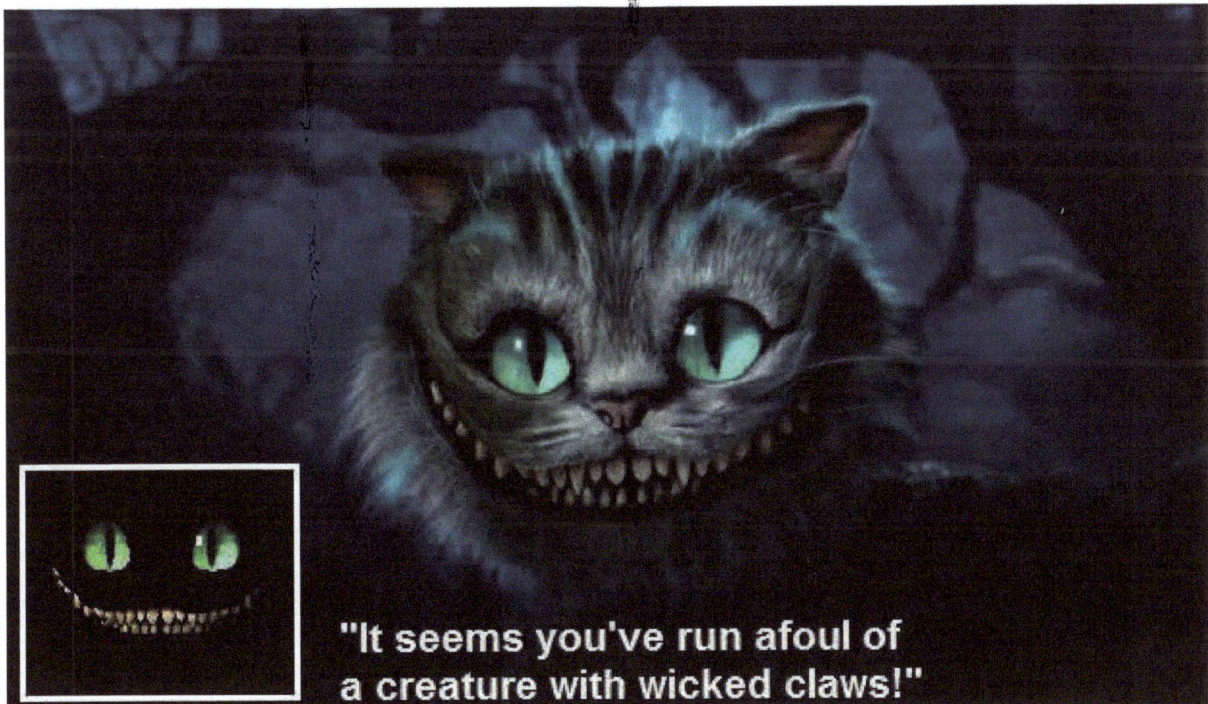

"It seems you've run afoul of a creature with wicked claws!"

Fictional fantasy becomes a reality with the creation of glow in the dark cats, the "smiling Cheshire Cat" comes to life in the genetic engineering labs of science!
https://www.walldevil.com/748653-cheshire-cat-wallpaper.html and http://tvtropes.org/pmwiki/pmwiki.php/Main/CheshireCatGrin

165

Glow in the dark kitties! What's not to love? We've got glow in the dark cat fur and claws, so where's that glow in the dark toothy smile?

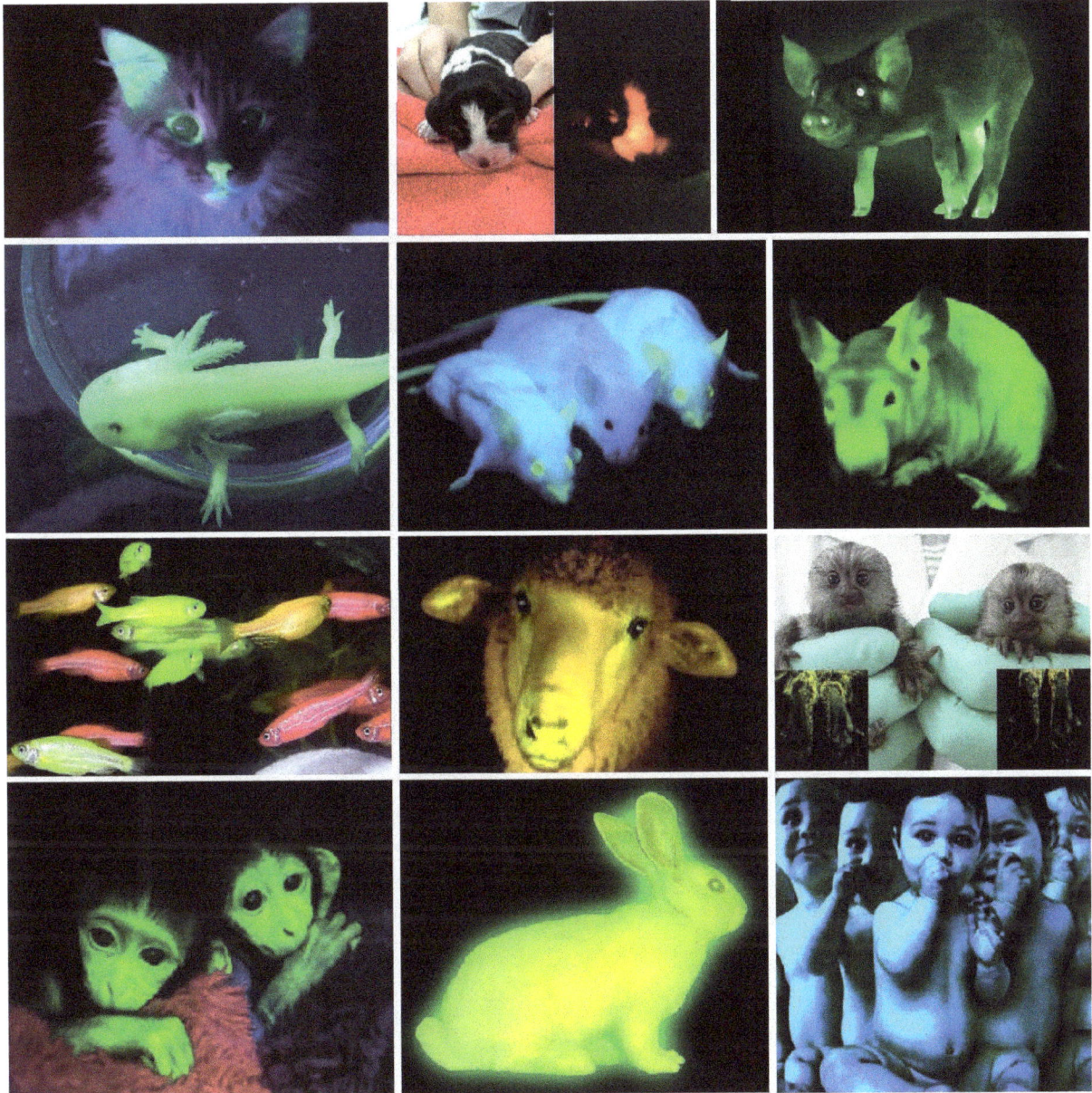

**Fluorescent animals for medical advancements in fighting human diseases
or designer glow in the dark pets and possibly humans?**
Google Images

Cloning, however, has been going on in the natural world for thousands of years. A clone is simply one living thing made from another, leading to two organisms with the same set of genes. In that sense, identical twins are clones, because they have identical DNA. Sometimes, plants are self-pollinated, producing seeds and eventually more plants with the same genetic code. Some forests are made entirely of trees originating from one single plant; the original tree spread its roots, which later sprouted new trees. When earthworms are cut in half, they regenerate the missing parts of their bodies, leading to two worms with the same set of genes. However, the ability to intentionally create a clone in the animal kingdom by working on the cellular level is a very recent development.

The first cloned animals were created by **Hans Dreisch** in the late 1800's. Dreich's original goal was not to create identical animals but to prove that genetic material is not lost during cell division. Dreich's experiments involved sea urchins, which he picked because they have large embryo cells, and grow independently of their mothers. Driesch took a 2 celled embryo of a sea urchin and shook it in a beaker full of sea water until the two cells separated. Each grew independently and formed a separate, whole sea urchin.

In 1902, another scientist, embryologist **Hans Spemman**, used a hair from his infant son as a knife to separate a 2-celled embryo of a salamander, which also grows externally. He later separated a single cell from a 16-celled embryo. In these experiments, both the large and the small embryos developed into identical adult salamanders. Spemman went on to propose what he called a "fantastical experiment" -- to remove the genetic material from an adult cell, and use it to grow another adult. In this way, he theorized, he would be able to prove that no genetic material was lost as cells grew and divided.

There were no major advances in cloning until November of 1951 when a team of scientists in Philadelphia working at the lab of **Robert Briggs** cloned a frog embryo. Rather than simply breaking off a cell from an embryo, the team took instead the nucleus out of a frog embryo cell and used it to replace the nucleus of an unfertilized frog egg cell, completing the "fantastical experiment" of nearly 50 years before. Once the egg cell detected that it had a full set of chromosomes, it began to divide and grow. This was the first time that this process, **nuclear transplant** was ever used, and it continues to be used today, although the method has changed slightly.
http://library.thinkquest.org/20830/Frameless/Manipulating/Experimentation/Cloning/long doc.htm

Dolly the Sheep

Dolly the sheep was the first successfully cloned mammal born on 5 July 1996 at the Roslin Institute near Edinburgh, Scotland. Her birth was only announced seven months later and was heralded as one of the most significant scientific breakthroughs of the decade. Dolly the Sheep, 1996-2003 from the Science Museum, London

The cell used as the donor for the cloning of Dolly was taken from a mammary gland, and the production of a healthy clone, therefore, proved that a cell taken from a specific part of the body could recreate a whole individual. As Dolly was cloned from part of a *mammary gland*, she was named after the famously curvaceous country western singer **Dolly Parton**.

They used the technique of somatic cell nuclear transfer, where the cell nucleus from an adult cell is transferred into an unfertilized oocyte (developing egg cell) that has had its nucleus removed. The hybrid cell is then stimulated to divide by an electric shock, and when it develops into a blastocyst it is implanted in a surrogate mother.

In the previous year, the same team had produced cloned sheep from the embryonic cells, but this was not seen as a breakthrough since, adult cloned animals had been produced from embryonic tissue as long ago as 1958, using cells from a South African frog *Xenopus laevis*.

Dolly the sheep, the world's first cloned mammal
http://www.cnn.com/2017/02/24/health/dolly-the-sheep-cloning-anniversary/

On 14 February 2003, Dolly was euthanized because of a progressive lung disease and severe arthritis. Her death raises questions of a premature aging because she came from an adult sheep six years in age and sheep normally live only 11 to 13 years. **" Dolly the sheep clone dies young", BBC News, Friday, 14 February 2003**

After cloning was successfully demonstrated through the production of Dolly, many other large mammals have been cloned, including horses and bulls, a mouflon (a form of wild sheep) and in 2005, a dog by the name Snuppy from Korea.

The door was open now to all manner of cloning possibilities and as a viable tool for preserving endangered species. Genetic engineering and gene manipulation was now mainstream science.

Scientists in January2009, scientists from the **Centre of Food Technology and Research of Aragon**, in Zaragoza, northern Spain announced the cloning of the Pyrenean ibex, a form of wild mountain goat, which was officially declared extinct in 2000. Using DNA from skin samples kept in liquid nitrogen the scientists managed to clone the Ibex from domestic goat egg-cells. The newborn ibex died shortly after birth due to physical defects in its lungs. However, it is the first time an extinct animal has been cloned. It has also increased the possibility that in the future it will be possible to reproduce long-dead species such as woolly mammoths and even dinosaurs. http://en.wikipedia.org/wiki/Dolly_the_sheep

Tetra, the Rhesus Monkey

Our last example of cloning is definitely setting the stage for the eventual cloning of humans from the embryonic stage to full adulthood and the subject, this time, is a simian, a rhesus monkey. Oregon researchers say they have cloned a monkey by splitting an early-stage embryo and implanting the pieces into mother animals. The technique has so far produced only one living monkey, a bright-eyed rhesus macaque female named Tetra, now 4 months old. **Professor Gerald Schatten**, a researcher at the Oregon Health Sciences University in Portland who led the research, said four more animals are on the way. *"This is essentially the method of Brave New World,"* said **Ronald M. Green**, an ethicist at Dartmouth College. *"This opens the prospect of mass identical replication."*

Tetra, the rhesus monkey is different from Dolly the sheep, which was produced by Scientists at Scotland's **Roslin Institute** using a process called nuclear transfer -- taking the nucleus out of an adult cell and using it to reprogram an unfertilized egg. Some scientists argue that animals like Dolly are not 100 percent clones because they have genetic material both from the adult cell they were taken from, and from the egg that is hollowed out to make the clone. Tetra was produced by a technique called "embryo splitting." Here's how it works:

- An egg from a mother and sperm from a father are used to create a fertilized egg.
- After the embryo grows into eight cells, researchers split it into four identical embryos, each consisting of just two cells.
- The four embryos are then implanted into surrogate mothers. Schatten said that in effect, a single embryo becomes four embryos, all genetically identical.

Tetra, the first cloned rhesus monkey the final stepping stone to human cloning

In the case of their experiment, three of the embryos didn't survive. The fourth, Tetra, was born 157 days later. Her name means "one of four." Tetra isn't the first monkey to be cloned but she is the first using the embryo-splitting technique. *"Researchers clone monkey by splitting embryo" CNN - January 13, 2000*

Cloning seems to have created such controversy and raised so many ethical questions in the minds of not only scientists and governments but also, in the minds of the public, not the least are the questions of moral and spiritual concerns raised by fundamentalist Christians and other religious groups particularly, with the latest allegations of human cloning. **Human cloning** is the one forbidden area that raises the specter of mankind playing god with the strong possibility of altering genetically, the evolution of his own destiny.

Stem cell research and cloning from stem cells is the new avenue in this new branch of medical science. Red flags were being raised by everyone concerned because at the rate at which DNA research and genetic engineering are moving forward in understanding the DNA makeup of every life form on the planet and the scientific community's acquisition of new knowledge appears to be leaving the moral majority choking and coughing in its dust.

Human cloning is the creation of a genetically identical copy of an existing or previously existing human. The term is generally used to refer to *artificial* human cloning; human clones in the form of identical twins are commonplace, with their cloning occurring during the natural process of reproduction. There are two commonly discussed types of human cloning: therapeutic cloning and reproductive cloning. Therapeutic cloning involves cloning adult cells for use in medicine and is an active area of research. Reproductive cloning would involve making cloned humans. A third type of cloning called replacement cloning is a theoretical possibility, and would be a combination of therapeutic and reproductive cloning. Replacement cloning would entail the replacement of an extensively damaged, failed, or failing body through cloning followed by whole or partial brain transplant.

The various forms of human cloning are controversial. There have been numerous demands for all progress in the human cloning field to be halted. Most scientific, governmental and religious organizations oppose reproductive cloning. Serious ethical concerns have been raised by the future possibility of harvesting organs from clones. Some people have considered the idea of growing organs separately from a human organism - in doing this; a new organ supply could be established without the moral implications of harvesting them from humans. Research is also being done on the idea of growing organs that are biologically acceptable to the human body inside of other organisms, such as pigs or cows, then transplanting them to humans, , a form of xeno-transplantation. There has even been a discussion among scientists to clone humans without heads to the adult stage so that human body parts could be harvested to treat the seriously ill or near terminal patients.

If what we have seen in the world of mainstream science is possible through government funding and private donations, imagine what can be achieved technologically in most branches of science that fall under the umbrella of National Security when you throw almost unlimited financial resources into the covert black programs of the military industrial complex! Scientific breakthroughs can occur in a very short time span in what would take 30 to 50 years (in some cases a hundred years or more depending on the specific technology) to appear in the everyday world of science, the same things in the secret black world of the military industrial complex can be achieved in 5 to 10 years or less!

Keep in mind that in the world of UFOs and Extraterrestrials, incredible breakthroughs in reverse engineering of alien technology, human and ET cloning programs and DNA manipulation and engineering were accomplished in the early 50's. Such programs were undertaken on the same level as the **Manhattan Project** during the Second World War to develop the first atomic bombs. In fact, the level of secrecy and the intensity for research and development surrounding alien technology has been many tens of levels higher than the Manhattan Project.
https://www.youtube.com/watch?v=Vvj78a7CP2k and
https://www.youtube.com/watch?v=D5iSm6auHi4 and
https://www.youtube.com/watch?v=Tz8HxNfIG8Q

In the black covert world of science, cloning has gone far beyond any mainstream science programs or projects that the general public is currently aware of. Not only is human cloning possible, it is a reality that pales almost into insignificance when compared to the cloning of Extraterrestrial biological entities, otherwise known as programmable life forms!

CHAPTER 27

PROGRAMMABLE LIFE FORMS (PLFs)

The "Greys" are the Extraterrestrial species that are most often reported by people around the world. Reports from individuals who have been in contact with the Greys vary on their intentions: some say they are kind beings here to guide us through our evolution, while others believe they are cruel invaders intent on ruling the planet. One thing is certain, most people who have come in contact with Greys strongly "believe" to the point of paranoia that these ET beings are actively abducting people on a regular basis. The Greys are reported to be abducting individuals, taking these humans on board their craft, performing invasive medical procedures such as extracting eggs or semen in order to create grey-human hybrids as proselytized by **Budd Hopkins, David Jacobs,** and **John Mack** and their supporting ilk.

The Greys are the Extraterrestrial species that are alleged to have crash-landed in Roswell, New Mexico in July 1947. Also, it is widely believed among researchers that the Grey species entered into an agreement with the U.S. government to provide us "hardware" in the form of advanced technology for "software", in the form of human beings.

The alien-human abduction phenomenon has created a very deep and ever-widening gulf among UFO/ETI researchers like no other aspect of Ufology before it. Over recent years, this chasmal disagreement as to the nature of the alien agenda currently being played out by the visiting Extraterrestrial presence to our planet has divided Ufologists into two distinct polarized camps.

On one side are Ufologists who believe that the so-call alien abductions of humans is a malevolent agenda against all of humanity which will eventually see most of mankind replaced with a new more intelligent species which is based upon an alien-human hybrid created ostensibly by the "Greys"!

On the other side of this chasmal disagreement, are the proponents that essentially feel that it is too early to know what the alien agenda is all about, however, they favour the benevolent agenda of ETs or at least the impartial neutrality of visiting ETI as there has been no open demonstration of alien hostility towards the people of Earth!

In late October 2010 in Rio Rico, Arizona, **Dr. Steven Greer** held a three day working **CSETI** seminar with over two hundred interested people from around the world in attendance who came to hear guest speakers talk on UFOs, ETs and related subjects as well as participating in night time UFO sightings and ET contact. During one of these seminar afternoons of lecturing, this author had the opportunity to ask him about the alien abduction scenarios that seem to occur so frequently in the United States and his thoughts about the subject matter.

He replied that besides reverse engineering of alien spacecraft since the late 40s, the US military industrial complex and its subsidiary deep black projects had also been working on associated alien technology found on board the retrieved spacecraft. But unknown to much of the world of Ufology, whose main concentrated efforts of investigation had been on the UFO crafts themselves and the abduction phenomenon, that they have completely overlooked the medical science of alien cloning and DNA decoding and manipulation of dead and still living ETIs found at saucer crash sites.

This typical Grey alien image is frequently portrayed in the news media and thus, in the public consciousness. It is a deliberate campaign of disinformation by the military industrial complex to influence the public's perception with a negative mindset of alien species and abductions
https://www.pinterest.com/lawnees/aliens/

Dr. Greer had been told by an insider, who wishes to remain anonymous to the public, that the black world besides making incredible breakthroughs in the early 50's in reverse engineering of crashed ET spacecraft but, had also made amazing medical breakthroughs not only from cloning Extraterrestrial life forms but, in the ability to marry these cloned beings with advanced computerization so that they were programmable. In essence they were synthetic or cybernetic life forms that were completely under the control of humans!!! They are **Programmable Life Forms (PLFs)** or as the British military and intelligence refer to as **Programmed Generated Life Forms (PGLFs).** With the use of remote control brain implants and a type of computerized brain and other computerized or digitized organs, etc. the **military industrial complex (MIC)** had in its possession an army of thousands of terrestrial-engineered alien beings that were essentially described as small 3 to 5 foot grayish- white beings with large heads and large black eyes!

Does all this sound familiar to most abduction researchers and Ufologists? Could this be the reason why this type of ET being is being propagandized over the news media and through

174

pseudo TV documentaries and movies? Is the MIC trying to setup a negative fear-based mindset in the general public using this particular ET xeno-type being for some future false flag alien invasion scenario? https://www.youtube.com/watch?v=aYk_nPoIu8E

Author's Rant: Now on a subjective level, I always suspected that this was the case. What was all too commonly reported particularly in the 60 s and 70s up to the present time was the reported sightings of small alien "Greys" particularly in, near or around landed ETs spaceships or in reported cases of alien abductions.

I had remote viewed this particular aspect of these small Grey ET beings and their human masters and have intuitively known that such beings were an artificial life form created on this planet but, this was merely a subjective piece of information which finally, had now been confirmed by someone well respected in the field of UFO/ETI research, namely Dr. Steven Greer.

However, this was not enough of a proof for me. There had to be some physical evidence or an account reported by one of the alien abductees to back up this story of terrestrially created extraterrestrials. Fortunately, years earlier in 1996, I came across a story about Darrel Sims, the self-styled "Alien Hunter" in the British magazine *"Encounter", issue 6, 1996* where Darrel Sims tells of such an event occurring:

Darrel Sims - "Alien Hunter"

Darrel Sims besides being an abductee has become controversial in his methods of abduction research which employ methods of fighting back against ET intrusions and abductions based upon his former years of training as a CIA agent. Sims who is the Chief of Abduction Investigations and Director of Physical Investigations for the Huston UFO Network uses techniques of hypnotherapy as a part of his investigations for people involved in the abduction phenomenon.

He is a Master Practitioner in **Neuro Linguistic** programming, a Master **Hypno-therapist** and a Master **Hypno-Anesthesiologist**. He has had 25 years experience in the research of UFOs and believing that abductions by aliens is real require physical evidence which he has acquired in the form of "implants" from various abductees.

One of his methods of obtaining information on the alien abductors is by hypnotizing people who have experienced re-occurring abduction events in their lives with a post-hypnotic suggestion or "re-programming" procedure to awaken when they are being abducted by their ET captors and try to recall as much vital data and establish a more active contact with the beings. This has resulted in various degrees of success to the point that the alien "Greys" are perplexed and have become aware of Sims and his some of his investigative methods.

This brings us to one of his more intriguing cases which will prove my point about these artificial life forms as not being true Extraterrestrial biological entities. **Laura Wilson** and her husband, Steve Wilson were constantly being visited by aliens and were at the breaking point. Laura sought out the help of Sims, telling him that her nocturnal intruders threatened her to stop visiting Darrel or there would be repercussions and to prove their point they surgically removed one of her fingers and then re-attached. They had also threatened her next of kin, if she didn't cease and desist her efforts to get outside help. (***This type of threat has military covert intelligence written all over it***). [My bold italics added].

Darrel Sims, the Alien Hunter
http://www.blogtalkradio.com/ufo_radio/2011/05/31/derrel-sims-alien-hunter

Through a re-programming procedure, Laura's subconscious was programmed to react to any abduction sequence with hostility. He gave her a post hypnotic suggestion which would be effective and then her conscious mind had to be erased of any knowledge of the suggestion as Sims is certain that the aliens could read our conscious minds telepathically.

On April 5th 1993, Laura's abduction and moment of truth had arrived. Two small Grey beings came through the wall of her bedroom. Laura remembers thinking to them, "come closer, come closer," not having any fear at that moment. When one of the beings got close to the bed, she lunged forward and grabbed the black protective eyepiece half peeling it off. She felt mucus from the eye land on her arm. The aliens retreated but returned two weeks later.

This time five "Greys" and one was described as being Brown came through the wall. As the being came closer, Laura grabbed the being around the waist with her left arm and with her right hand she pulled at the eyecup and removed the whole piece. She described the alien being's skin as being a rough texture.

When questioned by **Darrel Sims** as to whether the alien felt any pain from the injury, she stated, "No, however, the alien did seem to be bothered by the light from the garage and shielded his eye with one hand. When asked if she could see under his eye, she replied, "Yes, it's like white scanning dots behind a red screen. The lights…the lights, they leave a trial."

The Brown ordered the other ETs to put a helmet on her head which would sedate her down while they scrambled to retrieve the eyecup. The injured Grey was then floated out and the rest of the aliens left through the wall in a confused state.

Fearing that they may return which they did, she remained awake as they came for her son, but this time she quickly ordered them to leave and to her surprise they turned around and left.

What we see from this abduction account which probably hasn't yet occurred consciously to Darrel Sims is the very mechanical or cybernetic aspect of the alien life forms. The eye described by Laura was one of electronic circuitry like a medical monitoring device or oscilloscope in nature. This is an indication of their artificiality and may point to an advance programmable robot or cyborg creature. Now one case by itself proves nothing but, it points the way to another possibility which may not be Extraterrestrial in origin, but terrestrial or manmade in its creation. It is a plausible theory which must be seriously considered in the investigation of the whole UFO/ETI phenomenon, especially with regard to the whole alien abduction scenario. We need to research other cases similar to this one, to see if this is merely one of a kind account or strong evidence that the military industrial complex is further head in its reversed alien technology development or if this something that is truly alien in nature.

Author's Rant: I'm betting my money on Dr. Greer's research, that these beings are PLFs controlled by a covert human intelligence that may have had its inspirations and origins from an Extraterrestrial source! Read the next account below which confirms from another eye witness that PLFs are real.

**Are some "Grey" ETs really manmade Programmable Life Forms (PLFs),
a type of cyborg (part machine, part living creature)?** (c) Terry Tibando

As it turns out more witnesses are stepping out into the public spotlight with accounts and testimonies that state their involvement and witnessing these **PLFs (PGLFs)** in secret underground military complexes both in the US and in Britain.

Barry King - Testimony About Britain's PGLF Program

Barry King, who was a Security Enforcement Officer at AL-499 Peasemore Base, Berkshire in 1979 - 1981was one witness who worked at a top secret British AFB within the British military intelligence complex. In September 2000 he gave a resume and interview for CSETI team members and other UFO researchers over the **Programmable Generated Life Forms**, made by the NSA run underground Genetics Engineering lab 200ft under Peasemore, Berkshire, England.

His information comes via **British Intel** and his own experiences as part of a **Population Control programme** experimentee. He has two implants and has been experimented on in the **"Trip Seat"** as part of human mind control tests, for the ultimate total control of mankind.

Barry King was again, later interviewed inside the **RAF Bentwaters Base,** Suffolk.

King originally released this information in early 1994, via THE VOICE files, and initially in a 2 hr video with another investigator /experiencer in BASES 1. He was further questioned, in BASES 2, which includes **Larry Warren, Bill Uhouse** and more. **Matthew Williams** also grilled Barry about the veracity of his claims. Most notable are the **Montauk Project** books and videos.

Barry King, a former Security Enforcement Officer at Peasemore Base, Berkshire, UK whistleblower to the British government's involvement in genetic engineering of alien-like life forms used for MiLabs and Psi-War Exercises
https://www.youtube.com/watch?v=p7JF0aGdaI0

It was during the early 1990s **Barry King** released claims of alien mind control bases in the Berkshire countryside under the village of **Peasemore.** Linked with claims were the mind control experiments on the Greenham Common women protesters, near Hungerford, where the mind controlled massacre occurred in the 90s, thus linking Peasemore as the source of so called manmade alien Greys. Using a genetic population survey, people were selected at the age of 2 or younger, for these mind control tests and developments for a **Super Race** which form part of the video interviews: Bases 1, Bases 2, and the **CSETI Disclosure Program** release.

King describes an incident where he was startled to see a very tall, powerful looking, greenish-brown mottled-skin reptilian being, who was being escorted by two NSA operatives past him while he was walking in a corridor that was part of the underground complex in Berkshire. He

stated that these reptilian beings and the short Grey beings were created by humans and controlled by a central computer system via computerized implants.
https://www.youtube.com/watch?v=p7JF0aGdaI0 and
https://www.youtube.com/watch?v=8Wv-Igj8E7U and
https://www.youtube.com/watch?v=vx5F0r84wwg and
https://www.youtube.com/watch?v=eUF8NJIYrJY

An artist rendition of a Reptilian ET being or could it represent a genetically engineered creature created by human scientists as described by Barry King, to resemble an ET being for some possible future false flag alien invasion?
https://www.pinterest.com/quclark/shapeshiftersreptilianother-beings-living-on-earth/

Their purpose was to carry out **MiLabs - (Military Abductions)** of humans around Peasemore and throughout the British countryside under the operational control of the NSA. Its current project codename is **Puppetmaster** which grew out of a 1950s project called **Mannequin.**

In a startling revelation of information, King leaks out that *60% to 70% of the alien abductions globally are MiLab* and that some of the governments involved in this genetic engineering and MiLab program are the UK, Canada and the USA!!!!! No doubt other countries are also involved, particularly countries that are friendly or in alliance with the USA, like Brazil, Australia, Israel, and most of Europe, etc.

180

**Is this a "Peeping Tom" ET being looking through someone's window
or is it a PLF simulating an alien ET encounter or a hand puppet?**
https://www.youtube.com/watch?v=eG9furbZBH8

King states that besides adults, infants or young child are also subjected to mind control where in a laboratory called the **"Trip Seat"** are induced with psychotropic drugs, hypnosis and sensory deprivation which makes them believe that they are experiencing an alien abduction right there in the laboratory!!!

King goes so far to conclude that the 1980s UFO events at Bentwaters- Rendlesham AFB was a MiLab scenario, more precisely it was a **Psi-War Exercise** carried out on military personnel where the "hardware and software" were brought in from Peasemore.
https://www.youtube.com/watch?v=okacAlE2DQ8

Alien Abductions

In the early years of investigation of the UFO phenomenon, the general consensus amongst most researchers was mixed as to what UFOs represented. Theories ranged from either a misunderstood atmospheric or terrestrial phenomenon, to some advanced recent breakthrough in aerial technology developed by the Russians. The possibility of some new unknown cryptozoological life form missed by scientists like the Sasquatch or the Loch Ness monster was considered, and some researchers contemplated the rare possibly that these aerial objects could be a vanguard of remotely controlled alien vehicles from another planet or star system.

Careful research into the voluminous UFO eye witness accounts over many decades reveal the presence of alien life forms either, associated in close proximity to a landed UFO or spacecraft. Currently, in the field of Ufology, the acceptance that UFOs or Extraterrestrial spacecraft are piloted by Extraterrestrial life forms is now a given; what is not clearly understood was the reason why they were here, what is their agenda, and what is their intentions.

The terms *alien abduction* or *abduction phenomenon* describe "subjectively real memories of being taken secretly and/or against one's will by apparently nonhuman entities and subjected to complex physical and psychological procedures." People claiming to have been abducted are usually called "abductees" or "experiencers." http://en.wikipedia.org/wiki/Alien_abductions

In **Martin Cannon's** lengthy monograph, *"The Controllers: A New Hypothesis of Alien Abduction,"* presents the problem of alien abduction as "a dizzying number of individuals, who claim that travellers from the stars have scooped them out of their beds, or snatched them from their cars, and subjected them to interrogations, quasi-medical examinations, and "instruction" periods. Usually these sessions are said to occur within alien spacecraft; frequently, the stories include terrifying details reminiscent of the tortures inflicted in Germany's death camps. The abductees often (though not always) lose their memories of these events; they find themselves back in their cars or beds, unable to account for hours of "missing time." Hypnosis, or some other trigger, can bring back these haunted hours in an explosion of recollection – and as the smoke clears, an abductee will often spot a trail of similar experiences, stretching all the way back to childhood." https://www.youtube.com/watch?v=DZtil_ZsrOE

He then astutely observes an odd fact: "Many abductees, for all their vividly-recollected agonies, claim to love their alien tormentors. That's the word I've heard repeatedly: "love." There is "absurd internal contradictions" and "askew logic" to the abduction stories and a "discontinuity of emotional response" to the abduction scenario. The medical violation uses "vile and inhuman tactics of medical examination" yet, these victims of alien intervention are ascribed with the "wisdom of the ages and the beneficence of the angels." These ETs seem to know much about our society going about their business while remaining "undetected by the local authorities and the general public; they communicate in human tongue; they concern themselves with details of the percipients' innermost lives, yet they remain so ignorant of our culture as to be unaware of the basic moral precepts concerning the dignity of the individual and the right to self-determination."

Here, we will need to review some of the more classic abduction accounts to understand the initial eye witness experience as reported to the news media and to the Ufologists that investigated these cases. Then we will provide alternative and plausible explanations as to whether these were legitimate and authentic reports of real events or a misunderstood experience based on the limited understanding of the abductee of his immediate situation and the inadequate investigation methods by Ufologists. In all cases, the events were very real to the witness but, are most likely reported with a preconceived understanding, much like the biblical account of **Ezekiel's "Wheel** within a Wheel" event.

The first interspecies sexual relationship with intelligences from the stars began with the **Antonio Villas Boas** case of 1957 but, it was the **Betty and Barney Hill** case of 1961 that set the tone for establishing possible alien motives. These well publicized cases were followed very quickly with similar reports from people who recounted such experiences while still mentally conscious but, the majority of cases were remembered while under a state of hypnotic recall. Abduction reports seem to follow a consistent trend in which the intended victim of the experience was either physically or astrally (in a "out of body" state) taken on board a strange disc shape craft against their will by small alien beings.

Many abductees describe aliens as grey humanoids, known as "Greys."
http://harnois75.deviantart.com/art/Grey-Alien-123705422

These ET beings are often described as being grayish-white in complexion, approximately four to five feet in height, thin or skinny bodies with large craniums and large black featureless, almond eyes, basically humanoid in overall appearance. These beings have been popularized in the UFO culture as the "Greys", an obvious racial appellation which seem to continue the long standing tradition amongst humans of that despicable prejudice known as racism!

Few mainstream scientists believe the phenomenon literally occurs as reported. However, there is little doubt that many apparently stable persons who report alien abductions are sincere. According to **Dr. John Mack** after investigating over 800 abductee cases maintained that the majority of abductees do not appear to be deluded, confabulating, lying, self-dramatizing, or suffering from a clear mental illness. There is some psychopathology indicated in some isolated alien abduction cases which is to be expected (this author knows firsthand of one probable case that fits this description), however, abductees appear to be a rational and sane as the next person and this has been proven by "assessment by both clinical examination and standardized tests.

What is interesting is that an entire subculture has developed around the subject, with support groups and a detailed mythos explaining the reasons for abductions: The various aliens (Greys, Reptilians, "Nordics" and so on) are said to have specific roles, origins, and motivations.
http://en.wikipedia.org/wiki/Alien_abductions

Contrary to what some people may think, contactees are not abductees, as the experiences between the two is like night and day. Contactee experiences are considered positive, spiritually up lifting, life changing for not only the person in contact with Extraterrestrial beings but generally positive to all who accept the contactee and his message, even if at the best of times, the alleged contact experience is a hoax or a delusion. On the other hand, the abductee experience is usually negative, frightening, traumatic, and oftentimes forcefully intrusive on many levels besides the medical-physical aspect that is well documented in the UFO literature.

These abduction experiences are akin to humans abductions, but in the case of alien abductions, these experiences were not hostile or violent like so many human abductions, as care was taken to cause the least amount of harm to the abductee as possible.(which may be open to argument and personal perception by the victim) however, they were certainly unsolicited and intrusive. The orientation of most abductions seems to be toward forced medical examinations which emphasizes their reproductive system wherein, skin, ovum and sperm samples being the chief reason for abduction experience.

Abductees sometimes claim to have been warned against environmental abuse and the dangers of nuclear weapons. Consequently, while many of these purported encounters are described as terrifying, some have been viewed as pleasurable or transformative. In all cases the human victims were always return to their home or cars wherever the place of abduction was initiated.
http://en.wikipedia.org/wiki/Alien_abductions

Once again, mainstream scientists, mental health professionals, critics and skeptics overwhelmingly doubt that the phenomenon occurs literally as reported and instead attribute the experiences to deliberate "deception, suggestibility (fantasy-proneness, hypnotizability, false-memory syndrome), personality, sleep phenomena, psychopathology, psychodynamics and environmental factors.

Consequently, from these witness reports, eager UFO researchers possibly looking to establish themselves in a more favourable light among their UFO peers jumped upon the similarities of human abductions and this new type of abduction, thus, resulting in a new category of close encounters to explain the new experience, that of the **Close Encounters of the Fourth Kind**, the **Alien Abduction**.

Further case reports became epidemic throughout America which told of women becoming pregnant induced by ET medical intervention and then to the amazement these women and their doctors would then find themselves no longer pregnant. Research indicated that the ETs had retrieved the fetus after the first trimester. These fetuses were the product of human and alien conception. These alien hybrids were characterized with both alien and human traits, a new race of beings that were under the control of ETs. These types of cases, which although, were world-wide but, in much smaller numbers were essentially an American phenomenon. It would seem that hundreds of thousands of these alien abductions were perpetrated upon the American population, the unfortunate victims of a so called sinister alien agenda. Many abductees through this process of abduction and hybridization claim to have a child of alien- human genetic design floating around somewhere in space. The alien agenda therefore, according to **Dr. David Jacobs, Budd Hopkins, Whitley Strieber,** and **Dr. John Mack,** et al is the eventual replacement of all humans on this planet by alien hybrids, despite the relative paucity of corroborative evidence.

The abduction scenario has become so entrenched in the mythology and the mind set of Ufology that like so many controversial theories has divided researchers into opposing camps in their investigations of the UFO and ET phenomenon. If this is simply, a natural process of the investigative methodology of something new then, one can forgive and dismiss the initial differences, the eventual results will inevitably be conclusive and indisputable. However, as was already stated earlier, the data base in Ufology is faulty and corrupted. Thus, the end result is an inconclusive and disputable outcome due mainly to poor and improper research methods, faulty analysis and inadequate sample testing of the data leading to inconclusive proof that will only justify the means and methods often to fulfill an egotistic personal agenda.

Abductions can be viewed in similar fashion to the methods used by professional debunkers which is to select the data you wish to justify and prove your point while ignoring any and all data and facts that may prove contrary and which may get in the way to establishing your conclusions. Thus, in the immortal words of the great bard, **Shakespeare**: *"There is something rotting in Denmark!"* Demark in this case being Ufology and the "something rotting" being the abduction scenarios!

A serious revisit is required into the whole abduction scenario to make sense of it and to understand why these silly and absurd leaps of illogic constantly seem to be crippling the credibility of Ufology from the hastily reached conclusions of incredulity. There are certainly many other possible and more logical explanations that don't involve an alien agenda. This is certainly the place and time to exercise **Occam's Razor** to reach a more reasonable and satisfactory explanation to the whole abduction phenomenon.

We must establish what current theories have been accepted, what factors come into play, what real evidence has been collected and what associated circumstances exist in the current model of abductions and then compare it to the theories, facts, and evidences that have been sadly overlooked or ignored. In the abduction scenario, the old adage, *"not everything is what it appears to be"* still holds true and doubly so, throughout the field of UFO/ETI research.

Before leaving this controversial section, it bears repeating once more and that is the other possibility of why abductions are occurring. Allowing for the misinterpretation of alien abductions as being the covert agenda of military intelligence in the form of MiLabs, what if the abduction events as claimed by the abduction victims is true, based upon the limited ability to comprehend what is really taking place. ***Is it possible that ET Intelligence may be engaged in sampling and salvaging as much of the Earth's biological life forms as possible, including humans because of a potential doomsday countdown to planetary destruction at humanity's own hands?!***

CHAPTER 28

THE HUMAN PRESERVATION PROGRAM HYPOTHESIS –
ETI SAMPLING BIOLOGICAL SPECIMENS
INCLUDING HUMAN DNA

In this next chapter, for the perhaps, for the first time anywhere, this author propositions a new hypothesis that the reader should give serious consideration to, as it offers an explanation for the motivation behind the alien agenda and the specious alien abductions phenomena.

Is it probable that the so-called alien abductions scenario that people report every year has nothing to do with abducting humans, particularly when humans are always returned to their place of origin in most cases? Is there another motive to why people report being spirited away for short periods (hour or two) of time which may actually reveal the real alien agenda of why they are here?

There are numerous accounts that have been reported of little Gray-like ET beings (the "Greys") that are collecting plant, rock and animal bio-samples. One such example occurred in Valensole, France back in 1965 in a farmer's (**Maurice Masse**) lavender fields as recorded by **Aime Michel** and Jacques Vallee. **Michel, Aime. "The Valensole Affair"; Flying Saucer Review 11, 6.** and **Confrontations: A Scientist's Search for Alien Contact by Vallee, Jacques; 1990; New York; Ballantine Books**

Here is a list of accounts where people have witnessed ET beings walking about on Earth collecting flora, fauna, and soil samples:

- 1948. Svastika Ontario, Canada
  Source: http://www.ufoinfo.com/humanoid/humanoid1948.shtml
  Source: Jacques Vallee, Passport to Magonia
  Disc|ET|Force-Field|Vegetation

A Mr. Galbraith observed a disc-like object land on the ground; an entity left the object and collected samples of vegetation. The object transmitted a "force field" which pushed the witness to the ground. The same witness again observed an object on the ground, three entities who smiled at him, were observed outside the craft. A police patrol saw a light in the nearby woods, but couldn't approach it because of an "invisible wall".

- 1948. Campo Grande, Brazil
  Source: http://www.ufoinfo.com/humanoid/humanoid1948.shtml
  Source: FSR Vol. 4 # 1
  ET|Soil|Landing Mark

While strolling near a creek one morning, Mr. Ottaviano A. Souza Bueno observed the silent descent of a round luminous object on one of the banks of the stream. There emerged from it 3 creatures of less than medium height, with extremely rapid movements, who started collecting samples of soil, using a tube which made holes in the ground. After the UFO had left, the witness went to the spot and found that all the holes were perfectly square. He took a sample and had it analyzed; the result was, Silica, 61% Aluminum, 19%, Magnesium, 11%, Iron and other trace elements, 9%. **http://nawewtech.angelfire.com/soil.html**

- Summer 1948. Hemer Sauerland, Germany
  Source: http://www.ufoinfo.com/humanoid/humanoid1948.shtml
  Source: Illobrand Von Ludwiger, Mufon 1993 Symposium Proceedings|
  Dome|ET|Landing Mark|Soil|Speed|Whining sound

A shepherd was guarding his flock of sheep in an isolated grassy area when suddenly the animals scattered in a panic. He heard a rushing sound and saw a large metallic domed object emerge from what appeared to be an artificial fog and land nearby. The witness approached the object and touched it; he felt a strong electrical shock and was knocked unconscious. The witness later woke up and found himself in a different field surrounded by several short entities, described as having large heads, large slanted almond shaped eyes, narrow mouths, and small noses, they had bulging foreheads with short stubby hair. They all had boxes on their chests with tubes protruding from it. At times they appeared to grasp the tubes and breathe into it. The beings gesticulated and spoke among themselves in an unknown language. Nearby sat the object still encased in a mist. Next to the craft stood five other humanoids that began collecting soil samples and placing them into large containers. Finally, all the humanoids re-entered the object, which took off at high speed emitting a loud whining sound. Burnt spots were found at the scene where the object was said to have landed.

- June 17, 1950. Hasselbach, Germany
- July 9, 1952. Hasselbach, Germany
  Source: http://www.ufoevidence.org/Cases/case715.htm
  Source: Kathimerini newspaper (Athens, Greece), 1952, found in declassified CIA files, Oscar Linke.
  Disc|Hum|Landing Mark|Rotation|Soil|Whistling sound|ET|Silver Suit

I left my motorcycle near a tree and walked toward the spot which Gabriella had pointed out. When, however, I reached a spot about 55 meters from the object, I realized that my first impression had been wrong. What I had seen were two men who were now about 40 meters away from me. They seemed to be dressed in some shiny metallic clothing. They were stooped over and were looking at something lying on the ground.

- 1953. Lermarken Varmland, Sweden
  Source: http://www.ufoinfo.com/humanoid/humanoid1953.shtml
  Source: Sven Olof Svensson
  ET|Landing Legs|Soil

17-year old Lyyli Nilsson was cycling to a nearby village when she felt something warm on her body and soon saw an object which folded out a couple of legs and landed on a ridge. Four persons, wearing strange "equipment" on their backs, like divers, came out. They wore some kind of helmet, which seemed to be fastened to their light colored overalls. They started to dig in the ground and the witness thought they were taking samples. Frightened the witness fled on her bicycle. **http://nawewtech.angelfire.com/soil.html**

- 1953 (early). Burragorang Valley New South Wales, Australia
Source: http://www.ufoinfo.com/humanoid/humanoid1953.shtml
Source: Rex Gilroy, Nexus Vol. 2 # 17.
ET|Soil

Following sightings of low flying objects in the area, a lone hiker at a remote location saw from a distance and hiding behind some scrub brush a group of man-like figures wearing shiny white "space suits" like outfits. The figures appeared to be searching the ground with strange metallic devices, apparently collecting soil samples. Soon after this incident, three young hikers were reported missing in the same area and to this date never located.
January 12, 1953. Santana Dos Montes, Brazil

- Source: http://www.ufoinfo.com/humanoid/humanoid1953.shtml
Source: Merseyside UFO Bulletin Vol. 6 # 3
Dome|ET|Soil

While driving over a hayfield, Mauricio Ramos Bessa saw a bright metallic object, smaller than a Volkswagen, maneuvering 5 ft above the ground. It had a flat bottom and a rounded top. He drove to within 6 ft of it. A door opened and 2 men 5-5.5 feet tall jumped out; they wore "brilliant lead" colored suits, with a shining ball on each of their square-toed shoes. A third man looked out from the doorway. One of the men held a cylinder about 5" long and 1 1/2 "thick, which he pushed into the ground and withdrew; it seemed to become shorter and rounder. Then the men walked backwards to the craft, which was now closer to the ground. The witness developed a headache so severe that he did not observe the departure of the vessel; when the pain ceased suddenly, he saw it was gone.

- April 22, 1953. Ano Ilioupolis East Athens, Greece
Source: http://www.ufoinfo.com/humanoid/humanoid1953.shtml
Source: Panayotis Skayannis
Hut|ET|Soil

The next day, news was spread that in the nearby fields of Saint Nicholas, an "aluminum hut" came down from the sky and landed. It had "windows" through which a shepherd saw incredibly ugly dwarfs inside. Among them, there was a dwarf with the head of an animal. In a few minutes, a door opened on the "hut" and a dwarf came out, holding something like a golden plate in his hands. Then, the strange woman Mrs. F had seen the previous day made her appearance. The dwarf bowed to her, and she filled the "plate" with dirt. They both went into the "hut" and the latter took off and vanished in the sky with a loud bang.
http://nawewtech.angelfire.com/soil.html

- June 28, 1953. Molfetta, Bari, Italy
Source: http://www.ufoinfo.com/humanoid/humanoid1953.shtml
Source: Il Giornale dei Misteri # 277

Disc|Turret|ET|Ground samples|Hum|Speed|Sulfur smell

Sallustio Salvemini was walking by some nearby prehistoric ruins when he noticed a small luminous object that was descending quickly to earth. It was a bright orange color as it landed about 50 meters away. The light dimmed and he could now see that it was a classic disc-shaped, flat on the bottom with a turret-like structure on top, about 2 meters in height. As its brightness diminished, the witness saw a five-foot tall figure emerge from an opening on the object. The figure wore a shiny greenish coverall, with a gold colored belt, which emitted a lantern type light; it also wore a metallic helmet. The humanoid moved quickly around the terrain, apparently picking up ground samples and placing them inside the object. At this time the witness heard a loud humming sound coming from the turret and the light on the humanoid's belt became brighter. The witness experienced a tremendous headache as the humanoid quickly entered the object, which shot away into the sky at incredible speed. A sulfur like smell remained in the area after the object left.

- May 1954. Kensington Lake, Detroit, Michigan
  Source: http://www.ufoinfo.com/humanoid/humanoid1954.shtml
  Source: http://www.ufoevidence.org/sightings
  Disc|ET|Landing Legs|Speed|Vegetation

The 5-year old witness remembers seeing a large saucer-shaped object with a round top and three skinny telescoping legs that only came out when it was about to land, it emitted no sound. It was white metallic in color. 3 beings came out of the object through a door-drop down stairway on the sandy beach land grass covered area where it landed. The three beings looked tall, thin with long arms, and large oversized heads with large black eyes, they collected samples of sand, vegetation, water and communicated telepathically with the witness telling her, "Do not be afraid, we will not harm you" before embarking back into the ship. They seemed gentle and never bothered her; they wore white suits with golf-ball like indentation design.

- July 1954. Cold Blow, Lydden, Kent, England
  Source: http://www.ufoinfo.com/humanoid/humanoid1954.shtml
  Source: Chris Rolfe, UFOMEK
  Mushroom|Hum|ET|Hover|Soil|Speed

Harold Carpenter, an employee at a local state, over a period of several weeks had kept hearing a very loud, deep humming noise, which lasted for about two or three minutes. The noise always seemed to occur between four and five o'clock in the morning. One morning, very curious, Carpenter rose from his bed, quickly got dressed and quietly stepped outdoors. With the humming noise still audible, and seeming to emanate from the southwest, Harold set off to walk to a wooded area known locally as Sunny Calvert. Determined to track down the source of the humming noise, and having come this far Harold clambered down a deep slope to where the sound seemed to be coming from. Harold peered down and was shocked to see a strange looking device. It resembled a mushroom without a stem. The object was gray in color and seemed to be hovering about eight to ten feet above the ground. Harold came to a vantage point amongst the trees where he could actually see the underside of the craft. It was definitely hovering above the ground. It was about 15 to 20 ft in diameter across the base, which housed what looked like a lip on the inside bottom. Even more startling was the sight of up to five "peculiar looking beings" gathered underneath the craft on the surface below. The beings were about four to five feet tall, and carrying what looked like "pound jam jars" with a handle in one hand and an implement resembling a pair of tweezers in the other. The beings used these to pick up bits of twigs and leaves before placing them in the clear glass looking jars. As the beings wandered around picking up various samples, all kept very close to the craft, which hovered silently above. The humming sound had now stopped. All wore the same one-piece gray; flexible looking suits that stretched over their feet. It also covered their heads much like a Balaclava. Their faces were exposed and Harold could see that these were human like but totally expressionless, with normal human like eyes. If they had any ears, the one-piece suit hid them from view. They had a snout-like nose, resembling that of a "pig." Their mouths were similar to humans, and so were their hands, until Harold noticed they didn't appear to have thumbs. While observing in fascination, Harold had inched closer and closer, until he suddenly realized that he was only to within 10 ft of the craft. At this moment one of the beings suddenly looked up and with the same expressionless face, stared directly at him. Without so much as a sound or movement, it somehow alerted the others to Harold's presence, for the rest suddenly stopped what they were doing and slowly began to retreat to the craft, which had descended to hover barely two feet of the ground. One by one the beings disappeared underneath the craft by means of its lip. The craft then began to rise, at which point Harold saw one of the beings climb through a type of "gate" housed on the inside of the lip which then closed. The craft slowly continued to climb, moving slightly to one side towards a clearing among the trees. It gradually gained altitude and speed. It suddenly emitted a flash of light and departed at incredible speed. **http://nawewtech.angelfire.com/soil.html**

- Summer 1954. St. Priest-Taurion, Haute-Vienne, France
  Source: http://www.ufoinfo.com/humanoid/humanoid1954.shtml
  Source: G. Magne, http://ufologie.net/1954
  Disc|ET|Landing Legs|Vegetation

The witness, Marie-Louise was indulging in her daily walk after having suffered a partial stroke, 3 years earlier. The day of the observation she was walking on a dirt road about 500m from her house near the small farm of La Croze. Suddenly at the edge of an alley of tall spruces and a meadow, the witness saw a very shiny flattened hemispheric object. She described the object as flattened in shape, but round, and very brilliant, the color red was dominant but she could also distinguish yellow and orange, juxtaposed and spread out in strips of light. The object did not touch the ground as it seemed to rest on a kind of tripod. It was about 1m to 1.50m in height, at the level of the fir trees near the branches about the size comparable to that of a large car. The witness then saw two small beings occupied in collecting something around the meadow. Amazed, the witness did not dare to move from the dirt road, it didn't appear that the entities had noticed her. The witness described the beings as very small, 1 m or less, with a very bulky head (perhaps a helmet?) and of a metal gray color. After a few minutes, the humanoids penetrated under the UFO apparently carrying some vegetation with them. The machine rose without noise, and while trembling, to the summit of the spruces. There it very quickly took off disappearing from sight.

- November 1, 1954. Cennina, Italy
  Source: http://www.ufoevidence.org/cases/case703.htm
  Source: Sergio Conti, excerpt from FSR (Sept/Oct 1972)
  Spindle|ET|Landing Feet|Landing Mark|Plant Samples

Rosa Lotti nei Dainelli. Approaching Rosa Lotti, who was now terrified, they snatched out of her hands the bunch of carnations and one of her black stockings. When she remonstrated timidly, the one who seemed the older of the two handed some of the flowers back to her, but kept five of them. Then, having examined the structure of the flowers with an air of curiously, and laughing the while, he wrapped them in the stocking and threw them into the "spindle" through the small opening.

- November 4, 1954. Amacuro Delta area Venezuela
  Source: http://www.ufoinfo.com/humanoid/humanoid1954.shtml
  Source: Apro Bulletin 11/54
  Sphere|ET|Hover|Soil

Fernando de Moya, 39, hunting on his launch, saw a luminous sphere hovering 6 ft above the ground. The 2 Indians with him fled in terror, but he approached and saw 2 3-foot tall figures with round heads & large eyes running back & forth to the sphere, with little jumps, picking up various things. They entered through one of 2 windows in the UFO. When they saw De Moya, they jumped into the object, which ascended vertically.

- November 4, 1954. Pontal Brazil
  Source: http://www.ufoinfo.com/humanoid/humanoid1954.shtml
  Source: Olavo Fontes, Apro Bulletin 7/57
  Disc|ET|Foliage samples|Wobble

Jose Alves was fishing in the river Pardo near the location. It was a quiet night; the spot was deserted. Suddenly he saw a strange craft approaching with a wobbling motion, and it landed so near him that he could have touched it. It had the shape of "two washbowls placed together" and was between 10 and 15 feet in diameter. Too terrified even to escape, he watched three little men in white clothing and close-fitting skull caps emerge. Their skin appeared to be quite dark. They gathered samples of grass, herbs, and leaves of trees. One filled a shiny metal tube with water from the river. They then re-entered and the object rose swiftly silently and vertically and vanished. Jose Alves had never heard about flying saucers and thought he had seen "some kind of devils".

- November 14, 1954. St Maudan, France
  Source: http://www.ufoinfo.com/humanoid/humanoid1954.shtml
  Source: LDLN # 325 Type: B
  Boat|ET|Fruit

A 12-year old girl sees a luminous boat shaped craft land on a nearby field. Three hairy humanoids emerged from the object and began to gather apples from a nearby orchard. No other information. **http://nawewtech.angelfire.com/soil.html**

- November 14, 1954. Ortonovo, La Spezia, Italy
  Source: http://www.ufoinfo.com/humanoid/humanoid1954.shtml
  Source: Maurizio Verga "When Saucers Came to Earth"
  Cigar|ET|Rabbit

A 48-year old farmer, Amerigo Lorenzini, was placing some grass in a rabbit hutch just outside of his home. All of the sudden, Lorenzini heard a swishing sound (like a flock of swallows make). Interspersed with a sort of a metallic noise, coming from above him. Gazing up, Lorenzini was blinded by a dazzling light, and then he eventually passed out until the light went out. At this point, the farmer observed a cigar shaped object which had landed in a nearby clearing. The object was surrounded by a luminous halo. A hatch opened on one side of the object, and a group of smallish creatures emerged from it. The beings moved towards the kitchen garden of Lorenzini's house---showing no signs of indecision as if they were familiar with the location. Lorenzini was taken aback by what he saw, but he retreated to the house door. From there, nearly hidden from sight, he got a look at the very small beings wearing a metal diving suit like uniforms that covered even their faces. The creatures spoke with one another in low-pitched voices, paying no care to the farmer. They approached the rabbit hutch and seemed interested in the rabbits within. Lorenzini interpreted their gestures as indicative of their intent to carry all the rabbits away. Lorenzini went into the house, retrieved a rifle, went back outside and aimed the rifle at the creatures. He pulled the trigger, but the rifle did not fire. Lorenzini attempted once more, but he was still unsuccessful. At this point, he had to drop the rifle because it had become too heavy to hold. He became frightened and tried to shout---however, he was unable to accomplish even this. The farmer observed the creatures seizing all the rabbits, returning to the object and then entering it. The cigar-shaped craft took off very rapidly in the direction of Avenza, leaving in its wake a very bright trail, even more, intense than the halo that had surrounded the object while it was on the ground. Lorenzini managed to pick up the rifle and to fire it successfully, but the object was already long gone. Although in shock Lorenzini called out to some people, to whom he pointed out the emptied rabbit hutch.

- November 28, 1954. Caracas, Venezuela. (18)
- November 28, 1954. Petare, Venezuela.
  Source: http://www.ufoinfo.com/humanoid/humanoid1954.shtml
  Source: Apro Bulletin, 11/54
  Sphere|ET|Hairy|Soil

Gustavo Gonzalez, Jose Ponce. Ponce saw 2 others emerging from the bushes with soil in their hands. They jumped up into the sphere and shone a blinding light at Gonzalez.

- December 9, 1954. Linha Bela Vista, Rio Grande do Sul, Brazil
  Source: http://www.ufoinfo.com/humanoid/humanoid1954.shtml
  Source: Hanlon, The Humanoids
  Disc|ET|Hover|Speed|Plant samples

Farmer Olmira da Costa e Rosa was cultivating his crops of French beans and maize at this place, 2 ½ miles from Venancio Aires when he heard something 'like a sewing machine', and animals in a nearby field panicked. He then saw an object, shaped like a bowl or explorer's hat, cream colored and enveloped in a smoky haze. It was hovering just off the ground, and three strange looking men were there, one inside the craft, one examining a barbed-wire fence, and one close to the farmer. In astonishment, the farmer dropped his hoe, and the man smiled, approached, picked up the hoe, examined it carefully and handed it back to him. He then bent down plucked a few plants and walked back towards the machine, with the others. The craft rose slowly to about thirty feet, then accelerated, and flashed away towards the west at high speed. The farmer, almost completely illiterate, had never heard of 'flying saucers'. He was able to study these men at close range and in great detail. They were of medium height, broad-shouldered, with long blond hair blowing in the wind. With their extremely pale skin and slanted eyes, they were not normal looking by Earth standards. Their clothing consisted of light brown coverall garments fastened to their shoes, which were heelless. He concluded that they must be aviators from some foreign country. **http://nawewtech.angelfire.com/soil.html**

- December 11, 1954. Linha Bela Vista, Rio Grande do Sul, Brazil
  Source: http://www.ufoinfo.com/humanoid/humanoid1954.shtml
  Source: Hanlon, The Humanoids
  Disc|ET|Hover|Oscillating Motion|Plant sample

The farmer Pedro Morais heard a commotion among his fowls and looking around for a hawk, beheld an object 'that had a bottom like an enormous polished brass kettle'. It was hovering, with an oscillating motion, and making a noise like a sewing machine. Its upper part 'looked like the hood of a jeep'. In a nearby cultivated field, he next noticed two small human shaped figures. He could see no faces, for they seemed to be enveloped in a kind of yellow sack from head to toe. Indignant at this trespass on his crops, he headed for them. One of them began to run towards him, while the other raised his arm in what seemed to be a warning gesture to keep away. One of them then knelt down and plucked a tobacco plant from the ground, and both entered the craft, which vanished from sight in a few seconds.

The farmer, totally illiterate, had never heard of flying saucers or science fiction and thought the creatures were ghosts. When told that the Brazilian government was anxious to get one of these little men dead or alive he vowed he would shoot one if he got the chance.

- December 15, 1954. Near Campo Grande Brazil
  Source: http://www.ufoinfo.com/humanoid/humanoid1954.shtml
  Source: Col. Adil de Oliveira, Chief of Brazil's Air Force Intelligence
  Sphere|ET|Hover|Landing Gear|Soil

The witness, fishing in a river near his home, saw an unusual craft land a few hundred feet away. His dog began to howl. Through the telescopic sight of his gun, he could see a small sphere revolving around a larger one, which hovered 6 ft above the ground. At the bottom was a 3-ball landing gear. 3 occupants, human looking but quite small, emerged & came down to the ground; they were agile, with rapid movements. One held a luminous basket, the other a metal tube with a cone shaped end; with those they collected calcareous material from the river's edge, sucking it up into the tube & discharging it into the basket. They then got back into the object, which took off at high speed. The witness found square holes in the ground.

- December 19, 1954. Valencia Venezuela
  Source: http://www.ufoinfo.com/humanoid/humanoid1954.shtml
  Source: Apro Bulletin 1/55
  Disc|Beam|ET|Hover|Rocks

18-year old jockey Jose Parra, out on a training run, came upon 6 very hairy little men loading rocks into a disc-shaped craft which was hovering less than 9 ft from the ground.
http://nawewtech.angelfire.com/soil.html

- April 24, 1964. Tioga City, New York (19/P.104)
  Ellipse|ET|Soil

Gary T. Wilcox. Suddenly, however, two small humanoids appeared standing next to the craft. They were dressed in seamless uniforms with hoods covering their faces. Each carried a tray that appeared to be heaped full of earth that had been removed from his field.

- November 11, 1965. Mogi-Guacu, Brazil
  Source: Flying Saucer Occupants by Coral and Jim Lorenzen
  ET|Foliage Gathering|Glow|Foilage

Dario Anhaua Filho. The small figures picked up twigs and branches and appeared to examine a mare which was standing by the fence.

- April 14, 1967. Near Melville, New York
  Source: http://www.nicap.org/occupants_hall.htm
  Source: http://www.ufoinfo.com/humanoid/humanoid1967.shtml
  Source: Mark Rodeghier, UFO Reports Involving Vehicle Interference
  Sphere|Glow|ET-Robot|Interference|Soil|Whirring

A motorist saw a glowing object overhead, suddenly his car engine stalled, and a smaller circular metallic object landed beside the road. His car radio, which had been turned off, began to broadcast in a strange language. A tiny metallic robot-like figure appeared in the doorway of the object. It then dug up some dirt and placed it inside the craft. The doorway then closed, the object then turned a bright red color and rose into the sky emitting a whirring sound. It appeared to join the large glowing craft overhead.

- August 3, 1967. Caracas, Venezuela
  http://www.nicap.org/occupants_hall.htm
  Disc|ET|Hover|Glow|Gravel samples

One 3-ft tall being emerged from hovering luminous disc, took gravel samples.

- August 23, 1967. Joyceville, Ontario Canada
  Source: http://www.nicap.org/occupants_hall.htm
  Disc|ET|Landing Legs|Specimens

Stanley Moxon. Three 3- to 4-ft tall beings, white garb, translucent helmets from disc with three legs, gathered specimens. **http://nawewtech.angelfire.com/soil.html**

- August 29, 1967. Cussac, France
  Source: http://www.ufoevidence.org/cases/case705.htm
  Source: Joel Mesnard and Claude Pavy, excerpt from FSR, Sept/Oct 1968
- Sphere|Animal|ET|Landing Legs|No Seams|Soil|Sulfur Smell|Whistling Noise

Francois Delpeuch (13) and his sister Anne-Marie (9). One of the little beings is bending over and seems to be busy with the soil, while another, holding in one hand an object which reflects the sun (which Francois likened to a mirror), is waving his hands, apparently making signs to his companions.

- April 1970. Near Madison, Wisconsin
  Source: http://www.ufoinfo.com/humanoid/humanoid1970.shtml
  Source: Eric Norman, Saga Magazine May 1973
  Abduction|ET|Landing Legs|Specimens

Raymond Shearer. Under hypnotic regression, later, it was revealed that he had been taken onboard the UFO and had had an extensive conversation with the occupants, who were human-like and wore white coverall type uniforms with black belts, and who were "collecting specimens."

- June 1, 1972. Near Molong, New South Wales, Australia
  Source: http://www.ufoinfo.com/humanoid/humanoid1972.shtml
  Source: NUFORC

The main witness along with her mother and 2 of her friends was driving to the village of Collage. As she sat in the back seat she looked out the window to her right and noticed a bright shiny object, very large. She thought that if you were not looking in that direction you would never have seen it. It was saucer shaped with a dome on top, alongside of it on the left was a very tall person dressed in an outfit which glowed the same as the ship, another similar figure was walking towards the road (the craft was about 30 meters away from fence line, which was 5 meters from the road) and another one was standing on the right side of the object near a dead tree, all had the same clothing on and helmets with black visors. She described the aliens as being very tall, approximate 8 ft in height. As the car drove by the witness noticed that they seemed to be collecting items from the ground. She also noticed that the one walking towards the road seemed to have turned his helmet and look at the vehicle as it drove off. The witness never did tell anyone else in the car what she had seen.

- August 25, 1972. Fort Atkinson, Wisconsin
  Source: http://www.ufoinfo.com/humanoid/humanoid1972.shtml
  Source: Apro Bulletin, Nov. Dec. 1973

Steve Cleveland, a carnival worker from Eau Claire, called radio station WLS in Chicago to report that as he had been sitting on his suitcase outside of Fort Atkinson, hoping to hitch a ride to Eau Clare, he had seen "a huge ship" come down in a nearby field, out of which came two beings about 5 ft tall. They took some samples, & got back in the object, which then lifted off. He said he watched the incident for 15 minutes. **http://nawewtech.angelfire.com/soil.html**

- January 12, 1975. North Bergen, NJ.
  Source: http://www.nicap.org/occupants_hall.htm
  Disc|ET|Hover|Ground

George O'Barski. Eight to ten 3'/i-ft. (3 1/2 ft?) -tall beings with dark garb, helmets, emerged from underside of low hovering disc with lighted windows, took samples from ground.

- February 14, 1975. Petite-He, Reunion, France
  Source: http://www.nicap.org/occupants_hall.htm
  Disc|Dome|Beam|ET|Injury|Ground

Antoine Severin. Four 1- to 1.2-meter-tall beings, metallic white garb, helmets, from domed disc with portholes, directed light beam at witness, who felt a force pulling him, later had headaches and impaired vision; beings poked in ground as if gathering samples.

- July 1977. Jussaral, Brazil
  Source: http://www.ufoinfo.com/humanoid/humanoid1977.shtml
  Source: Bob Pratt, UFO Danger Zone
  ET|Ground|Landing Mark

Local villagers reported seeing an object come down to the ground, then they saw two or three small figures come out from under it and get something off the ground. The light from the object was as bright as the sun. After the object left, marks were found on the ground.

- 1978. Northeast of the island of Grenada
  Source: UFO Digest - Sir Eric Gairy: Prime Minister of Grenada His UN UFO Meeting And His E.T. Secret by Wesley H. Bateman
  Crash|Debris|ET|Specimens

Sir Eric Gairy: Prime Minister of Grenada. The reef mentioned above is located to the northeast of the island of Grenada. The fact that Sir Eric said that one of the alien metal boxes that was found on the beach with the alien body contained specimens of fish and sea plant life does bring to mind the possibility that the alien met with a disaster after visiting this unique marine reef. Or maybe he was part of an expedition to restock the reef.

- October 8, 1978. Ord, NE
  Source: http://www.nicap.org/occupants_hall.htm
  Disc|Beam|Buzz|ET|Hover|Plant samples

Deanne Kearns. Two 5 to 6 ft. tall beings descended in light beam from hovering UFO shaped like one plate inverted on top of another, loud buzzing sound; beings collected earth and plant samples. **http://nawewtech.angelfire.com/soil.html**

- January 1992. Alegrete, Brazil
  Source: http://www.ufoinfo.com/humanoid/humanoid1992.shtml
  Source: GEPUC Brazil
  Globe|ET|Soil|Speed|Ground samples

A luminous globe shaped craft landed in a wooded area near a police station. Four humanoids emerged from the object. These were described as 1.45 meters in height, very large heads, as in proportion to their bodies, short arms, and huge staring eyes. They wore tight fitting metallic objects. The humanoids collected ground samples and vegetation and then re-entered the object, which took off at high speed.

- January 31, 1996. Near Romatambo Peru
  Source: http://www.ufoinfo.com/humanoid/humanoid1996.shtml
  Source: UFO Roundup Vol. 1 # 15
  Collosal|ET|Soil|Speed

Silvia Bedoya. These were described as 3 feet, 3 inches tall, with large oversized heads, long spindly arms and short bandy legs. They wore gunmetal gray helmets and matching one-piece coveralls. The humanoids took out transparent containers and, ignoring the shepherds, collected many samples, like soil, grass, mountain weeds, & water from the nearby river. Soon the humanoids entered the large object and all the objects shot away into space @ high speed.

- March 7, 1997. Magog, Quebec, Canada
  Source: http://www.ufoinfo.com/humanoid/humanoid1997.shtml
  Source: Canadian UFO Survey 1997
  Disc|ET|Hum|Soil|Landing Trace

A witness encountered a humming silver disc-shaped object on the ground. Three short beings only 30 cm tall were seen next to the object. The beings wore silvery-white inflated suits, with round helmets and appeared to be collecting items from the ground around the object. Ground traces and electromagnetic effects were reported. No other information.

- Summer 2000. Svetloye, Kaliningrad region, Russia
  Source: http://www.ufoinfo.com/humanoid/humanoid2000.shtml
  Source: Denis Rozhnov, "Komsomolskaya Pravda in Kaliningrad" February 27, 2004, # 37.
  Disc|Animal|ET|Landing Legs|Soil

Several meters from the disk three entities appeared to be taking soil samples. One of the aliens who was apparently the leader was about 2 meters in height and was standing apart from the other two, which were much shorter than he was and were actually collecting the samples. The humanoids were bending down in ways utterly impossible for normal humans; their bodies appear to be much flexible. **http://nawewtech.angelfire.com/soil.html**

- July 1, 2001. Paracambi, Rio De Janeiro, Brazil
  Source: http://www.ufoinfo.com/humanoid/humanoid2001.shtml
  Source: Thiago Luiz Ticchetti EBE-ET
  Sphere|Cupola|ET|Speed|Vegetation

Odete Fortini. On the center of the object a door opened and two humanoids about 1 meter in height stepped out. They seemed to be interested in the vegetation around the object. After a few minutes of investigating the area, they re-entered the object, which became bright again and left at high speed.

- August 1, 2001. Fortaleza, Ceara, Brazil
  Source: http://indianamufon.homestead.com/2001Archive.html
  Source: Thiago Luiz Ticchetti, EBE-ET
  Disc|ET|Landig Trace|Speed|Ground samples

There were five geography students as witnesses. Suddenly a small door opened and two short humanoids came out. The object and the humanoids were about 200 meters away from the witnesses. The little men seemed to be collecting ground samples when they became aware of the witnesses presence. They ignored the witnesses and after several minutes reboarded the object, which took off at high speed. Deep ground traces were found at the landing site.

- December 11, 2009. Monahans, Texas
  Source: http://mufoncms.com/
  Oval|Animal|ET|Glow|Hover|Interference|Speed|Ground samples|

As we stood there in total shock, my brother next to me expressing his disbelief, we saw two strange looking "Men" appear from around the object. They at first didn't notice us, they were busy running around and gathering objects from the ground, from the trees. They didn't seem to notice us at first. They looked to be about the size of my oldest nephew, about 4 feet tall, but they looked very skinny like they had no meat on their bones. Their heads were large and their arms were long, skinny and hung down around their knees.
http://nawewtech.angelfire.com/soil.html

It obvious from the above examples that this sample collecting from the multi-strata of the biosphere wasn't just because the life forms on Earth were different, unlike anything on their own planet, although, this may be possible, but rather, these Extraterrestrial intelligences were interested in preserving the uniqueness of the rich biodiversity of this planet. In much the same way that Earth scientists have been collecting seeds and storing and preserving them in temperature controlled seed vaults around remote locations on the planet. So too, have ETI been doing the same but, on a much grander scale which includes animal life and human DNA. This then would account for the unusually large number of people throughout the United States and in other countries around the globe who have reported personal accounts of alien abductions.

When our biologists fly over different parts of the country, they may shoot and tranquillize a small population of animals and attach radio transmitters to them so that they may be easily located at some future time. This is not to say we are being treated in the same manner as the

198

animals we tranquillize and tag but, there are those Ufologists like **Hopkins** and **Jacobs** who believe that this is exactly what is happening to some people. The question then, is why do ETs who come to this planet utilize the same technology that we employ upon animal life forms? Why is it that not all humans who have an ET encounter with another intelligent species, never undergo the traumatizing experience of a medical examination, etc. unless, these types of abductions which are often repeated upon some of the same individuals are in reality a part of the military abductions and false flag covert agenda that is being perpetrated upon an unsuspecting populace?

Add into this scenario is the overflights of military bases and nuclear weapons installation sites by Extraterrestrial spacecraft as publically disclosed by **Robert Hastings**, former US Air Force officer at the **National Press Club** in Washington back in September 2010. Though Hastings doesn't appear to know the reason for these over-flights by UFOs, he states that their presence is very real and of major concern to the Department of Defence both here in the US, but also, in countries like Russia, Britain, France, China, India, etc. <http://www.ufohastings.com/>

There is a real concern among Extraterrestrial Intelligences that our immature human race has armed itself with **weapons of mass destruction (WMDs)** and with a climate of political instability that raises its ugly head at any moment on the planet, the ETI fear a real potential for global destruction that could wipe out all life upon this planet., thus their need to preserve the indigenous life forms on this planet including humanity!

Dissecting the Alien Abduction Scenario

From the number of different Extraterrestrial beings that are visiting the Earth which vary from insect-like beings to humanoids, reptiles, humans (who could be our cosmic cousins), to exotic and robotic beings, very few species are actually involved with any abductions of humans. Apart from the occasional insect-like and human type interventions or partnership with other humanoid beings, it appears, that it is the humanoid type beings described as the Greys that are behind the abduction of humans.

The **abduction scenario** can be broken into precise well-orchestrated events although, not all phases have been routinely reported in each abduction case:

1. **Capture**. The abductee is forcibly taken from terrestrial surroundings to an apparent alien spacecraft.
2. **Examination**. Invasive medical or scientific procedures are performed on the abductee.
3. **Conference**. The abductors speak to the abductee.
4. **Tour**. The abductees are given a tour of their captors' vessel.
5. **Loss of Time**. Abductees rapidly forget the majority of their experience.
6. **Return**. The abductees are returned to earth. Occasionally in a different location from where they were allegedly taken or with new injuries or disheveled clothing.
7. **Theophany**. The abductee has a profound mystical experience, accompanied by a feeling of oneness with God or the universe.
8. **Aftermath**. The abductee must cope with the psychological, physical, and social effects of the experience.

First, the methodology of abductions or **capture** was initially reported as occurring by coercing or the outright physical capture of humans from their homes or from their vehicles, wherever the opportunity presented itself to the ETs. This method was soon refined and the method of abduction became the extraction of the astral body from the physical form whereby, people would describe leaving their bodies while accompanied by small alien beings, the Greys. The consciousness astral body would be floated through a wall or window or up through the roof and on board their spacecraft.

Second, in a physical abduction where the human victims are no longer in control of their psycho-motive abilities and once on board the spacecraft, the **examination** procedure begins where most often the abductees are stripped of their clothing. They are then, either in the physical or astral state placed upon a horizontal platform or table which was described as being one continuous functional part of the ship, as if it grew out of the floor of the craft. Here a medical examination and various procedures were performed upon the abductee. The individuals were examined from head to toe, but primarily, the reproductive system seems to be the main focus of the medical examination

As stated before, skin samples were taken leaving small red dot-like scars or small scoop marks on various areas of the body which may eventually disappear over a period of time or not and of course, ova would be taken from females and sperm from males. Other procedures may be the insertion of a number of tracking devices in different parts of the body e.g. hands, feet, head, and lower abdomen areas to keep tabs on and location of the abductee.

Depending on the person claiming the abduction experience, individuals may be abducted many times during their lifetime or over generations in the same family. Many women described becoming mysteriously pregnant for the first trimester and then, mysteriously losing the fetus but, not through any intentional abortive event. Women who were abducted multiple times described seeing a hybrid child who is human but also, alien in appearance much like those paintings that were painted back during the seventies of the very large blue-eyed children with slightly larger heads perhaps, with very light blond or wispy hair. ***Do you think the artist was trying to tell us something of the subject matter?*** [My bold italics] These women would tell stories of how these small Grey beings would present these small hybrid children to these women to be held and create some type of bond between them or an acceptance of them.

Third, after the medical procedures, the individual in **conference** fashion between human and ET would sometimes be counseled or shown different things related to Earth's near future. Abductees sometimes claim like their predecessors, the contactees before them to have been warned against environmental abuse and the dangers of nuclear armament.

Fourth, a **tour** is given to the abductee almost as a reward for their unsolicited corporation in the abduction event by viewing various parts of the spacecraft, or a trip and a glimpse of outer space, etc.

Fifth, in most cases, if not in all, the memories of the event would be erased or masked so that the individual would not remember anything then, the human would then be physically or astrally **returned** to their place of origin.

200

Sixth, if a car was the point of departure which was en route from one location to another location, then the abductee may have experienced **"missing time"** which they could not account for.

Seventh, by the time the abductees are returned to their homes, many report the **theophany** or **mystical union**, a sense of understanding and forgiveness toward the kidnappers for what had happened to them and sometimes, associated feelings of oneness with God or the universe. One would see this as a parallel experience with human abductions or kidnappings that are commonly known as the Stockholm Syndrome (disambiguation), where the hostage identifies irrationally with their kidnappers such as in the case of Patty Hearst and the Symbionese Liberation Army.

Eighth, in the **aftermath** of the abduction event, many of these purported encounters are described as terrifying, some have been viewed as pleasurable or transformative. Many abductees were initially in fear and panic when confronted by their alien abductors, then feeling dread and anger for their personal safety during the intrusive medical examination and finally relief and curiosity as the ETs would try to calm them down and explain a few of their mysteries or counsel and admonish them about potential dangers that threaten humanities existence. Often abductees would report black unmarked military helicopters overflying their homes after an abduction experience, sometimes in harassment type fashion.

Most importantly, what is truly remarkable about these abduction stories is that although, some memories can be recalled consciously, most of them were recounted or recalled while under hypnotic regression by either a professional psychologist or by some UFO researcher trained in hypnosis. Over a number of hypnotic therapeutic sessions, the full story of the abduction event would eventually be disclosed.

An entire subculture has developed around the subject, with support groups like **Budd Hopkin's** *"Intruders Foundation"* and a detailed mythos explaining the reasons for abductions such as certain Extraterrestrial species involvement, hybridization and duplicitous government involvement with cover-ups of the alien agenda. Now, there are many finer details of the abduction event but, at this time only the salient points of abduction will be examined here.

According to some, but not all UFO researchers, a working abduction theory based on the raw data, speculation, analysis and deductions, leads Ufologists to an inescapable conclusion, that we are being manipulated against our wills in an alien agenda to produce a race of hybrid beings for the purpose of co-existence with us or to eventually replace us as the dominant species on this planet.

This particular perception has been ballyhooed and promoted through Ufology, the news media, and the Hollywood entertainment industry as the main reason why ETs are visiting our planet and the main reason why ETs are viewed with suspicion, fear, and outright racial indifference and prejudice.

Quite frankly, this type of alien-human interaction as claimed by many abductees is for the most part, negative and traumatic and would literally scare the be-Jesus out of most people, who know little to nothing about the ET intelligence subject matter.

It should be understood that an abduction is a very specific type of criminal act committed by one or more humans against another human and is associated with the act of kidnapping. Abduction and kidnapping are terms that are commonly used interchangeably, even though, they have similar meaning, they are actually different criminal acts. A person is kidnapped by force or by fraud and is usually held prisoner against their will as a hostage i.e. a political hostage for example, or for ransom. The outcome is frequently the release of the individual if ransom or hostage agreements are met but, this may not always be the case. When the person is not returned because the kidnapper's ransom conditions were not met, the hostage may be taken away beyond any hope of rescue, living out their life as a prisoner or possibly meeting an untimely death by murder then, the act of kidnapping becomes an act of abduction.

When an abduction occurs much in the same way as a kidnapping, it may be more secretive or coercive by which the victim is lead away, however, the outcome is in most cases is very negative. As a rule, the abductee is never returned to their home or to their loved ones. The motive is not for any political or financial gain but, for some other reason which may never become known.

Thus, in the mind of this writer, aliens who return their victims to their homes after a so-called abduction have in reality performed only an act of kidnapping against the will of the individuals. If or when a true alien abduction is perpetrated against some hapless human, who happens to be in the wrong place at the wrong time, ***they are never returned to their home, safe and sound. The individual is gone for good***, for reasons that are unknown and which will remain a mystery, unless, the alien perpetrators have decided to reveal their intentions which may become a retelling of a contactee account. This then is a real abduction event with a very sad ending.

Even, if the person is gone for good, never to be seen again, can it be said he was abducted and met an untimely demise or could the individual have gone willingly because he was given an unprecedented opportunity to participate in something extraordinary, a once in a lifetime opportunity? In other words, can we be sure that what would appear to us as an abduction, a negative event, be exactly that or something else? Here then, is the flaw in human thinking! We perceive all things in an anthropocentric paradigm. We associate human qualities and attributes to all things which may be fine for most things here on planet Earth but, we are dealing with alien life forms which as far as we know, do not behave within that same anthropocentric paradigm. After all, we have yet to greet and establish a meaningful relationship with any Extraterrestrial visitors to our planet so; we cannot hope to know the mind or the intentions of another interstellar intelligence until that historic day finally arrives.

There are as many opposing theories to the abduction phenomenon as there are with the cattle mutilations phenomenon. The **Extraterrestrial Hypothesis** is an obvious theory supported by many researchers in Ufology which is dismissed by academics as being pseudoscientific. Skeptics and mainstream scientists in typical fashion reject reports of people literally being kidnapped and then, subjected to forced medical examinations by non-human creatures. Reports of such abduction events could not be occurring because in the minds of the skeptics and scientists, the alternative of accepting the possibility of alien abducting humans is one that upsets their paradigm of reality. The comfortable notion that we are the only intelligence in the universe or that ETs are visiting the planet leaves scientists powerless to do anything to prevent or stop abductions. It's the notion of ET visitations that upsets the whole socio-geo-political structure on the planet (or is it merely the status quo of the economic power structure of a few nations or a few wealthy corporate elitist?), at least that is the current theory held by the skeptics.

The Skeptic's Viewpoint

"Skeptics propose theories and explanations that are more down to earth than the **ET Hypothesis** that are centered on known psychological processes that can produce subjective experiences similar to those reported in abduction claims. Skeptics are also likely to critically examine abduction claims for evidence of hoaxing or influence from popular culture sources such as science fiction.

Psychological Hypothesis

Some of these **psychological hypotheses** included hallucination, temporary schizophrenia, epileptic seizures and parasomnia—near-sleep mental states (**hypnogogic states**, night terrors and sleep paralysis). **Sleep paralysis**, in particular, is often accompanied by hallucinations and peculiar sensation of malevolent or neutral presence of "something," though usually people experiencing it do not interpret that "something" as aliens. Occasionally the abduction phenomenon is also theorized to be a confused memory of past events (such as sexual abuse)."
http://en.wikipedia.org/wiki/Alien_abductions

From the psychological perspective, **Dr. Michael Persinger** argues that most of the features of the abduction phenomenon can be explained as the manifestation of measurable functions of the human brain. Persinger explores the UFO and ET visitation experiences from the perspective of modern neuroscience. The average visitation experience attributed to an alien entity he states is indistinguishable from average mystical or religious experience attributed to gods and to spirits. His research has been to isolate those areas of the brain and those electromagnetic patterns within the brain that are involved with the general visitation experience.

He goes on to argue that "Nearly every basic element of mystical, religious, and visitor experience has been evoked with electrical stimulation" of test subjects' brains. Individuals with some forms of epilepsy often experience vivid hallucination, and Persinger suggests that the same areas of the brain are activated in these individuals as in those who experience extraordinary visitations.

This does not indicate that all people who report visitor experiences associated with UFOs are undiagnosed epileptics or that the phenomena will cease when with this particular medication. Instead, it indicated that well-formed and meaningful experiences, attributed to alien sources and sufficient in magnitude to disrupt the person's sense of self and adaptability, can be associated with periods of electrical activity that can be affected by treatments not typically associated with these types of experiences. Persinger suggests that while underlying neurological factors give the experience its basic form, how such events are interpreted is shaped by cultural factors: "Because human brains are more similar than they are different, the themes of these experiences have been and remain remarkably similar across space and time. The details are simply punctuation from the person's culture."
http://en.wikipedia.org/wiki/Perspectives_on_the_abduction_phenomenon

Recreational Drugs Hypothesis

Other theories include that some alleged abductees may possibly be mentally unstable or under the influence of recreational drugs. This is the **Recreational Drugs Hypothesis.**

Author Rant: I know of such a case in Vancouver where the individual has some mental problems, uses recreational drugs and claims that ETs have abducted him on more than one occasion yet, this individual would also exhibit lucid, rational and acceptable social behaviour when he wasn't high.

Paranormal and Mythos Hypothesis

In explaining the **Paranormal and Mythos Hypothesis**, the late astronomer, **Carl Sagan** points out the somewhat paranormal aspects and similarities in his book *"The Demon-Haunted World"* and in a minor piece in Parade Magazine (1993) that the alien abduction experience is remarkably similar to tales of demon abduction common throughout history. These are similar to the stories and reports that Jacques Vallee tells in his book *"Passport to Magonia"* of **elves, faeries,** and **leprechauns,** who would interact with human and manipulate them but, who would also, appear to alter their appearance and "modus operandi" always staying several steps ahead of any human investigation.

"Various authors, including Jacques Vallée and John Mack, have suggested that the dichotomy 'real' versus 'imaginary' may be too simplistic; that a proper understanding of this complex phenomenon may require a re-evaluation of our concept of the nature of reality. While some corroborated accounts seem to support the literal reality of the abduction experience, others seem to support a psychological explanation for the phenomenon's origins. Jenny Randles and Keith Basterfield both noted at the 1992 MIT alien abduction conference that of the five cases they knew of where an abduction researcher was present at the onset of an abduction experience, the experiencer "didn't physically go anywhere." http://en.wikipedia.org/wiki/Alien_abductions

Hypnosis has often been criticized as being frequently unreliable as most alien abductees recall their abductions through hypnosis thus, the vast majority of evidence for alien abduction is based on memories 'recovered' through hypnosis. It has been demonstrated that false memories are often very easily created and that hypnosis can unintentionally aid in confabulation. Some abductees, however, report vivid, detailed accounts without hypnosis.

Budd Hopkins counters this notion by stating that not all psychotherapists consciously or unconsciously implant abduction accounts by hypnosis, who are otherwise, viewed by skeptics as "manipulative practitioners who 'believe in' the reality of such events. Case in point is the Hill abduction account, where psychologist Simon, who hypnotized the Hills was an avowed skeptic and remained so at the end of the treatment, felt strongly that the Hill memories could not be literally true even when the Hills stubbornly held to their hypnotically recovered accounts. Therefore, Hopkins concludes the bias of the hypnotist had nothing to do with the content of their hypnotic recall.

From what we have seen so far, there are real physical, psychological and psychical or paranormal elements associated with the abduction phenomenon. Something real is going on and it is interacting with the abductees on many levels with many hypotheses overlapping or intersecting with each other. It would seem that everyone who investigates this subject has part of the answer but not the whole answer or the "big picture." When a subject matter of this importance and controversy comes to the attention of the general public and mainstream science, all possibilities and theories need to be considered and scrutinized in the search for truth. Sometimes, the truth we discover may be unsettling and not what we want to hear at that moment but, it will be undeniable and not easily covered up or suppressed.

As was already stated, "There is something rotting in Denmark!" Something is fishy and not all the evidence adds up, nor does the evidence of omission seem to justify the current perception of alien abductions.

Military Abductions (MILABs) Hypothesis

An alternative theory to the current alien abduction model and the other various hypotheses is that of **Military abductions (Milab),** where the military industrial complex or some branch of it is behind most of the abductions scenarios using sophisticated mind control techniques to create or hoax an alien threat. The Milab model presents a strong argument in contrast to the alien abduction hypotheses and is certainly convincing many UFO researchers and the general public that not everything as it appears and a closer scrutiny of this hypothesis is well warranted.

Given the fact the military industrial complex is implicated in the whole UFO/ETI phenomenon, is it possible that their duplicity extends into all manner of covert activity and advanced technology that is associated with the black world of **"unacknowledged special access projects" (USAPs)?** We know that the privatized industrial sector has abundant financial resources and the advanced technology to develop and build most of the weapons for the U.S. military forces, employing some of the most brilliant minds on the planet in their "think tanks" and R&D programs. When the military forces from the Army, Navy or Air Force or from any of the intelligence agencies, whether independent or as a department of the arm forces require a new or improved weapon ("big toys for big boys"), the war and weapons industrialists are only too happy to supply them with what they need.

Could an arm of military intelligence (there's an oxymoron!) in order to fulfill some agenda, abduct people and perform the same type of medical examinations and procedures as described and attributed to ET Intelligences? The answer is a resounding yes! Even, when we consider the Betty and Barney Hill case, the technology already existed, right down to the actual medical biopsy procedures and the implantation of electronic tracking devices.

Sedating a person with a tranquilizer dart or a gas would be no problem in order to transport the individual to a medical centre. **Psychotropic (hallucinogenic drugs)** along with "**screened memories**" from hypnosis would give the illusion of having an out of body experience and silent stealth type helicopters would easily convince a drug induced abductee and anyone else for that matter, that they were on board an alien spacecraft, all this could easily be staged. People with the appropriate expertise along with props, costumes, staging areas, whatever resources would be required to convince the abductee that he had been kidnapped by aliens would be used by this covert intelligence group.

This basic level of mind control where people and equipment necessary to carry out a hoaxed alien abduction to convince someone that his reality has undergone a severe transformation would be considered in the estimation of the military intelligence, a crude theatrical and amateurish attempt. When you have almost limitless financial resources and a scenario that is critical to an overall master plan, you as the military industrial complex do not do anything half way or amateurish. You use the best equipment that technology has to offer and if you are fortunate enough to get your hands on actual alien technology, you would use it leaving no doubt in the mind of the abductee that he had experienced a genuine alien encounter and abduction scenario. The military industrial and military intelligence sectors are professional in everything they do, leaving nothing to chance. So if they are going to carry off an alien abduction scenario that would bring the house down with ovations then, you know that they will go all out to get every detail of the staged event as realistic as possible. Common amongst abduction claimants is the fact that one or more members of a family has involvement in the military, or has military friends or associations or possibly a residence near a military base, all of which makes selecting, targeting and "conditioning" of intended victims, convenient for the black op abductors.

We have shown that most of the conditions of the abduction scenario can be reasonably explained with conventional theatrics, the use of drugs, various medical technologies and the use of stealth aircraft. But, what about those reports which stated in no uncertain terms that real, physical, small ET entities known as the Greys were involved in the abduction event?

Even within the Milab Hypothesis, there are elements of absurdity where specific statements are taken as factual evidence but are based on pure speculation from the accounts of a few hypnotically regressed "alien abductees" like Katherina Wilson. We need to be very careful when examining such evidence gleaned from hypnosis as to whether to accept such evidence as factual or dismiss outright because of its subjectivity and loose logical associations with other UFO or ETI events.

The so-called factual evidence that I am referring to are such stories that speak of a hierarchy of alien species such as the "shapeshifting reptoids" or "reptilians" as the masterminds behind human abductions for the purpose of human DNA experimentation and manipulation, cloning and time travel using the "Greys" and the "**Nordic Pleiadians**" (our human cosmic cousins) as "proxies" for the covert operations. Add to this mix, a dash of zealous religious overtones like Christian fundamentalist intervention to "save" female abduction victims who are perceived as "SINNERS" because they have used psychic abilities and/or received such powers from "demon" ETs and you have the makings of a great science fiction novel.

More importantly, behind all this gobbly-gook, all the ingredients of a military intelligence "mind control programming event" are in place. You mix elements of true science developments and breakthrough research together with a lot of pseudoscience and technobabble and outright disinformation, misinformation and lies and the perceived mess is one more area of UFO/ETI research that bogs down any real understanding of this whole phenomenon. **The Controllers: a New Hypothesis of Alien Abductions; Martin Cannon; 1990**

Alien abduction researchers will claim that the evidence supports the analysis and conclusions that aliens are hostile and this perception plays right into the Military Industrial complex's agenda without the MIC doing much to motivate this kind of conclusion. We must not forget that propaganda and disinformation are the main stable of military tactics and is as old as man himself but, has been developed into a fine art in the current modern age. Its usage is in effect constantly in all areas of national security and military strategy and is fully deployed into the whole subject of the UFO/ETI.

Are there real so-called abduction events *(A more preferable terminology would be the word: ET encounters)*? It would be ludicrous to think that all ETs have our best interests at heart, but to label them "hostile, bad and demonic" is a pure, prejudicial human emotion of anthropocentric thinking. We need to look beyond such thinking and examine only the hardcore evidence, not some wild speculation of disjointed events.

CHAPTER 29

FAMOUS ALIEN ABDUCTION CASES

We are going to examine some of the more well-known and famous "so-called" abduction cases as well as being the most controversial in the files of Ufology. "So-called" because they appear to be human abductions by alien beings but, new research has brought serious evidence to bear on their authenticity as to whether these are true alien abductions. Re-examination of the evidence is required and by applying a little diagnostic logic with some insightful or intuitive perception, there is a strong probability that a genuine ET contact experience may have taken place. The problem, however, could be a mistaken interpretation of the circumstances leading to a popular but, faulty conclusion of the event, that being the probability that most alien abductions are Milab scenarios carried out by "ET type beings" to give the "feel or impression" of a genuine Extraterrestrial encounter.

In the mind of the human experiencing the event, they would not know the difference having no previous point of reference with such an experience to make a decisive and conclusive judgment. In other words, without a point of reference with something completely unknown or in this case alien, it would be easy and understandable to project an anthropocentric thinking and reasoning onto the behaviour of alien who is not behaving necessarily in a human fashion and therefore, the behaviour of the alien would not conform to the rational or expectations of a human being. They are after all, non-human, but alien in thinking and behaviour. And thus, we cannot interpret their actions as benevolent or malevolent without a period of observation and interaction, the Travis Walton Incident is a classic example of misinterpretation of events due to limited human understanding and anthropocentric projection of human values onto another intelligent species.

Travis Walton Experience

The Travis Walton abduction case is the best-known account of an alleged alien abduction and no other abduction case has generated more controversy in the field of Ufology, in mainstream science, and amongst skeptics. Furthermore, it still remains one of the very few alien abduction cases with corroborative eyewitnesses, and one of few abduction cases where the time allegedly spent in the custody of aliens plays a minor role in the overall account. **Jerome Clark; The UFO Book: Encyclopedia of the Extraterrestrial; Visible Ink; 1998; ISBN 1-57859-029-9**

There hasn't been an abduction story before or since, that has begun in the manner as related by Walton and his co-workers and the case is uniquely singular where the abductee disappeared for days on end with police squads, family, and friends searching desperately, without success. It is a typical **"Close Encounter: Fourth Kind" (CE4)** ... which bucks the trend so much that it worried some investigators; others defend it staunchly." **Jenny Randles and Peter Houghe; The Complete Book of UFOs: An Investigation into Alien Contact and Encounters; Sterling Publishing Co, Inc, 1994; ISBN 0-8069-8132-6**

Travis Walton (born April 23, 1957) is an American logger who claims to have been abducted by a UFO on November 5, 1975, while working with a logging crew in the Apache-Sitgreaves National Forest in Arizona.

As the report goes Travis, his work boss and best friend, **Mike Rogers** and their fellow work crew of **Ken Peterson, John Goulette, Steve Pierce, Allen Dallis and Dwayne Smith** were contracted by the **United States Forest Service** to brush cut and thin out the undergrowth from a large 1200 acre area near Turkey Springs, Arizona. They typically would work long shifts from 6 a.m. until sunset to fulfill their contract agreement with the Forest Service. On November 5, just after 6 p.m., Rogers and his crew finished their work for the day and were piled in Rogers' truck driving back to Snowflake, where they all lived.

Travis Walton
http://ufocongress.com/2010-convention-speaker-spotlight-travis-walton/

"It wasn't long as they began the drive home, that Travis saw a bright light from behind an upcoming hill which he thought was the setting sun but quickly realized that the sun had already set about a half hour earlier. Then, it occurred to Travis and the crew who now also, saw the bright light that perhaps it was some hunters' camp fire or even, a small plane has crashed and hung up in the trees. As they drove closer past a thicket of trees to an unobstructed area, they saw a large golden flying saucer hovering below the tree line above a clearing and shining with a yellowish brilliance. The flattened disc was around 8 to 10 feet thick and 15 to 20 feet in diameter, like to two large pie pans facing each other, lip to lip at the rim with a small white dome or cupola on the top of the craft. There were darker strips of a dull silver sheen that divided the glowing areas into panel-like sections. The overall yellowish light from the surface of the craft had the luster of hot metal.

Rogers' truck slowed to a stop, and almost immediately Walton being a risk taker and throwing all caution to the wind, leapt from the truck and ran toward the disc hovering about 30 yards away. The other men began shouting at Walton to come back but, in a committed and brazen yet, slow approach, he continued toward the disc. He was nearly below the object in a half crouched, somewhat cringing fashion observing the detail of the alien craft when Travis perceived a soft mechanical sound coming from the craft which suddenly began to increase in volume similar to a very loud turbine. The disc then began to wobble from side to side, at which point Travis thinking the saucer was about to streak away, decided it best to get the hell out of there.

"Walton began to rise from his crouched position to move away from the object when his crewmates witnessed a tremendous foot-wide blue-green ray of light emanate from the bottom of the craft striking Travis on the head and chest! He arched backward, his arms and legs outstretched, and appeared to rise a foot off the ground, and was hurled backward through the air for 10 feet, all the while caught in the glow of the light. His right shoulder hit the earth, and his body sprawled limply over the ground." **Travis Walton "Fire in the Sky"; Marlowe & Company Publishing, 1996; ISBN 1-56924-840-0**

Travis recalls, "I saw and heard nothing. All I felt was a numbing force of a blow that felt like a high voltage electrocution. The intense bolt made a sharp crack or popping sound. My mind sank quickly into feeling blackness. From the instant, I felt that paralyzing bolt, I did not see, hear, or feel anything more. I didn't see what hit me. I was totally oblivious to all that happened after that."

Thinking that his best friend was dead, Rogers drove away quickly in sheer panic over the rough road, afraid that the disc was chasing the truck. After about a quarter-mile, the truck skidded off the road and Rogers stopped. After some discussion, the crew claims they decided to go back to the site and find Walton.

"When they returned to the area where they saw Walton last, the disc was gone, and his co-workers said they searched for Travis for a half hour but found no sign of him. Allen thought that Travis had been killed. Ken saw the blue ray engulf him all over. Dwayne exclaimed that he thought poor Travis had been disintegrated like he got hit by lightning or something but, Steve saw him hit the ground in one piece. John swore that it sure knocked the hell out of him. Ken reported that it was like a grenade had gone off in front of Travis blowing him backwards." **Travis Walton; The Walton Experience; 1978; Berkley Publishing Corp; ISBN 425-03675-8**

Rogers's crew piled back into the truck and drove into Heber, Arizona to the police and related to them what had happened. Police, Sheriff Officers, their Deputies, volunteers and some of the work crew returned to the site of the incident. "By the morning of November 6, many officials and volunteers had scoured the area around the scene where Travis went missing. Still, no trace of him was discovered, and police suspicions were growing that the UFO tale was concocted to cover up an accident or homicide." http://en.wikipedia.org/wiki/Travis_Walton

"Saturday morning, Rogers and Duane Walton arrived at Sheriff Gillespie's office "explosively angry" because they had returned to the scene and found no police there. By that afternoon, police were searching for Travis with helicopters, horse-mounted officers, and jeeps." **Jerome Clark; The UFO Book: Encyclopedia of the Extraterrestrial; Visible Ink, 1998; ISBN 1-57859-029-9**

Walton's alleged abduction cause a media sensation internationally, with many news reporters, ufologists and the curious travelling to Snowflake. Accusations began to fly around town as the police investigation was stepped up with the prime focus of suspicion aimed squarely at the woodcutters as having done away with Travis. Still others like Snowflake Town Marshall, Sanford Flake felt that the entire affair was a prank engineered by Travis and his brother, Duane. His reasoning was based on earlier statements that the pair had made when interviewed by Fred Sylvanus, a Phoenix UFO investigator and also, with Rogers and Duane, that police were only doing a half-hearted search, statements that would return to haunt them, when seized upon by critics, like Philip Klass.

On Monday, November 10, all of Rogers' remaining crew took polygraph examinations administered by Cy Gilson, an Arizona Department of Public Safety employee. His questions asked if any of the men caused harm to Travis (or knew who had caused Travis harm), if they knew where Travis's body was buried, and if they told the truth about seeing a UFO. The men all denied harming Travis (or knowing who had harmed him), denied knowing where his body was, and insisted they had indeed seen a UFO.

Excepting Dallis (who had not completed his exam, thus rendering it invalid), Gilson concluded that all the men were truthful, and the exam results were conclusive. Clark quotes from Gilson's official report: "These polygraph examinations prove that these five men did see some object they believed to be a UFO and that Travis Walton was not injured or murdered by any of these men on that Wednesday". If the UFO was hoaxed, Gilson thought, "five of these men had no prior knowledge of a hoax".

Following the polygraph tests, Sheriff Gillespie announced that he accepted the UFO story, saying "There's no doubt they're telling the truth." Flake, however, remained unpersuaded. **Jerome Clark; The UFO Book: Encyclopedia of the Extraterrestrial; Visible Ink, 1998; ISBN 1-57859-029-9**

Just before midnight on Monday, November 10, **Grant Neff**, Travis' brother –in-law gets a phone call from what he believes is a prankster who speaks in a weak voice, "This is Travis. I'm at a phone booth at the Heber gas station, and I need help. Come and get me." In total disbelief, Neff is about to hang up when the caller in near hysterics screams "It's me, Grant ... I'm hurt, and I need help badly. You come and get me." Immediately Neff and Duane Walton drive to the gas station and find Travis collapsed in the second of three telephone booths wearing the same clothing that he wore when he disappeared. He seemed thinner and to have not shaved in the time he was absent.

Travis was shaken to the very core of his being; he was fearful and repeatedly mumbled on about strange beings with terrifying eyes. He was stunned to learn he'd been absent for nearly a week even though, he thought he'd been gone only a few hours.

Word of Travis's return had leaked out to the public and had reached **Coral Lorenzen**, Director of APRO, the civilian UFO research group. She had promised Duane that she could arrange an examination for Travis by two medical doctors at Duane's home. Duane agreed, and the exam began at about 3:30pm Tuesday.

The medical examination revealed that Travis was essentially in good health, but they did note two unusual features: A small red spot at the crease of Travis's right elbow that was consistent with a hypodermic injection, but the doctors also noted that the spot was not near a vein;

Analysis of Travis's urine revealed a lack of acetones. This was unusual, given that if Travis had indeed been gone for five days with little or no food as he insisted (and as his weight loss suggested), his body should have begun breaking down fats in order to survive, and this should have led to very high levels of acetone in his urine. Also of note, which is more difficult to explain is the absence of bruises, which one might expect in the wake of Travis's alleged beam-driven collision with the ground.

Travis later speculated that he'd gotten the mark on his elbow in the course of his logging work; critics would speculate that the mark showed where Travis (or someone else) had injected drugs into his system, however, medical doctors found no sign of it. Travis later noted that he'd been an amateur boxer and had rarely bruised even after rough matches; he also noted that in his logging duties, he and others had taken some painful bumps and falls which had not left significant marks. This, of course, may raise an inconsistency in that Travis suspects that a minor puncture wound could still be visible after five days, while he simultaneously insists that being roughly tossed some ten feet would leave no bruise or abrasion.

Sheriff Gillespie upon learning of Travis's return through the news media was angered, even though, he had come to believe Travis' UFO story following the polygraph exams. Travis felt that Gillespie needed to know what had happened during the five days he was gone.

Travis reported that after approaching the UFO near the work site, the last thing he remembered was being struck by the beam of light. When he woke, Travis said he was on a reclined bed. A bright light shone above him, and the air was heavy and wet. He was in pain, and had some trouble breathing, but his first thought was that he was in a normal hospital.

Arizona
Department of Public Safety

R. O. BOX 6638

Phoenix, Arizona 85005

13 November 1975

JAMES J. HEGARTY.
DIRECTOR

PLEASE REPLY ATTENTION OF

Sheriff Marlin Gillespie
Navajo County Sheriff's Office
300 Navajo Boulevard
Holbrook, Arizona

Dear Sheriff Gillespie:

On 10 November 1975 a polygraph examination was administered to John E. Goulette, Dwayne D. Smith, Kenneth E. Peterson, Michael H. Rogers, Jeff S. Pierce and Allen M. Dalis.

The purpose of these examinations was to determine whether or not any of the above listed people were deliberately trying to conceal a criminal act i.e.: aggravated assault and/or homicide in which Travis Walton was the victim.

The relevant questions asked and the answers given are as follows:

Question #1 - Did you cause Travis Walton any serious physical injury last Wednesday afternoon?

Question #2 - Do you know if Travis Walton was physically injured by some other member of your work crew last Wednesday?

Question #3 - Do you know if Travis Walton's body is buried or hidden somewhere in that Turkey Springs area?

Question #4 - Did you tell the truth about actually seeing a UFO last Wednesday when Travis Walton disappeared?

Results of the polygraph test (page 1)

213

Sheriff Marlin Gillespie
Page Two
13 November 1975

Copy 2-9-76 [illegible]

Each of the six men answered NO to questions # 1, 2, and 3 and they each answered YES to question #4. The test results were conclusive on Goulette, Smith, Peterson, Rogers, and Pierce. The test results on Dalis were inconclusive.

Based on the polygraph chart tracing, it is the opinion of this examiner that Goulette, Smith, Peterson, Rogers, and Pierce were being truthful when they answered these relevant questions.

These polygraph examinations prove that these five men did see some object that they believe to be a UFO and that Travis Walton was not injured or murdered by any of these men, on that Wednesday (5 November 1975). If an actual UFO did not exist and the UFO is a man-made hoax, five of these men had no prior knowledge of a hoax. No such determination can be made of the sixth man whose test results were inconclusive.

The charts obtained will be maintained in the Polygraph Section of the Department of Public Safety.

Sincerely,

C. E. Gilson

C. E. Gilson
Polygraph Examiner
Department of Public Safety

CEG:jg

Results of the polygraph test (page 2)

As his faculties returned, Travis says he realized he was surrounded by three figures, each wearing a sort of orange jumpsuit. The figures were not human but humanoid a head, two legs, two arms and hands with five digits on each. Travis described them as similar to the so-called "Greys" which feature in some abduction accounts: "shorter than five feet, and they had bald heads, no hair. Their heads were domed, very large. They looked like fetuses ... They had large eyes — enormous eyes — almost all brown, without much white in them. The creepiest thing about them were those eyes ... they just stared through me." Their ears, noses and mouths "seemed real small, maybe just because their eyes were so huge." **Jerome Clark; The UFO Book: Encyclopedia of the Extraterrestrial; Visible Ink, 1998; ISBN 1-57859-029-9**

"Afraid for his safety, Travis manages "despite his 'weakened' condition, 'aching body' and 'splitting pain in his skull' as a result of being hit with a burst of energy like lightning earlier, manages to muster enough strength to get to his feet, and shouted at the creatures to stay away. He grabbed a glasslike cylinder from a nearby shelf and tried to break its tip to create a makeshift knife, but found the object unbreakable, so instead waved it at the creatures as a weapon. The trio of creatures left him in the room.

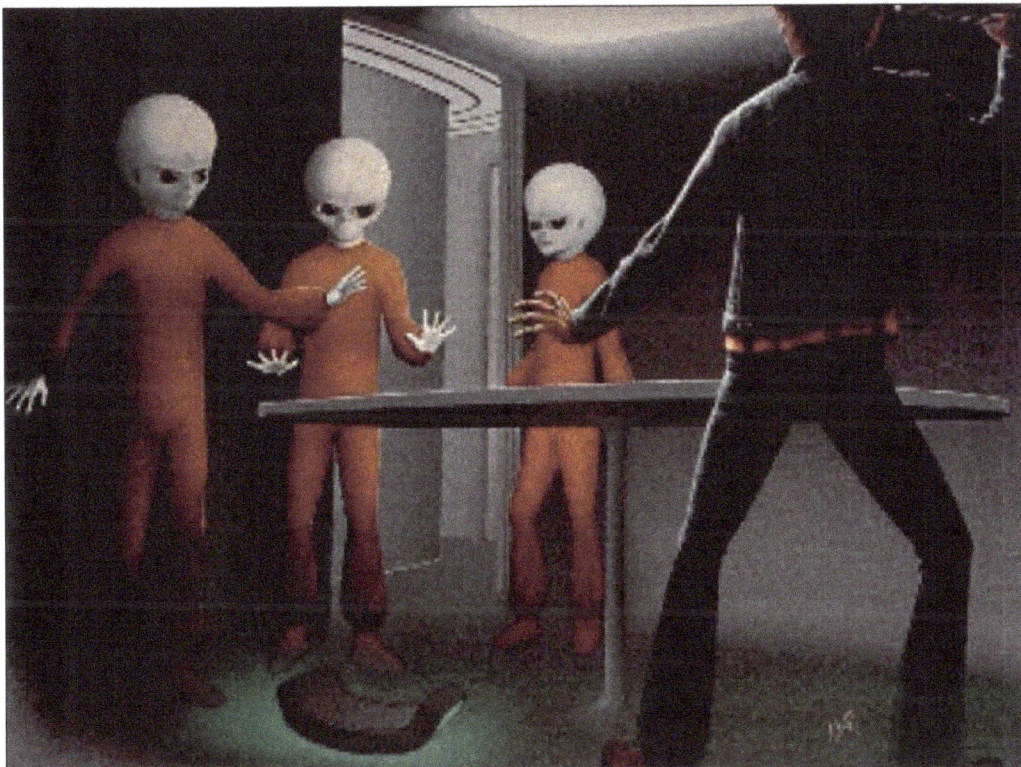

Walton awakens and is startled by strange alien beings standing over him
http://www.ufocasebook.com/Walton2.html

Travis then left this "exam room" via a hallway, which led to a round, spherical room with only a high-backed chair placed in the room's center. Though he was afraid there might be someone seated in the chair, Travis says he walked towards it. As he did, lights began to appear in the room. The chair was empty, so Travis says he sat in it. When he did, the room was filled with lights, similar to stars projected on a round planetarium ceiling.

The chair was equipped on the left arm with a single short thick lever with an oddly shaped molded handle atop some dark brown material. On the right arm, there was an illuminated, lime-green screen about five inches square with black lines intersected at all angles.

When Travis pushed the lever, he reported that the stars rotated around him slowly. When he released the lever, the stars remained at their new position. He decided to stop manipulating the lever since he had no idea what it might do.

He left the chair, and the stars disappeared. Travis thought he had seen a rectangular outline on the rounded wall — perhaps a door — and went to look for it.

Just then, Travis heard a sound behind him. He turned, expecting more of the short, large eyed creatures, but was pleasantly surprised to see a tall human figure wearing blue coveralls with a glassy helmet. At the time, Travis said, he did not realize how odd the man's eyes were: larger than normal, and a bright gold color.

Travis in the navigation room by the pilot's control seat

Travis says he then asked the man a number of questions, but the man only grinned and motioned for Travis to follow him. Travis also said that because of the man's helmet he might have been unable to hear him, so he followed the man down a hallway which led to a door and a steep ramp down to a large room Travis described as similar to an aircraft hangar. Travis says he realized he'd just left a disc-shaped craft similar to the one he'd seen in the forest just before he'd been struck by the bluish light, but the craft was perhaps twice as large.

In the hangar-like room, Travis reported seeing other disc-shaped craft. The man led him to another room, containing three more humans — a woman and two men — resembling the helmeted man. These people did not wear helmets, so Travis says he began asking questions of them. They responded with the same dull grin and led him by his arm to a small table.

Once he was seated on the table, Travis says he realized the woman held a device like an oxygen mask, which she placed on his face. Before he could fight back, Travis says he passed out.

Two human type aliens approach Travis; the female has a portable gas mark and Travis falls unconscious when the mask is placed over his face

When he woke again, Travis says he was outside the gas station in Heber, Arizona. One of the disc-shaped craft was hovering just above the highway. After a moment, the craft shot away, and Travis stumbled to the telephones and called his brother-in-law, Grant Neff. He thought that only a few hours had passed." **Travis Walton; Fire in the Sky; Marlowe & Company Publishing; 1996; ISBN 1-56924-840-0** and **Travis Walton; The Walton Experience; Berkley Publishing Corp., 1978; ISBN 425-03675-8**

After hearing Travis's story, Gillespie speculated that Travis may have been hit on the head and drugged, then taken to a normal hospital where he had confused the details of a routine exam with something more spectacular. Travis dismissed this, noting that the medical examination had found no trace of head trauma or drugs in his system. Travis told Sheriff Gillespie that he was willing to take a polygraph, a truth serum, or undergo hypnosis to support his account. Gillespie said that a polygraph would suffice, and he promised to arrange one in secret to avoid the growing media circus.

Duane and Travis then drove to Scottsdale, Arizona, where a meeting with APRO consultant James A. Harder had been arranged. Harder hypnotized Travis, hoping to uncover more details of the missing five days. Clark writes that "Unlike many other abductees, however, Walton's conscious recall and unconscious 'memory' were the same, and he could account for only a maximum of two hours, and perhaps less, of his missing five days. Curiously ... Walton encountered an impenetrable mental block and expressed the view that he would 'die' if the regression continued." **Jerome Clark; The UFO Book: Encyclopedia of the Extraterrestrial; Visible Ink, 1998; ISBN 1-57859-029-9**

McCarthy then administered the polygraph, which remains mired in controversy. Travis asserts McCarthy behaved unprofessionally, while McCarthy insists Travis failed the polygraph and tried to cheat. Travis states that McCarthy was biased, confrontational and aggressive in the manner by which he administered the polygraph questions.

After completing the exam, McCarthy determined that Travis was lying. Clark quotes from McCarthy's official report: "Based on his reaction on all charts, it is the opinion of this examiner that Walton, in concert with others, is attempting to perpetrate a UFO hoax and that he has not been on any spacecraft". Later, McCarthy would assert that "sometimes Travis would hold his breath, in an effort to 'beat the machine." **Jerome Clark; The UFO Book: Encyclopedia of the Extraterrestrial; Visible Ink, 1998; ISBN 1-57859-029-9**

The Waltons, APRO and the National Enquirer (who financed the medical examination and polygraph tests) then agreed to keep the results of this polygraph a secret, due in large part, they insisted, to doubts about McCarthy's methods and objectivity. Eight months later, when word of this decision was made public, there would be more charges of deception and cover-up. Travis would later take and pass two additional polygraph exams, though the suppressed results of the first exam would shadow him and earn mention in nearly every discussion of the case to the present. http://en.wikipedia.org/wiki/Travis_Walton

This is where the story stands today in the files of Ufology, full of controversy, biases, and lies from both the witnesses and the investigators, a story muddy with too many mistakes and harsh critique.

In 1993, Walton's book was adapted into a film, *Fire in the Sky*, directed by Robert Lieberman and starring **D. B. Sweeney** as Travis Walton and **Robert Patrick** as Mike Rogers. Clark writes that the film found "Moderate success, mixed reviews, and Ufologists' complaints about its inaccuracies and exaggerations." **Jerome Clark; The UFO Book: Encyclopedia of the Extraterrestrial; Visible Ink, 1998; ISBN 1-57859-029-9**

Especially inaccurate was the portion of the film detailing his time on the UFO; it bears almost no resemblance to the original narrative. Screenwriter Tracy Tormé even sent letters to many ufologists, claiming that the changes were requested by studio officials, and apologizing for making such substantial alterations to Walton's narrative. **Jenny Randles and Peter Houghe; The Complete Book of UFOs: An Investigation into Alien Contact and Encounters; Sterling Publishing Co, Inc, 1994; ISBN 0-8069-8132-6**

Dr. Greer concurred with the general complaints leveled at Tracy Tormé stating the sinister nature of the ETs as portrayed in the movie was grossly inaccurate and reflected a fear factor that was aimed at the general public to ensure and re-enforce the stereotypical negative mindset that all Extraterrestrial Intelligences are bad and to be avoided if possible, at all costs. The nature and reality of ET Intelligence is really something quite different than what is currently being milled out to the public from the Hollywood movie industry.

The ETIs are in fact very wary of humans, given our tendency for hostility and aggression to anything foreign or different to our anthropocentric sensibilities. Everything recorded or reported, strongly reflects the non-hostile nature of ETs down through the history of mankind's interaction with them. The truth is that ETI are very passive and almost neutral in their behaviour towards us, a behaviour which is more of a non-interfering but, concerned observer.

Author's Rant: I had an opportunity briefly to meet Travis Walton at the Whole Life Expo in Seattle back in 1994 and expressed to him my personal support and conviction that his story was exactly as he described it. I discovered that Dr. Greer had also interviewed Travis and had offered him another possible explanation, from a medical perspective as to what may have happened to him during his five days of missing time.

Dr. Greer suggests that at the point where Travis was near the craft observing its structure and based on his and the rest of Roger's work crew's eyewitness testimony, it would appear that the craft's motives and intentions were not hostile as were reported by investigating ufologists but, rather a matter of an untimely and unfortunate scientific principal being displayed with Walton in the wrong place at the wrong time. Namely, the flying saucer was releasing a high electrical discharge of energy much like a lightning bolt which is how Travis and his work buddies described it when it hit Travis, who merely was nearest conduit point for grounding the discharge! Since the saucer was above him and the source of the energy discharge, it would seem logical that Travis was hit in the head and chest area which is of lower resistance than the earth (he may have also had something metallic on him or his clothing), those areas of the body being closest to the disc.

Luminosity and Distance as a Ratio of Power Generation

In order to understand the energy generation potential from a flying saucer which could discharge an energy bolt in the direction of Travis Walton sending him backward off his feet, we need to first consider several other UFO cases which Jacques Vallee investigated. He makes a brilliant deduction from examining an important piece of the evidence from the UFO phenomenon that is rarely contemplated by Ufologist until now, that of distance and luminosity as a ratio of the power output generated from flying saucers.

"His first case is the Fort-de-France event where the French Navy is anchored for a layover at the island of Martinique in late September 1965 after a joint naval exercise with the US Navy in July for the recovery of a Gemini capsule near Bermuda. Three hundred witnesses from the submarines *Junon* and the *Daphne* and from the weather observatory observed a large UFO arrive slowly and silently from the west, flew to the south, made two complete loops in the sky over the French vessels, and vanished like a rapidly extinguished light bulb. The UFO was the colour of a fluorescent tube, about the same luminosity as a full moon at an estimated distance of 10 kilometers south of the ships. It moved from west to east with a whitish trace similar to the glow of a television screen. The distance as agreed by the witnesses would represent a disc of ninety meters in diameter. Based on this information and a simple mathematical calculation, results in an energy output from the disc of 2.3 megawatts (MW).

To put this into perspective, the energy output of a lawn mower is 3 kilowatts (KW), a car engine will range between 75 and 1000 KW or (I MW), a commercial airliner will reach 50 to 150 MW; while the energy from a nuclear power plant is measured in thousands of megawatts.

The second case occurred between Rives and Grenoble, France where a UFO was witnessed by a French Physicist "Dr. Serge" (not his real name). On his way home from the Grenoble airport, he observed in the sky a luminous disc as bright as the full moon. Stopping his car, he got out to watch the slightly flattened object (with an aspect ratio of 0.9) with an angular diameter of twenty arc minutes (the full moon has an angular diameter of thirty minutes). The object was white in the centre and bluish-white at the periphery and was surrounded by an intense green halo about two to three arc minutes thick. It flew toward east-southeast stopped for about 10 seconds, changed course and flew in front of the Le Taillefer Mountains 36 kilometers away and finally disappeared behind Le Neron Mountains, nine kilometers away.

Reconstruction of the evidence indicated that the object flew at an altitude between 1500 to 2500 feet which give a diameter between 6 to 20 feet and approximate speed of 3600 miles per hour. Assuming a luminosity as bright as a full moon, its energy level in the visible part of the spectrum would be a modest 15 KW.

The third case took place near Arcachon, France on June 19, 1978, which was initially investigated by **GEPAN**, the French government's official UFO task force (also known as **SEPRA**). In this case, a UFO triggered photocells that control the lights of a whole town. From the distance and the threshold level of the cell, it is possible to derive another estimate of the energy of the object. The town of Gujan-Mestras was the actual location of the sighting where a number of citizens, in particular, two young men in their late teens while repairing a turn signal light on their car heard a powerful rumble like an earthquake and looking up reported seeing an oval object, red and surrounded by white flames flying at an altitude of 11,000 feet approaching them quickly. One of the young men found it hard to breathe and fainted whereupon the object flew off in a different direction

Another witness, a thirty-five year old restaurant manager saw a very bright orange ball hover 1000 feet over La Reole, a neighbouring community. An additional witness, a student who lived in Gujan confirmed that when he was outside, the town lights died just before midnight;

concurrently, he had heard a low rumble that scared him and flashes of light above the pine trees but below the cloud ceiling.

GEPAN investigators interviewed all the witnesses and concluded that they all had seen the same object. Allowing for expected human errors, there was an agreement in time, duration, distance, trajectory, sound, and luminosity parameters. The manager of the town utility department was also interviewed; he showed the investigators the location of the photo-electric cells that control the street lights. Naturally, when these cells are exposed to a light that exceeds their threshold, they assume that daylight has arrived and they turn off the system. The result of the analysis bracket the distance between the cells and the UFO as 135 meters and 480 meters, or roughly 400 and1500 feet. This yields an energy level between 160 kW and 5 MW.

The fourth and final case happened in Haynesville, Louisiana on December 30, 1966, when a professor of physics and his family were driving north on Highway 79 between Haynesville and the Arkansas border. His wife called his attention to a red-orange glow appearing through and above the trees ahead to their left. As they continued down the highway, the light became a luminous hemisphere, pulsating regularly, ranging from dull red to bright orange with a frequency of two seconds. When the car came to a point about one mile from the source of the light, it suddenly brightened to a blinding white, washing out the car headlights and casting sharp shadows. The burst of light was so intense that the physicist shielded his eyes and it even woke his children in the back seat. After a few seconds, the light returned to its red-orange appearance.

Later investigation of the area revealed that the trees had a blacking or burning of the bark in the direction pointing to the centre of the area as if exposed to an intense source of radiated energy. From an estimate of the energy required to produce the depth of the burn it would be possible to estimate the power of the source, however, this was never done. Going back to the initial report of the physicist, he that the car headlights are a known 150 watts and of a complicated mathematics and since the clearing in which the object was observed from the road was 1800 feet, the energy output was determined at 500 MW. This figure, which is a minimum estimate, places the object in the energy range of a small nuclear reactor; the actual energy figure may be orders of magnitude beyond the original calculations. ***"The light emitted may be a side effect of the propulsion mechanism of the UFO, much in the same way that carbon monoxide is a side effect in the exhaust of an automobile engine."*** [My italics added]. Jacques Vallee; Confrontations: A Scientist's Search for Alien Contact; Random House, Inc., 1990; ISBN 0-345-36501-1

The Walton Experience Continued

Returning back to the Walton account, we see from the above examples, that UFOs or more accurately, Extraterrestrial spacecraft have as a component of their propulsion system, an energy output generated in extremely high orders of magnitude that are equivalent to a flying nuclear reactor. Their side effect is the emission of light in every colour of the spectrum and with the intensity of a noonday sun. It would not be a far stretch of the imagination that such high energy generation spills over into other areas of the electromagnetic spectrum from Long Waves to Radio/TV to Microwaves and from Infrared to Ultraviolet and from X-rays to Gamma rays and Short Waves.

Given the possibility that these ET crafts are discharging radiation across the spectrum when they fly and hover, any approach by a living creature including humans is bound to cause probable injury or even death. It would seem **Travis** was fortunate that he survived due to the quick actions of his alien benefactors and caregivers.

Natural terrestrial lightning almost always emanates from the ground up toward the sky which means that if you are struck by lightning while in the wrong place and time like near a tree, you would be hit feet first with the lightning discharging through your body and out your hands or head first or simultaneously.

The bolt from a lightning strike as is often reported by eyewitnesses will in most cases knock a person off his feet and even send him flying backwards or off to the side from the position in which he was struck. Lightning injuries result from three factors: electrical damage, intense heat, and the mechanical energy which these generate.

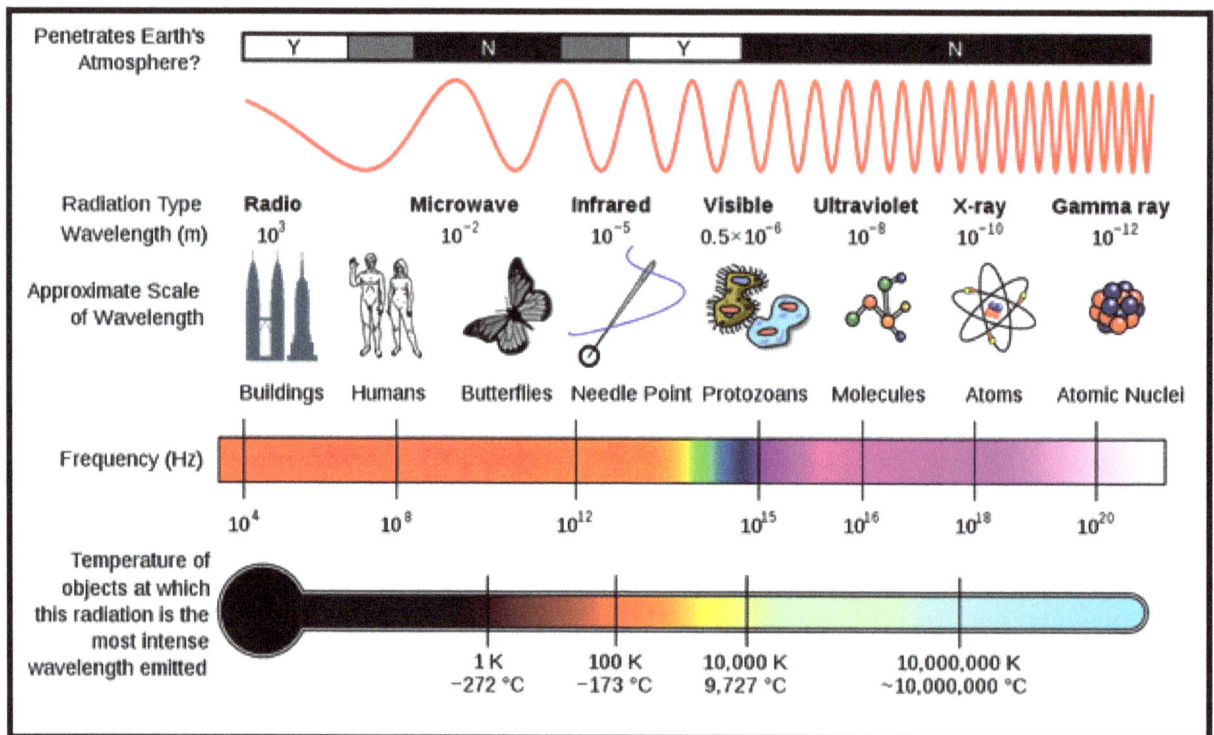

The Electromagnetic Spectrum. Although some radiations are marked as *N* for no in the diagram, some waves do in fact penetrate the atmosphere, although extremely minimal compared to the other radiations
http://earthsky.org/space/what-is-the-electromagnetic-spectrum

Dr. Greer suggests further, that **Travis Walton's** work buddies in a complete state of panic and hysteria drove off down the dirt road out of the area leaving what they presumed was a now deceased Travis and very fearful that the saucer was going to follow them and do harm to the rest of them.

It was obvious at this point that the ETs aboard the craft realizing what had accidently occurred and being responsible for the cause of bodily harm to this poor and somewhat foolish human who had gotten too close to their craft, rescued him and carried him on board their spacecraft for medical attention.

Dr. Greer states that from first-hand experience with victims of lightning strikes, you have to response quickly to the injuries suffered by the individual as it really is a matter of a life and death situation. The intense heat generated by a lightning strike can burn tissue, and cause lung damage, and the chest can be damaged by the mechanical force of rapidly expanding heated air. Lightning can affect the brainstem of the patient, which controls breathing. If a victim appears lifeless, it is important to begin artificial resuscitation immediately to prevent death by suffocation, and should be continued until the brainstem recovers, this then ensures the chances of survival. The intense electrical energy can cause a loss of consciousness; it is also speculated that the **EMP (Electromagnetic Pulse)** created by a nearby lightning strike can cause cardiac arrest. The atmospheric expansion by lightning (the cause of "thunder") can cause concussive and auditory injury at extremely close range.

A doctor's immediate response will be to diagnose the most critical injuries which are the circulatory system, the lungs, and the central nervous system. There may also, be an immediate cardiac arrest or myocardial infarction and various cardiac arrhythmias, either of which can be rapidly fatal as well. Loss of consciousness is very common immediately after a strike. Amnesia and confusion of varying duration often result as well. Experiments on sheep show that a lightning strike to the head of a victim enters through the eyes, ears, nose, and mouth, converging at the brainstem, which controls breathing. Many unconscious victims, who appear lifeless, die of suffocation.

Dr. Greer also points out that *"Rhabdomyolysis",* the rapid breakdown of skeletal muscle due to injury to muscle tissue or its destruction leads to the release of the breakdown products of damaged muscle cells into the bloodstream; some of these, such as myoglobin (a protein), are harmful to the kidney and may lead to acute kidney failure. Treatment is with intravenous fluids, and dialysis or hemofiltration if necessary.

The most reliable test in the diagnosis of rhabdomyolysis is the level of creatine kinase (CK) in the blood. This enzyme is released by damaged muscle, and levels above 5 times the "upper limit of normal" (ULN) indicate Rhabdomyolysis. Initial and peak CK levels have a linear relationship with the risk of acute renal failure: the higher the CK, the more likely it is that kidney damage will occur. CK levels rise after 12 hours of the initial damage, remain elevated for **1–3 days** and then fall gradually.

Because internal cellular damage will have probably occurred in some of the organs and muscles, it will take approximately three to five days to flush through the liver (the main filtering organ) and kidneys out of the body. At this critical stage, liver damage and renal failure could develop in the filtering and flushing out of dead cells and bodily fluids in which case, further medical treatment may be required. It is thus, important to keep the patient under observation for that prescribed time period to ensure complete recovery.

Now, isn't it curious that the time period of Travis Walton's disappearance is exactly five days?! That is long enough to ensure that he would make a full recovery, including from any other associated complications, even if as he reported, he was groggy and weak upon regaining consciousness.

Walton sits up on the single pedestal table and correctly deduces that he is in some sort of hospital room, although unawares at first that he is on board a flying saucer. His blurry sight clears and the fog in his mind disappears and to his amazement and horror, he discovers that there are three strange alien beings standing over him.

It would seem that to any rational and perceptive person that Travis' "doctors" albeit, alien in appearance were professional in their duties to their patient, obviously concerned for his welfare and may even be thought of as kind and caring. Hardly, what you would describe as being diabolically sinister with ulterior motives of abduction and an agenda to perform bizarre medical experiments upon humans. And how does Travis repay his alien caregivers? He grabs the nearest object, a long plastic-like tube which he tries to break the end off into a jagged point but, it is unbreakable and uses it as a weapon to physically and verbally threaten the small ET caregivers!!! Now, who let this "ape" loose on board the ship?! It's no wonder that we are probably viewed by some Extraterrestrial civilizations as primitive and barely out of the trees!

Travis gets up and staggers out of the medical room and finds his way into the navigation room where he is permitted to fly the spacecraft. Whether this was an intentional act by the ETs to help settle Travis' mind as to where he was or one which they were unaware of his immediate whereabouts, it appears no harm came to him or the ETs. He is suddenly greeted or confronted by a very human looking blond male alien wearing a helmeted spacesuit and is gently but, firmly lead off the now landed saucer into a hangar where he sees other saucer craft parked. He is taken down a short corridor to another room where he seated in a comfortable chair beside a black slab, single pedestal table. He is then approached by another male and a female without helmets, who get him to lie down on the table and a mask of some kind with gas is placed over his nose and mouth by the female and Travis quickly blacks out into unconsciousness. He awakens beside a road near the town of Heber looking up to see the departing ET spacecraft above him flying off, back into space.

As mentioned earlier abductions are absolute in their finality, their outcome has no happy ending for the victim. The Walton case demonstrates beyond a shadow of a doubt that his safe return to terra firma, family and friends cannot be remotely perceived or justified as an abduction event. Waltons events are real and this is was makes them so different from all other alleged abduction cases as it does not fall into the category of deliberate kidnapping, has nothing to do with strange medical or sexual reproduction examinations for the purpose of harvesting human sperm. No message of warnings or admonitions and no new scientific information was imparted and it certainly wasn't a "catch and release" program for the sport of it. This was an act of the highest spiritual and moral respect in preserving the life of another intelligent life form.

As of this writing, (Dec. 2014) Travis Walton as predicted by **Dr. Greer** has now come out on the lecture tour circuit and has altered the conclusion of his story which is more in line with the ETs trying to save Travis from a life-threatening situation caused by a careless act on the part of

a human coming too close to a hovering ET spacecraft that was powering up!
https://www.youtube.com/watch?v=Nwy-A0OtQIA

The Schirmer Abduction Case

The next abduction case is that of Police officer **Sgt. Herbert Schirmer** which received a high credibility status by most UFO researchers because of the type of witness that Schirmer represented, namely a police officer who was well respected by his fellow officers and the community at large in which he worked, until his UFO sighting on that fateful night. Skeptics, however, debunked his claim of abduction based on the fact that he was the only witness to the event and could not offer any proof of his claim, other than that of his written log statement of the event.

Officer Herbert Schirmer who was on routine late-night patrol in the town of Ashland, Nebraska, on December 3, 1967. While driving his patrol car along State Highway 6 going west, at about 2:20 a.m., he noticed a group of lights near the ground which he thought were that of a semi-trailer off the road. He approached to investigate, only to see the thing - whatever it was - take off and disappear in the night sky. When he returned to the police station at 3:00 a.m., he wrote in the station log, "Saw a flying saucer at the junction of highways 6 and 63. Believe it or not!" He went home that morning with a splitting headache and an inexplicable red welt on the side of his neck. Apart from this, however, Schirmer had no immediate sense that anything else had happened. **Ralph Blum and Judy Blum; Beyond Earth, Man's Contact With UFOs; Bantam Books, Inc; 1974**

Though only 22 at the time, Schirmer was so respected in Ashland that he was named Chief of Police a short time later. But only two months after becoming police chief, he resigned from the force. He said he couldn't get the UFO encounter out of his mind. He was experiencing terrible headaches and found it difficult to concentrate on his work.

The Condon Committee was in the midst of its UFO study at that time and hearing about Schirmer sent U.S. Air Force investigators from Boulder, Colorado to interview him. A preliminary investigation turned up the fact that 20 to 25 minutes seemed to be unaccounted for in his log of December 3, he had no memory of events between the takeoff and landing of the UFO. It was suggested that Schirmer undergo hypnotic regression to see if he could remember more details of the incident. UCLA scientists placed Schirmer into a deep trance and under hypnotic recall, a startling story emerged.

In early 1968, shortly after his UFO experience, among those who spoke with Schirmer was **Dr. R. Leo Sprinkle**, who was director of counselling and testing at the University of Wyoming at that time. Dr. Sprinkle told The Province that during the interviews Schirmer appeared "sincere and telling the truth as he saw it."

"He was aware of a loss-of-time experience (the missing 25 minutes in his radio log) and said he sensed someone coming toward him. At that time he didn't have any memory of being taken aboard." **Unidentified Flying Object (the Schirmer Abduction case) Source: James Spears, Vancouver Province (BC, Canada), Mar. 20, 1976**

Later, Schirmer was put under hypnosis to delve into the missing 25 minutes. Dr. Ron Katz, of the University of California (Los Angeles) department of anesthesiology, told The Province that Schirmer had submitted to a number of hypnotic trances and drew sketches of what he saw while in a trance. The drawings were made in 1974.

Generally, he said, subjects under hypnosis do not attempt to fabricate stories. *The one exception,* said Dr. Katz*, might occur if a subject had been previously hypnotized to "remember" a series of events under subsequent trances.* [Italics added]. Unidentified Flying Object (the Schirmer Abduction case) Source: James Spears, Vancouver Province (BC, Canada), Mar. 20, 1976

The time was 2:20 am when he saw the UFO, first as flashing red lights, then as a football-shaped object. He tried to call the Wahoo Sheriff office but his radio was dead. "The craft was moving over the highway north about 40 feet in the air. It seemed that some great force was pulling the patrol car up the embankment towards where the craft was beginning to land."

Sgt. Herbert Schirmer
http://www.ufoevidence.org/cases/case659.htm

Schirmer saw a red-orange glow on the bottom of the craft and saw a hatch open. "It was then that a shape of a man came out and walked over to the front of the patrol car. Then a second man came out of the craft and walked over to the car. The first man stood in front of the patrol car and, holding a small box-like object, pressed something. A green mist came out, spraying all over the patrol car. Schirmer found he couldn't draw his revolver. Without knowing why, he rolled down his window. "The other man walked over to the driver's side of the patrol car and reaching inside pressed a silver object against my neck, directly under my left ear. I felt a tingling sensation go through my whole body."

Seconds later, he found himself standing outside his car. The being asked, "Are you the watchman over this place?" Schirmer couldn't answer. The being asked, "Would you shoot at a space ship?" "No, sir," Schirmer said. The being motioned for him to come aboard the craft.

Schirmer described the beings as four and a half to five feet tall, with heads somewhat narrower and longer than an average human. Their skin looked grayish white; their noses were fairly flat and their mouths looked like slits that didn't move. Their eyes were slightly slanted, though not overly large; the eyelids did not blink. They wore silvery-gray jumpsuits that included a tight-fitting headpiece with a small antenna on the left side. On the left breast, they wore an emblem of a winged serpent. http://www.ufoevidence.org/Cases/CaseSubarticle.asp?ID=661

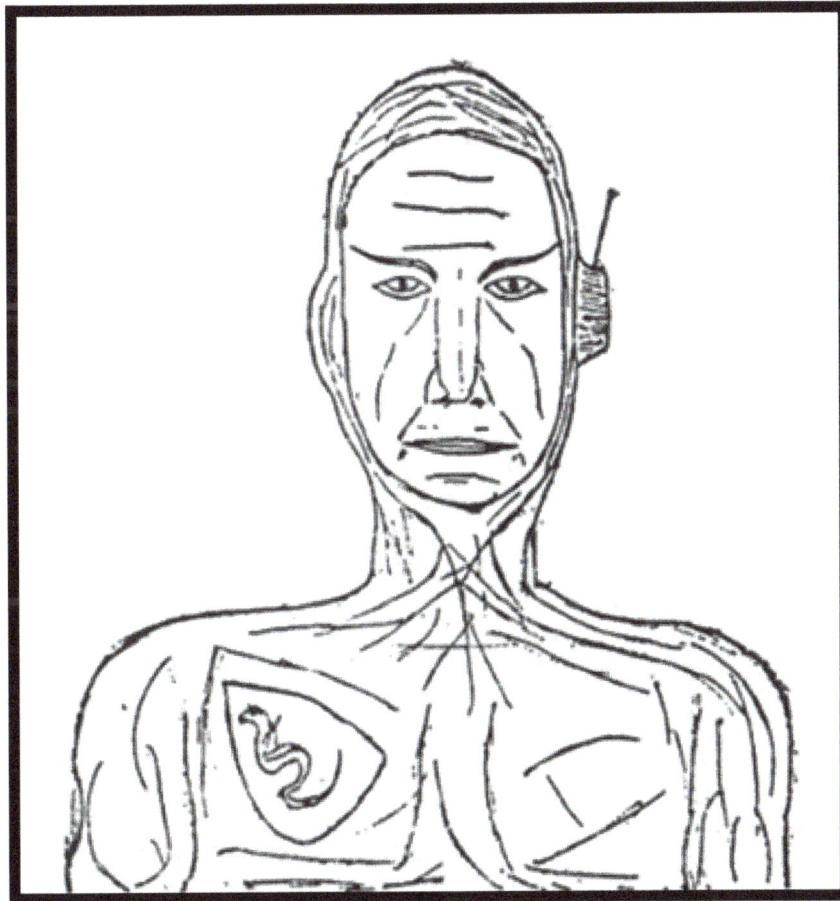

A drawing of the crew leader made by Schirmer indicates the beings look similar to humans, and not like the "Gray" type often linked with more recent abduction accounts. (credit: Wendelle Stevens)
http://www.ufoevidence.org/cases/case659.htm

During the next fifteen minutes of earth time, the beings communicated both audibly and telepathically as they showed Schirmer around their craft. He recalled being shown a smaller craft that was parked on board. He was told it was used for surveillance. He was told that these beings had been watching the human race for a very long time and were engaged in what he called a "breeding analysis program." They explained that they purposely contacted people at random so as not to reveal too much of a pattern. They want to keep us confused, Schirmer said, so we won't get too upset as we gradually get used to their presence. He felt sure they weren't hostile.

The grey-skinned man told Schirmer how the craft operated, showed him the power source and computers, and answered questions on the purpose of the craft's visit. Under hypnosis, Schirmer has drawn detailed sketches of the power plant of the craft, which the spaceman said was based on "reversible electromagnetism."

Schirmer on board the UFO was shown the alien solar system, "This is where we are from a galaxy next to yours." He also saw pictures of the craft's mother ship on the screen. (credit: Hesemann)

http://www.ufoevidence.org/cases/case659.htm

He saw a room circled with canister-shaped objects, with wires connecting them. In the middle of the room was an object, glowing red, like a spinning football. Schirmer said he was conveyed, by a glass elevator, to another level of the craft and saw a TV-like screen which the spaceman manipulated. He showed Schirmer what his own solar system "in a galaxy next to yours," looked like, and saw pictures of the craft's mother ship on the screen. By the screen, he said, was a map of a sun with six planets "with writing beside each planet. The language was one that I could not read."

Among the answers: The craft landed to extract power from a nearby hydro transmission line. 'He said: Look watchman. And I saw a blue bolt of light, through a porthole. It was like a beam and it hit the transformer on a power pole approximately 200 feet from the craft. There was a bright flash of fire and the blue bolt came back to the craft. "Then the man said that one of the reasons they landed was to extract electricity from power stations. He also said that they have been observing us for a long time."

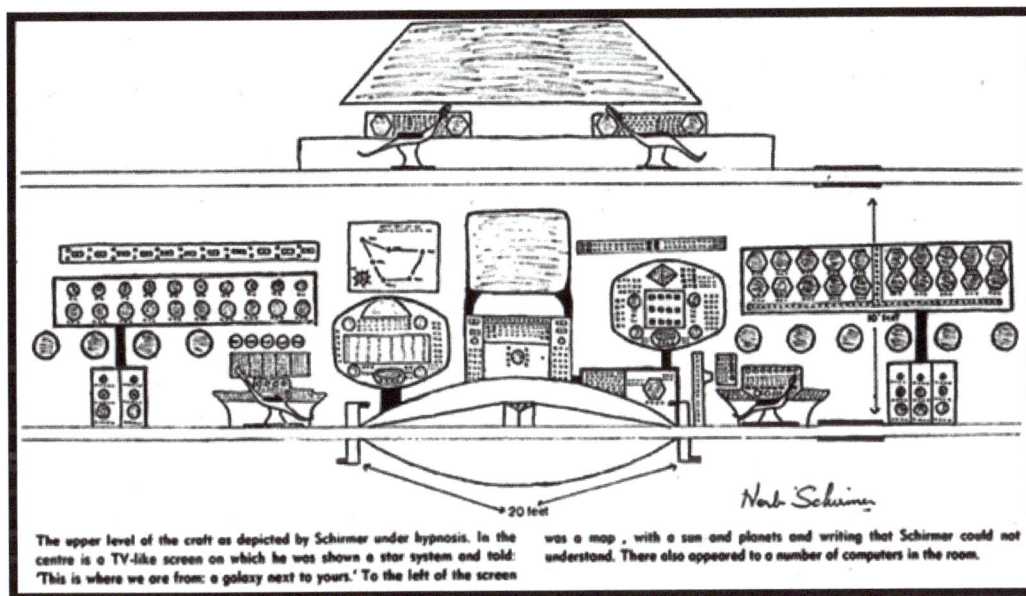

The Navigation Room depicted by Schirmer while under hypnosis. In the centre is a TV-like screen on which he was shown a star system and to the left was a star map, with a sun and planets and writing that Schirmer could not understand, also there appeared to be computers in the room.

http://www.ufoevidence.org/cases/case659.htm

One of the men also told the patrolman that spacecraft would visit him again twice in his lifetime. "Someday, Watchman, you will see the universe," but it isn't true. I never saw anything again." **Unidentified Flying Object (the Schirmer Abduction case)** *Source: James Spears, Vancouver Province (BC, Canada), Mar. 20, 1976*

They told him that he wouldn't remember most of what happened during the encounter and that they would visit him again.

Herbert Schirmer was returned to his car and then watched as the craft flew away. The welt on his neck remained as a hint that an alien being had touched him. After recalling these events under hypnosis, he did not return to the police force. Without doubt, his encounter on December 3, 1967, changed his life forever.

The fact that this was not really an abduction in the true sense when compared to other abduction scenarios, I feel lends creditability to Sgt. Schirmer's case. His lapse of conscious memory or missing time and his recall through regressive hypnosis strongly suggests that the event was a kidnapping of sorts and maybe not even that. Considering that here was a police officer who was armed and was ready to pull out his revolver to defend himself was no doubt standard police conditioning necessary in an unpredictable and singularly unique situation. Little wonder that the Extraterrestrial Intelligences on board the craft were not going to take chances with an Earth being who was ready to become aggressive and potentially dangerous. He needed to be disarmed both mentally and physically which explains the green mist or fog sprayed into and over the patrol car and the statement by the ETI to Schirmer, "Would you shoot at a space ship?"

It would seem that Schirmer was the recipient of ET knowledge and not the victim of some medical examination and harvesting of human DNA, skin, and sperm samples, even though, the ETI were forthright with their intentions that they were here for surveillance of the human race and a "breeding analysis program." Schirmer's description of the ET beings did not fit the typical "Grey" alien beings so often described by other abductees. These humanoid beings were a different racial group all together and obviously from another star system and another galaxy as was explained to Sgt. Schirmer. Given the position of authority, his truthfulness and dependability in his community, even though, Schirmer had no witnesses to his reported UFO/ETI encounter, he is considered by many UFO investigators as a credible witness.

This case has the air of respectability, but for one annoying little fact which tends to crop up too often in abduction accounts wherever, the witness is shown a *"star map"* of the so-called home world or star system in which the ET beings originate from as well as the out of place piece of evidence in which one little alien wore a *"cap or hat"*. This star map information event is similar to the **Barney and Betty Hill case** (which is currently suspected as one of the first *"staged MiLab events"*) and may actually be *"mental triggers"* or *"cues"* for memory recall designed to aid the abductees to remember some aspect of the event, rather than an information exchange or a goodwill gesture by ETs for any inconvenience forced upon the poor human victims. Jacques Vallee correctly points out that "the map makes no sense as a navigational aid." Apart from the inaccuracy in scale of the **Fish star map** interpretation from the Hill case, alien spaceships would surely navigate the same way we guide our own spacecraft: via computers and telemetry. The validity of the Fish interpretation and the Schirmer case and any similar case become irrelevant as the point here is that any such chart would have NO value to an interstellar star-traveller unless it was for no other purpose than for esthetics or art appreciation. **Jacques Vallee; Dimensions; 1988; Ballantine Books; ISBN 0-345-36002-8**

His case is listed in the **Condon Committee of the University of Colorado Report on UFOs** as Case No. # 42 and after numerous psychological tests conducted from University Colorado's Clinical Services, committee members concluded to the Condon project staff that they had no confidence that the trooper's reported UFO experience was physically real."
The U.S. Air Force; Dr. Edward U. Condon – Project Director; Scientific study of Unidentified Flying Objects; Bantam books, Inc; 1969

230

Pascagoula Incident

The **Pascagoula Abduction** or **Pascagoula Incident** occurred in 1973 when coworkers **Charles Hickson** and **Calvin Parker** claimed that they were abducted by aliens while fishing near Pascagoula, Mississippi. The case earned substantial mass media attention, and is, along with the earlier Hill Abduction, among the best-known cases of alien abduction. "It was the first time in any major UFO encounter that the witnesses' testimony was recorded so swiftly, and on tape, barely three hours after Calvin and Charlie saw the flashing blue light."
http://www.nicap.org/hicksontape.htm

On the evening of October 11, 1973, 42-year-old **Charles Hickson** and 19-year-old **Calvin Parker** — co-workers at a shipyard — were fishing off a pier on the west bank of the Pascagoula River in Mississippi. They heard a whirring/whizzing sound, saw two flashing blue lights, and reported that a domed, cigar-shaped aircraft, some 30 to 40 feet across and 8 to 10 feet high, suddenly appeared near them. The ship seemed to levitate about 2 feet above the ground.

Charles Hickson and Calvin Parker
http://www.nydailynews.com/news/national/man-1973-ufo-incident-turned-life-upside-article-1.1482818

A door opened on the ship, they said, and three creatures emerged and seized the men, floating or levitating them into the craft. Both men reported being paralyzed and numb. Parker fainted due to fright. They described the creatures as being roughly humanoid in shape and standing about five feet tall. The creatures' skin was pale in color and wrinkled, and they had no eyes that the men could discern, and slits for mouths. Their heads also appeared connected directly to their shoulders, with no discernible neck. There were three "carrot-like" growths instead - one where the nose would be on a human, the other two where ears would normally be. The beings had

231

lobster-like claws at the ends of their arms, and they seemed to have only one leg (Hickson later described the creatures' lower bodies looking as if their legs were fused together) ending in elephant-like feet. Hickson also reported that the creatures moved in mechanical, robotic ways. Hickson described being grabbed by the robot-like creatures and was floated aboard their craft, Parker had passed out or fainted and he too was floated onto the ET craft.

On the ship, Hickson claimed that he was somehow levitated or hovered a few feet above the floor of the craft, and was examined by some kind of instrument which Hickson had never seen before. When pressed by the interviewer that perhaps, it was like an X-ray machine

Hickson said it wasn't like an X-ray machine. *"There ain't no way to describe it. It looked like an eye. Like a big eye. It had some kind of an attachment to it. It moved. It looked like a big eye. And it went all over my body. Up and down. And then they left me."* **Ralph Blum and Judy Blum; Beyond Earth, Man's Contact With UFOs; Bantam Books, Inc; 1974**

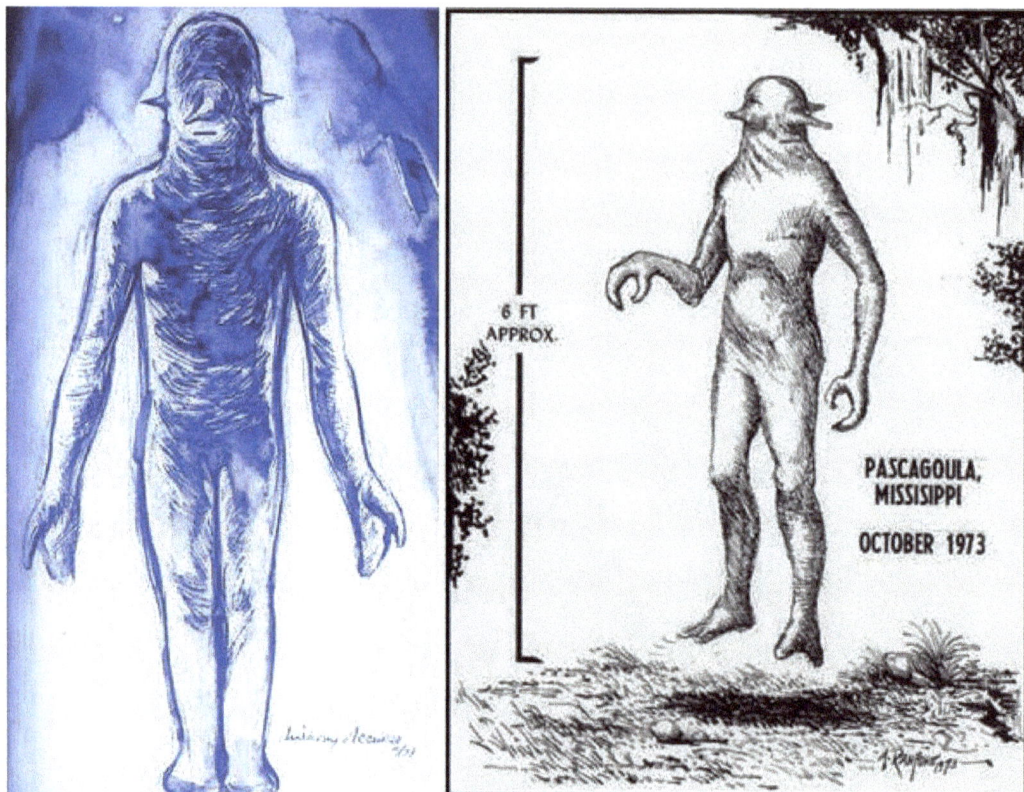

Two artist impressions of the Pascagoula, Mississippi ET being which Charlie Hickson and Calvin Parker as having clawed hands and fused feet
http://www.ufoevidence.org/cases/case97.htm and https://bizarreandgrotesque.com/2015/10/17/the-1973-pascagoula-alien-abduction/

Parker could not recall what had happened to him inside the craft, although later, during sessions of hypnotic regression he offered some hazy details. The men were released after about 20 minutes and the creatures levitated them, with Hickson's feet dragging along the ground, back to their original positions on the river bank.

Both men said they were terrified by what had happened. They claimed to have sat in a car for about 45 minutes, trying to calm themselves. Hickson drank some whiskey. After some discussion, they tried to report their story to officials at Keesler Air Force Base, but personnel told them the United States Air Force had nothing to do with UFO reports (Project Blue Book had been discontinued about four years before) and suggested the men notify police.

Within days, Pascagoula was the center of an international news story, with reporters swarming the town. **Professor James A. Harder** (a U.C. Berkeley engineering professor and **APRO** member) and **Dr. J. Allen Hynek** (an astronomer formerly with **Project Blue Book**) both arrived and interviewed Parker and Hickson. Harder tried to hypnotize the men, but they were too anxious and distracted for the procedure to work--Parker especially so.

Hynek withheld ultimate judgment on the case but did announce that, in his judgment, Hickson and Parker were honest men who seemed genuinely distressed about what had occurred. Both of the witnesses then took polygraph tests, and both passed without a problem. Harder and Hynek both believed that the two tormented men were telling the truth. Clark, Jerome, The UFO Book: Encyclopedia of the Extraterrestrial; Visible Ink Press, 1998

Before leaving this bizarre case to return back home in Chicago, UFO researcher **J. Allen Hynek** would proclaim that "... there was definitely something here that was not terrestrial." http://ufos.about.com/od/aliensalienabduction/p/pascagoula.htm

On October 21, 2001, Associated Press reporter **Natalie Chambers**, reported that a new witness to the now legendary **Pascagoula UFO Incident** had been discovered. Chambers identified this individual as **Mike Cataldo**, a retired Navy chief petty officer.

Cataldo saw the UFO through the windshield along with companions, Ted Peralta and Mack Hanna and thought at first that it was a shooting star which appeared to move from right to left. It then came down into a marshy, tree-lined area and seem to hover in the area for about 30 seconds. It was spinning and had blinking lights on the top of it all around its edge, all the way around it in a circle. These were blinking lights arranged on it just like you would tape lights to the side of a cake pan. Cataldo affirmed that there was a definite structure to be seen, not just blinking lights.

"This thing was like a whitish-gray colored sailor hat or a tambourine, and it was less than half a mile away and looked as big as any big American airliner I've ever seen. And then it just shot away, almost like it was just suddenly gone." *[This description of the shape of the craft is different than the description that Hickson and Parker witnessed.]* (My italics added). http://www.nicap.dabsol.co.uk/newwitness.htm

Cataldo said that shortly after the sighting he parted company with Ted Peralta and Mack Hanna and changed cars. Then while en route to Ocean Springs as darkness was setting in, he saw the same object a second time at about the same distance away, watching it for another 45 seconds to one minute before it again shot off in the same manner as before.

Mike Cataldo said that there were other motorists in the area during each episode and that some of them had slowed down to look at the UFO.
http://www.ufocasebook.com/pascagoulanewwitness.html

A subsequent investigation by Joe Esterhas of *Rolling Stone* uncovered some additional information, leading to much skepticism about the abduction claim. The supposed UFO landing and abduction site was in full view of two twenty-four hour toll booths, and neither operator saw anything that night. Also, the site was in range of security cameras from nearby Ingalls Shipyard, and the cameras additionally showed nothing that night.

The question to be answered here is: was this a genuine encounter with a strange spacecraft and its even, stranger robotic-like alien entities that took Hickson and Parker on board their craft for a brief but, mentally terrifying medical examination? Did Cataldo, Ted Peralta and Mack actually see the same object that abducted Hickson and Parker or did they see some other UFO which coincidently was in the same area at the time? *(note also, the previously mentioned difference in description of the object by the first-hand witnesses and the second-hand witnesses).* Why is it that the two 24 hour toll booth operators and the security cameras saw nothing or recorded nothing unusual at the time of the alleged abduction?

Allagash Abductions – The Weiner Twins, Rak and Foltz

The Allagash case is perhaps the first and best well-documented multiple-witness UFO abduction event in the abduction literature. The witnesses comprising twin brothers **Jack and Jim Weiner**, their friend **Chuck Rak** and their guide, **Charlie Foltz** (aka. the "Allagash Four") stated that the incident started on August 20, 1976, when the group of four men, all in their early twenties, ventured on a camping trip into the wilderness near Allagash, Maine.

 They say their first day went by without incident. However, on their second night, they noticed a bright light not far from their campsite which they first passed off as being a helicopter or weather balloon, but later noticed it displayed a strange quality of light. Suddenly, the object imploded and disappeared. The following day went by without incident, as the first. The men were unlucky in their fishing so they decided to try it at night. They set camp on the shore of Eagle Lake on August 20. As darkness settled, the men built a blazing campfire which they expected to burn for several hours to be used as a beacon for their campsite while out on the lake. They then headed out in a canoe.

After a short time, Rak noticed the bright light they had seen two nights before in the distance above the tree line. He called the others' attention to it. They watched the object intently and noticed it appeared to be much larger this time and made no sound. Foltz grabbed a flashlight and began flashing an SOS.

Suddenly, a bright beam of light shot out from the bottom of the craft and it quickly made its way towards the men. All men, minus Rak, began to paddle furiously back towards shore. Rak seemed entranced by the object as it closed in on them. Suddenly, the light enveloped the canoe and the four men.

The Allagash Four plus one. Left to right: Charlie Foltz, Chuck Rak, and the Weiner twins. The man sitting on the ground came out of the woods, stayed for a while and left, and never gave his name. (From Raymond Fowler's book: The Allagash Abductions)

Where did that strange woodsman, "the Park Ranger" go in this colour photo which is virtually identical to the black and white photo above? It is impossible to get two identical photos without some minor movement in people's positions

The next thing the men knew, they were back on shore at their campsite. They stood at the edge of the water and stared blankly at the craft, which was hovering no more than a few dozen feet from them. After watching for several minutes, the craft suddenly imploded as it had done two nights previous, and reappeared over the treetops on the other side of the lake. It then shot upwards into the sky.

The men suddenly all felt exhausted and decided to sleep for the night. The large fire they had made only minutes previous was now a pile of burnt embers. Without much conversation following the unusual incident which just took place, the men went to sleep. The next morning, the men spoke little of the incident and packed their belongings to move to a new campsite.

About 12 years after the incident, twins Jim and Jack suddenly began to have similar nightmares of being in a "medical examination room". They also dreamt of strange creatures with large heads and large black eyes. The twins told Chuck Rak and Charlie Foltz of their dreams and were surprised to learn the other men were having similar nightmares. The four men all claimed the same sequence of bizarre events occurred to them despite little to no discussion of the incident amongst them.

Under separately conducted hypnotic regressions, the men recalled being abducted from the canoe; they were levitated by the light up to and through the craft and finally being inside the craft that they had seen at Eagle Lake. While inside, they were subjected to several medical tests that have now become familiar to most researchers in the abduction phenomenon where blood, urine, sperm and skin samples were taken. During these tests, the beings telepathically informed the men that they would cooperate and that they would not be harmed. Each of the men described the beings as resembling large insects with "bug-like" eyes, although, there does appear to be differences as their drawings and descriptions indicate.

Jim and Jack Weiner's drawings resemble each other's closely with slight variations; Charlie Foltz's drawings seem to depict the usual "Grey" being that we've come to known but again, some similarities and minor variations with the two Weiner brother's drawings. Chuck Rak's description and drawing depict a being that looks more like an embryonic chicken head that was elongated from the top of the head to the bottom of the chin area. Even the description of the ETs hands was different depending on who was retelling their story under hypnotic regression. Some say they had four fingers, two of which were opposable to the other two fingers, like the human opposable thumb which permitted the palm area to close together like a sea-clam opening and closing. Chuck thought that the aliens only had three fingers that were functional, while a smaller fourth finger was vestigial. The differences in memory can be chalked up to mere differences in opinion and perception or the fact that there may have been more than one alien species on board the spacecraft. **"The Allagash Abductions"; Raymond E. Fowler; 1993; Wild Flower Press; ISBN 0-926524-22-4**

Jack Weiner has stated that, over the years, he has seen strange lights in the skies near his house and dreams where the aliens had re-appeared to him warning him, "not to tell."

It is unusual to find four men who are all artist in their own right, all sharing passion for art fishing, camping, and canoeing the great outdoors but who've also, experienced an extraordinary alien contact event together.

Author's Rant: I had an opportunity to meet and interact with Chuck Rak in June 1996 at Crestone, Colorado during a CSETI week long seminar hosted by Dr. Steven Greer and his late partner, Shari Adimak. I had asked Chuck to do a caricature drawing of me on the last day of the CSETI event, taking the opportunity to get an amusing sketch of myself but, more importantly, to talk to Chuck further about his Allagash Lake abduction story.

He told me that up to that very last day in Crestone, he felt that the abduction event that he and his friends had experienced may not have been exactly as recalled under hypnotic trance. That he felt that Raymond Fowler, author of "The Allagash Abductions" was trying desperately to make the pieces of the story, all fit together. Chuck had thought that the event may have been "staged", that it may have been a military intelligence operation but, he had no way of knowing for sure. I also questioned him about the strange bald-headed man with the mustache wearing the loafer shoes, who appeared out of nowhere just as the four men were packing up to leave their campsite. He was photographed with Chuck and his artist buddies standing on the lake shoreline, while he sat on the ground in front of them. Who was he and who took the photograph? Chuck was unable to identify who this person may have been whether he was a "park ranger" or another camper from nearby who happen to just be in the area.

Some researchers have said that he was a "park ranger" of the area, but to date, that information has not been substantiated. On this matter, Chuck has no idea who he was or why he showed up out of the blue when he did. From the two photographs above, it is obvious that in the colour photo, the "park warden" is missing from the picture perhaps erased by "photoshop" or by some similar film altering process for whatever reason. Was this stranger another camper who happen to be nearby during the time of the abduction event? Did he witness what had happened to the "Allagash Four" or as Chuck Rak suspected, that the abduction event was a staged covert military event and that this stranger was a part of what took place and merely showed up to see the aftermath effects on the abductees? We will have to wait and see if further investigation into this case will produce new informational leads.

https://www.youtube.com/watch?v=Da9UKjvdvJk

CHAPTER 30

MIND CONTROL AND ABDUCTIONS

Martin Cannon and MKULTRA

The strongest scenario for alien abductions was proposed by **Martin Cannon**, who argues that memories of alien abductions might, in fact, have been created in the "abductees" by a secret government mind control program, such as **MKULTRA**. In a lengthy paper entitled the ***"The Controllers,"*** Martin Cannon builds an exceptionally persuasive discourse that most abduction cases emanate chiefly from that alphabet soup known as the intelligence community. (Much of what follows comes from the extensive research and writings of Martin Cannon's thesis and papers on this mind control subject). Cannon states that "an unprejudiced overview of abduction reports in the popular press and the less-familiar material on mind control will demonstrate a striking correlation. Once other abduction researchers have been educated in the ways of **MKULTRA,** they may note similar patterns and that the abduction enigma contains within it many sub-mysteries that slide into the mind control scenario then, they can begin to write a revisionist history of the phenomenon.

"The with surprising ease, even elegance — mysteries which fit the E.T. hypothesis as uncomfortably as a size 10 foot fits into a size 8 shoe. As we have seen, the **MKULTRA** thesis explains the reports of **abductee intracerebral implants** (particularly reports involving nosebleeds), unusual scars, "telepathic" communication (i.e., externally induced intracerebral voices) concurrent with or following the abduction encounter, allegations that some abductees hear unusual sound effects (similar to those created by the hemi-synch and cognate devices), haywire electronic devices in abductee homes, personality shifts, "training films," manipulation of religious imagery, and missing time. Needless to say, the thesis of clandestine government experimentation readily accounts for abductee claims of human beings "working" with the aliens, and for the government harassment that plays so prominent a role in certain abductee reports." **The Controllers: a New Hypothesis of Alien Abductions; Martin Cannon; 1990**

All theories of UFO abductions except the skeptical ones put out by people like Philip Klass; include some aspect of the concept of mind control. Now, if people's minds are being controlled, and this technology is certainly in existence, then the question to be asked: can we trust the participant's reports of what they are seeing even, in terms of the UFO's that they claim to be seeing, but in particular, the nature of the abduction experience itself?

Drawing from a very old example of hypnotism, Cannon states, "it was a very common practice, going back many, many decades, to see if somebody was truly under a deep hypnotic state, they would suggest to the person that there was a small, black dog in the room, and he's coming up to you, and would you pet him. The subject would often, actually see the dog as being physically real as they would see any normal dog that may come from an animal shelter. Now, if the human brain can be tricked to that extent, then is it not possible that the ET's that people are seeing are of an exactly like substance to that dog?"

"The government's involvement in mind control operations would have people believe that this was something that they were doing back in the 50's and the 60's, and it was all to play catch up with the Russians who had this huge lead in the field, but then, they stopped doing it around 1963 because they never really found anything to it. It's all a lie!"

Cannon says that these programs went very far beyond this. The U.S. were far ahead of the Russians in psychic research and mind control as mentioned in a memo in Cannon's possession, where **Allen Dulles** (former CIA Director) admitted as much to the **Warren Commission**.

Mind control goes back to World War II, possibly to the 30's, in fact, occult groups have been doing experimentations with what they call electronic mind control, going all the way back to the 19th century (the 1880s). In fact, there isn't a single technology of mind control that doesn't go back to that time." **UFO CONTACT CENTER INTERNATIONAL GROUP MEETING; 1988; Speaker: Martin Cannon; Subject: UFOs and Mind Control**

There is increasing evidence that various departments of the U.S. intelligence community including the **Central Intelligence Agency (CIA), Defense Advanced Research Projects Agency (DARPA),** the **Office of Naval Intelligence (ONI),** the **National Security Agency (NSA)** and the **National Reconnaissance Office (NRO)** even, the **National Aeronautics and Space Administration (NASA),** the **Atomic Energy Commission (AEC)** and most branches of the Defense Department are involved in mind control covert projects such as **Artichoke, Bluebird, Pandora, MKDelta, MKSearch** and the now infamous **MKUltra.**

"Spy- chiatrists" as Cannon calls them can be found working behind the scenes in college campuses, CIA sponsored institutions and in prisons. Cannon tells of laundry lists of mind altering techniques employed by these spy agencies that make the everyday mind-altering street drugs found in most cities' poorer districts look like a simple candy popping addiction. These mind altering and behaviour altering techniques include erasure of memory, hypnotic resistance to torture, truth serums, post-hypnotic suggestion, rapid induction of hypnosis, electronic stimulation of the brain, non-ionizing radiation, microwave induction of intracerebral "voices", and a plethora of other disturbing technologies.

The need to control people, particularly masses of people and whole societies is an aspect of human behaviour that is thousands of years old, In ancient times, the overt nature to control people was accomplished through war, political and religious conflict and internal turmoil pitting one political or religious faction against another in order to achieve the desired result or agenda of the party-politic or the economic group usually lead only by a few individuals who seek power. Kings and political rulers from many countries and the ecclesiastical orders from many of the world's major religions of the day have been at the heart of most of the world's problems, ever since civilizations first started springing up. Power, control, ideology, and greed have been the source of corruption for anyone and any nation that have arisen to seize the scepter of power and control. The annals of human history are littered with such accounts. Hundreds of millions of people have died because of one man's need to seek leadership or exercise control over others, either in his own country or over the peoples of other nations.

Yet people are attracted to and repelled by authority. ***"On one hand, they have recognized that it is essential to the achievement of any degree of peace, order, and freedom. On the other, they have been repeatedly disillusioned by the harm it has eventually inflicted on these same cherished achievements. It has seemed impossible either to live with authority or to live without it."*** "The Power of the Covenant, Part One and Two"; National Spiritual Assembly of the Baha'is of Canada; 2000; Baha'i Canada Publications; ISBN 0-88867-101-6

Today, these same basic attitudes and behaviours continue to be the main reason and motivation behind the hidden agendas of political leaders and governments. The only difference in this day and age is the use of advanced methodologies and technology to manipulate the minds of others without their knowing it.

We will look at a few of these mind control techniques especially those that are closely associated with cases of alien abduction. We will discover that the "advanced alien technologies" used on abductees which seem so futuristic has an unmistakable terrestrial origin that dates back to the 19th century (**Electronic Stimulation of the Brain – ESB** was conducted in 1898 by a professor of psychology in Straussbourg). The public literature on this research has most certainly been sanitized from the original research literature. In the black world of covert technology, there is almost unlimited funding for researchers and their **USAPs (Unacknowledged Special Access Projects),** *beyond the oversight and control of the government and unstrained by peer review where major breakthroughs can be achieved more rapidly than by scientists in the outside world.*

Most of us all know of someone and of the illegal drugs that are sold on our city streets to these poor drugs addicts whose lives are so damaged and dependent on their next fix that they can no longer function as normal contributing members of society. Yet, it is these very same drugs that were developed and used during World War II by both sides of the war to act as a "truth drug" against captured enemy soldiers and prisoners of war such as used in Dachau on the Jews, Slavs, and gypsies. Mind-altering drugs in combination with hypnosis as used by the Nazi scientists proved to be efficacious in producing the desired compliance in the subjects. Interestingly, the results from these tests were made available to the United States after the war, along with other heinous experiments during **Operation Paperclip (Project Paperclip),** where many Nazi scientists were brought over to the U.S. and allowed to continue their work but under tight supervision and control of the ever-watchful eye of the American military.

We will examine later, just how far the scientific and political influence of this Nazi "incursion by invitation" via Operation Paperclip extended into the psyche of the American Government.

The CIA's involvement in mind control required a cover for their programs, Project Bluebird and Artichoke by using propaganda to convince the world that the Communist Bloc were devising insidious methods to control and shape the human will. This propaganda ploy has become a standard tactic to deflect unwarranted investigation into their secret deep black programs. The CIA studied and employed hypnosis, conditioning, brain implants and even ESP and when MKUltra was leaked to the public in the 1970s, the public's focus was on drug and ESP experimentation while the CIA's focus was on the development of psycho-electronics or psychotronics.

One of the most interesting pieces of an alleged alien abduction is the intracerebral (brain) implants that often show up on X-rays and MRI scans of many abductees. Abductees often describe an operation where a needle is inserted up through the sinus cavity and into the centre of the brain with the audible reported crunch or cracking sound. Abduction specialists and ufologists attribute the intracranial incursions as the handiwork of interstellar scientists for the purposes of tracking, monitoring and possible control of human emotions, thoughts, and behaviour. Cannon points out that researchers have failed to familiarize themselves with the advances in terrestrial medical technology which in this and other aspects of UFO and ETI research points to a glaring short-sightedness in investigative methodology.

Miniature electrode implants into the brain such as the **"stimoceiver"** invented in the late 50's to early 60's by neuroscientist Jose Delgado, may also be positioned in other subcutaneous parts of the body such as the hands, arms, legs, abdomen and groin or reproductive areas. The fact that such implants have been found in almost every conceivable part of the body, as claimed by UFO abductees, suggests that "alien implants" are placed in humans in an almost hide and seek fashion for the UFO researcher and medical specialists to try and discover.

The stimoceiver is a miniature FM radio receiver and transmitter which enables the subject to be controlled by an outside operator. This device not only worked in animals but also, in humans to the degree that emotional and behavioural responses were easily elicited in such "extremes of temperament as rage, lust, fatigue, etc." "A variety of effects included pleasant sensations, elation, deep thoughtful concentration, odd feelings, super relaxation, coloured visions and other responses such as fainting, fright or "floating." The term "coloured vision" clearly indicating remotely induced hallucinations which may be controlled by an outside operator.

"Floating feelings" are also, recognizable to Ufologists when abductees describe the opening minutes of their experience of floating or flying aboard the alien spacecraft. Humans could, in essence, be controlled like robots via two-way communication from a computer with these type of intracerebral implants. ***God forbid that such a notion would even become a reality!*** [**My italics added**]. Ralph Blum and Judy Blum; Beyond Earth, Man's Contact With UFOs; Bantam Books, Inc; 1974

Author's Rant: As a teenager growing up in North Bay, Ontario in the mid-60s, I watched a lot of television; I was addicted to its influence. Television back then was a media where the public could be both entertained and educated, sometimes at the same time. Science advancements with new technology were often broadcast over the airwaves and like many viewers, I received some of my education right off the TV set. The news media would frequently inform the viewing audience that Science was doubling every two years and that it wouldn't be long before it would double every two months or every two weeks or less.

It was during those teenage years of television viewing that a news piece was aired, that discussed and demonstrated in a somewhat frightening manner, a rhesus monkey with a brain implant (see below) showing how neuropsychologists could control the behaviour and emotions of these monkeys. I was at that time, a witness to the historical beginning of a potential and inevitable hijacking of our future and that mind control in one form or another was to become an important tool in the shaping of that future. Today one would

have to wonder how far science had advanced with this type of technology, as the popular press and news media of the time stopped reporting on such things after the late 60s.

Rhesus Monkey in restraints experiencing electrode stimulation to its brain in an experiment by José Delgado, infamous mind control researcher. (Google Images)

The quandary in this whole subject matter, much to the chagrin of the critics who state that such intracerebral implants didn't exist back then, which has been demonstrated to the contrary is that scientists have had the capability to create brain implants like the ones seen in abductee MRI scans. In fact, advancements have continued far beyond implants and ESB effects can now be achieved with microwaves and other forms of electromagnetic radiation with or without the use of electrodes.

So why, if we accept at face value the stories of UFO abduction are the "advanced aliens" using outdated old terrestrial technology, a technology that is probably by our own Earth standards rendered obsolete if it hasn't been already? Cannon recalls the "charming anachronistic old flash Gordon TV serials where swords and spaceships clashed continually," then adds whimsically, "Do they also watch black and white television on Zeta Reticuli?"

The imagination and ingenuity of humans appear to know no boundaries, as this same stimoceiver was attached to the tympanic membrane of a cat, effectively turning the cat's ear into a simulacrum of a microphone. The assistant would whisper into the cat's ear and Delgado could hear the words over a loudspeaker in another room. It seems that not only is our paranoia justified that our phones may be "bugged," but now, so is Kitty in on the spying business.

Cannon posits that Delgado's experiments could also make a subject's ear into a loudspeaker beside turning it into a microphone, which may explain the "voices that abductees claim to hear and indeed such ear implants were used to allow the "therapist" to communicate with the subject. Keep in mind these are not simply hearing devices for the impaired but miniature radio transceivers which we have all now become familiar with in many spy movies, but in the 50's and 60's this was cutting edge technology. Today this type of technology has undergone many generations of upgrading and improvement.

Because of its ability to control by turning on and off emotions, to re-enforce certain compliant behaviour and affect memory (all the aspects currently found in UFO abduction cases), it was even used at one point to effect a cure for homosexuality. According to Joseph A. Meyer of the **National Security Agency (NSA),** there is potential use in controlling "socially troublesome persons" like criminals and "urban dwellers" from New York's Harlem district, which is, of course, a black community. The technology is not limited to animals and the socially challenged but even children could be implanted and tagged as a way to keep track of them for health and safety reasons that are numerous and for the sake of "peace of mind" for the parents. We should not forget the "benefits" of implanting a rice-size tracking implant into our senior citizens, whose health and well-being is of prime concern to their children and to society in general. The usages are virtually unlimited in a modern day society. Remember, all this "mind management system" could be easily controlled and monitored by computers from a secure location at considerable distance from the subjects.

Through R & D and the limitless financial resources of the Military Industrial Complex, brain implants were superceded by more exotic forms of mind control, "ESB effects could now be elicited with microwaves and other forms of electromagnetic radiation, used with and without electrodes." However, Hypnosis remained at the heart of mind control and no actual physical contact was required between the "hypnotherapist" and the victim (abductee) when the stimoceiver or similar device was used. This practice became known as remote hypnosis and was first used in the early 1930s by parapsychological means, by L. L. Vasiliev, Professor of Physiology at the University of Leningrad. In the U.S., the CIA has mastered a technology called **"Radio Hypnotic Intracerebral Control – Electronic Dissolution of Memory" (RHIC – EDOM).** Thus, these techniques can remotely induce hypnotic trance, deliver suggestions to the subject and erase all memory for both the instruction period and the act which the subject is asked to perform. **The Controllers: a New Hypothesis of Alien Abductions; Martin Cannon; 1990**

The microminiaturized offspring of the stimoceiver technology interestingly involves the use of intramuscular implants which is not so dissimilar to the "scars" mentioned by abductees in Budd Hopkins "Missing Time." These implants are stimulated to induce post-hypnotic suggestion. In other words, EDOM is nothing more than missing time itself – the erasure of memory from consciousness by blocking the synaptic transmission in certain areas of the brain.

Cannon goes on to state that other fellow mind control researchers like Walter Bowart and James Moore have found that RHIC – EDOM techniques for sending radio signals to certain parts of the brain of an abductee/victim can trigger emotional responses such as anger or rage for any apparent reason or for no reason or cause at all, at the discretion of the controller. Even the controller can conceal a small EDOM generator- transmitter on his person which can transmit a small electrical charge plus an ultra-sonic signal tone by merely shaking hands or by a slight touch of the person, which will cause a time disorientation of the person affected. **The Controllers: a New Hypothesis of Alien Abductions; Martin Cannon; 1990**

It should now start to become obvious that to any unbias investigator that a terrestrial explanation for alien abductions which employs RHIC – EDOM technology resolves much of the mysterious "missing time" scenarios. Even, "alien implants, both intracerebral and intramuscular, etc. are explained. The reference to "recurring hypnotic state," which can be "reinduced automatically by the same radio command" may also account for the "repeater" abductees who claim to have on-going sessions of "missing time" and abduction – even while a bed-time mate sleeps undisturbed. Cannon cautions the reader and investigator with a caveat that the RHIC – EDOM technology may only be in the development stage but also states that the subject is taken seriously by most people in the intelligence community and a lot of research and development has occurred since the late 70s when the CIA was questioned on this matter.

It should be noted that MKULTRA may have investigated and even used RHIC – EDOM in its earlier days but according to Lawrence and Moore it appears to be a product of Military research. The CIA program of MKULTRA which was officially terminated in 1963 begat other programs of **MKNAOMI, MKACTION,** and **MKSEARCH**, etc. that may have continued with RHIC – EDOM but, the main focus of the CIA was on psychoelectronics.

"Science marches on" to better and improve the life of society and at least this adage is certainly true in the field of mind control of society's citizens. Such "mind machines" to promote creativity, stimulate learning and alter consciousness i.e., to produce a drugless, more natural high, could also induce "Out- of-Body Experiences" (OBEs). These OBEs gave the percipient that ability to mentally travel to another location while leaving the body behind and at rest. Shaman and certain cultures globally have practiced this ability since ancient times, without the use of mechanical devices but perhaps, sometimes with drug compounds or with meditative exercises to induce this state of being for either personal or religious reasons.

Hemi-synch headphones playing slightly different frequencies in each ear causes the brain to calculate the difference and produces a binaural beat rhythm which "entrains" the brain to this beat. This can cause the subject's EEG to either speed up or slow down and because the brain has a beat of its own, certain frequencies are associated with a relax, alert state (Alpha – 8-13 cycles per second), agitation and intense mental concentration (Beta – 14-30 cycles per second), a hypnogogic state (Theta – 4-7 cycles per second), and sleep (Delta - .5-3.5 cycles per second). It didn't take a rocket scientist but, a perhaps a scientist in neuropsychology on the payroll a government or military-funded project to see the potential of "Brain Entrainment" especially if the subject has a suitably entrained brain making him more responsive to suggestions and vivid hallucinations.

"Stereophonic sound" is one of the effects that UFO abductees describe preceding many "encounters" particularly if a stimoceiver is implanted in the ear canal of the abductee in which to transmit a certain frequency. Many inexpensive devices are on the market for personal consumer use but in the hands of government scientists engaged in clandestine bioelectric research projects, such devices in combination with other previously described techniques can produce mind-boggling and mind-blowing effects. **The Controllers: a New Hypothesis of Alien Abductions; Martin Cannon; 1990**

244

One final piece of technology needs to be discussed before we leave this subject and that is the use of microwave beam radiation. Most people are familiar with a microwave oven, which cooks your meals with high energy microwaves resulting in at least in this writer's opinion a meal that is this side of a rubbery cadaver-type consistency particularly when "nuking" meat! But in the context of mind control, we are speaking of **extremely low frequencies (ELF)** and ultrasonics.

Have ELF microwaves been used on an unsuspecting American public? According to most senators, congressional representatives and state representatives, they have in their possession many "wavie" files. Wavies are people who claim they are the victims of clandestine bombardment with non-ionizing radiation – or microwaves. Most cases are typically dismissed but there are too many for all to be the mere imaginings or ravings of the socially challenged or psychologically deranged. Cases come in from all strata of society from the poor and transient to the working class and the professional. The Wavies report sudden changes in psychological states, alteration of sleep patterns, intracerebral voices ("voices in the head"), and other sounds and physiological effects.

The effects of low-level microwaves on the mind range from metabolic changes, alter brain functions, disrupt behaviour patterns. Pulsed microwaves can create leaks in the blood/brain barrier, induce heart seizures, and behavioural disorganization. A **Rand Corporation** scientist reported that microwaves could cause insomnia, fatigue, irritability, memory loss, and hallucinations. Needless to say, a plethora of emotional and behavioural effects could be inducted into the subject from pleasant positive states to extreme negatively hostile states. Microwaves acted much like "hemi-synch" devices by entraining the brain to theta rhythms.

But all these effects become secondary as Cannon points out that microwaves could also produce booming, hissing, buzzing and other intracerebral static or sounds! In fact, it was demonstrated that spoken words could be transmitted and "heard" via a pulsed-microwave analog of the speaker's sound vibrations. What we see here is that by means of low frequency microwaves, a modulated electromagnetic energy i.e., **"radio frequencies" (RF)** could be transmitted **"directly into the subconscious parts of the brain"** without employing any technical devices for receiving or transcoding the message and without the person exposed to the influence having a chance to control the information input consciously.

In other words, by using an RF carrier frequency carried across a microwave beam using a broadband dispersal pattern or a pinpoint beam aimed e.g., either in a room full of people or at a selected "target" in that room, that person would hear a spoken voice in his head not knowing who or where it came from. He would believe, if it were the intentions of the controller of the microwave device, that he was hearing the "voice of God or any god of his choice!" **The Controllers: a New Hypothesis of Alien Abductions; Martin Cannon; 1990**

Interestingly, modern psychology has now acknowledged that people hearing voices in their head may not necessarily be diagnosed as schizophrenic but actual victim of "psychotronic targeting." As the literature and the case, reports support this technology. *I personally know of someone who worked in the President Jimmy Carter Administration, who stated that he was such a victim of psychotronic targeting.*

A targeted subject could be driven crazy with "voices" or be delivered undetectable instructions as a programmed assassin. *"Shades of a real-life Manchurian Candidate scenario!"* [**My italics added**]. In such a case, who would listen to the ravings of such a person when electronically-induced hallucinations parallel exactly the classical signals of paranoid schizophrenia and/or temporal lobe epilepsy?

There has been much speculation with some supporting evidence that suggests that **Lee Harvey Oswald,** the assassin of **President John F. Kennedy** in Dallas, Texas in 1965, was publically executed before his trial by **Jack Ruby.** It is believed that Ruby was a *"programmed assassin"* of the CIA or of some other intelligence agency who was programmed to kill Oswald in order to prevent him from revealing his part in the planned presidential assassination and his CIA connection. Before Oswald's untimely death, he told a reporter that he was just a *"patsy"* suggesting that there was a conspiracy ordered from higher up in the echelons of officialdom.

Then, there is the assassination of **Senator Robert Kennedy**, brother of the late President John Kennedy, on June 5, 1968, in Los Angeles, California by Palestinian immigrant Sirhan Sirhan, the evidence demonstrates Sirhan's odd mechanical or robotic, dissociative behaviour which is an indication of mind control programming.

With the tiny electrode "stimoceivers" implanted in an animal like a monkey or even a human, they can act as amplifiers of the electromagnetic effect. Kathleen McAuliffe, "The Mind Fields," OMNI Magazine, February 1985.

The critics may counter that any microwave device and the power from it, when aimed at someone by some clandestine operator from a low flying helicopter into an abductee's bedroom or from a truck passing alongside a subject's car, would fry or cook *(your choice of the eventual outcome)* [my italics] the victim before the operator downloaded his thoughts.

That's fair criticism, but the device is more than likely built with a multi-function capability much like the **Star Trek Phaser** from the same TV show, where the Phaser could be set for kill or stun, for wide dispersal or pinpoint accuracy. Military and intelligence agencies expectations are always high when it come to weapons development and of course, money is no barrier, so it would be no different with such brain implants and microwave technology.

All the pieces of the puzzle of the alien abduction scenario are now in place! "Once an abductee has been implanted and if we are to trust the regressive hypnotic accounts of abductees at all, the first implanting session may occur in childhood – the chip-in-the-brain would act as an intensifier of the signal. Such an individual could have any number of "UFO" experiences while his or her bed partner doses comfortably." The Controllers: a New Hypothesis of Alien Abductions; Martin Cannon 1990

The enigma of the abduction phenomenon is a riddle within a mystery, a multi-layered conundrum of confusion surrounded with poorly acquired field data which the abduction research of Ufology has paraded and trumpeted as if there were no other viable conclusion or perspective on the subject matter.

To date, Martin Cannon's theory offers perhaps, the best explanation of the alien abduction scenario in the field of Ufology, as it answers all the oddities and unusual circumstances surrounding the alien abduction phenomenon. I suspect that most of the other aspects of abduction, such as seeing and handling hybrid children or seeing glass containers with humans or Grey ETs in liquid suspension are most probably, "window dressing" to convince the abductees of a "real-time ET experience."

Cannon has demonstrated that the MKULTRA thesis explains the reports of abductee intracerebral implants with its associated nosebleeds, unusual scars and strange pattern marks on the body, mysterious "telepathic" communication induced externally during or following the abduction encounter, reports from abductees that hear unusual sound effects much like hemi-synch and cognate devices, strange and bizarre electronic devices in or near the abductee's homes, personality shifts, "training films," manipulation of religious imagery as in the case of *"The Betty Andreasson Affair"* by Raymond Fowler and missing time. Covert programs involving government experimentation easily accounts for abductee claims of human beings "working" with aliens as well as government harassment are main factors in most abduction cases. The Controllers: a New Hypothesis of Alien Abductions; Martin Cannon; 1990

First Brain Implants

Rauni-LeenaLuukanen-Kilde (born 1939) was the provincial medical officer of the Finnish Lapland Province with a doctorate in medicine from 1975 until a car accident in 1986, which took away her ability to continue her work and career. Since then she has been best known for her UFO contacts and related thoughts. According to Kilde, the United States and other nations have been involved in covert **microchip mind control** research.
http://www.wikidoc.org/index.php/Rauni-Leena_Luukanen-Kilde

According to **Kilde**, "brain electrodes were inserted into the skulls of babies in 1946 without the knowledge of their parents," electrical implants were used in the 1950s, and, two decades later, the first brain implants were surgically inserted in 1974 in...
Ohio".http://www.examiner.com/article/secretly-forced-brain-implants-a-brief-history and Microchip Implants, Mind control, Cybernetics

Technological Advances

Technological advances have enabled researchers to continuously improve upon the size and quality of the implants, Kilde asserts, stating that the initial implants were one centimeter in size, but later shrank to "the size of a grain of rice and, whereas early implants were made of silicon, later ones were manufactured from gallium arsenide. "Today," Kilde says, the implants "are small enough to be inserted into the neck or back, and also intravenously in different parts of the body during surgical operations, with or without the consent of the subject," and, using low-frequency radio waves and satellites, "it is now almost impossible to detect or remove them".
http://www.wikidoc.org/index.php/Rauni-Leena_Luukanen-Kilde and Microchip Implants, Mind control, Cybernetics

Location and Tracking Capabilities

Every newborn baby could be implanted with implants that would enable government officials to locate and track individuals at all times, anywhere, and Sweden has authorized the implantation of prisoners with such devices, Kilde claims. According to a May 1995 article in *The Washington Post*, the British government has implanted **Prince William** when he was 12 years old so as to be able to pinpoint his location if he were ever to be kidnapped by targeting his microchip "with a specific frequency". [Microchip Implants, Mind control, Cybernetics](#)

Martin Anderson, a former top advisor to **U.S. President Ronald Reagan**, apparently shares Kilde's concern with the possible abuses of such technology. In the October 11, 1992, issue of *The Washington Times* Anderson declared that an "identification system" manufactured by the **Hughes Aircraft Company** uses a "syringe implantable transponder" that allows "identification using radio waves" by implanting "a tiny microchip the size of a grain of rice... under the skin." This microchip, he declares, can "be injected simultaneously with a vaccination or alone" and "contains a 10 character alphanumeric identification code that is never duplicated." The microchip can be read by passing "a scanner... over the chip." The scanner "emits a 'beep'" as the identified individual's "number flashes in the scanner's digital display."

Military Applications

In addition to locating individuals anywhere in the world at any time, brain implants have military uses, Kilde says. For example, "The U.S. **National Security Agency's (NSA)** 20 billion bits/second supercomputers could now 'see and hear' what soldiers experience in the battlefield with a **remote monitoring system (RMS)"**. Furthermore, by electronically manipulating the brain, enemy soldiers can be made to hallucinate, hearing and seeing whatever remote operators want them to perceive. This technology, she says, was used to monitor and record the thoughts and emotions of U. S. astronauts. [Microchip Implants, Mind control, Cybernetics](#)

Kilde's contentions are seconded by an otherwise unidentified "group of scientists" who report that a program known as *"the 'Phoenix Project'*, which had to do with Vietnam veterans, employed a chip called *'the Rambo Chip'*," which "would actually cause extra adrenaline flow". http://www.israelect.com/reference/WillieMartin/NEWS-35.htm

Cognitive and Emotional Manipulation

Microchip mind control technology has other, equally nefarious uses, Kilde contends: without their knowledge or consent, a person "can be manipulated in many ways," such as:

- having emotional life changed
- being made "aggressive or lethargic"
- having "sexuality. . . artificially influenced"
- having "thought signals and subconscious thinking" monitored
- having dreams "affected" or "induced". [Microchip Implants, Mind control, Cybernetics](#)

248

"Zombified" Assassins

Such technology can also be used to "**zombify**" individuals into assassins, Kilde states, so as to make them act against their own conscience and values. She says that such manipulation has been used in the United States to transform unsuspecting individuals into assassins who have no knowledge or memory of having committed murder, and "the latest supercomputers are powerful enough to monitor the whole world's population," so no one is immune to microchip mind control operations. Microchip Implants, Mind control, Cybernetics

Reportedly, the chief **Central Intelligence Agency (CIA)** "mind control researcher, **Dr. Sidney Gottlieb**," who headed the **MKULTRA** behavior-control program, "told the Agency director that "research findings by scientists demonstrated that humans could finally be programmed to attack and kill on command". http://www.israelect.com/reference/WillieMartin/NEWS-35.htm

American Psychiatric Association Involvement

According to Kilde, the **American Psychiatric Association (APA)** is a silent partner with government mind control researchers, using their *Diagnostic Statistical Manual IV*, which is "printed in 18 languages" to provide a basis for labeling victims of microchip mind control as paranoid schizophrenics. Microchip Implants, Mind control, Cybernetics

Certainly, a serious review and analysis of the abduction data, along with more insightful research into terrestrial medical technology and the intelligence community's practices and methodology in mind control techniques would no doubt bring about a revisionist rewriting of the historical facts behind alleged alien abductions.

Betty and Barney Hill Abduction Experience Re-Visited

Another revisionist perspective on alien abductions can be made for the famous **Betty** and **Barney Hill** case where "retro-technology" appears to have been employed by Grey visitors from the "home world" of **Zeta Reticuli**. The full account of this case can be found in **John G. Fuller's** book: "The Interrupted Journey" (New York: Dell, 1966) and will not be recounted again here, as this case was discussed earlier in the section: **The Beginning of the Modern Age of UFO Sightings.**

At one point in the alleged UFO abduction, Betty Hill had a needle inserted into her navel by one of the ET examiners, informing her that this procedure was a test for pregnancy. Rather hastily and rashly, some Ufologists have assumed that Betty's "pregnancy test" is evidence of advanced alien technology, since her 1961 account predates the official announcement of amniocentesis, which does make use of a needle inserted in the navel. Other, much less invasive tests for pregnancy than amniocentesis are now used, like a home kit for peeing on a test strip of paper to see a colour change in acidity of the urine. Amniocentesis is still used on occasion to gather information on the fetus but why would a highly evolve interstellar race visiting our planet to perform pregnancy tests not use a more advanced technology instead of one that is frequently used by Earth medical doctors back in the sixties?

Most medical breakthroughs undergo years of testing before official disclosure of their discovery is made known publicly. In Betty Hill's time finding medical volunteers for new medical procedures was a major obstacle, unlike today's more open-minded society. Perhaps, Betty was the unwilling test subject in an "external and objective stimulus" for a new medical procedure carried out by humans, ten years ahead of its time as oppose to encountering Extraterrestrials (reputedly a "billion years ahead of us") using science from eons before their own time?

Barney and Betty Hill circa 1966
http://bswett.com/1963-09BettyAndBarney.html

Delving deeper in the Hill case we find that Betty Hill's aliens seemed to have no grasp of basic human concepts (such as how we measure time) — yet they knew enough about us to speak English fluently and had even mastered our slang. Martin Cannon asks in his thesis on mind control: *The Controllers* if these were real aliens, or humans engaging in theatricals (and occasionally muffing their lines)? For that matter, why did Betty Hill originally recall her abductors as humanoid, only later describing them as aliens?

Intriguingly, Barney hill had quite a few friends in the Air Force intelligence at Pease Air Force Base (through church functions) and it is suggested that the Hills were singled out as candidates for possible black ops activity given their relationship with some members of the intelligence community and for "the clandestine services to satisfy a number of itches with one scratch."

On an internet Blog Talk Radio program called **"The Black Vault,"** Joe Montaldo interviewed **Jim Marrs** on the 12th of October, 2008 about alien abductions and **Milab (Military Abductions)** with regard to the Hill abduction case. Montaldo stated that information gleaned from the actual Hill hypnotic regression tapes of which there were 17 tapes and from the two books originally published don the case indicated that the hills were actually abducted while still in Quebec, Canada. Montaldo posits that there were three alleged abduction accounts which somehow became mixed and muddled together into one account which most Ufologists are familiar with in the current UFO abduction literature.

The first account Betty talks about her dreams, in the next account version Betty talks about what happen while under hypnotic regression and in the third account of her experience she talks about her waking thoughts. All three versions of the account are different.

When Barney recalls the UFO experience, he always talks about that it was people dressed in black pants, black shirts and black "duck-bill" caps which appears strangely reminiscent of World War II black Nazi uniforms. Today, most Ufologists would describe this as Black Ops fatigue uniform but back then, they would not have known this at that time. Even Betty confirms this clothing attire detail during her regression sessions,

It appears that from the regression tapes that an actual alien abduction scenario took place in Canada according to Joe Montaldo research and when the Hills crossed the border into the States, they were re-abducted again by Black Ops and re-programmed. Some of these black uniformed people were described as one being a black American cop and another was said to be an Irish cop by his scent. When they drove along the road into New Hampshire, they were in typical terrestrial procedure stopped and pulled over by five people on the of the road, then they were walked or escorted to the spacecraft or whatever kind of vehicle was used and then after the experience they were escorted back to their car.

They couldn't remember what had happened or where they were the night before because their memories of the actual event had been removed and a new memory implanted into both Betty and Barney with unique differences in recall to both of them. This may explain why their memories were re-combined into one event but were, in fact, two separate events in two different countries. According to Montaldo, the details of the account had to be sorted out to make sense of the whole abduction event which is obviously a lot different than the standard story of the Hills that we have all in the UFO field have come to know.

Cannon adds that if the memory is only comprehensible as an example of **"Artificially Induced Hyperamnesia"** in other words Betty Hill was **"directed"** to store the memories of the pregnancy test and the star chart (which we have already discussed earlier) within her subconscious. Thus the star map, even if it is astronomically correct was nevertheless, a prop to convince her and the general public of the validity of the event. **The Controllers: a New Hypothesis of Alien Abductions; Martin Cannon; 1990**

Before we leave this abduction section completely, we need to look at one other aspect of mind control technology which may be considered the "crowning jewel in the crown" of mind control.

One has to wonder to what extent this type of technology would be used and in what fashion. And more importantly, when. It would seem from an outsider's perspective, that the American public is a nation of unsuspecting guinea pigs, who are at the mercy of one or more clandestine intelligence agencies seeking to improve on some new and more bizarre medical experiment or procedure. **Haarp,** it would appear, seems to be that ultimate mind control experiment that is destined to be played out in the very near future, if we are to accept all the theories, conspiracies, and warnings from those research investigators who study such things. From the title of Jeane Manning's and **Dr. Nick Begich's** the book: *"Angels Don't Play This Haarp: Advances in Tesla Technology"* this is definitely, one Haarp that angels don't play, however; the minions of evil certainly wouldn't hesitate to play it!

HAARP (High Frequency Active Auroral Research Program)

The **High Frequency Active Auroral Research Program (HAARP)** is an ionospheric research program jointly funded by the US Air Force, the US Navy, the University of Alaska and the **Defense Advanced Research Projects Agency (DARPA)**. Its purpose is to analyze the ionosphere and investigate the potential for developing ionospheric enhancement technology for radio communications and surveillance purposes (such as missile detection). The HAARP program operates a major Arctic facility, known as the HAARP Research Station, on an Air Force owned site near Gakona, Alaska.

**HAARP (High Frequency Active Auroral Research Program) research station
in Gakona, Alaska operating agency office of naval research and
the air force research laboratory for ionospheric research**
https://en.wikipedia.org/wiki/High_Frequency_Active_Auroral_Research_Program

The most prominent instrument at the HAARP Station is the **Ionospheric Research Instrument (IRI),** a high-power radio frequency transmitter facility operating in the high frequency (HF) band. The IRI is used to temporarily excite a limited area of the ionosphere. Other instruments, such as a VHF and a UHF radar, a fluxgate magnetometer, a digisonde and an induction magnetometer, are used to study the physical processes that occur in the excited region. http://www.haarp.alaska.edu/haarp/factSheet.html. **Retrieved 2009-09-27**

The project is funded by the Office of Naval Research and jointly managed by the ONR and Air Force Research Laboratory, with the principal involvement of the University of Alaska. Many other universities and educational institutions have been involved in the development of the project and its instruments.

According to HAARP's management, the project strives for openness and all activities are logged and publicly available. Scientists without security clearances, even foreign nationals, are routinely allowed on site. The HAARP facility regularly (once a year on most years according to the HAARP home page) hosts open houses, during which time any civilian may tour the entire facility.

HAARP approaches the study of the ionosphere by following in the footsteps of an ionospheric heater called EISCAT near Tromsø, Norway. There, scientists pioneered exploration of the ionosphere by perturbing it with radio waves in the *"2-10 MHz range"*, [my italics added for emphasis] and studying how the ionosphere reacts. HAARP performs the same functions but with more power, and a more flexible and agile HF beam.

Some of the main scientific findings from HAARP include:

1. Generation of very low frequency radio waves by modulated heating of the auroral electrojet, useful because generating VLF waves ordinarily requires gigantic antennas
2. Production of weak luminous glow (below what you can see with your eye, but measurable) from absorption of HAARP's signal
3. Production of ultra low frequency waves in the 0.1 Hz range, which are next to impossible to produce any other way
4. Generation of whistler-mode VLF signals which enter the magnetosphere, and propagate to the other hemisphere, interacting with Van Allen radiation belt particles along the way
5. VLF remote sensing of the heated ionosphere

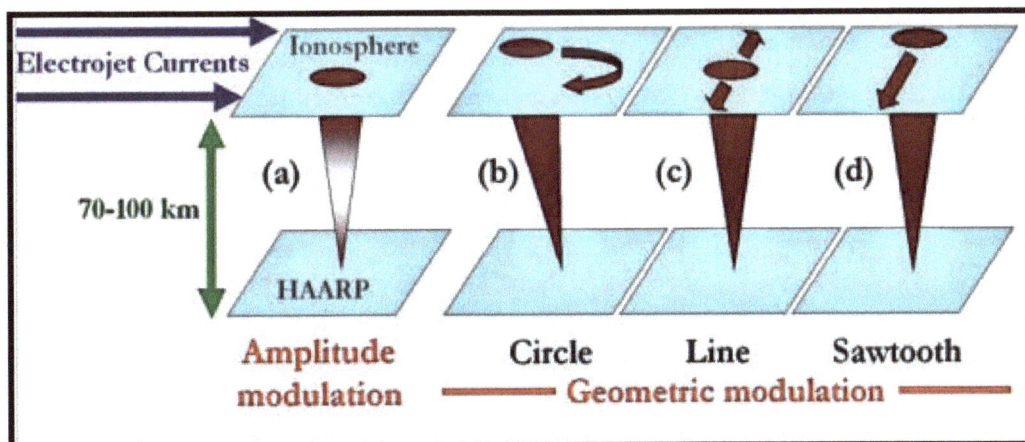

ELF/VLF waves with geometric modulation
https://chemtrailsplanet.net/2014/10/27/naval-research-lab-confirms-haarp-can-create-climate-change/

Research at the HAARP includes:

1. Ionospheric heating
2. Plasma line observations
3. Stimulated electron emission observations
4. Gyro frequency heating research
5. Spread F observations
6. Airglow observations
7. Heating induced scintillation observations
8. VLF and ELF generation observations <http://www-star.stanford.edu/~vlf/publications/2008-03.pdf> (See diagram below)
9. Radio observations of meteors.
10. Polar mesospheric summer echoes: PMSE has been studied using the IRI as a powerful radar, as well as with the 28 MHz radar, and the two VHF radars at 49 MHz and 139 MHz. The presence of multiple radars spanning both HF and VHF bands allows scientists to make comparative measurements that may someday lead to an understanding of the processes that form these elusive phenomena.
11. Research on extraterrestrial HF radar echoes: the Lunar Echo experiment (2008).[5][6]
12. Testing of SS-Spread Spectrum Transmitters 2009
13. Meteor shower impacts on the ionosphere
14. Response and recovery of the ionosphere from solar flares and geomagnetic storms
15. The effect of ionospheric disturbances on GPS satellite signal quality

Yet, the black world of the military industrial complex is always looking to get "the maximum bang for their buck" and "to kill two birds with one stone" in their research projects. The **HAARP project** is just such a multi-functional device as can be seen from the examples above but, its science is based on early 1900s Tesla technology.

Haarp allows for communication between submarines and military ground bases and to date, there are HAARP antenna bases located in Alaska, Norway, Arecibo in Puerto Rico and in a few in Russia. There may possibly be others in South America, Australia, and Africa but these have not been confirmed. The HAARP base in Alaska is the largest one with a near 40 hectare square base, where radio frequencies can be focused coming out of the energy field or "Ionic Heater" as a cyclotron resonance which appears as a corkscrew energy beam in the atmosphere. Such a beam was seen over Norway in 2010 that was incorrectly assumed to have originated in Russia, the video of which can be found on the **YouTube** website. This same beam can also be inversed into a narrow beam and projected out into space with corkscrew end closest to the earth or ground base.

Besides punching holes some 50 to 60 miles up in the upper atmosphere, it appears to have another potential use, as it relates to our topic at hand. Recall that HAARP can send out ELF radio waves between 2 - 10 Hz into the atmosphere and even over the horizon.

This little fact is what makes the HAARP program so frightfully insidious in its concept and potentially harmful in its usage. Haarp can bath the Earth in **ELF (Extremely Low Frequencies)** *radio waves designed to exactly match the same frequency at which that the* *human brain operates!* [My italics added]. Thus, Haarp can potentially be used to create huge mental dysfunction across vast portions of the planet. Haarp therefore could be the ultimate mind control device, if it were operated in the hands of unscrupulous black op or rogue agents from either, the military or intelligence community or both, in order to facilitate some diabolical agenda on a nation or all of humanity. https://www.youtube.com/watch?v=VkMhDT3P7mc

A Torsion Spiral over Norway emanating from the European Incoherent Scatter Scientific Association (EISCAT), an ionospheric heating facility, capable of transmitting over 1 GW Effective Radiated Power (ERP), near Tromsø in Norway
https://www.pinterest.com/starseednetwork/ufos/

The European Incoherent Scatter Scientific Association (EISCAT),
a similar HAARP-like device near Tromsø in Norway
capable of generating high frequency radio waves
https://redice.tv/news/was-norway-s-haarp-facility-eiscat-responsible-for-the-norway-spiral

Author's Rant: It is my personal belief that this is exactly the ultimate reason for Haarp's creation. From the all diverse data on mind control technology of which I have only skinned the surface of the subject, there is associated with this technology growing evidence that there is a hidden agenda by one or more covert quasi-governmental groups, some with an eschatological agenda that is strongly fundamentalist Christian in origin and extremely fanatical with bizarre cultish overtones, and there are other rogue groups who are motivated by pure greed and the need to seize control over most of the planet's resources with total disregard of the environmental damage to the Earth or the consequences to the rest of humanity!

This mind control technology along with other reversed engineered alien technology has little to nothing to do with an alien invasion from the stars but rather, instead, has more to do with a planetary coup d'état of the Earth's governments under a rogue **One World Order** government based on materialism, power and control of earth's citizenry.

Frederick Valentich - A Genuine Abduction Case or Military Interference?

Before we leave this topic on abductions and mind control, one last case needs to be examined as it is a true abduction case in every aspect and sense of the word. This is the now famous but, tragic account of **Frederick Valentich**, a 20-year-old, who on 21 October 1978 disappeared in

256

unknown circumstances while piloting a Cessna 182L light aircraft over the Bass Strait to King Island, Australia during a 127 mile (235 km) training flight.

Valentich advised Melbourne air traffic control that he "was being buzzed by a UFO with four bright lights" about 1,000 feet (300 m) above him. He made several reports, including that the engine was running roughly, and his last message was "it is not an aircraft." No trace of Valentich or his aircraft was ever found, and a Department of Transport investigation concluded that the reason for the disappearance could not be determined. Clark, Jerome (1998), *The UFO Book: Encyclopedia of the Extraterrestrial.* Visible Ink, ISBN 1578590299

Valentich, a pilot with a Class Four instrument rating and 150 hours flight experience had earlier filed a flight plan from Moorabbin Airport, Melbourne, to King Island in the Bass Strait. His departure from the airport went routinely and the first hour of flight was eventful. Visibility was good and winds were light.

At 5:00pm, Valentich contacted Melbourne Flight Service Officer, Steve Robey and requested information on other aircraft at his altitude (below 5000 ft, 1524 m) and was told that there was no known traffic at that level. Valentich then reported that he could see a large unknown aircraft which appeared to be illuminated by four bright landing lights. He was unable to confirm its type but said that it had passed about 1,000 feet (300 m) overhead and that it was moving at high speed. Valentich then reported that the aircraft was approaching him from the east and voiced the opinion that the other pilot might be purposefully toying with him.

About 5 minutes later, Robey asked Valentich to confirm his altitude and that he was unable to identify the aircraft. Valentich confirmed his height and began to describe the aircraft, saying that it was "long", but that it was traveling too fast for him to describe it in more detail. Valentich stopped transmitting for about 30 seconds, during which time Robey asked for an estimate of the aircraft's size. Valentich then came back saying that the aircraft was "orbiting" above him and that it had a shiny metal surface and a green light on it. This was followed by 28 seconds of silence before Valentich reported that the aircraft had vanished. There was then a further 25-second break in communications, after which Valentich returned and wondered if he was being tailed by a military aircraft. Paul Norman, (1996), "The Frederick Valentich Disappearance", Victorian U.F.O. Research Society Inc. (2007-04-27)

Robey attempted to gather more information about the unidentified aircraft and its location; Valentich reported that it was now approaching from the southwest. Twenty-nine seconds later, at 5:12 pm Valentich reported that he was experiencing engine problems and was going to proceed to King Island. There was brief silence then he came back to say "it is hovering and it's not an aircraft." This was followed by 17 seconds of unidentified noise, described as being "metallic, scraping sounds," then all contact was lost.

A search and rescue alert was initiated at 19:12. Valentich failed to arrive at King Island by 5:33 pm, and a sea and air search was undertaken by sea, with two RAAF P-3 Orion airplanes conducting an aerial search over a seven-day period and continued until 25 October 1978.

A slick of fuel was discovered on the sea roughly near where Valentich had last radioed Robey, but analysis proved that the slick was not aviation fuel. No trace of the aircraft was found. The aircraft was equipped with four life jackets and an emergency radio beacon and was designed to stay afloat for several minutes.

An official **Department of Transport (DOT)** investigation was launched into Valentich's disappearance. It lasted two weeks but was unable to determine the cause. DOT findings on the incident read "The reason for the disappearance of the aircraft has not been determined", and that it was "presumed fatal" for Valentich.

Analysis by **Dr. Richard F. Haines**, a former researcher with NASA-Ames and Associate Professor of Psychology at San Jose State University, described the metallic, scraping sounds," as "thirty-six separate bursts with fairly constant start and stop pulses bounding each one," and said that there were "no discernible patterns in time or frequency." The significance of the sounds, if any, has remained undetermined.

After news of Valentich's disappearance became public, a number of individuals came forward claiming to have witnessed unusual activity in the area. Among the 20 or so accounts given, people claimed to have seen "an erratically moving green light in the sky" and in one instance witnesses, located about 2 km west of Apollo Bay, Victoria, stated that they saw a green light trailing or shadowing Valentich's plane and that he was in a steep dive at the time.

According to ufologists, these accounts are especially significant as most were recorded several years prior to the 1982 release of transcripts in which Valentich had described the object above him as having a green light.

According to an Associated Press report, Guido Valentich, the missing pilot's father, said "*he hoped his son had been taken by a UFO and had not crashed. The fact that they have found no trace of him really verifies the fact that UFOs could have been there.*" Guido Valentich also told the AP that "*his son used to study UFOs 'as a hobby using information he had obtained from the air force. He was not the kind of person who would make up stories. Everything had to be very correct and positive for him.*"

Whenever, a potential UFO case obtains this kind of sensational press coverage, the "UFO vultures" come out for the "feeding frenzy" to give either credence to the event or dispute and debunk the account for anything other than a genuine UFO report.

Frederick Valentich, a photo of a UFO that was taken 20 minutes before Valentich's encounter and disappearance with a UFO and a plaque dedicated to him by his family

The debunkers view the Valentich disappearance as an error in human judgment that tragically resulted in the untimely death of a young pilot before his time. His plane, the debunkers believe crashed into the sea due to a lack of fuel which was not the case as he had more than enough fuel to reach his destination, or they believe he was the victim of drug dealers being a drug runner himself according to UFO skeptic, **Philip Klass**. All of which is sheer nonsense to those who knew him.

On the flip side of the argument, Ufologists are certain that he met with foul play by "planet hopping" extraterrestrials who were looking to get their kicks with an abduction of an unsuspecting human being who was in the wrong place at the wrong time.

This unconventional argument may be closer to the truth although one cannot rule out the possibility of a mechanical problem with his Cessna 182 aircraft or that he became disoriented at sea, either situation meant he would have gone down into the sea where his aircraft would have eventually sunk below the waves. However, no oil or fuel slicks in the sea were found that could be attributed as coming directly from his aircraft although, an oil slick was found in the sea near the last known location of where the aircraft was reported however, it was determined not to have come from the Valentich's Cessna but from some other source.

This would seem to be the most plausible cause, except for the fact that the events and circumstances leading up to Valentich's disappearance remain unexplained. The strange aircraft with the green light that he reported was possibly, an extraterrestrial vehicle, which in turn either abducted Valentich or caused the destruction of his plane in some inadvertent fashion.

Speculation that a UFO was involved has been fueled by a number of factors, including Valentich's last transmission, in which he described the aircraft shadowing him as "hovering" and "not an aircraft", the unexplained sounds that were heard at the end of his transmission, and a rash of UFO reports from the area.

If we give credence to the alien abduction scenario, then as stated earlier in this section on abductions, a real abduction does not have a happy outcome. This is not a case of mere kidnapping and eventual release by the abductors. The Valentich disappearance is a real abduction event. Because we do not know what became of Valentich, even if it was extraterrestrial in origin, we cannot assume that he died from this mysterious circumstance.

It is just as likely, that his outcome could have been positive as oppose to being negative, where as an amateur UFO researcher, Frederick Valentich may have gotten a first-hand, up-close and in your face, real life UFO encounter where he got his free ride to the stars or he is now working for the ETs in some exotic program that we can only imagine. Any outcome is possible. It would be nice to think that as a strong, intelligent young man, the ETs offered Valentich a once in a lifetime opportunity and he probably accepted their offer, but we'll never know sure.

AIRCRAFT ACCIDENT - INTERIM REPORT

| M116 | 783 | 1047 | FOLIO |

1. THE ACCIDENT

Location Between Cape Otway and King Island	Date 21.10.78	Time (Local) 1912	Zone EST
Aircraft Type Cessna 182L	Registration VH-DSJ	Owner C. Day, 35 Reserve Road, Beaumaris. Vic.	
Operator Southern Air Services, Moorabbin Airport, Mentone.	Flight From Moorabbin	To King Island	Purpose Travel	Class of Operation Private
Damage to Aircraft Unknown			

2. PERSONS INVOLVED

Name	Role	Description of Injuries	Probable Cause	T. 462 Forwarded to
Frederick VALENTICH	Pilot	Unknown		

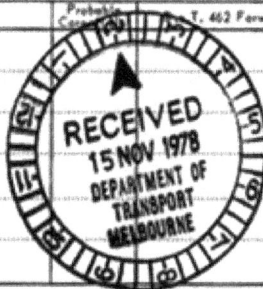

RECEIVED
15 NOV 1978
DEPARTMENT OF TRANSPORT
MELBOURNE

3. SEQUENCE OF EVENTS

At 1723 hours EST on 21 October 1978 a flight plan for VH-DSJ (Moorabbin to King Island and return, ETD 1745 hours EST, time interval to King Island 69 minutes), was submitted at Moorabbin Briefing Office. The aircraft was refuelled to capacity at 1810 hours EST and subsequently departed Moorabbin for King Island at 1819 hours EST. The pilot reported Cape Otway at 1900 hours EST and at 1906 hours EST asked if there was any traffic below five thousand. After being advised there was no known traffic, he reported what seemed to be a large aircraft below five thousand. He could not identify the type but described it as having a green light, being metallic like and all shiny on the outside. After describing the movements of the aircraft, its disappearance and reappearance, the pilot stated that it was hovering on top of him and that it was not an aircraft. He also reported at approximately 1912 hours EST that the engine was "rough idling" and declared that he was continuing to King Island. The final transmission from the pilot of the aircraft's callsign followed by 17 seconds open microphone was made at 1912:28 hours EST. Last light King Island was 1918 hours EST but the pilot had not requested aerodrome lighting. Flight Service King Island was notified and lighting was arranged but the aircraft failed to arrive there.

A search was initiated but no trace of the pilot or aircraft was found. The pilot was the only known person on board.

4. MATTERS ARISING

Details of what the pilot reported seeing prior to his disappearance have been referred to the RAAF for investigation.

Signed P.R. GRAHAM for Director	Designation A.S.I.W.	Date 14.11.78	Regional Reference V116/783/1047

D.o.T. 1573 (Rev. 3/77)

Department of Transport Interim Report - Note: Matters Arising
(Courtesy of www.ExplorOz.com)

261

TIME	FROM	TEXT
:46	FSU	D D DELTA SIERRA JULIET what type of aircraft is it
:50	VH-DSJ	DELTA SIERRA JULIET I cannot affirm it is four bright it seems to me like landing lights
1907:04	FSU	DELTA SIERRA JULIET
:32	VH-DSJ	MELBOURNE this (is) DELTA SIERRA JULIET the aircraft has just passed over over me at least a thousand feet above
:43	FSU	DELTA SIERRA JULIET roger and it it is a large aircraft confirm
:47	VH-DSJ	er unknown due to the speed it's travelling is there any airforce aircraft in the vicinity
:57	FSU	DELTA SIERRA JULIET no known aircraft in the vicinity
1908:18	VH-DSJ	MELBOURNE it's approaching now from due east towards me
:28	FSU	DELTA SIERRA JULIET
:42		// open microphone for two seconds //
:49	VH-DSJ	DELTA SIERRA JULIET it seems to me that he's playing some sort of game he's flying over me two three times at a time at speeds I could not identify
1909:02	FSU	DELTA SIERRA JULIET roger what is your actual level
:06	VH-DSJ	my level is four and a half thousand four five zero zero
:11	FSU	DELTA SIERRA JULIET and confirm you cannot identify the aircraft
:14	VH-DSJ	affirmative
:18	FSU	DELTA SIERRA JULIET roger standby
:28	VH-DSJ	MELBOURNE DELTA SIERRA JULIET it's not an aircraft it is // open microphone for two seconds //
:46	FSU	DELTA SIERRA JULIET MELBOURNE can you describe the er aircraft
1909:52	VH-DSJ	DELTA SIERRA JULIET as it's flying past it's a long shape // open microphone for three seconds // (cannot) identify more than (that it has such speed) // open microphone for 3 seconds // before me right now Melbourne
1910:07	FSU	DELTA SIERRA JULIET roger and how large would the er object be
:20	VH-DSJ	DELTA SIERRA JULIET MELBOURNE it seems like it's stationary what I'm doing right now is orbiting and the thing is just orbiting on top of me also it's got a green light and sort of metallic (like) it's all shiny (on) the outside
:43	FSU	DELTA SIERRA JULIET

D. o. T. Summary report page 2
(Courtesy of www.ExplorOz.com)

262

6. RELEVANT EVENTS (cont'd)

TIME	FROM	TEXT
:48	VH-DSJ	DELTA SIERRA JULIET // open microphone for 5 seconds // it's just vanished
:57	FSU	DELTA SIERRA JULIET
1911:03	VH-DSJ	MELBOURNE would you know what kind of aircraft I've got is it (a type) military aircraft
:08	FSU	DELTA SIERRA JULIET confirm the er aircraft just vanished
:14	VH-DSJ	SAY AGAIN
:17	FSU	DELTA SIERRA JULIET is the aircraft still with you
:23	VH-DSJ	DELTA SIERRA JULIET (it's ah nor) // open microphone 2 seconds // (now) approaching from the southwest
:37	FSU	DELTA SIERRA JULIET
:52	VH-DSJ	DELTA SIERRA JULIET the engine is is rough idling I've got it set at twenty three twenty four and the thing is (coughing)
1912:04	FSU	DELTA SIERRA JULIET roger what are your intentions
:09	VH-DSJ	my intentions are ah to go to King Island ah Melbourne that strange aircraft is hovering on top of me again // two seconds open microphone // it is hovering and it's not an aircraft
:22	FSU	DELTA SIERRA JULIET
:28	VH-DSJ	DELTA SIERRA JULIET MELBOURNE // 17 seconds open microphone //
:49	FSU	DELTA SIERRA JULIET MELBOURNE

There is no record of any further transmissions from the aircraft.

The weather in the Cape Otway area was clear with a trace of stratocumulus cloud at 5000 to 7000 feet, scattered cirrus cloud at 30000 feet, excellent visibility and light winds. The end of daylight at Cape Otway was at 1918 hours.

The Alert Phase of SAR procedures was declared at 1912 hours and, at 1933 hours when the aircraft did not arrive at King Island, the Distress Phase was declared and search action was commenced. An intensive air, sea and land search was continued until 25 October 1978, but no trace of the aircraft was found.

7. OPINION AS TO CAUSE

The reason for the disappearance of the aircraft has not been determined.

Approved for publication		(A.R. Woodward) Delegate of the Secretary	Date 27.4.1982

D. o. T. Summary report page 3
(Courtesy of www.ExplorOz.com)

263

CHAPTER 31

THE CORRUPTION OF THE UFO DATABASE REQUIRES A "GOOD PHYSICIAN" TO CURE IT

In reviewing this section, we've seen the strengths and weaknesses in the field that has been termed "Ufology". We've seen how the general public has responded to strange aerial phenomenon that does not fit in with the standard conventionality of things that fly and which are terrestrial in origin and design. We've seen how these reported sightings were investigated by amateur researchers who arose to do the research work in which mainstream science refused to get their hands soiled in the controversy that surrounded this extraterrestrial phenomenon. Yet, a few courageous scientists stayed true to the creed of scientific investigation of the unknown with the hopes of advancing and expanding the frontiers of scientific knowledge and understanding. Often these few rare scientists placed their necks, their careers and their reputations on the chopping block of public censorship and ridicule usually instigated by the new media desperate for a sensational news story and a higher percentage of the public audience. In most cases, skepticism from professional debunkers, outright condemnation from fellow scientists and loss of tenure from the hallow halls of academia often stalked and hounded the efforts of these amateur researchers and professional scientists.

We've seen how the global science community has dismissed the reports and claims of those who have asserted that the earth is being visited by extraterrestrial beings as the ravings of the deluded, of unscientific, untrained minds bordering on social dysfunctionality and mental paranoia. In the astute minds of science, UFOs and ETs simply do not exist and if they did, scientists would be the first to know about it, and based on current scientific understanding and knowledge, they cannot get here from there. Yet! Take these same scientific naysayers away from the earshot range of the general public, the new media, and their fellow scientists and they will tell you a different story! They will openly admit in a one on one interview or private discussion that they know that the Extraterrestrial Hypothesis is real, that UFOs or alien spacecraft and alien intelligences are real and have historically and are still currently visiting the Earth, today. These same scientists are afraid to come out of the closet in a public disclosure because of their positions and reputations that are so intrinsically tied into their professions, whether government, military, intelligence, societal or in the private industrial sectors of society.

Sixty years of investigative research by members of many well known UFO organizations with some of the most prominent and largest organizations particularly in America being infiltrated by various military and government intelligence agencies resulting in their dysfunction and/or dissolution. With such deliberate wholesale dismemberment of these publicly influential UFO organizations like **APRO, NICAP**, and **CUFOS**, any movement toward public disclosure by these organizations was thwarted by agencies within the government or the military. The resultant effluent of any continued UFO investigation has been mired down by the endless and needless internal squabbling, bickering, "back-stabbing" of researchers' reputations and ego power-tripping over leadership, the enforcing of personal theories, and the direction in which UFO and ETI research should go.

Every manner of the paranormal, pseudo-scientific, and crypto-zoological encounter is lumped together in the waste basket of Ufology making any true understanding and sense out of it, practically impossible. To any discerning eye looking into Ufology it would appear that Ufology is going to hell in a handbasket, while the riders on board for the trip whistle a happy tune, oblivious to where the final destination will take them.

Skeptics and professional debunkers are right about one thing in their argument against Ufology, in its current state, with the exception of a few professional researchers, there is not much that is scientific in their approach to solving the alien enigma.

The field of Ufology and the public loves, so it seems a good mystery. We see wild speculations and pet theories surrounding the cattle mutilation phenomenon and the abduction phenomenon (which has taken on a separate life of its own), tossed back and forth as the only acceptable models of UFO/ETI understanding. All of which leaves the general public and the scientific community feeling confused or duped into buying the erroneous data that comes from Ufology or buying the lie that comes from the military industrial complex. Neither choice is acceptable in the search for and the disclosure of truth.

Into this quagmire of Ufology enters the free enterprising, entrepreneurial UFO businessman whose only real motive is to exploit the potentially profitable and commercial aspects of this phenomenon. The UFO entrepreneur sees such opportunities as serving not only his own interests but, as a "contribution" to the basic understanding of the UFO phenomenon. To the outsider whether, UFO investigator or the general public, the "cottage industry" of UFO conventions and conferences are the only sources of information where one can distribute, sell, listen to or pick up the latest UFO information, all in one place.

There is really nothing wrong with this commercialization of the UFO/ETI mystery as there are paranormal conventions and seminars, fantasy and role playing conferences, movie, TV and comic conventions, sex oriented and lifestyle conventions, in fact, there are conferences, seminars and conventions on nearly every aspect and interest one can imagine. So, it is not surprising that UFOS and ETs fit into this dazzling panoply of unique interests.

To compound the existing problems, forces from within and from without have overtly and covertly usurped the direction and influence of most large UFO organization. These internal and external forces which are very human and not alien by nature instigate changes by splintering and creating off-shoots of UFO interest with new age UFO-based religions that go nowhere in contributing to an advancement of society's spiritual condition or like great magicians of deception bring about negative mindsets with distrust and prejudice toward any Extraterrestrial Intelligence coming to the planet, thus distracting the focus and attention away from any real serious investigation or any conclusive resolution to the ETI question.

The database of Ufology is so corrupt and contaminated with sightings of lights in the night sky from misidentification of conventional objects and aircraft, to the spurious claims from reports of alleged alien intervention in the mutilations of cattle and farm animals, to the outright hoaxing and fabrication of ET encounters as claimed from the "abduction scenario accounts by "wannabes" to actual encounters with strange entities mistaken for real ETs (namely the

"Greys"), but in reality are the product of the military industrial complex that are **"Programmed Life Forms" (PLFs)** to carry out a specific operation or agenda that it becomes difficult to sort the "signal from the noise" in genuine data accumulated from such voluminous evidence over the years.

The ratio of "signal" from real hardcore data to the "noise" from misidentified, hoaxed, disinformation, misinformation, lies, denials, and of false claims requires sorting by a very good computer software program to pull out the pertinent data in order to make sense of it. The right questions need to be asked, other viable alternatives need to be considered, an understanding and investigation into current technology both public, military and even privatized must be undertaken where possible and where accessible. The data collected from the many fields of science on this UFO subject needs to be analyzed and compared to the hardcore and statistical data already compiled by various UFO organizations and from the torrent of files that are now being released by many of the world's government. All perceptions and viewpoints need to be considered and only those do not lead to speculation should be kept and added to the overall body of first-hand evidence, so that irrefutable conclusions can be drawn or extrapolated with only a small percentage error. In essence, then, the UFO database needs to be overall and re-evaluated.

It could be said, that if the study and the evidence of UFOs and ETI were a religion, and that religion was corrupt or impotent from forces both inside and out, then it would be in need of a Divine Physician to cure it!

When there have been over sixty years of UFO and ETI research and investigation, the questions that need to be answered by members in the field of Ufology are: "What have you deduced, what is your assessment and conclusion from your investigations"? "Are Extraterrestrial beings real and are they visiting the planet at this time"? "Have you determined conclusively the nature of their agenda"? "Does the government, the military or the intelligence communities know anything about UFOS and the nature of ETI"? "Do they possess retrieved alien spacecraft and captured alien beings live or dead"? To any of the above questions, "What do you plan to do now with all the information and data that you've accumulated over the years? What is your next strategy for further investigations or do you have a plan or agenda in place for open contact with these Extraterrestrial visitors?

These would be the questions a divine physician would ask and perhaps no adequate cure could fix the current problems with Ufology. What may be needed now, more than any time since UFO research began is a new direction based on a proactive first-hand approach instead of a passive second-hand investigative approach.

The passive approach to UFO research has served our understanding well up to this current time when we have been able to "sift the chaff from the wheat", but now, we need a fresh proactive approach to move forward to the eventual day of disclosure.

Unofficial public disclosure has already occurred with Dr. Steven Greer's Disclosure Project's witness testimony event that came from high-ranking government and military officials, scientists and astronauts at the National Press Club in Washington, D.C. back in 2001. To this end, Dr. Greer is a pioneer in the new frontier of UFO research and is considered the "Father of UFO and ETI Disclosure."

With the release of thousands of new UFO files from many governments around the world and the Disclosure Project campaign to inform not only the powers that be, besides the general public, the world now waits expectantly to hear official government disclosure on the subject of the Extraterrestrial visitation. Many billions of people in the Unites States, Canada, Britain, Europe, the Middle and the Far East expect the **"Truth Embargo"** on UFOs and ETI to be lifted in the very near future. In fact, I have been told by Dr. Greer that he had a personal consultation with a few members from a G8 nation that had approached him to help them in establishing contact with the Extraterrestrial Intelligences and in bringing official disclosure to the world. This nation has taken this whole matter very seriously and feels that America's leadership in this matter is too slow or complacent for whatever reason and thus, this European nation will probably make a pre-emptive move toward official disclosure.

Many Ufologists feel that this may backlash negatively toward the U.S. unless, the **President Obama** administration make the first move in disclosure which isn't likely to happen given the political climate and other pressing government priorities currently weighing on the U.S., only time will tell the story and that may be a lot sooner than most people will suspect.

CHAPTER 32

DO YOU RETURN YOUR Ph.D. OR DOCTORATE
IN ALIEN ABDUCTION RESEARCH WHEN
YOUR WORK IS FAULTY?

Frequently the question comes up as to why there are no recognized courses or classes that teach the subject of Ufology as either a part of science or sociology or preferably a separate course unto it. The answer may surprise you. Until recently, within the last decade, no one would touch the UFO subject or even giving it legitimacy because of its controversial nature, and it's unproven evidence for reality. No self-respecting scientist would taint his reputation, his stature or tenure with an investigation into a non-reality subject that is still considered a source of ridicule, let alone even entertaining the notion of teaching a course on the phenomenon.

There are some non-accredited courses being offered either in senior high school as an extra-curricular activity like football or gymnastics or as a night course program for adults and sometimes they are offered as a part of a major course in university. They are even offered to the general public through a UFO or paranormal organization as a one day or evening program or as a week-long session.

Those who ventured into this arena of controversy did so, contrary to the warnings and advice given by their fellow colleagues, families, and friends. But undeterred by their admonitions, these courageous scientists, these seekers of truth and knowledge pursued a higher truth (pun intended) that they felt deserved a more serious consideration than what was given from the community of mainstream science.

Dr. Ronald Leo Sprinkle

Dr. Ronald Leo Sprinkle (born August 31, 1930) is an American psychologist. He studied at the University of Colorado and earned his Ph.D. at the University of Missouri.

During his time as a professor at the University of Wyoming, he was considered the first academic figure to investigate tales of supposed alien abduction; his involvement in this field began in the 1960s.

Sprinkle became convinced of the phenomenon's actuality and was the first to suggest a link between abductions and so-called cattle mutilations. In 1980 he founded the Rocky Mountain Conference on UFO Investigation. In 1989 he was forced to resign his tenure from the University of Wyoming when it became public knowledge that he claimed to have been abducted by aliens as a child. Bryan, C.D.B.; (1995); *Close Encounters of the Fourth Kind: Alien Abduction, UFOs, and the Conference at M.I.T.,* New York; Alfred A. Knopf; ISBN 0-679-42975-1

Dr. Ronald Leo Sprinkle
http://www.ovnis-usa.com/2010-4.htm

It appears that most universities at that time and right up to the current time do not tolerate controversy of any kind unless it adds to the prestige of the university and their board of regents. The study of UFOs and alien beings visiting the Earth would be one subject that doesn't fit in with university studies or the protocols of behaviour especially when you're abducted by aliens who are from the stars and not south of the border. Had Dr. Leo Sprinkle defended his position, the outcome in '89 may have been given a clear message to all universities to be careful in their due diligence with regards to faculty expulsion when one of their faculty members enters into a controversial field of research or makes states about their personal life.

Dr. John E. Mack

Dr. John E. Mack *(Repeated from Part One – Famous Psychologists in the Field of Ufology).* (October 4, 1929–September 27, 2004) unlike many scientists who are afraid to tackle this UFO issue, the late psychiatrist never hesitated to step into the unknown in the search for scientific truth and understanding. Mack was an American psychiatrist at the **Harvard Medical School** and a Pulitzer Prize-winning biographer, and a leading authority on the spiritual or transformational effects of alleged alien abduction experiences which he discusses in his book: *"Abduction – Human Encounters with Aliens"*, 1994, by Macmillan Publishing Co. As stated earlier, alien abductions seem to have taken on a life of its own as an adjunct to the UFO phenomenon and Dr. John Mack was no stranger to the subject and its psychological implications as attested by the respect garnered for his work from the few other brave scientists in the field of Ufology.

Dr. John E, Mack
http://www.montaguekeen.com/page68.html

Born in New York City, Mack received his medical degree from the Harvard Medical School (Cum Laude, 1955) after undergraduate study at Oberlin (Phi Beta Kappa, 1951). He was a graduate of the Boston Psychoanalytic Society and Institute and was Board certified in child and adult psychoanalysis.

Mack advocated that Western culture required a shift away from a purely materialist worldview (which he felt was responsible for the Cold War, the global ecological crisis, ethno-nationalism and regional conflict) towards a transpersonal worldview which embraced certain elements of Eastern spiritual and philosophical traditions.

The dominant theme of his life's work has been the exploration of how one's perceptions of the world affect one's relationships. He addressed this issue of "worldview" on the individual level in his early clinical explorations of dreams, nightmares, and teen suicide, and in his biographical study of the life of British officer **T. E. Lawrence (Lawrence of Arabia),** for which he received the Pulitzer Prize in biography in 1977.

Mack's interest in the spiritual aspect of human experience has been compared by the New York Times to that of a previous Harvard professor, William James. Like James, Mack became controversial for his efforts to bridge spirituality and psychiatry.

This theme was taken to a controversial extreme in the early 1990s when Mack commenced his decade-plus study of 200 men and women who reported recurrent alien encounter experiences. Such encounters had been reported since at least the 1950's (the account of Antonio Villas Boas) and had seen some limited attention from academic figures (Dr. R. Leo Sprinkle perhaps being the earliest, in the 1960s). Mack, however, remains probably the most esteemed academic to have studied the subject.

Mack initially suspected that such persons were suffering from mental illness, but when no obvious pathologies were present in the persons he interviewed, Mack's interest was piqued.

Following encouragement from long time friend Thomas Kuhn (who predicted that the subject might be controversial, but urged Mack to simply collect data and temporarily ignore prevailing materialist, dualist and "either/or" analysis), Mack began concerted study and interviews. Many of those Mack interviewed reported that their encounters had affected the way they regarded the world, including producing a heightened sense of spirituality and environmental concern.

In 1994 the Dean of Harvard Medical School appointed a committee of peers to review Mack's clinical care and clinical investigation of the people who had shared their alien encounters with him (some of their cases were written in Mack's 1994 book "*Abduction*"). In the same BBC article cited above, Angela Hind wrote, "It was the first time in Harvard's history that a tenured professor was subjected to such an investigation."

Dr. Mack is one of those courageous scientists, who put his reputation and career on the line for what he believes was the right course of action. Dr. Mack held tenure at the prestigious Harvard University, who when the faculty of deans heard of his involvement in UFO and abduction research attempted to have Mack fired and disbarred from teaching at Harvard. However, Dr. Mack fought back and won his case against Harvard University who were forced to let Mack continue teaching at Harvard. No doubt Harvard viewed his involvement into UFO investigation as a pseudo-science, a waste of his time and effort and unworthy of his illustrious career, feeling ultimately that such tarnishment on Dr. Mack also, reflects ignominiously on their university's prestige.

It is apparent that there is still much resistance to the topic or interest of UFOs, ETI, and alien abductions in many of the universities in America and no doubt in Canada and elsewhere on the planet. As a course by itself, there is no such program offered anywhere but, as a part of another course such as sociology or physics and most likely in an astronomy class, there doesn't seem to be much in the way of opposition as it doesn't form the basis of any course be taught.

David Jacobs

David Jacobs is an American historian and Associate Professor of History at Temple University, specializing in twentieth century American history and culture. He is well known in the field of Ufology for his research into alleged alien abductions and UFOs. He was the first to initiate a

UFO course at Temple University as a part of his Sociology class and in fact was probably the only person anywhere in the United States to receive a Ph.D. in History from the University of Wisconsin-Madison in 1973 based on his dissertation on the controversy over unidentified flying objects in America namely, that of alien abductions. **Biography: David M. Jacobs";
International Center for Abduction Research.**
http://www.ufoabduction.com/biography.htm. Retrieved 2009-11-14.

A revised edition of his dissertation was published as *The UFO Controversy in America* by Indiana University Press in 1975. His current research interests "involve a delineation of the role of anomalous experiences in personal and cultural life." For over 25 years Jacobs has taught a course on "UFOs in American Society."

In recent years Jacobs has argued publicly that the evidence from his research, using hypnosis with alleged alien abductees, shows that alien-human hybrids are engaged in a secret program of infiltration into human society with the final goal of taking over the Earth. He asserts that some of his research subjects are teaching these hybrids how to blend into human society so that they cannot be differentiated from humans, and that this is occurring worldwide.

Dr. David Jacobs
http://www.ufoevidence.org/topics/DavidJacobs.htm

Jacobs approaches the subject of abductions, hybridization and the "hidden alien agenda" in a forthright manner employing the scientific method of research and believes unquestionably in his conclusions.

With that said, however, his critics have taken Jacobs to task questioning the validity of hypnotic regression methods which seem to rear its "ugly little head" whenever the subject of UFOs and alien abductions is the topic of discussion. Faulty memory retrieval is a typical argument when hypnotists "lead" the patient in a regression session and sleep paralysis is also cited as a cause of alien abduction events, most notably, the late **Carl Sagan** and **Susan Clancy** have been his chief critics.

Of late, more criticism has been levelled at Jacobs from within the UFO community on his unwavering position to the whole alien abduction and conquest of the human race by subterfuge using alien-human hybrids to slowly replace humanity on the planet. The main alien race behind this plot is course our now very familiar "Greys".

Joe Montaldo from **"The Black Vault – UFOs Undercover"** and from his website: **ICAR (International Center for Abduction Research)** as well as **Jesse Randolf** from **Ufonaut Radio, and Alejandro Rojas** from **Open Minds Radio,** (these are **Blog Talk Radio** podcast programs) along with many of their guest speakers do not agree for the most part with **Jacobs** abduction results. They find that the evidence does not support the data and that much of Jacobs conclusions appear to be speculation or extrapolation of the raw data in order to make sense of the reports which many come from hypnotically regressed witnesses claiming an abduction experience. http://www.ufoabduction.com/

We already explored and seen that abductees can and have been given screened memories from military intelligence in order to dig out the truth behind the ETI agenda or to fabricate a whole abduction episode that has nothing to do with aliens but some other covert human agenda or program.

Even, **Budd Hopkins**, the foremost proponent of the abduction scenario is strangely silent on the whole abduction phenomenon as of late. It is suspected that he too is firmly entrenched in his own theories of the alien agenda, much like Jacobs choosing to ignore the many thousands of accounts that report positive encounters with Extraterrestrial Intelligences.

Author's Rant: It would appear to me that it would be a terrific blow to both these men's egos and their reputations which were built upon their many years of research as well as supporters of their work to discover that their findings and conclusions were faulty, based on incorrect or poor assessments of the data. Yet, that is exactly what appears to be the case and only time will play out the final conclusive proof on the whole abduction scenario.

In defence of these men and particularly Jacobs, it is alright to be wrong, to have made an error or two in judgment (big or small), to have concluded incorrectly and even to have built a reputation on the promotion of the "evidence" as long as you recognize your mistakes, corrected them and then learn something from your mistakes. To stubbornly not admit your error is to perpetuate an ongoing misunderstanding of the ETI phenomenon which may never be resolved clearly or accurately in which case, you should be stripped of your doctorate degree!

I feel that Hopkins, Jacobs, et al have not asked the correct questions or investigated other possibilities that may produce similar answers or conclusions but is not necessarily related to the ET Question. It is like a doctor who is trying to do a diagnosis on a patient who is complaining about chest pains. The doctor had better know his medical knowledge on this sort of problem because there are many things that will give a patient chest pains. Some are related to other things and some are not, like a torn pectoral muscle or heartburn or a heart attack but, whatever is the cause, he better be right about it because one poor diagnosis may kill the patient if the doctor is wrong.

People interested in the UFO and ETI subject should check their schools and colleges course curriculum as well as community newsletters to see what type of UFO classes may be offered. The increase in public interest in the subject of UFOs and ETI has spurred many to seek a satisfying answer to one of the biggest mysteries of all time. It is my hope that this textbook may offer an accurate foundation of basic knowledge in this subject matter.

CHAPTER 33

UFOLOGISTS FROM AROUND THE WORLD

Earlier in this section, we looked briefly at some of the major players in the field of UFO and ETI research. These are UFO researchers from the past and to the present time, who have made a profound impact and advancement to the understanding of Ufology. There are also, various UFO organizations (whether defunct or still operational) who sounded the clarion call for public and government's attention to the need for disclosure. Into this chaotic rabble of Ufology, a few courageous scientists placed their reputations and careers on the line, who were willing to do some professional research work in a field that is highly charged and controversial.

The response by governments worldwide and particularly by the G8 nations, including the US Military and the alphabet soup of government intelligence agencies was to standby in silent observation, afraid to give verbal or financial support for a legitimate research and investigative program into the phenomenon for reasons of political career suicide or because the military had already concluded through their own commissioned investigation that there was nothing to be gained and no threat to national security and any further research would be a waste of the taxpayers' dollars. In support of the military action to UFOs, the various intelligence agencies deemed that any reality to the phenomenon, the public didn't have a need to know.

The American news media who were steadily being viewed with suspicion of not performing their duty as the "watchdogs" of society or as the "Fifth Estate" to report all news unbiasedly and without "sanitizing" or "spin-doctoring" were either, no longer reporting any new accounts of UFO sightings, determining that the public had lost interest in such news or they reported such events with a "tongue in cheek" or ridicule factor playing up the humorous side of the phenomenon.

However, the UFO phenomenon and in particular the ETI unswervingly continued their agendas and unintimidated by any human response or threats, all to the consternation of UFO investigators and particularly to the **Military Industrial Complex (MIC).** UFO investigators have searched relentlessly for that one reported account that would reveal the whole picture and shine the limelight of respectability on the subject matter, hopeful that it would lead to eventual disclosure. While the MIC and the intelligence community were duplicitous in a cooperative agenda to do everything in their power to ensure the truth of UFOs and ETI never sees the light of day.

A Who's Who of Professional UFO Investigators and Researchers

Currently, who then are the public's major players in Ufology? To understand the major players in this ever evolving phenomenon, we need to see who's who. Initially, in considering the many investigators in Ufology, this list appears surprisingly small when considering the worldwide nature of the phenomenon. The cream of the crop consists of UFO researchers, who are well known in Ufology and by the public at large and have been briefly discussed elsewhere in this book.

They are the husband and wife team, Jim and Coral Lorenzen of APRO; **Major Donald Keyhole** of NICAP; Dr. J. Allen Hynek, astronomer and the U.S. Air Force's chief UFO investigator and spokesperson and also, founder of CUFOS; Dr. Jacques Vallee, French mathematician and world renown UFO researcher; Dr. Stanton T. Freidman. nuclear physicist and foremost authority on the famous Roswell, New Mexico saucer crash; **Dr. Steven M. Greer**, International Director of CSETI, the Disclosure Project and **The Orion Project**, and probably the world's foremost authority on Extraterrestrial life: Timothy Good, world famous British UFO investigator and a leader in UFO research; and **Grant Cameron**, a well respected Canadian UFO researcher and renowned for his research work on US Presidents and their UFO connection.

A more in-depth look into Ufology worldwide reveals the following list of UFO researchers and ETI specialists, however, it is not currently a complete or detailed listing, as many new investigators have joined the vanguard of Ufology but, their work and contributions are as yet, not publicly recognized. Still, others have retired from the field to pursue other interests or family life while others have unfortunately passed away. I have only chosen those people who are either, well known in their own country, who have made a significant contribution or impact in the understanding of the UFO phenomenon, or who have attained international recognition and status.

It is also, obvious that many of the well-known Ufologists appear to come from the U.S.A., Canada, the U.K., this could be that these countries give more exposure and publicity than other countries or simply other investigators are not known in the western hemisphere. There are of course many premiere investigators from other European countries, Central and South American countries, and lesser known UFO researchers from the Middle East, India, China and other countries of the Far East.

 Famous contactees were not included in this list, as I felt that they are not UFO researchers in the true sense of the word but, more like people for whom investigations and case reports were written about them and their alleged contactee claims. Likewise, individuals who had famous documented UFO and /or ETI cases who publicly recounted their experiences on a lecture tours, like the many witnesses from the Roswell saucer crash and retrieval or the Travis Walton case and other similar UFO accounts were also, not included as they have not made on-going investigations into any other UFO cases than their own. Military personnel and officers were also, omitted from the list unless of course, they undertook full-time investigation in the public UFO domain. Needless to say, private or corporate industry and the military would have their own UFO investigators where information obtained from field research would in all probability, not be shared with the general public or with UFO organizations.

Many UFO investigators, who may be well known in their own cities, towns or in their states and provinces, may not be as well known as other publically lime-lighted investigators who appear to be known nationally and internationally. The list below indicates both those who are known and those who less well known but, whose research work is just as invaluable to the overall understanding of the UFO/ETI phenomenon as those current luminaries of Ufology. Consider then, this as a short list of who's who, with an apology to all those thousands of UFO researchers not mentioned by this author.

ASIA – Indonesia

J. Salatun was Indonesia Air Commander and is a retired Major General. He is known as a founder of Indonesia's National Institute of Aeronautics and Space in 1963.

Salatun observed UFO phenomena in 1974. Two of his famous works are *"Revealing the Secret of Flying Saucer"* in 1960, and *"UFO, One of The Present Riddles"* in 1982. He established **SUFOI (Indonesian UFO Study)** in the 1980s. In a letter published in 1974 he wrote, "I am convinced that we must study the UFO problem seriously for reasons of sociology, technology, and security." [1]

Nur Agustinus, (1966 -), is a writer, an Indonesian ufologist, director of BETA-UFO Indonesia and has a degree in psychology, but what he enjoys most is hunting UFOs, although he has never seen one, or an alien for that matter.

"I believe that we're not the only ones [in the universe]," he said.

His interest in extraterrestrials started back in the 1970s when he first came across tales of natural mysteries linked to astronomy and the wider universe.

"When I was a kid I always liked to look up at the sky and all the stars," Nur said.

https://www.youtube.com/user/nuragustinus

So, together with friends who shared the same interest, he founded **Beta-UFO** (Beta stands for

Benda Terbang Aneh or Unusual Flying Objects) to investigate reports and document alleged UFO sightings in Indonesia.

Beta-UFO claims to be the largest UFO community in Indonesia.

"Our missions are the documentation, investigation, research, public education and scientific study of UFOs for the benefit of humanity. We collect and investigate claims of alleged UFO sightings in Indonesia," Nur said.

The group has a mailing list of 1,000 and 80 to 100 active members, who meet once a year in Jakarta for an annual gathering. At other times they share information and news through the group's Web site, as well as investigating reported UFO sightings.

"We have a special group of UFO hunters," Nur said. The group sets up camp at spots where many reports of UFOs have originated.

"They'll stay for several days with their cameras and video recorders ready in case a UFO appears," Nur said.

"During the 1950s and 1960s, there were many reports of UFOs from eastern Indonesia. But now the reports come from all over Indonesia, with most coming from Jakarta, Bandung, and Cirebon." [2] http://www.ufocasebook.com/2009/jakarta.html

Kuwait

Sanad Rashed is a Kuwaiti novelist/author/Ufologists and paranormal Researcher and a member of SETI, who is one of the very first Kuwaiti writers to attempt writing Horror/Science Fiction novels with all Kuwaiti characters and events that happen in Kuwait and all around the World. He also writes periodical articles for journals and web-based magazines. [3]

EUROPE - Finland

Rauni-Leena Luukanen-Kilde (born 15 November 1939 in Värtsilä, now in the Republic of Karelia) was a provincial chief medical officer in Finnish Lapland Province with a doctorate in medicine from 1975 until a car accident in 1986, which took away her ability to continue her work and career. Kilde received her medical doctor degree in 1967 from Univ. of Turku, Finland. She is also the author of many books about UFOs, alien abductions, mind control and other topics, which have been published in 6 different languages inside and outside of Finland. She has spoken openly about the hiding of the UFO evidence and other conspiracy theories like mind controlling. She has taken part in numerous UFO conferences.

Experiments involving the implantation of microchips in the human brain, she says, have been conducted without subjects' knowledge or consent. Kilde first published these astonishing allegations in "Microchip Implants, Mind Control, and Cybernetics" in the 36th-year edition of the Finnish-language journal SPEKULA (3rd Quarter, 1999), a publication of Northern Finland medical students' and doctors' club at the Oulu University.

Kilde is the author of another article, "My 100 Encounters with Space Aliens," in which she claims that extraterrestrials have saved her life three times. Kilde reports having seen her first UFO—yellow sphere hovering 60 to 90 feet above a field in Rovaniemi, Lapland"—in August 1986. She claims to have met three kinds of aliens: One that is "three feet tall, has a huge head and big black eyes, but no nose or teeth"; A second that "is like the first but has a large nose"; A third that "is about 12 feet tall, with a very small head and large dark glassy eyes" and "wear lab coats, gloves and hoods over their heads".[4]

France

Marcel Griaule (1898 – 1956) was a French anthropologist known for his studies of the Dogon people of West Africa, and for pioneering ethnographic field studies in France. Between 1928 and 1933 Griaule participated in two large-scale ethnographic expeditions -- one to Ethiopia and the ambitious Dakar to Djibouti expedition which crossed Africa. On the latter expedition, he first visited the **Dogon**, the ethnic group with whom he would be forever associated. During these trips, Griaule pioneered the use of aerial photography, surveying, and teamwork to study other cultures. In 1938 he produced his dissertation and received a doctorate based on his Dogon research.

He died in 1956. Griaule is remembered for his work with the blind hunter Ogotemmeli and his elaborate exegeses of Dogon myth (including the Nommo) and ritual. His study of Dogon masks remains one of the fundamental works on the topic. A number of anthropologists are highly critical of his work and argue that his claims about Sirius and his elaborate accounts of cosmic eggs and mystic vibrations do not accurately reflect Dogon belief.

He died in 1956. Griaule is remembered for his work with the blind hunter Ogotemmeli and his elaborate exegeses of Dogon myth (including the Nommo) and ritual. His study of Dogon masks remains one of the fundamental works on the topic. A number of anthropologists are highly critical of his work and argue that his claims about Sirius and his elaborate accounts of cosmic eggs and mystic vibrations do not accurately reflect Dogon belief.

In 2007, Laird Scranton offered new evidence in support of Griaule and Dieterlen's Dogon cosmology by demonstrating abiding parallels between **Dogon cosmology** as evoked by a Dogon granary and classic Buddhist cosmology - expressed in terms that closely match Griaule's - as derived from a ritual aligned stupa. Such parallels support the notion that Griaule's Dogon cosmology represents a known, legitimate classic cosmological form.
http://en.wikipedia.org/wiki/Marcel_Griaule [5]

Yves Sillard (Engineer, former director of CNES, 1976–1982, head of the new French government UFO investigation organization, GEIPAN). The French Government, and its official **French national space agency (CNES)** will restart its official UFO study program. The French government has had an official program to study UFOs for several decades, called GEPAN / SEPRA. It was closed recently, but after an audit, it will now be restarted, with a new director, Yves Sillard.

This program is an official French government project and is conducted under the official French space agency CNES (the equivalent of NASA in the US). The new director was a former director of CNES and has numerous other high positions and achievements to his credit. In an interview with him, he seems to express a very strong support for the subject of UFOs and advocates the need for serious, rigorous scientific study of the phenomenon. Mr. Sillard's most recent position was Assistant Secretary General for Environmental and Scientific Affairs for NATO. [6]

Germany

Michael Hesseman is a cultural anthropologist, an internationally published author, filmmaker, journalist, science writer, and Germany's most prolific UFO expert. His never- ending quest for UFO information has taken him around the world.

Born in Duesseldorf/Germany, he studied History and Cultural Anthropology at Goettingen University. He lives in Düsseldorf and Rome.

His books ("UFOs: A Secret History", "The Fatima Secret" and "Beyond Roswell") are international bestsellers, published in 16 countries with a total print run of over a million copies. Hesemann produced several video documentaries ("Ships of Light"; "UFOs: The Footage Archives"; "UFO Abductions: A Global Phenomenon" and "UFOs: The Secret Evidence").

He has worked as an advisor for several TV programs as an advisor for TV stations and programs in the US (including 'The X-FILES', 'SIGHTINGS', 'ENCOUNTERS', for Paramount, Disney, The Fox Network and others), UK (Channel 4, MTV), Japan (NTV), Mexico (Televisa, TV Azteca), Germany (ARD, RTL, SAT 1, PRO SIEBEN, KABEL 1, TV NRW), Poland (TVN) , Italy (Rai Due) and the Vatican (TelePace).

Hesemann was the first western researcher who published the KGB Files on "Anomalous Aerial Phenomena", officially released to him. For his research, Hesemann travelled over 700.000 miles and visited 44 countries. He has lectured at international conferences, universities and been a guest on several radio shows in the US and Mexico.

From 1979-2000 he published and edited the "Magazine 2000", Europe's largest magazine on mysteries of the past and presence, which came out in German and Czech language with a print-run of 100.000 copies.

He is an Associate Member of the prestigious "Society for Scientific Exploration", the "World Explorer Club" and the "Israel Exploration Society". He received the Honorary Membership of the Italian C.U.N. and was honoured with the Hungarian "Colman VonKeviczky Medal" for his research on Extraterrestrial Intelligences and the UFO Phenomenon. He lectured at international conferences in 25 countries and at over 30 universities, at the Vatican Lateran University, the Pontifical University LUMSA, both in Rome, and on invitation by the Ministries of Culture and

Transportation of the Republic of San Marino, the Human Potential Society, the Humanist Society at the United Nations and the United Nations Recreational Council. As a journalist, he is accredited at the Holy See Press Office.

Hesemann was the first western researcher who published the KGB Files on "Anomalous Aerial Phenomena", officially released to him. He is the principal investigator of the Carlos Diaz case in Mexico and one of the main investigators of the Billy Meier case in Switzerland, the British and international crop circle phenomenon (on which he wrote three books) and the Fatima Secret, for which he received international attention. He uncovered the truth on the controversial alleged Roswell Autopsy Footage and exposed international fraud cases like the alleged **Lesotho Crash of 1995** or the "**Adrain Case**" in Miami, Florida.

For his research, he travelled over 700.000 miles and visited 44 countries of all five continents. He managed to interview the Bedouin discoverer of the Dead Sea scrolls on camera just a year before he passed away and documented the hidden story behind the archaeological discovery of the 20th century. He interviewed US astronauts **Gordon Cooper** and **Edgar Mitchell**, several Russian cosmonauts and Ex-Generals as well as high-ranking prelates of the Roman Catholic Church.

With official permission from the **Vatican**, Hesemann investigated the history of the alleged relic of the inscription of the cross of Jesus, hidden in Rome for nearly 1700 years. It was dated by seven Israeli experts for comparative palaeography -the established method to date inscriptions- into the 1st century, indicating its authenticity.

On December 17, 1998, Hesemann was personally received by **H.H. Pope John Paul II** to present the results of his study on this important relic. Later, in a letter written by his personal secretary, John Paul II expressed Hesemann his "admiration and appreciation for your laborious research".

Hesemann's latest research covered the history of the **Holy Grail**, the **Chalice of the Last Supper**, which he claims he has located in a chapel in Spain. He continues his studies on Christian Relics, Marian Apparitions, Miracles, Historical Topics, Extraterrestrial Life and the UFO Phenomenon. On his latest research, he reports regularly under "Current News" on this website.

Hesemann is a regular guest on several radio shows in the US and Mexico, including "DREAMLAND" with Whitley Strieber, the "21st CENTURY SHOW" with Bob Hieronymus, the "HILLY ROSE"-Show and, in Mexico, "TERCER MILLENNIO" with Jaime Maussan. He has numerous books in the German language:

Findet der Weltuntergang statt (The Doomsday Chronicles), 1984Unser Gott ist die Erde (The Aztec Heritage), 1988
UFOs: Die Beweise (UFOs: The Evidence), 1989
UFOs: Die Kontakte (UFOs: The Contacts), 1990
Botschaft aus dem Kosmos (The Cosmic Connections), 1992
Geheimsache UFO (UFOs: A Secret History), 1994

UFOs: Neue Beweise (UFOs: New Evidence), 1994
Jenseits von Roswell (Beyond Roswell), 1996
Kornkreise (Crop circles), 1996
UFOs ueber Deutschland (UFOs over Germany), 1997
Geheimsache Fatima (The Fatima Secret), 1997
Die kommende Weltkrise (The Coming World Crisis), 1998
Die Jesus-Tafel (The Jesus-Title), 1999
Die stummen Zeugen von Golgatha (The Silent Witnesses of Golgatha), 2000
Das Fatima-Geheimnis (The Fatima Secret - Revised Version), 2002
Die Kornkreis-Chroniken (The Crop circle Chronicles), 2002
UFOs: Besucher aus dem All (UFOs: Visitors from Space - a Picture Book), 2002
Die Entdeckung des Grals (The Discovery of the Grail), 2002

And his other books in the English language:

The Cosmic Connections, 1994
Beyond Roswell (with Philip Mantle), 1997
UFOs: The Secret History, 1999
The Fatima Secret, 2000

His Videos include: Mystery of the Crop Circles, 199; UFOs: The Secret Evidence, 1992; UFOs and Area 51: Secrets of the Black World, 1994; UFOs: The Contacts, 1996; UFOs: The Footage Archives, part 1-5, 1998-2000; UFO Abductions: A Global Phenomenon, 1998; Ships of Light, part 1-2, 2000-2001; and The Face of Christ 2002. [7]
http://www.ufoevidence.org/researchers/detail93.htm

Hungary

Daniel Tarr (1974-), a cultural anthropologist and cyber-guru, who formulated the Extraterrestrial Energyzoa Hypothesis as an alternative explanation to the extraterrestrial hypothesis (**ETH**) in 2006.[8]

Endre Kriston (Founder and director of the **(RYUFOR) Foundation, and the Hungarian Center for UFO Studies & Fund**) [9]

Italy

Enrico Baccarini is a journalist, writer, and researcher. He is an award winner in Experimental Psychology with a focus on the anthropological world.[10]

Monsignor Corrado Balducci, (born May 11, 1923, in Italy died September 20, 2008, in Italy), was a Roman Catholic theologian of the Vatican Curia, a close friend of the pope, long time exorcist for the Archdiocese of Rome, and a Prelate of the Congregation for the Evangelization of Peoples and the Society for the Propagation of the Faith. He has written several books about the subliminal messages in rock and metal music, diabolic possessions, and extraterrestrials.

Monsignor Balducci often appeared on Italian TV to talk about satanism, religion, and extraterrestrials. Balducci has said that extraterrestrial contact is real.

In relation to the teachings of the Catholic Church, he has stressed that extraterrestrial encounters "are not demonic, they are not due to psychological impairment, and they are not a case of entity attachment, but these encounters deserve to be studied carefully." He denies the rumor that the Vatican follows reported incidents of extraterrestrial encounters from its Nunciatures in different countries.

Balducci's opinions received a wider context in 2008 when the Vatican's chief astronomer José Gabriel Funes also discussed the possibility of extraterrestrial life. As God's power is limitless, it is not only possible but also likely that inhabited planets exist. [11]

Paola Leopizzi Harris (Italy/Europe/Vatican) is an Italo-American photojournalist and one of the very few exceptionally qualified female investigative reporters in the field of extraterrestrial related phenomena research. She is also a widely published free-lance writer, especially in Europe. She has studied extraterrestrial related phenomena since 1979 and is on personal terms with many of the leading researchers in the field. From 1980-1986 she assisted Dr. J. Allen Hynek with his UFO investigations and has interviewed many top military witnesses concerning their involvement in the government truth embargo.

In 1997, Ms. Harris met and interviewed **Col. Philip Corso** in Roswell, New Mexico and became a personal friend and confidante. She was instrumental in having his book *The Day After Roswell*, for which she wrote the preface, translated into Italian. She consequently brought Colonel Corso to Italy for the editorial group *Futuro*, publisher of *Il Giorno Dopo Roswell*, and Corso was present for many TV appearances and two conferences. She returned to Roswell in the summer of 2003 for the American debut of her book, *Connecting the Dots; Making Sense of the UFO Phenomenon.*

Because of her international perspective on extraterrestrial related phenomena, Paola has consulted with many researchers about the best avenues for planetary disclosure with emphasis on the "big picture" and stressing the historical connection. She is a close friend of Monsignor Padre Corrado Balducci and assisted in filming the Italian witnesses, including the Monsignor, for the Disclosure Project for the May 9, 2001, press conference. She was instrumental in bringing to Italy Robert Dean, Dr. Steven Greer, Linda Moulton Howe, **Dr. Richard Boylan**, **Russell Targ, Travis Walton, Derrell Sims, Helmut Lamner, Michael Lindemann, Nick Pope, Bill Hamilton, Ryan Wood, Carlos Diaz** and Dr. John Mack. Her new non-profit association, Starworks Italia, will continue to bring speakers to Italy and promote disclosure and exo-political dialogue worldwide.

She has a regular column in Area 51 *UFO Magazine, has written for Nexus, Australia, Notizario UFO* and *Dossier Alieni*, among others publications.

Paola lives in Rome and Boulder, Colorado and has a Masters degree in Education. She teaches history and photojournalism and On-line classes in Exopolitics for **Dr. Michael Salla's**

Exopolitics Institute for which she is the International liaison director. [12]
http://www.paolaharris.com/bio.htm

Roberto Pinotti, (1944 -), was born in Venice, Italy, in 1944. He is a Sociologist, journalist and author and the main Italian ufologist. According to his website, he has a degree in political science and applied social sciences from the University of Florence, a professional journalism qualification, and a degree in Business Administration and Management Science.

Since 1960 he has searched for any available documentation or information about the UFO phenomena. Formerly an officer of the NATO Italian Army III Missile Brigade, near Venice He was the only Italian ufologist to participate in the French *Groupement d'Étude des Phénomènes Aérospatiaux Non-identifiés* (GEPAN). He organized the first "International UFO Symposium", working for the Republic of San Marino. He is also a writer for the *Rivista Aeronautica* of the Italian Air Force. Pinotti is also a proponent of several international rules of conduct for post-contact with extraterrestrial intelligent beings.

http://www.robertopinotti.com/index_ing.htm

He has a proposed solution for Enrico Fermi's paradox (briefly: "If extraterrestrials do exist, why haven't they been detected?"). His proposal is that humans live in a sort of Indian reserve under an agreement he calls "The Big Game" (Il Grande Gioco) between Earth governments and E.T. beings in order not to reveal themselves. As a sociologist, Pinotti explains this arrangement could be aimed at preventing the anomie that might result from a breakdown of traditional assumptions and values.

In 2000 he published material regarding the so-called "Fascist UFO Files", which dealt with a flying saucer that had crashed near Milan in 1933 (some 14 years before the Roswell, New Mexico crash), and of the subsequent investigation by a never mentioned before Cabinet RS/33, that allegedly was authorized by **Benito Mussolini**, and headed by the Nobel scientist **Guglielmo Marconi**. A spaceship was allegedly stored in a hangar in Vergiate.[13]

Poland

Zbigniew Blania-Bolnar (born 17 October 1948 , died 1 October 2003) was a Sociologist, one of the best-known ufologists in Poland. He lived in Lodz and was a researcher of **CE III** event in Emilcin in 1978.

Zbigniew was the first researcher interested in the case of John Wolski, a farmer of Emilcin who had seen alien beings that he claimed that he met in the forest of Emilcin which has been classified as a CE III encounter with UFOs. There is a monument in the forested area in memorial to the 1978 *event in Emilcin* to the resident of the village, John Wolski,

The identification and development of this case it took Zbigniew several years of research which resulted in a book on this UFO subject and this is probably the most thoroughly documented case of the "Middle Polish Stage III Meeting" (CE III) of UFOs. He was responsible for the creation *Experiencing UFO* project, which was to search for any evidence of the existence of extraterrestrials. [14] http://cezary_kwiatk.republika.pl/blania.htm

http://www.emilcin.com/artykuly,2328.html

Romania

Doru Davidovici (1945–1989), was an aviator and writer. Born in Romania in a Jewish family, Doru Davidovici became one of the most loved Romanian fiction writers in the 1980s. During the communist years, his books gave an unusual sense of liberty and new horizons by describing the experience of flying, and the closeness it forged - both between pilots and between pilots and their machines. The plane is seen by Davidovici not simply as a machine that enables one to fly but as an actual character, with its own personality and almost with its own soul. Beside his narrative work, Doru Davidovici is known for his essay on the UFOs, *My colleagues from the unknown*. Here Davidovici regards, once again through his pilot eyes, the UFOs and the issues raised by their presumed existence. [15] **http://en.wikipedia.org/wiki/Doru_Davidovici**

Spain

Fernando Jiménez del Oso (Madrid, on July 21, 1941 - Madrid, on March 27, 2005), psychiatrist and Spanish journalist, specializing in topics of mystery and parapsychology, Ufology, the director of magazines (*Año Cero*, *Enigmas* and *Más Allá*) and the television programs *The Door of Mystery*, *Beyond*, *Point of Meeting* and *The Magic Spain* between others.

Juan José Benítez, born September 7, 1946, in the city of Pamplona, Navarre, is a Spanish author. He received a journalism degree from the University of Navarra in 1965 and started his work as a journalist in the newspaper *La Verdad* of Murcia in January 1966.

Over the next few years, he became increasingly interested in the UFO phenomena, initiating his extensive investigation on such topics. After holding jobs at several other newspapers, he eventually left journalism in the late 1970s in order to fully immerse himself in the study of UFO phenomena.

http://literatura.wikia.com/wiki/Juan_Jos%C3%A9_Ben%C3%ADtez

During his three decades as an author, he has published more than 50 books, including investigative reports, essays, novels, and poetry, and sold more than nine million copies worldwide, over five million of which belong to the *Caballo de Troya* saga.

He has also directed a TV documentary series called *Planeta Encantado* (Enchanted Planet), in which he travels to 17 different countries in order to render his interpretation of some of the great unsolved mysteries of past history.

He now resides in the coastal town of Barbate, Andalusia, with his wife and two dogs. [16]
http://en.wikipedia.org/wiki/Juan_Jose_Benitez

Sweden

Clas Svahn, (1958 -), Chairman of **UFO-Sverige (UFO-Sweden)** group since 1991. Vice chairman of Archives of UFO Research. Co-editor of UFO-Sweden's magazine "UFO-Aktuellt". Clas has published several books in the genre and contributed to even more.

Håkan Blomqvist, (1952 -), Co-founder and board member of **Archives for UFO Research** since 1973. University studies at Stockholm University: history of religion, philosophy, ethnology. Special interests: Contactee cases, psychological and religious aspects of the UFO phenomenon. Building of archives for the preservation of UFO history.

Anders Liljegren, (1950 -), One of the founding members of UFO-Sweden in 1970, and co-founder and board member of Archives for UFO Research since 1973. Involved in the study of several Swedish close encounters, particularly the Domsten case and the Mariannelund humanoid case.

http://www.afu.se/afu2/?page_id=223

Switzerland

Erich Anton Paul von Däniken (born April 14, 1935) is a controversial Swiss author best known for his claims about extraterrestrial influences on early human culture, in books such as *Chariots of the Gods?* published in 1968. Däniken is one of the main figures responsible for popularizing the paleo-contact and ancient astronaut hypotheses.

Däniken is a co-founder of the **Archaeology, Astronautics and SETI Research Association (AAS RA),** and designed the theme park, Mystery Park in Interlaken, Switzerland that first opened on 23 May 2003. His 26 books have been translated into more than 20 languages, selling more than 60 million copies worldwide, and his documentary TV shows have been viewed around the world.

http://wheniwasbuyingyouadrinkwherewereyou.blogspot.ca/2016/04/31st-year-n-105-thursday-14-april-2016.html

Building on previous works by other authors (including Italian Peter Kolosimo, who was later critical of Däniken), Däniken claimed that intelligent extraterrestrial life exists and has entered the local solar system in the past and that evidence of this past contact is abundant.

He also claims that human evolution may have been manipulated through means of genetic engineering by extraterrestrial beings. The evidence that Däniken has put forward to support his paleo-contact hypotheses can be categorized as follows:

- Artifacts have been found which are alleged to represent a higher technological knowledge than existed at the times when they were manufactured.
- In ancient art throughout the world, themes are observed which can be interpreted to illustrate astronauts, air and space vehicles, non-human but intelligent creatures, and artifacts of a high technology.
- Origins of religions might be a reaction to contact with an alien race by primitive humans.

Candida Mammoliti is a Swiss-Italian UFO researcher, president of the "Centro Ufologico Nazionale della Svizzera Italiana (CUSI, Italian Switzerland's UFO centre). [17]

United Kingdom

After the end of the cold war, and the collapse the Soviet empire, Good became the first Western Ufologist to be interviewed on Russian television. He was also invited to the Pentagon in 1998 and at the headquarters of the French Air Force in 2002 to discuss UFOs and other related matters and has acted as a consultant for several U.S. Congress investigations.

http://www.bbc.co.uk/birmingham/content/articles/2008/07/10/timothy_good_feature.shtml

He is a British Ufologist and professional violinist. He was born in London and educated at The King's School, Canterbury before being granted a scholarship to the Royal Academy of Music, where he studied the violin.

In addition to his background in classical music, Good grew up with a strong interest in aviation and space which became an interest in UFOs in 1955 after he encountered the writings of Major

Donald Keyhoe. In 1961, after reading a book by U.S. Air Force intelligence officer Captain Edward J. Ruppelt, Good began conducting his own research into the UFO phenomena, eventually becoming a well-known authority on the subject.

Timothy Good is probably best known among Ufologist's for his groundbreaking book "Above Top Secret" Which is an eye-opening look at the structure of sensitive top secret info that is kept under wraps in various Governments around the world. According to ufologist Timothy Good (in his books *Alien Liaison* and *Alien Contact*), after **Jackie Gleason's** death, his wife reported that one day in 1973 Gleason had come home extremely shaken. He confided to her that because of Gleason's interest in UFOs, **U.S. President Richard Nixon**, who was a friend of his, had arranged for him to view bodies of extraterrestrials at Homestead Air Force Base, Florida, under conditions of extreme secrecy. Gleason had found the experience very troubling. [18]

Nicholas Pope (born 19 September 1965) is a 25 year serving former employee of the British Government's Ministry of Defence. He joined the MoD in 1985 and he is most notable for having served as a regular term between 1991 and 1994 as that agency's official first point of contact and investigator of UFO reports and sightings, in the department then known as **Sec(AS)2a (Secretariat of the Air Staff).**

In addition to his main duties, he was given the more minor task of documenting the UFO phenomenon, mirroring the work done in the US by the now defunct Project Blue Book. Although most of the cases could be explained as misidentifications of known objects and phenomena, a hard core of sightings defied any conventional explanation.

https://www.aircraft-info.net/2014/05/ex-mod-expert-nick-pope-believes-aliens/

He was the Ministry of Defence official responsible for researching and investigating UFOs, alien abductions, crop circles, cattle mutilation and other strange phenomena. His involvement in UFO phenomena within the MOD led to his self-description as the "British Fox Mulder" from the hit television series, *The X-Files*.

As a result of the cases that he worked in Pope became a believer in UFO phenomena, and that UFOs raised serious defence and national security issues He came across numerous instances where UFOs had reportedly been tracked on radar, leading to jets being scrambled. There were also cases where there had allegedly been near-misses between UFOs and civilian aircraft. All this led him to believe that an extraterrestrial explanation for some sightings such as the so-called Cosford Incident could not be ruled out.

In November 2006, he resigned from his post at the MOD, criticizing the MOD and saying the government's "X-Files have been closed down." Although no longer carrying out this sort of work for the government, he continues his research and investigation in a private capacity.

Pope has written four books in total as of January 2009: *Open Skies, Closed Minds, The Uninvited, Operation Thunder Child, Operation Lightning Strike* and hundreds of articles relating to Ufology. All of his written works required clearance by the Ministry of Defence prior to publication; this is the case whenever any MOD employee writes a book and does not mean that the MOD in any way endorses his material [19]

Jenny Randles (born **Christopher Paul Randles** on October 30, 1951) is a British author and former director of investigations with the **British UFO Research Association (BUFORA)**, serving in that role from 1982 through to 1994. Randles was assigned male at birth.

Randles specializes in writing books on UFOs and paranormal phenomena. To date 50 of these have been published, ranging from her first *UFOs: A British Viewpoint* (1979) to *Breaking the Time Barrier: The race to build the first time machine* (2005). Subjects covered include crop circles, ESP, life after death, time anomalies and spontaneous human combustion.

According to her publishers, Simon & Schuster, Randles studied physics and geology at university, has written articles for *New Scientist*, and has sold more than 1.5 million copies of her fifty published books. It was stated in 1997 that her books had been published in 24 countries.

Randles is a regular contributor to the magazines *Fortean Times* and *The Skeptic* and she was editor of Northern UFO News (a 12-20 page A5 journal detailing UFO activity within Northern England) from 1974 up to 2001. [20]

William Francis Brinsley Le Poer Trench, 8th Earl of Clancarty, and 7th Marquess of Heusden (September 18, 1911–May 18, 1995) was an Irish peer, as well a nobleman in the Dutch nobility. He was the fifth son of the 5th Earl of Clancarty and Mary Gwatkin Ellis.

Clancarty was known to be a prominent ufologist and was a firm believer in flying saucers, and in particular, the **Hollow Earth Theory**

He claimed to know a former U.S. test pilot who said he was one of six persons present at a meeting between President Eisenhower and a group of aliens, which allegedly took place at Edwards Air Force Base on April 4, 1954. Clancarty reported that the test pilot told him "Five different alien craft landed at the base. Three were saucer-shaped and two were cigar shaped... the aliens looked something like humans, but not exactly."

From 1956 to 1959 Clancarty edited the *Flying Saucer Review* and founded the *International Unidentified Object Observer Corps*. In 1967, he founded *Contact International* and served as its first president. He also served as vice-president of the British UFO Research Association (BUFORA). Clancarty was an honorary life member of the now defunct *Ancient Astronauts Society* which supported the ideas put forward by Erich von Däniken in his 1968 book Chariots of the Gods?.

In 1975 he succeeded to the earldom on the death of his half-brother, giving him a seat in the British Parliament. He used his new position to found a UFO Study Group at the House of Lords, introducing *Flying Saucer Review* to its library and pushing for the declassification of UFO data.

Four years later he organized a celebrated debate in the House of Lords on UFOs which attracted many speeches on both sides of the question. In one debate, Lord Strabolgi, for the Government, declared that there was nothing to convince him that any alien spacecraft had ever visited the Earth.[21]

NORTH AMERICA - Canada

Stanton T. Friedman is a Nuclear Physicist-Lecturer having received his BSc. and MSc. Degrees in physics from the University of Chicago in 1955 and 1956.
He was employed for 14 years as a nuclear physicist by such companies as GE, GM, Westinghouse, TRW Systems, Aerojet General Nucleonics, and McDonnell Douglas working in such highly advanced, classified, eventually cancelled programs as nuclear aircraft, fission and fusion rockets, and various compact nuclear power plants for space and terrestrial applications.

He became interested in UFOs in 1958, and since 1967 has lectured about them at more than 600 colleges and 100 professional groups in 50 U.S. states, 9 Canadian provinces, and 16 other countries in addition to various nuclear consulting efforts. He has published more than 90 UFO papers and has appeared on hundreds of radio and TV programs including on Larry King in 2007 and twice in 2008, and many documentaries.

He is the original civilian investigator of the **Roswell Incident** and co-authored "Crash at Corona: The Definitive Study of the Roswell Incident." "TOP SECRET/MAJIC," his controversial book about the Majestic 12 group, established in 1947 to deal with alien technology, was published in 1996 and went through 6 printings. An expanded new edition was published in 2005. Stan was presented with a Lifetime UFO Achievement Award in Leeds, England, in 2002, by UFO Magazine of the UK. He is a co-author with **Kathleen Marden**

(Betty Hill's niece) of a book in 2007: "Captured! The Betty and Barney Hill UFO Experience." The City of Fredericton, New Brunswick, declared August 27, 2007, Stanton Friedman Day. His new book "Flying Saucers and Science" was published in June 2008 and is in its 3rd printing.

He has provided written testimony to Congressional Hearings, appeared twice at the UN, and been a pioneer in many aspects of Ufology including Roswell, Majestic 12, The Betty Hill-Marjorie Fish star map work, analysis of the Delphos, Kansas, physical trace case, crashed saucers, flying saucer technology, and challenges to the **S.E.T.I. (Silly Effort To Investigate) cultists**. He has spoken at more MUFON Symposia than anyone else.

Stanton T. Friedman is a dual citizen of the USA and Canada and lives in Fredericton, NB, Canada. [22]

Jennifer Jarvis, UFO researcher from St. Catherines, Ontario. A member of CSETI, her ongoing observations can be found at **http://orbwatch.com**

Since the summer of 1993, when Jarvis first saw the crop circles in England, her interest in unusual and inexplicable phenomena increased proportionally to its complexity. In 1994, her attention was moved to the direction, of a strange incident occurring near Silbury Hill, close to the stone circle of Avebury, in Wiltshire.

Witnesses testify that they saw strange light phenomena in the proximity of Silbury Hill, and apparently, these light phenomena were moving in an intelligent and directed manner. The witnesses mentioned a project that had taken place in the vicinity of Silbury Hill only days before. This had been a meditative project directed by two gentlemen from Germany, in conjunction with an organization called CSETI - The Center for the Study of Extraterrestrial Intelligence.

Jennifer became a member of the CSETI organization in 1995 and felt that CSETI's mandate was a sincere effort to work towards world, and maybe, universal peace. In June of 1996, in the beautiful mountains of the Sangre de Cristo, and the spectacular San Luis Valley of Southern Colorado, Jarvis attended a week of training led by Doctor Greer and Ms. Adamiak.

Returning from Colorado, Jarvis formed a CSETI branch group of interested and like-minded people together to form a small working group and use the "contact protocols" developed by Dr. Greer.

Synchronicities and messages via lucid dreams initiated the newly formed group one Spring cold night to the north shores of Lake Ontario which lead to the group's first golden orb sighting. From that point on Jennifer Jarvis has photographed and videotaped many orb lights going into and coming out of Lake Ontario and her website illustrates her research work on this phenomenon. [23] http://orbwatch.com

Martin Jasek was born in the Czech Republic in 1963. In 1970 his family immigrated to Canada and settled in Edmonton, Alberta. He attended the University of Alberta where he obtained his Bachelor's degree in Civil Engineering in 1986 and his Master's degree in Water Resources Engineering in 1992. In 1993 he moved up to Whitehorse Yukon to work for Indian and Northern Affairs Canada, Water Resources Division.

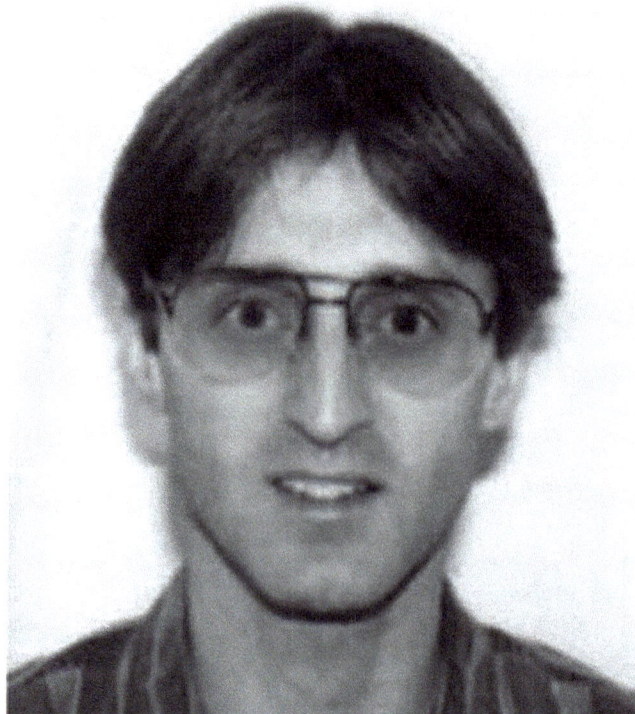

http://www.ufobc.ca/yukon/martin.htm

298

It was August 1995 when his interest in the UFO subject from early childhood was rekindled by a television program. Although slightly skeptical, he started to read everything he could about the subject. In 1996 he started collaboration with his neighbour, Lorraine Bretlyn who had a lifelong interest in the subject. In the spring of 1997, he teamed up with **UFO*BC** and has become their Yukon Representative. In July of 1997, he traveled to Roswell, New Mexico to attend the 50th anniversary of the legendary UFO crash.

In 2002 he moved to the Vancouver lower mainland and joined **UFO*BC** as a director but still remains active with the **UFO Yukon Research Society** based out of Whitehorse. Martin, UFO*BC, investigated and created a comprehensive report on the **Little Fox Lake UFO Incident** December 11, 1996.[24]

Paul Kimball, (1967-) film producer / director, and paranormal commentator / researcher, whose documentaries include *Stanton T. Friedman is Real* (2002), *Do You Believe in Majic* (2004), *Aztec: 1948* (2004), *Fields of Fear* (2006), *Best Evidence: Top 10 UFO Cases* (2007) and *The Island of Blood* (2009).

Kimball is an outspoken ufologist who isn't afraid to speak his mind, often critiquing other UFO researchers with his own perspective on the UFO phenomenon and could be considered as "part of the new young guard of Ufology", a term referred to by "Ufonaut Radio" host Jesse Randolf on "Blog talk Radio." [25]

http://www.rense.com/1.mpicons/redstar.html

Chris Rutkowski, an astronomer has been studying reports of UFOs [or PFTs - "pesky flying things", as he sometimes likes to call them] since the mid-1970s. A writer and editor for science and technology publications, he also writes about his investigations and research into UFO cases.

The **Ufology Research of Manitoba** to which he belongs publishes, along with Geoff Dittman, **'The Canadian UFO Survey'**, an annual compilation of UFO Reports from across Canada which has been operating since 1998.

Chris has had five books published: 'Visitations?' (1989) and 'Unnatural History' (1993), the latter of which is in its third printing. 'Mysterious Manitoba' (1997), co-authored with Dave Creighton and Brian Fidler, 1997, 'Abductions & Aliens - What's Really Going On' published by Dundurn, 1999 and *The Canadian UFO Report* with Geoff Dittman

His day-job is with the Public Relations Department of The University of Manitoba in Winnipeg.

The Canadian Federal Government currently (2007) directs all UFO sightings to *Chris Rutkowski* of Ufology Research of Manitoba.[26]

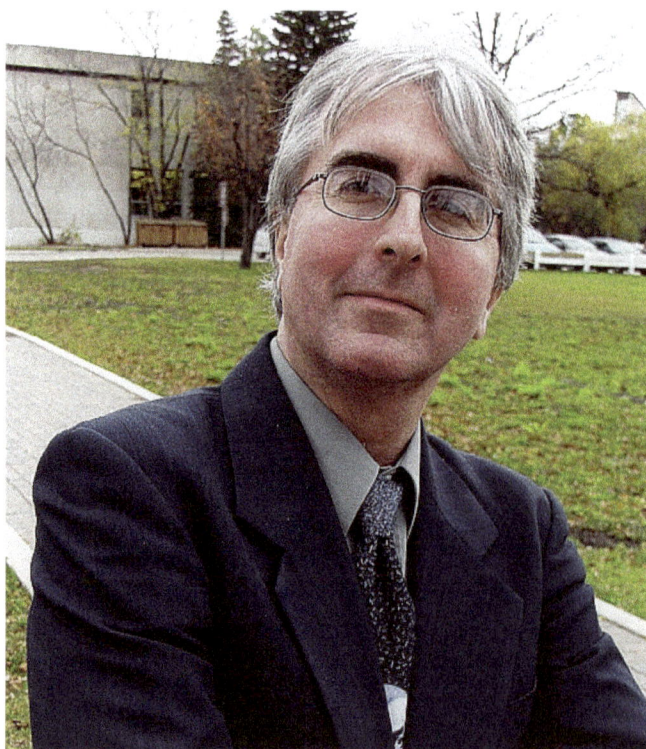

http://calgaryherald.com/news/local-news/fortney-kananaskis-ufo-sighting-a-reminder-of-sightings-past

David Sereda, born in Edmonton, Alberta, Canada on August 21, 1961, became a permanent Resident of the United States of America with his family in 1964 and grew up in the State of California. David Sereda's first aspiration in life was to become an astronaut. He kept a scrapbook of every Apollo mission while growing up in Berkeley and San Francisco, California in the 1960s.

In approximately 1968, while David was on his way home from elementary school in Berkeley, California, he and his friend Tommy witnessed a large, classic disc-shaped UFO hovering in a clear blue sky with hundreds of witnesses shouting with excitement in the streets all pointing upwards in utter amazement. After 20 minutes to a half an hour of witnessing this UFO, David said he saw it literally disappear in one second into another invisible dimension, a subject that would take him until the year 2000 to understand. With a permanent and affected memory, David grew up as a UFO enthusiast never living in doubt of the phenomena that has swept the world since the **Roswell incident in 1947**.

His interest in space, religion, philosophy, astronomy and science led him on his career in related fields. He has worked deeply in high technology, on environmental and humanitarian issues and as a professional photographer for over 20 years. He has studied world religion, science, physics and paranormal psychology for over 25 years on virtually hundreds of issues.

From 1995 through 2001, at the request of Martyn Stubbs (who video cataloged over 400 hours of live NASA broadcast missions) and Michael Boyle (Professional Photographer associate of Mr. Stubbs) David Sereda conducted a deep scientific investigation into the scientific community at NASA and outside of the agency into 1990s space shuttle mission video footage of Unidentified Flying Object Phenomena. The results of that investigation led to the development of the book and documentary film, ***"Evidence, The Case For NASA UFOs."*** [27]

http://www.latest-ufo-sightings.net/2010/09/ufo-congress-2010-david-sereda.html

Wilbert Brockhouse Smith (1910–1962) was a Canadian electrical engineer, radio engineer, ufologist and contactee. Wilbert Smith epitomises like no other UFO researcher, the intelligence, the insightfulness, the open-mindedness and the respected trust by his peers and his fellow scientists, by government and military officials, both in Canada and the United States, especially at a time in history when UFO secrecy and cover-up by the military industrial complex was tightening its grip on public disclosure.

Smith is generally known as an official with the Department of Communications (DOC) and the head of a Canadian Government UFO project, Project Magnet.

Wilbert Smith held both a B. Sc from the University of British Columbia in 1933, and M. Sc. in Electrical Engineering in 1934, and was Senior Radio Engineer, Broadcast and Measurements Section, Department of Transport (DOT). Smith was internationally recognized for his work in radio communications as an engineer at several commercial radio stations before co-founding the **Canadian Association of Broadcast Consultants** and he held a number of patents.

He had been responsible for engineering aspects of everything concerning the use of radio in Canada, including equipment standards, radio relay systems, broadcast facilities and interference studies. He was in charge of establishing a network of ionospheric measurement stations throughout Canada, and he often represented Canada at international conferences. [37]

Another notable achievement was the 1947 establishment of stations to measure the ionosphere.

At his death, Smith was the Superintendent of Radio Regulations Engineering for the DOT, an organization responsible for licensing broadcast facilities, setting equipment standards and performing other acts of commercial radio regulation.

Smith died on December 27, 1962, of intestinal cancer. Though the fact was not disclosed until well after his death, Smith claimed that, in the late 1940s, he received what a friend described as "mental messages from space people" Smith related this claim to very few people during his life.

In 1950, Smith's interest in UFOs was further piqued by a book written by retired U.S. Marine Corps **Maj. Donald E. Keyhoe,** claiming that flying saucers were real and their existence being covered up by the U.S. government, and by **Frank Scully**'s book *Behind the Flying Saucers*, which claimed that several crashed UFOs had been recovered by U.S. officials and they operated on magnetic principles.

The Scully story is now widely regarded as a hoax perpetrated on Scully by two con-men. However, according to a memorandum located by **Arthur Bray**[3], Smith met in 1950 with U.S. Defense Dept. physicist **Robert Sarbacher**, who confirmed the core elements of Scully's story as accurate, and who further claimed that American scientist Vannevar Bush was involved in a secret, high-level UFO research group known as "Majestic 12" or "MJ12" operating out of the **Defense Department's Research and Development Board (DRB)** looking into the modus operandi of the saucers.

The interview with Sarbacher was arranged by the Canadian embassy in Washington D.C. after Smith asked them to find out from U.S. officials whether the flying saucers were indeed real. Sarbacher, a consultant to the DRB, confirmed that they were and added that the subject was classified higher than the H-bomb. Contacted 33 years later, Sarbacher confirmed the interview had taken place and the substance of Smith's memo.

In a series of "discreet inquiries" in the United States Smith was able to discover five key points about flying saucers, based on what he was told about the American flying saucer program:

1) The matter of UFOs was the most highly classified subject in the US, rating higher than the H-bomb
2) Flying saucers exist
3) Their modus operandi is unknown, but concentrated effort is being made by a small group headed by **Dr. Vannevar Bush,** (Of 'MJ12' fame)
4) The entire matter is considered by US authorities to be of tremendous significance.
5) There were a number of other things including "mental phenomena" that were being studies because of their possible link to the saucers.
[SMITH BACKGROUND] by Grant Cameron, UFOROM (3 pages);
http://ufo-joe.tripod.com/gov/wbsebk

Immediately afterward, Smith lobbied the **Canadian Defence Research Board (DRB)** and the DOT to establish a UFO research group. The program began after Smith wrote a Top Secret memo to the Deputy Minister of Transport for Air services C.P. Edwards, and the Assistant Minister J.R. Baldwin.

The Top Secret memo described a possible new propulsion system powered by the earth's magnetic field. Smith felt that this was the principle being demonstrated by the flying saucers.

Project Magnet was the result. Founded in December 1950, it operated under DOT auspices and DRB assistance until 1954. Magnet was the official flying saucer investigation by the Canadian government. It was headed up for the four years it was operational by Wilbert Smith.

Smith wrote several reports for Magnet, concluding that UFOs were likely extraterrestrial in origin, and used principles of magnetism to travel. After Magnet's funding was ended in 1954, Smith was allowed to use Magnet's facilities at his own expense.

Also after Magnet's formal funding ended, Clark writes, Smith "became more open about his fascination with what many considered Ufology's fringe aspects." Smith met or corresponded with contactees like Frances Swan, and used séances to contact **"ufonauts"**. Smith was satisfied with results, which he thought confirmed his opinions about UFOs.

In 1957, Keyhoe asked Smith to join the advisory board of NICAP, the UFO research group Keyhoe headed, citing Smith's notable achievements in radio and electronics. Smith declined, largely because he disagreed with NICAP's skepticism towards cases where UFO occupants were purportedly witnessed. [28] **http://ufo-joe.tripod.com/gov/wbsebk.html**

Grant Cameron became involved in Ufology in May 1975 with personal sightings of an object which locally became known as **"Charlie Red Star"**. The sightings occurred in Carman, Manitoba about 25 miles north of the Canada-US border. Hundreds of other people sighted objects at the same time during a prolonged flap of sightings.

Over the next 18 months he had many sightings of large objects and small (monitor) objects in the area. He spent countless days in the area photographing a series of strange objects and interviewing hundreds of witnesses who were involved.

After composing a manuscript about the flap, he moved on to research the work of the late **Wilbert B. Smith**. Smith headed up the Canadian government flying saucer investigation known as Project Magnet which ran from 1950 to 1954. During two decades of work on Smith, Cameron was able to collect most of Smith's files and material written, along with 12 hours of audio related to Smith's work. He interviewed most of the associates around Smith who worked on the flying saucer investigation

Years of Smith research led to the discovery of former Penn State University president Dr. Eric Walker, who was identified by Dr. Robert Sarbacher as a key person inside the UFO cover-up. Cameron teamed up with T. Scott Crain to research and write a book titled "UFOs, MJ-12, and the Government", published by MUFON. The book summed up their three years of research on Dr. Walker's involvement.

In the past few years Cameron has turned his research interests to the involvement and actions of the President of the United States in the UFO problem. He has made 20+ trips to the National Archives and most of the various Presidential archives looking for presidential UFO material. One highlight his presidential UFO research was the chance to question Vice-president Dick Cheney on his knowledge of the UFO subject. Another highlight of the presidential UFO research was a FOIA to the White House Office of Science and Technology which yielded 1,000 pages of UFO documents from the Clinton administration. Many of these findings have been written up on the **Presidents UFO website: www.presidentialufo.com.**

At present Cameron is working on producing monographs for Dr. Walker, Wilbert Smith, the Presidential involvement in the topic of psychic phenomena, and a monograph looking at a possible disclosure pattern to try and explain the many actions of the American government, related to UFOs, during the last 50 years. He is also working on a detailed paper detailing the "64 Reasons the Government is Covering Up the ET Presence."

He has lectured widely in Canada and the United States on the 1975 Carman flap, the Canadian government's early investigations into flying saucers, UFO disclosure politics, the **Rockefeller UFO Initiative**, and the **Presidential UFO connection**.

In September 2005 he was denied access to the United States to lecture on UFOs. He produced a video lecture for the 2006 Ozark convention and plans to resume lecturing for free in the United States in 2008.

At present Cameron is awaiting almost 100 FOIA requests from the **Clinton Presidential Library** in Little Rock, Arkansas, related to the UFO related actions and policies inside the two Presidential terms of Bill Clinton. So far 12 UFO related requests have been released. Their release caused a stir among media outlets around America. [29]
http://www.presidentialufo.com/grant-cameron-biography

DEPARTMENT OF TRANSPORT
INTRA-DEPARTMENTAL CORRESPONDENCE

OTTAWA, Ontario, November 21, 1950.

	SUBJECT		OUR FILE
	Geo-Magnetics		(R.57.)

Downgraded to Confidential SEE MEMO 15/9/69 CR

MEMORANDUM TO THE CONTROLLER OF TELECOMMUNICATIONS:

For the past several years we have been engaged in the study of various aspects of radio wave propagation. The vagaries of this phenomenon have led us into the fields of aurora, cosmic radiation, atmospheric radio-activity and geo-magnetism. In the case of geo-magnetics our investigations have contributed little to our knowledge of radio wave propagation as yet, but nevertheless have indicated several avenues of investigation which may well be explored with profit. For example, we are on the track of a means whereby the potential energy of the earth's magnetic field may be abstracted and used.

On the basis of theoretical considerations a small and very crude experimental unit was constructed approximately a year ago and tested in our Standards Laboratory. The tests were essentially successful in that sufficient energy was abstracted from the earth's field to operate a voltmeter, approximately 50 milliwatts. Although this unit was far from being self-sustaining, it nevertheless demonstrated the soundness of the basic principles in a qualitative manner and provided useful data for the design of a better unit.

The design has now been completed for a unit which should be self-sustaining and in addition provide a small surplus of power. Such a unit, in addition to functioning as a 'pilot power plant' should be large enough to permit the study of the various reaction forces which are expected to develop.

We believe that we are on the track of something which may well prove to be the introduction to a new technology. The existence of a different technology is borne out by the investigations which are being carried on at the present time in relation to flying saucers.

While in Washington attending the NARB Conference, two books were released, one titled "Behind the Flying Saucer" by Frank Scully, and the other "The Flying Saucers are Real" by Donald Keyhoe. Both books dealt mostly with the sightings of unidentified objects and both books claim that flying objects were of extra-terrestrial origin and might well be space ships

...... 2

Smith's memo to the Dept. of Transport, Nov. 21, 1950

from another planet. Scully claimed that the preliminary studies of one saucer which fell into the hands of the United States Government indicated that they operated on some hitherto unknown magnetic principles. It appeared to me that our own work in geo-magnetics might well be the linkage between our technology and the technology by which the saucers are designed and operated. If it is assumed that our geo-magnetic investigations are in the right direction, the theory of operation of the saucers becomes quite straightforward, with all observed features explained qualitatively and quantitatively.

I made discreet enquiries through the Canadian Embassy staff in Washington who were able to obtain for me the following information:

a. The matter is the most highly classified subject in the United States Government, rating higher even than the H-bomb.

b. Flying saucers exist.

c. Their modus operandi is unknown but concentrated effort is being made by a small group headed by Doctor Vannevar Bush.

d. The entire matter is considered by the United States authorities to be of tremendous significance.

I was further informed that the United States authorities are investigating along quite a number of lines which might possibly be related to the saucers such as mental phenomena and I gather that they are not doing too well since they indicated that if Canada is doing anything at all in geo-magnetics they would welcome a discussion with suitably accredited Canadians.

While I am not yet in a position to say that we have solved even the first problems in geo-magnetic energy release, I feel that the correlation between our basic theory and the available information on saucers checks too closely to be mere coincidence. It is my honest opinion that we are on the right track and are fairly close to at least some of the answers.

Mr. Wright, Defence Research Board liaison officer at the Canadian Embassy in Washington, was extremely anxious for me to get in touch with Doctor Solandt, Chairman of the Defence Research Board, to discuss with him future investigations along the line of geo-magnetic energy release.

••••••• 3

Smith's memo to the Dept. of Transport, Nov. 21, 1950

I do not feel that we have as yet sufficient data to place before Defence Research Board which would enable a program to be initiated within that organization, but I do feel that further research is necessary and I would prefer to see it done within the frame work of our own organization with, of course, full co-operation and exchange of information with other interested bodies.

I discussed this matter fully with Doctor Solandt, Chairman of Defence Research Board, on November 20th and placed before him as much information as I have been able to gather to date. Doctor Solandt agreed that work on geo-magnetic energy should go forward as rapidly as possible and offered full co-operation of his Board in providing laboratory facilities, acquisition of necessary items of equipment, and specialized personnel for incidental work in the project. I indicated to Doctor Solandt that we would prefer to keep the project within the Department of Transport for the time being until we have obtained sufficient information to permit a complete assessment of the value of the work.

It is therefore recommended that a PROJECT be set up within the frame work of this Section to study this problem and that the work be carried on a part time basis until such time as sufficient tangible results can be seen to warrant more definitive action. Cost of the program in its initial stages are expected to be less than a few hundred dollars and can be carried by our Radio Standards Lab appropriation.

Attached hereto is a draft of terms of reference for such a project which, if authorized, will enable us to proceed with this research work within our own organization.

(W.B. Smith)
Senior Radio Engineer

Smith's memo to the Dept. of Transport, Nov. 21, 1950

WASHINGTON INSTITUTE OF TECHNOLOGY
OCEANOGRAPHIC AND PHYSICAL SCIENCES

DR. ROBERT I. SARBACHER
PRESIDENT AND CHAIRMAN OF BOARD

November 29, 1983

Answer from Dr. Sarbacher Received 12-5-83 Wm Steinman

Mr. William Steinman
15043 Rosalita Drive
La Mirada, California 90638

Dear Mr. Steinman:

I am sorry I have taken so long in answering your letters. However, I have moved my office and have had to make a number of extended trips.

To answer your last question in your letter of October 14, 1983, there is no particular reason I feel I shouldn't or couldn't answer any or all of your questions. I am delighted to answer all of them to the best of my ability.

You listed some of your questions in your letter of September 12th. I will attempt to answer them as you had listed them.

1. Relating to my own experience regarding recovered flying saucers, I had no association with any of the people involved in the recovery and have no knowledge regarding the dates of the recoveries. If I had I would send it to you.

2. Regarding verification that persons you list were involved, I can only say this:

John von Neuman was definitely involved. Dr. Vannever Bush was definitely involved, and I think Dr. Robert Oppenheimer also.

My association with the Research and Development Board under Doctor Compton during the Eisenhower administration was rather limited so that although I had been invited to participate in several discussions associated withthe reported recoveries, I could not personally attend the meetings. I am sure thatthey would have asked Dr. von Braun, and the others that you listed were probably asked and may or may not have attended. This is all I know for sure.

500 BRAZILIAN AVENUE   PALM BEACH, FLORIDA 33480   305-833-1118

1983 Sarbacher letter to researcher William Steinman

310

WASHINGTON INSTITUTE OF TECHNOLOGY

OCEANOGRAPHIC AND PHYSICAL SCIENCES

DR. ROBERT I. SARBACHER
PRESIDENT AND CHAIRMAN OF BOARD

November 29, 1983

*Answer
from Dr. Sarbacher
Received 12-5-83
Wm Steinman*

Mr. William Steinman
15043 Rosalita Drive
La Mirada, California 90638

Dear Mr. Steinman:

I am sorry I have taken so long in answering your letters.
However, I have moved my office and have had to make a
number of extended trips.

To answer your last question in your letter of October 14,
1983, there is no particular reason I feel I shouldn't or
couldn't answer any or all of your questions. I am delighted to answer all of them to the best of my ability.

You listed some of your questions in your letter of
September 12th. I will attempt to answer them as you had
listed them.

    1.   Relating to my own experience regarding re-
covered flying saucers, I had no association with any
of the people involved in the recovery and have no knowledge regarding the dates of the recoveries. If I had I
would send it to you.

    2.   Regarding verification that persons you list
were involved, I can only say this:

    John von Neuman was definitely involved. Dr.
Vannever Bush was definitely involved, and I think Dr.
Robert Oppenheimer also.

    My association with the Research and Develop-
ment Board under Doctor Compton during the Eisenhower
administration was rather limited so that although I had
been invited to participate in several discussions associated withthe reported recoveries, I could not personally
attend the meetings. I am sure thatthey would have asked
Dr. von Braun, and the others that you listed were probably
asked and may or may not have attended. This is all I know
for sure.

500 BRAZILIAN AVENUE   PALM BEACH, FLORIDA 33480   305-833-1118

1983 Sarbacher letter to researcher William Steinman

Chris Styles is an active UFO researcher who investigates select classic and current UFO incidents that have occurred in Atlantic Canada. He is best known for his work on the Shag Harbour Incident. In 1993 Chris received a modest research grant from the Washington based group, **The Fund for UFO Research** which helped underwrite the expense on an on-site research effort at Canada's National Archives in Ottawa.

In 1995 Chris directed an underwater search for physical evidence that may have remained undetected on the seabed of **Shag Harbour**. That effort was funded by Paramount Television and resulted in several segments for the syndicated UFO documentary program *Sightings*. Chris has also appeared in several US feature documentaries.

http://www.noufors.com/images/Who's%20Who%20in%20UFOlogy/

Chris served as a paid technical advisor with several Canadian UFO feature documentaries such as Ocean Entertainment's *The Shag Harbour Incident* and *Northern Lights*, a 2 hour feature production of Roadhouse Films. His most recent on camera appearance was in the US History Channel's 60 minute feature documentary *UFO Files: Canada's Roswell*, which was first broadcast in March, 2006.

In 2001 Chris co-authored *Dark Object* with fellow UFO researcher Don Ledger. He has also written several speculative papers on different aspects of UFO research and has presented at various international UFO symposia. He can be reached via E-mail at shagharbour@hotmail.com. [30] **http://www.shagharbourufo.com/**

Terry Tibando is an active independent UFO who has been quietly investigating and researching the UFO and ETI phenomenon since he was a teenager and has experienced UFO sightings and ET encounters dating back to early his childhood, over 60 years ago. With many first-hand experiences and close encounters with UFOs and ET beings, Tibando finds himself in

312

a rather unique and enviable position among Ufologists. Many UFO researchers spend years investigating the phenomenon, but very few have ever seen a UFO or experienced an ET contact encounter. This makes Tibando eminently qualified to understand the big picture of the UFO and ETI phenomenon, its motives and its agenda.

Tibando attended the University of Victoria studying astronomy and the sciences for two years before gaining employment in the BC Forestry Department and has been a Route Manager and Technician for 30 years in the Pest Control industry. Throughout his university education, it was always in the back of his mind that his study of sciences was to serve in his understanding of the world, the universe and primarily understanding the "other intelligent life in the universe"!

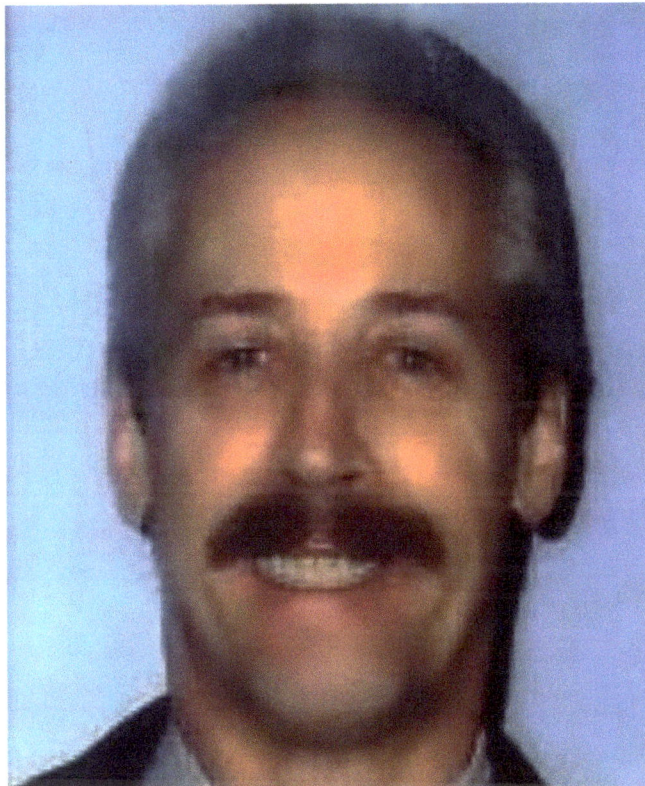

It was, however, his Extraterrestrial encounters that Tibando began to seriously study UFO /ETI phenomenon back in 1965 when he had a daylight sighting, very close to his home in Victoria, of a luminous disc-shaped craft that resembled a water chestnut which travelled in and out of the clouds of an overcast sky. This sighting however, was not Terry's first close encounter with other worldly visitors. His first experienced happen as a five year old child in November 1953 in St. Jeans, Quebec.

As an air force "brat" of a father in the **Royal Canadian Air Force (RCAF)** and a family constantly transferred from one air force station to another from Victoria to North Bay to St. Jeans, Quebec. It was at St. Jeans that Terry saw a flying saucer during the day with his mother and younger baby brother (James) while out for a walk. It flew over a military building, hovered momentarily and then flew off again. Terry asked his mother what the object was and she said it was just an airplane and with that explanation, Terry seemed satisfied that such "airplanes" were also disc-shaped, silver in colour with a dome on the top and a dark non-reflective flat bottom, similar to the photo of the **Passaic, New Jersey saucer of 1952.**

This sighting was merely a prelude to an ET encounter a few nights later when three little luminous beings "materialized and floated" through his bedroom wall. Awaken from sleep by the sudden presence of these strange little beings, Terry became frightened and called out to his mother several times to investigate their sudden appearance which he described as being like **"Casper the Ghost"** type beings! His mother saw nothing and reassured Terry that he was just having a "bad dream". However, these "luminous ghosts" were persistent, if not determined to accomplish whatever mission they had in mind for Terry and so, on their *third return*, Terry still in fright and not wanting to call for his mother again or re-awaken his brothers, simply pulled the bed covers up over his head and hid from these approaching ET "ghosts" until sleep over came him. At least that was the last conscious memory of the event. After that time, it seemed Terry had an expanded view of life in the universe and never doubted the existence of ETIs!

Terry recalls that even his father once told him and his family about an unusual event in 1951 on a visit back home to see his own parents in Toronto when he saw a flying saucer parked in a military hangar.

Corporal Tibando flew out of Victoria by military air transport and arrived in Moulton, Ontario, a well known RCAF airbase at the time. Upon stepping off the aircraft, Cpl. Tibando's attention was immediately focused on an aircraft hangar and noticed that its large doors were starting to open up. In that jaw-dropping moment, Cpl. Tibando witnessed a flying disc-shaped craft levitating and slowly moving out through the hangar doors to the outside in front of the hangar. The flying saucer hovered there for a few moments then, shot straight up into the sky to a low altitude between 500 to 1000 feet. It hovered there briefly then, quickly descended and moved back into the aircraft hangar from whence it came. Perhaps, this was a secret test flight in which other military personnel weren't supposed to have been present to observed it, such as the arrival of Cpl. Tibando, thus the hasty retreat of the disc back into the hangar.

Right up to the end of his life Cpl. Tibando was convinced that he had indeed seen a flying saucer in possession by the RCAF. As to how the Canadian military came into possession of the disc, whether this was the famous **Avrocar** or a genuine captured ET saucer remains unknown to this day.

In 1965 while attending high school, Terry became a member in **APRO (Aerial Phenomenon Research Group),** the world's largest UFO organization at the time and its affiliated organization, **CAPRO (Canadian Aerial Phenomenon Research Organization).**

Tibando offers a fresh perspective of the public's "everyman" knowledge and understanding of the UFO/ETI mystery, a sort of rational, intellectual "Joe Average" perception that goes beyond mere curiosity *(this textbook is a testament to that knowledge).* His assessment of the phenomenon was that the ETI presence was real, that there was a cover-up and suppression of knowledge on multiple levels being orchestrated by the **Military Industrial Complex.** He concluded that many UFO researchers had major pieces to the UFO puzzle but, because of an inability to consult together and with many **Big Egos** among researchers, Ufology appeared to be in a state of disintegration, confusion, chaos with a corrupt UFO database. This malaise made it impossible for anyone in Ufology to connect the dots in order to see the big picture until Dr. Steven Greer appeared upon the world's UFO stage.

In early 1992, Tibando joined the **CSETI (Centre for the Study of Extraterrestrial Intelligence)** organization headed by International Director, **Dr. Steven Greer.** Tibando discovered that here was someone else, who had reached the same conclusions as he had after many decades of research and had taken the next step in a pro-active investigation of the ETI phenomenon. Dr. Greer stated that a new paradigm in UFO research had been reached. After 50 years of analysis and conclusions from documented UFO case reports, the good doctor stated that it was possible for humans to initiate contact and communications with Extraterrestrial beings visiting the Earth utilizing a set of communication protocols.

Tibando shared much of the same common knowledge, experiences, and conclusions on the UFO/ETI phenomenon reached by Dr. Greer. and in 1993, Tibando formed and organized a branch of CSETI in Vancouver, B.C. becoming Canada's first CSETI working group.

Then in 1995, with the assistance of one of Greer's former team members, **Dr. Joseph Burkes**; the fledgling CSETI group was trained in the CE-5 Protocol Initiative.

Following this intitial training Tibando became the **CSETI Vancouver Coordinator** leading numerous teams of interested people in the UFO phenomenon out on field trips to establish human-initiated contact and communications with visiting ETI to our planet.

In mid-October 1995 the **Discovery Channel** from Toronto came to Vancouver to interview Tibando and the CSETI Vancouver team as a part their television **"Alien Week"** series which they were filming for next summer's viewing. Almost unexpectedly, two ET spacecraft showed up, one during the interview and one after the interview. The first object was a small intensely bright object which hovered about a mile away and seen by everyone present except for the Discovery Channel TV crew who unfortunately were unable to capture the object on camera, fortunately, however, the Vancouver CSETI team did video record it. The second object was a large triangular craft which showed up and paced the CSETI team's vehicles down the highway after the Discovery Channel had completed their interview.

In the summer of 1996 in Crestone, Colorado Tibando was trained in the **"Ambassadors to the Universe"** program by Dr. Greer, developing remote viewing capabilities and higher states of meditative consciousness (perquisites to human-initiated contact with Extraterrestrials life forms) as well as becoming a supporter and friend of Dr. Greer.

In June 1997, Tibando was a speaker at the **Bellingham UFO Group (BUFOG)** UFO lecture seminar filling in for **Dorothy Izzat** from Vancouver, who was unable to attend for health reasons. Other guest speakers were **Peter Davenport** from **NUFORC** and **Sharon Filip,** alien abduction researcher.

In September 2001, Tibando along with the **CSETI Vancouver** team and aided by **Dr. Alfred**

Webre sponsored Canada's first public UFO/ETI disclosure event as a part of Dr. Steven Greer's **Disclosure Project** lecture tour. Tibando facilitated and emceed the venue at the Simon Fraser University in Burnaby, B.C. with **Dr. Greer, Dr. Carol Rosin** and **Alfred L. Webre** as the guest speakers.

Tibando and **CSETI Vancouver** has also been interviewed by **BCTV** where serendipitously, an unusual physical high strangeness event occurring toward the end of the interview. An activated strobe light had mysteriously moved from its original location to a spot two hundred feet away without anyone having seen it moved. No deliberate pranks or hoaxes or even animals in the immediate area could account for this translocation of a fully functional light that was in full view of everyone present during the TV interview. It was, in fact, the TV crew who pointed out its final resting spot from the area in which it was originally located.

In 2012, **Terry Tibando** is a major donor and supporter of Dr. Greer's film documentary **"Sirius"** and hid current second movie "Unacknowledged", 2017 which have been successfully received globally (both may be viewed on Netflix) adding to the greater understanding of the UFO and ETI phenomenon as a part of Dr Greer's **Disclosure Project**.

Currently, Tibando is working on writing his first UFO book (this book). It will be a major 3300-page treatise, a distillation of the vast UFO database into a textbook format collating the best-known eyewitness accounts, the history of the phenomenon, the suppression of alien technology by the military industrial complex, and the hopeful future of humanity through the release of new energy generation systems. It also includes his personal experiences and insights on the subject matter. The book titled *"A Citizen's Disclosure on UFOs and ETI"* connects the dots and present the big UFO/ETI picture as it is currently understood in the field of Ufology.

Mexico

Jaime Maussan was born in 1953 in Mexico City and earned a B.A. degree in radio and television from the Miami University in Ohio. After one year at the National Autonomous University of Mexico (1972-73), he attended Miami University (Ohio) and graduated with a B.A. in Radio and Television. During his college years, he was Midwest Correspondent and sports reporter for Televisa (Mexico).

Since 1991 events taking place in Mexico have had extraordinary implications for exopolitics. Perhaps the greatest sightings flap in history has occurred just outside the borders of the United States. An entire nation was given an intense indoctrination into extraterrestrial-related phenomena. The Mexican military and civilian government became involved. No one is more qualified to present this history than Jaime Maussan.

The July 1991 solar eclipse, which passed over Mexico, was the beginning of a wave of mass sightings. Thousands of people across Mexico witnessed and video recorded night lights and daylight disks. Over the past dozen years Maussan has compiled over 5,000 videos and photos from such eyewitnesses and has presented this information around the world.

He has had a 25-year career in the media, during which he has received numerous awards, and is now General Director and Anchorman of "60 Minutes" Mexico. He produced 20 commercialvideos for "Programas de Investigacion", achieving leadership for an independent production company in Mexico. He also produces a radio program "Jaime Maussan, UFOs and other Mysteries", one of the most popular shows in Mexico.

He also works as an Investigative Journalist/Anchorman and General Producer of the TV show "Tercer Milenio" that is broadcast via the Televisa Network to all Latin American countries, Europe, the United States and also through the Sky Satellite System. In 1996, he was the producer of the feature radio program "Jaime Maussan, UFOs and Other Mysteries" transmitted through XEW Radio and the RASA Network, which became one of the most popular radio programs in Mexico.

Jaime has earned many awards for his films, radio and television shows both in North America and Europe. He received a special award at The Capitol Hill in March 1982, the Premio Ondas award in Barcelona in 1980, and the Global 500 award from the United Nations in 1990. He also received the National Journalism Award from the Journalists Club of Mexico in 1983, 1987 and 1993. He was presented the "AMPRYT" award for the TV program "60 Minutos" in 1990-91-92. His investigation into the still on-going Mexico UFO wave has garnered worldwide acclaim and he has been invited to speak at conferences around the world. [31]

United States

Stephen Bassett is a leading advocate in the nation for ending the 61-year government imposed truth embargo regarding an extraterrestrial presence engaging the human race. He is a political activist, lobbyist, commentator, and columnist. He is the founder of the **Paradigm Research Group (PRG)**, the Executive Director of the **Extraterrestrial - Phenomena Political Action Committee (X-PPAC)**, the creator of the **Paradigm Clock** and the executive producer of the X-Conference. His work has been covered extensively inside and outside the United States.

Since 1996 Bassett has assisted numerous organizations and initiatives working to 1) raise public awareness, 2) convene open Congressional hearings to take government witness testimony, and 3) end the **"truth embargo"**. He has appeared on hundreds of radio and television talk shows and in numerous documentaries speaking to millions of people about the implications and likelihood of "Disclosure" - the formal acknowledgement of the extraterrestrial presence by world governments.

In 1995 Bassett decided to bring a 15-year background in business development and consulting and a degree in physics into the issues surrounding extraterrestrial-related phenomena. His point of entry was a five-month stint working at the **Program for Extraordinary Experience Research (PEER)**, founded by the late **Dr. John Mack** in Cambridge, MA.

In July of 1996 he left Cambridge to set up an advocacy organization, Paradigm Research Group in Bethesda, MD, just outside of Washington, DC. The mission of PRG was to use all means possible to confront the United States government regarding its policy of a truth embargo on the events and evidence demonstrating an extraterrestrial presence engaging the human race and the formal acknowledgement of that presence. One of the first moves of PRG was for Mr. Bassett to register as a lobbyist advocating UFO/ET issues, most importantly Disclosure. He was the first person to ever do so.

In April of 1997, he attended the press and congressional briefings conducted in Washington, DC by **Steven Greer** and **CSETI**. During the July 1997 events in Roswell, New Mexico, he assisted **Stargate International** in launching the "Petition for an Open Congressional Hearing on Modern UFO Evidence." This is part of an ongoing process to bring the United States Congress formally into the process of disclosure. The petition continues to gather paper and e-mail signatures and will eventually be presented to the Congress at the appropriate time. In October 1997 he assisted **Frances Barwood,** city councilwoman in Phoenix Arizona, in dealing with an aggressive attempt by her political opponents to discredit and punish her for speaking publicly and in council session on the lack of a proper investigation of the **Phoenix UFO sighting** in March of 1997. Ms. Barwood won her recall election by a 2 to 1 margin.

At midnight on April 30, 1998, he published the **"Paradigm Clock"** to the **World Wide Web.** This website is designed to support the researchers/activists in the field. The Clock itself is a metaphor for citizens and the media to develop a feel for the process of disclosure underway. Then from early April of 1998 Mr. Bassett worked to develop media interest in the anomalies present in the photographic returns of the **Cydonia** region of Mars from the **Mars Global Surveyor**. Additional concerns included the fate of journalists who had done investigative pieces unfavorable to elements of the government and the secrecy reform legislation (S-712) introduced by **Sen. Patrick Moynihan.**

318

Early in 1999 Bassett began collaboration with **Robert Bletchman** and **Larry Bryant** of MUFON who had developed a UFO State Ballot Initiative targeting the 16 states permitting direct referendum. He developed a web subsite to represent this project. Missouri became the first test effort under the guidance of MUFON state director, Bruce Widaman. Then, in July after assisting Virginia MUFON in promoting MUFON's '99 International Symposium in DC metro area, he began a process for a more intense approach to Congressional members. Consequently, on July 13, 1999, he founded the **Extraterrestrial Phenomena Political Action Committee (X-PPAC)**, the first PAC in history to target the politics of UFO/ET phenomena and the government embargo of facts confirming the presence of extraterrestrial life forms in our world, now. Announced nationally on July 21 it will be promoted via its website at www.x-ppac.org and by ongoing guest media appearances.

In 2000 political initiatives were restrained by the pending change of government. Mr. Bassett continued to speak to the issue at conferences in Santa Clara, CA and Santa Fe, NM later that year.

Early in 2001 Paradigm Research Group and X-PPAC put their support firmly behind a major new political initiative titled "The Disclosure Project." Specifically, assistance was given to the holding of an important press conference on May 9, 2001, at the National Press Club in Washington, DC. At that time the testimony of 70 government and agency employees, 21 in

319

person, directly relating to extraterrestrial-related events and evidence was presented to the media. The totality of this testimony, compiled from 120 hours of interviews, essentially confirmed an extraterrestrial presence and a government embargo on the formal acknowledgement of this fact.

In April of 2004, he produced the 1st Exopolitics Expo - the X-Conference - held at the Hilton Washington, DC North Gaithersburg. This was followed by X-Conferences 2005, 2007 and 2008.

On October 1, 2008, PRG launched the Million Fax on Washington with the goal of directing thousands of letters, faxes and emails toward the transition headquarters of the new President Elect. The considerable correspondence generated all called for the new president to end the truth embargo, release all files, call for congressional hearings, and make available ET derived technologies to the public domain. Phase II of the Million Fax on Washington began on February 1, 2009, with the new destination the White House. [32] [33]
http://www.paradigmresearchgroup.org/stephenbassett.html

Theodore Richard Bloecher (born August 22, 1929) in Summit, New Jersey is an American ufologist. Researcher **Jerome Clark** described him as "highly regarded" for his scientific rigor, his detailed research, his efforts at balance and neutrality, and his reluctance to speculate beyond what is supported by given data.

He became interested in UFOs during the great UFO "flap" of 1952. In 1954 Ted co-founded **CSI (Civilian Saucer Intelligence)** of New York with researcher Isabel L. Davis (July 30, 1902 - June 19-20, 1984), research chemist Alexander Duff "Lex" Mebane (1923 - Dec. 4, 2004), Elliott Rockmore of Brooklyn, and Marilyn Feifer of the Bronx. Although CSI only lasted about 5 years (1954-1959) as a focused and effective group (it was never formally disbanded, but it finally ceased functioning altogether sometime in the mid-1960s), historians of Ufology have always regarded it one of the most significant of the early critically-minded civilian UFO groups. Bloecher was a Staff Member of NICAP (National Investigations Committee on Aerial Phenomena) from May 1968 through July 1969.

In 1967 NICAP published Bloecher's *Report on the UFO Wave of 1947*, a detailed compilation of and reference guide to the hundreds of flying discs reported across the U.S. in 1947. Jacques Vallee describes Bloecher's *Report* as "authoritative." Bloecher's other UFO publications of note include the "Humanoid Catalog" (co-written with David Webb) and *Close Encounter at Kelly and Others 1955* (with Isabel Davis).

320

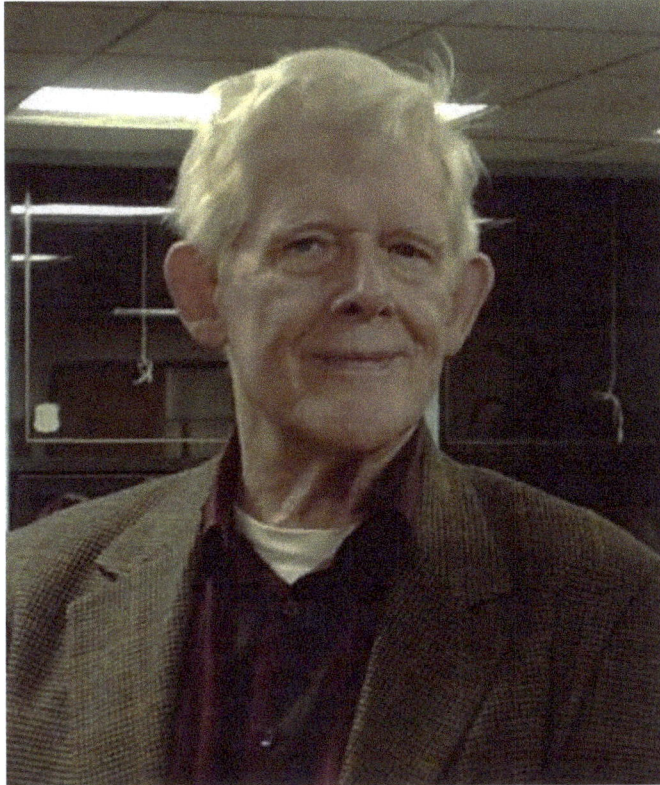

In the 1970s, Bloecher was associated with **MUFON (Mutual UFO Network)** and **CUFOS (Center for UFO Studies)**. In 1975, he co-investigated a series of UFO abduction reports with Budd Hopkins.

Bloecher retired from Ufology in 1984, donating many of his files to CUFOS.[34]

Jerome E. Clark, (Nov. 27, 1946 -) - UFO historian, author of the authoritative *UFO Encyclopedia* **Jerome Clark** (born November 27, 1946) is an American researcher and writer, specializing in unidentified flying objects and other anomalous phenomena; he is also a songwriter of some note.

Clark is one of the most prominent UFO historians and researchers active today. Although Clark's works have sometimes generated spirited debate, he is widely regarded as one of the most reputable writers in the field, and he has earned the praise of many skeptics. Clark's works have been cited in multiple articles in the debunking-oriented *Skeptical Inquirer*. Despite the fact that most contributors to the British periodical *Magonia* disagree with Clark's endorsement of the extraterrestrial hypothesis, they have nonetheless consulted his books for their articles, and have described his works as "invaluable" and described him as one of "Ufology's finest" and as "highly respected."[4] The skeptical RRGroup describes Clark as a rare "Bona fide UFO researcher." In his *Saucer Smear*, longtime ufologist James W. Moseley writes that Clark "is acknowledged ... as the UFO Field's leading historian."

Clark is also a prominently featured talking head on made-for-television UFO documentaries, most notably the 2005 prime-time U.S. television special *Peter Jennings Reporting: UFOs — Seeing Is Believing*, discussing the early history of the U.S. Military's UFO investigations (see also **Project Sign** and Project Grudge.) In addition to the **Peter Jennings** special, Clark has also appeared on episodes of NBC's *Unsolved Mysteries* television series and on the syndicated television series *Sightings*. In 1997 he was prominently featured on the A&E Network's documentary "Where Are All the UFOs?" which examined the history of the UFO phenomenon.

In the years since, Clark has championed a sort of open-ended agnosticism, choosing to focus on phenomena that are purported to have some degree of documentable support—whether physical evidence or reliably reported events. He has argued very cautiously in favor of the extraterrestrial hypothesis, not as proven fact but as a working hypothesis, choosing to focus on the UFO cases he regards as the most promising: multiple witnesses and/or UFO cases which are said to leave physical evidence.

Perhaps Clark's greatest accomplishment in the field of UFO studies came in the 1990s with the publication of his massive, award-winning *UFO Encyclopedia*.

The *UFO Encyclopedia* was first published by respected academic and reference books specialists Omnigraphics as a three-volume hardcover set in the 1990s. In 1997, Visible Ink published an abridged, mass-market trade paperback version under the title *The UFO Book*, and an updated two-volume hardcover edition of the *Encyclopedia* was published in 1998.

Backed by detailed research and extensive bibliographies, Clark's encyclopedia is widely regarded by most UFO researchers, and even many skeptics, as one of the best-researched and most credible publications on the often-controversial subject of UFOs; the Association of College and Research Libraries described the book as "the definitive work on the [UFO] subject for many years to come" while *Library Journal* notes that one of the judges for Clark's Benjamin Franklin Award declared the *UFO Book* (a condensed, mass-market version of the *UFO Encyclopedia*) "an exhaustive, non-judgmental look at the history of unidentified flying objects ... the writing is top notch and clear. [34]

Robert Dean, (1929 -) - Ufologist, reportedly read a document called *An Assessment* (1964), a NATO report on UFOs prompted by an incident on February 2, 1961, during which 50 UFOs allegedly appeared over Europe.

Mr. Dean's history includes forty years of research in the UFO field. He spent twenty-seven years of active duty in the US Army where he retired as Command Sergeant Major after serving as a highly decorated infantry combat veteran. He also served in Intelligence Field Operations and was stationed at **Supreme Headquarters Allied Powers Europe (SHAPE)**, the military arm of NATO. His **Cosmic Top Secret clearance** opened the way for his study of the highly controversial document of which he now speaks.

(c) Terry Tibando

Mr. Dean also spent fourteen years as an emergency services manager with the **Federal Emergency Management Agency (FEMA)** for the Arizona Pima County Sheriff's Department. He is the former Arizona Assistant Director and Pima County Director for the **Mutual UFO Network (MUFON)** and is a former member of the Center for UFO studies **(CUFOS)** and the **Ancient Astronauts Society**. He also served twelve years as a member of the board of Directors for the **Aerial Phenomenon Research Organization (APRO).**

Mr. Dean majored in ancient history, psychology, and philosophy at Indiana University, completed extensive studies in archeology and theology and holds the equivalent of a masters level degree in Emergency Management. Beyond extensive United States engagements, he has spoken all over the world to share his message, appearing at sponsored events in England, Switzerland, Italy, Japan, Germany, Brazil, Mexico, Australia, Hungary and Puerto Rico. He also has countless television, radio, video and news documentary appearances to his credit and has been honored with three lifetime achievement awards for his contribution to the field of Ufology.[35]

Richard M. Dolan was born in Brooklyn, New York, in 1962. He holds an MA in History from the University of Rochester and a BA in History from Alfred University. He earned a Certificate in Political Theory from Oxford University and was a Rhodes Scholar finalist. Prior to his interest in anomalous phenomena, Dolan studied US Cold War strategy, Soviet history, and international diplomacy.

In 2000, he published a 500-page study, *"UFOs and the National Security State, Vol. I: Chronology of a Cover-Up 1941-1973."* This is the first volume of a two-part historical narrative of the national security dimensions of the UFO phenomenon from 1941 to the present. Included are the records of more than fifty military bases relating to violations of sensitive airspace by unknown objects, demonstrating that the US military has taken the topic of UFOs seriously indeed.

http://www.noufors.com/twelve_government_documents_that_take_ufos_seriously.html

Apollo 14 astronaut Dr. Edgar Mitchell called Dolan's book "monumental," while Dr. Hal Puthoff, Director of the Institute for Advanced Studies at Austin, declared it to be "a must-read for serious students in the field."

Dolan has appeared on numerous television documentaries for the History Channel, Sci Fi Channel, as well as BBC and European networks. He appears regularly on radio stations throughout the U.S., including Coast to Coast, and has been a featured speaker at conferences internationally. In 2006, he hosted a six-episode series, Sci Fi Investigates, produced by NBC for the Sci Fi Channel.

Dolan has published numerous articles on anomalous phenomena, science, and the intelligence community. In 2003, he was a founding member of Phenomena, a magazine dedicated to leading edge issues pertaining to science and society.

Richard Dolan also continues to research and write volume two of UFOs and the National Security State, which is due for release in 2009. He lives with his family in Rochester, New York. [36]

Raymond E. Fowler was born in Salem, Massachusetts (1934 -) and received a B.A. degree (magna cum laude) from Gordon College of Liberal Arts. His career included a tour with the USAF Security Service and 25 years with GTE Government Systems. He retired early after working as a Task Manager and Senior Planner on several major weapons systems including the Minuteman and MX Intercontinental Ballistic Missiles.

Ray Fowler's contributions to Ufology are respected by UFO researchers throughout the world. His investigation reports have been published in Congressional Hearings, Military Publications, newspapers, magazines and professional journals in the U.S.A. and abroad. The USAF UFO Projects' Chief Scientific Consultant, Dr. J. Allen Hynek, has called Raymond Fowler, "An outstanding UFO investigator … I know of none who is more dedicated, trustworthy or persevering."

Ray served as chairman of the NICAP Massachusetts Subcommittee, an early warning coordinator for the USAF-contracted UFO Study at the University of Colorado and as a scientific associate for the Center for UFO Study. In later years he served as director of investigations on the board of directors of the Mutual UFO Network.

He also has appeared on hundreds of radio/TV shows in the U.S.A. since 1963 including Dave Garroway, Dick Cavett, Mike Douglas, Good Morning America, Unsolved Mysteries, Sightings and a number of Network and syndicated documentaries on UFOs. UFO Books by Ray Fowler include *"UFOs: Interplanetary Visitors, The Andreasson Affair, The Andreasson Affair - Phase Two, The Watchers, The Watchers II, The Andreasson Legacy, The Allagash Abductions, Casebook of a UFO Investigator, The Melchizedek Connection, UFOs: Interplanetary Visitors,* and his newest book, *Synchrofile"*. [37][38] **http://www.nicap.org/bios/fowler.htm**

Later in life Fowler wrote about being an abductee himself sharing this information, in his autobiographical book *UFO Testament: Anatomy of an Abductee*. During an interview with Rosemary Ellen Guiley, Fowler listed some of his abduction experiences which seem to correlate with other abductee testimony such as **Betty and Barney Hill** abduction and **Betty Andreasson Luca.**

Fowler's claim of being an abductee, and his UFO research as a whole, was not always welcome by his family members, because of their religious beliefs on the subject of UFOs[*UFO*

Testament: Anatomy of an Abductee]. Fowler's extensive investigations in the UFO field lessened after the publication of *The Watchers I* and *The Watchers II* , in which Fowler initially acknowledged his UFO abduction experiences.

Dr. Steven M. Greer (June 28, 1955 -), is an American physician, ufologist, author, lecturer and founder of the **Orion Project, The Disclosure Project,** and **Sirius Disclosure** websites. He is also as a proponent of openness in government, media, and corporations when it comes to advanced technologies that he and others believe to have been shelved and hidden from public awareness for reasons of profit and influence.

Greer enrolled at Appalachian State University in Boone, North Carolina, but in 1974 he "left traditional college at Boone to enter teacher training at Maharishi International University". As of 2007, Greer was licensed to practice medicine in Virginia and has also been licensed in North Carolina. He is a lifetime member of Alpha Omega Alpha, the nation's most prestigious medical honor society; Dr. Greer is an emergency physician and former chairman of the Department of Emergency Medicine at Caldwell Memorial Hospital in North Carolina.

Greer and his wife Emily have four daughters and reside in the Charlottesville, Virginia area. Dr. Greer has coined the term "close encounter of the fifth kind" (a.k.a. CE5) to describe human-initiated contact with extraterrestrials which are encounters that are "characterized by mutual, peaceful and sustainable bilateral communications rather than unilateral contact." Debunkers and skeptics have derogatorily referred to Dr. Greer as a new age "contactee" which is further from the reality than they would imagine. It would be safer and more accurate to state that he is a seeker of truth and a servant of disclosure of that truth.

Dr. Greer is one of those rare individuals who come along once in a lifetime to bring about a positive change in the collective consciousness of society with a genuine desire to see humanity advance spiritually and technically toward a greater civilization globally. There is perhaps, no other man in the field of Ufology who has done so much to promote and advance the cause of UFO and ETI research and disclosure, to contribute to the enlightenment and the education of both military and government officials and the public in general, to the development of environmentally clean and cheap alternative sources of free energy generating systems, all with insightful vision toward a good and hopeful future for all mankind. To that hopeful end, he continues to work tirelessly and diligently. He has been called a new renaissance man in the modern era of Ufology, a man who even has the attentive ear and respect of a sitting CIA Director. He has trained 10's of thousands of people worldwide as ambassadors to the universe in establishing human - ET relationships. He has attracted friends and worldwide supporters for his cause and he has also, unfortunately, attracted the attention of the naysayers, the detractors and the would-be evil perpetrators of a sinister world order regime.

Dr. Greer is the Founder and International Director of The **Center for the Study of Extraterrestrial Intelligence (CSETI),** an international non-profit scientific research and education organization dedicated to the furtherance of our understanding of extraterrestrial intelligence founded in 1990. CSETI's projects include the **CE-5 Initiative** and the **Disclosure Project**.

CSETI grew out of the **Project starlight Coalition** as a means to identify high-level witnesses, to developed a strategy to coordinate a high-level disclosure on the UFO/ETI subject, and to successfully to brief numerous world leaders. The wide consensus exists among senior political, scientific and military / intelligence figures for a public acknowledgement of the reality of UFOs and ETI.

The **Disclosure Project** is a nonprofit research project working to fully disclose the facts about UFOs, extraterrestrial intelligence, and classified advanced energy and propulsion systems. Over 400 government, military, and intelligence community witnesses have testified to their direct, personal, first-hand experience with UFOs, ETs, ET technology, and the cover-up that keeps this information secret.

In addition to being the Director of the Disclosure Project, he has also been supervising a world-wide search through purview of The Orion Project for alternative energy sources, specifically those known as zero-point or over-unity devices with the plan to identify and develop systems which will eliminate the need for fossil fuels.

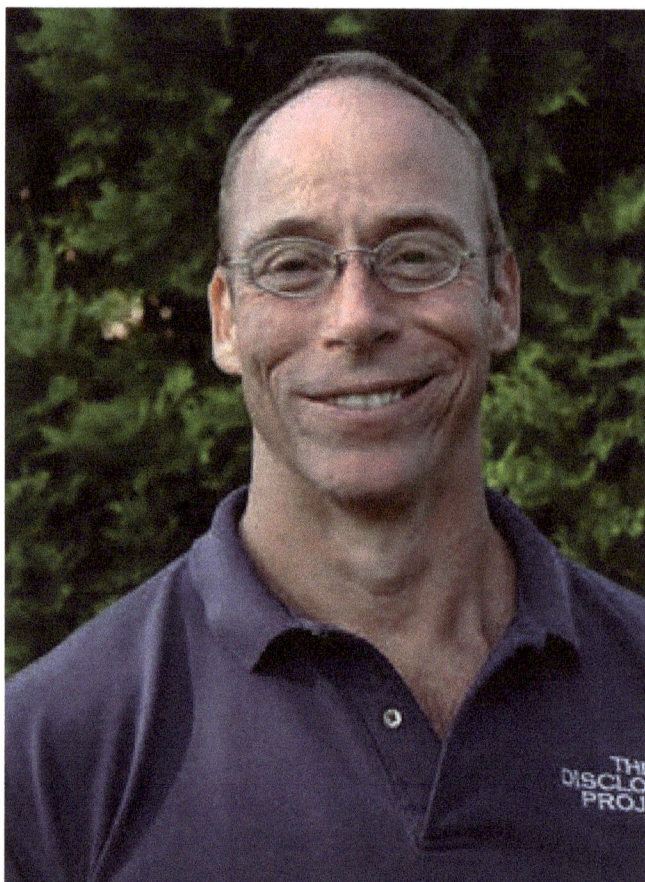

In 1990 Greer founded the Center for the Study of Extraterrestrial Intelligence, and offers how-to training on initiating contact with extraterrestrial intelligence. Greer also teaches the use of meditation techniques that he claims allow attendees to "remote view" locations and times (past and present), and develop "cosmic consciousness" and supernatural abilities such as precognition. Three years later, Greer founded the Disclosure Project, claiming evidence of extraterrestrial visits to Earth from the testimonies of over 400 first-hand witnesses from military, government, intelligence officials and the private industrial sector. Greer is also the founder of the Advanced Energy Research Organization (AERO) and The Orion Project.[8] Both seek funding for research into infinite energy systems and other free-energy and over-unity devices.

The Disclosure Project is an organization started by Greer in 1993 that asserts the position of the existence of a US government cover-up of information relating to unidentified flying objects (UFOs). The Project has adopted Greer's contention that UFOs are spacecraft piloted by intelligent extraterrestrial life, and that the United States government is keeping this secret. Greer also uses the project to disseminate documented evidence that the government has concealed advanced energy technologies obtained from the extraterrestrials by suppressing and hiding them in top secret "black projects" in order not to upset the global geopolitical power and energy sector financial status-quo and its oil industry "special interests".

328

Greer and the Disclosure Project call for congressional hearings of all data regarding UFOs, including the large amount of information they claim is being hidden, and for release of the technology they claim is being suppressed particularly free energy sources. He uses written statements and accounts from military personnel and defense industry employees as evidence for the various claims associated with the Disclosure Project.

The Disclosure Project has been well-received by UFO enthusiasts, with speeches from Greer and various other witnesses being presented at various UFO-themed conferences. Greer has held press conferences and embarked on a continuing series of lectures and television appearances trying to raise popular support. Mainstream media coverage of the group centered around a 2001 conference at the National Press Club , a public disclosure of the nature, the reality and implications of extraterrestrial visitors to our planet. Greer convened the conference with more than 100 other military and government witnesses offering testimony and arguing for investigations into a UFO and ETI phenomenon. Such arguments were met with by derision by skeptics and spokespeople for the U. S. Air Force who pointed out that there is no convincing evidence for the existence that UFOs are alien spacecraft

Greer has appeared on *Larry King Live*, CBS, BBC, NTV in Japan, *Sightings, Encounters*, Ancient Aliens on the History Channel, *Coast to Coast AM*, the Art Bell show, and the Armstrong Williams radio show and hosts an Internet radio segment called "Conversations with Dr. Steven Greer" on the World Puja Network.

Dr. Greer has written numerous four books to date, "Extraterrestrial Contact – The Evidence and Implications", "Disclosure", *"Hidden Truth - Forbidden Knowledge", and Contact: Countdown to Transformation"* as well as numerous papers: *The Foundations of Interplanetary Unity,, Disclosure and 9/11 – An Analysis, Disclosure and Transformation, Open Letter to PI-40, Unacknowledged, Abductions – Not All That Glitters is Gold, Extraterrestrials and the new Cosmology, The Crossing Point,* Who Are the Controllers, Animal Mutilations and the ETs and many more papers. His books and documents represent many aspects his own personal experiences with Extraterrestrial peoples and the unfolding of cosmic awareness since his childhood - from his sighting of a UFO at an early age, to his amazing near-death experience at age 17, to his unraveling of the secret cabal running the illegal transnational energy and UFO related projects, to his meetings with a CIA Director, US Senators, heads of state and royalty, to the CIA involvement in most of the major news media, newspapers and television broadcasting companies.

On May 9, 2001, as Director of The Disclosure Project, Dr. Greer presided over The Disclosure Project Press Conference from the National Press Club in Washington, DC. Over 20 military, government, intelligence, and corporate witnesses presented compelling testimony regarding the existence of extraterrestrial life forms visiting the planet, and the reverse engineering of the energy and propulsion systems of these craft. Over 1 billion people heard of the press conference through webcast and subsequent media coverage on BBC, CNN, CNN Worldwide, Voice of America, Pravda, Chinese media, and media outlets throughout Latin America. The webcast had 250,000 people waiting online the largest webcast in the history of the National Press Club.

Dr. Greer was the first UFO researcher to bring disclosure of the UFO secrecy and cover-up to the attention of the public and has been recognized with the appellation as "Father of UFO Disclosure" for his efforts.

On a personal note: Steven Greer has been a mentor, an inspiration, a teacher of wisdom and diplomacy, an intellectual giant among his peers, he is a friend and a spiritual brother with whom I share much in personal belief and experiences.[39]

Richard Harris Hall, (Dec. 25, 1930 - July 17, 2009), former assistant director of NICAP in the 1960s, former director of the **Fund for UFO Research** in the 1980s. He was a leading Ufologist and proponent of the extraterrestrial hypothesis to explain UFO sightings; he also wrote numerous books and articles dealing with the role of women in the American Civil War.

http://www.washingtonpost.com/wp-dyn/content/article/2009/08/22/AR2009082202196.html

Hall held a bachelor's degree in philosophy from Tulane University in New Orleans. He lived most of his life in the Washington, D.C. area.

From 1958 to 1969 he worked for the **National Investigations Committee on Aerial Phenomena (NICAP)**, one of the most prominent and influential UFO civilian research groups in American history. He began as executive secretary, and eventually became NICAP's assistant director. In this role Hall was both an eyewitness and participant to much of the early history of the UFO phenomenon in the United States. Working with NICAP director Donald Keyhoe, he helped lobby the United States Congress to hold public hearings and investigations into the UFO phenomenon. In 1964 Hall researched, edited, and wrote much of *The UFO Evidence*, a compendium of the best UFO sightings and incidents of the 1940s, 1950's, and early 1960s. A copy of *The UFO Evidence* was sent to every member of Congress in 1964, and the book is still regarded by many UFO researchers and historians as one of the best UFO books ever published.

Following Keyhoe's ouster as NICAP director in 1969, Hall left NICAP to work as a technical writer and editor. He continued to work in the UFO field. He served as the director of the Fund for UFO Research, which provides grant money to legitimate researchers working in UFO studies. He was also the editor of the *MUFON Journal*, the official publication of the Mutual UFO Network (MUFON), the largest civilian UFO group in America today. In 2001 he wrote a sequel to *The UFO Evidence*; it covered major UFO incidents from the mid-1960s through the 1990s. He was also the chief editor of the *Journal of UFO History*, which is published six times per year. Hall was a strong proponent of the theory that UFOs are extraterrestrial spacecraft from an advanced alien civilization, and he was an active member of the "UFO Updates" message board and website.

To supplement his income as a UFO researcher, Hall worked for many years as an abstractor-indexer for the Congressional Information Service in Bethesda, Maryland. A member of the Authors Guild, Hall also published numerous books and magazine articles dealing with the role of women in the American Civil War, and he maintained a strong interest in Civil War history through his life.

On the morning of July 17, 2009, Richard Hall died after a long battle with cancer, according to ufologist Jerome Clark and anomalist Patrick Huyghe. [40]

James Albert Harder, Ph.D., (Fullerton, California in 1926-2006) was a professor of civil and hydraulic engineering at the University of California, Berkeley. He is a professor emeritus there. His education includes BS (Caltech); MS, Ph.D. (U.C. Berkeley). He is a Fellow, AAAS; Life Member, ASCE; and Founding Member, Society for Scientific Exploration. Professor Emeritus, U.C. Berkeley.

Dr. Harder was perhaps best known as a prominent UFO researcher who has studied the subject for over 50 years, first becoming interested in 1952. He was Director of Research for the Aerial Phenomena Research Organization (APRO) from 1969-1982. APRO was one of the first civilian organizations to study the UFO phenomenon. When the U.S. House of Representatives Committee on Science and Astronautics held hearings on UFOs in 1968, he was one of six scientists asked to testify on UFOs before the committee. In a 1998 interview, Harder said the subject was generally treated with disdain by the scientific community, but he was still one of about 300 academics who were actively investigating the phenomenon.

Dr. James Harder also served as an electronics technician in the U.S. Navy and one of six scientists asked to testify on UFOs before the U.S. House of Rep. Committee on Science and Astronautics in 1968. Harder was the primary investigator on a number of classical UFO cases, mainly related to alien abductions, including the 1973 Pascagoula Abduction and the 1975 Travis Walton case. He also took over the Betty and Barney Hill abduction investigation and continued it for many years. According to Harder, in about 95% of abduction cases he's studied, abductees report the encounter as positive, benevolent, and/or enlightening. He also investigated the claims of legendary CIA remote viewer Pat Price (who allegedly died under suspicious circumstances in 1975). Based on his remote viewing, Price believed aliens had underground bases at four locations on Earth.

Harder had long been a strong advocate of extraterrestrial origins for UFOs or the Extraterrestrial hypothesis. He also firmly believed that the subject has been covered up by the U.S. government, which he thought was extremely worried about what is happening.

One of his more controversial statements, based primarily on hypnotic regressions on alien abductees, is that there is a "Galactic Federation" of aliens similar to our United Nations. There are perhaps as many as 57 alien species in this Federation (a number, he says, which frequently pops up in abductee recollections). Some have been visiting Earth and studying humans for a very long time, and are generally benevolent, he believes (though not always). Many communicate through telepathy, and, said Harder, can sometimes be channeled through subjects while they are hypnotized.

Harder had also applied his physical sciences and engineering background to the study of UFOs. In his Congressional testimony of 1968, Harder mentioned physical analysis of magnesium fragments found near Ubatuba, Brazil, said to have come from an exploded flying saucer. The magnesium was of very high purity. Harder conjectured that the lightweight metal, normally very brittle, might become exceptionally hard and strong if purified and made free from crystalline defects. If that were the case, it would be a very good metal for the construction of a flying device. Construction of such high-strength metals is now thought possible with insights gained from the emerging field of nanotechnology.

Another theory advanced by Harder's arose from a sighting of an oval UFO by a chemist named Wells Allen Webb near Phoenix. Webb was wearing Polaroid glasses and noticed three concentric dark rings around the object. Harder thought the observation might be explained by a very powerful magnetic field surrounding the object causing polarized light from the sun to be rotated, or the Faraday Effect. Exactly how this magnetic field might explain the object's propulsion was unclear, but he thought it might be connected with gravitomagnetism, an analog of electromagnetism, predicted from general relativity. Theoretically, a gravity-like field can be generated by a moving mass, but the effect is normally minuscule. Harder was again unsure how a practical gravitomagnetic force might be produced.[41]

Budd Hopkins (1931 - 2011), Budd Hopkins is a world-renowned artist, author, and pioneer UFO abduction researcher. Having investigated well over 700 cases, he now heads the Intruders Foundation, a nonprofit, scientific research and support organization. Budd first became interested in the UFO phenomenon when he and two others had a daylight UFO sighting near Truro, Massachusetts, in 1964. In 1975 he carried out his first major investigation which involved a UFO landing and occupant incident in North Hudson Park, NJ. Shortly thereafter, he began to concentrate on the investigation of the UFO abduction phenomenon, which led to the eventual publication of his findings.

Taken together, his three books, *Missing Time,* 1981, *Intruders,* 1987, and *Witnessed,* 1996, are widely regarded by researchers and skeptics alike as comprising the most influential series of books yet published on the abduction phenomenon. These works, Hopkins' lectures, and his other presentations have been responsible for bringing a number of other noted researchers-David Jacobs, John Carpenter, Yvonne Smith, and John Mack, among others into this extraordinary area of specialization. His documented discoveries have become the basis of later abduction investigations and research. IF, the Intruders Foundation, was founded in 1989 to further facilitate this important work. At present, with a dedicated team of fellow researchers and volunteers, IF is flourishing.

Budd Hopkins has long been considered Ufology's most visible figure. He pioneered and continues to lead the investigation into the most controversial aspect of the UFO phenomenon...the systematic abduction of human beings by UFO occupants. As the world's premier expert on this issue, he has worked with more than one thousand people who have reported abduction experiences over the past twenty years. These individuals come from all walks of life and include physicians, psychiatrists, attorneys, police officers, military personnel, political figures, personalities from the entertainment world, and even a NASA scientist.

A prolific writer and internationally respected painter, Hopkins has delivered hundreds of UFO lectures around this country and around the world. His groundbreaking first book, *Missing Time,* was the first work to compare a number of UFO abduction cases in order to isolate the patterns they revealed. His second book, *Intruders-The Incredible Visitation at Copley Woods,* was a New York Times bestseller and the basis for the popular 1992 CBS miniseries, Intruders, which has since been broadcast internationally. His widely acclaimed latest book is *Witnessed-The True Story of the Brooklyn Bridge UFO Abductions.*

Hopkins' goal has always been to bring an objective, dispassionate scientific intelligence to bear on the UFO abduction phenomenon. To this end, he founded the Intruders Foundation (IF) in 1989. IF is a nonprofit organization devoted to research and public education concerning this extraordinary enigma. They publish a respected journal, as well as offer a nationwide referral service for those wishing to explore their own suspected abduction experiences.

(c) Terry Tibando

Despite its extremely controversial nature, Hopkins' research has received serious commentary in such mainstream publications as *Time, Paris Match, The Washington Post, The New York Times, The New York Review of Books, Omni, People, and Cosmopolitan.* He has been a guest on hundreds of television and radio programs including *Nightline, Good Morning America, The Today Show, The Oprah Winfrey Show, The Tonight Show, Charlie Rose, Larry King Live, The Charles Grodin Show, Sally Jesse Raphael, The Geraldo Rivera Show, 20/20, 48 Hours, Unsolved Mysteries, Encounters, A Current Affair, Nightwatch, The Late Show, The Art Bell Show, Tom Snyder, The Laura Lee Show, Hieronimus & Company, Weekend Edition* (National Public Radio)*, Voice of America, Armed Forces Radio,* numerous BBC affiliates, and many other shows and forums.

Budd Hopkins first became interested in the UFO phenomenon when he and two others had a daylight UFO sighting near Truro, Massachusetts, in 1964. In 1975 he carried out his first major investigation, which involved a UFO landing and occupant incident in North Hudson Park, N.J. Shortly thereafter he began to concentrate on the investigation of the UFO abduction phenomenon which led to the eventual publication of his findings. [42]

Linda Moulton Howe, born in Boise, Idaho (January 20, 1942 -), is an American investigative journalist and documentary producer-writer-director-editor who is currently based in Albuquerque, New Mexico.

She entered the 1963 Miss Boise pageant for college scholarships and went on to win the 1963 Miss Idaho crown and scholarships, and participated in the Miss America Pageant that year in Atlantic City. Howe received her 1965 B.A. *cum laude* in English Literature from the University of Colorado, Boulder. In 1966, Howe was awarded the Stanley Baubaire Scholarship for her Master's Degree work at Stanford University, Palo Alto, California. She received her Master's Degree in Communication from Stanford University in 1968, where she produced a documentary film for the Stanford Medical Center and her Master's Thesis, "A Picture Calculus," at the Stanford Linear Accelerator.

Howe has devoted her documentary film, television, radio, writing and reporting career to productions concerning science, medicine, and the environment. Ms. Howe has received local, national and international awards, including three regional Emmys, a national Emmy nomination, and a Station Peabody award for medical programming. While Linda was Director of Special Projects at KMGH-TV, Channel 7, Denver, Colorado from 1978 to 1983, her documentaries included *Poison in the Wind* and *A Sun Kissed Poison* which compared smog pollution in Los Angeles and Denver; *Fire In The Water* about hydrogen as an alternative energy source to fossil fuels; *A Radioactive Water* about uranium contamination of public drinking water in a Denver suburb; and *A Strange Harvest* and *Strange Harvests 1993*, which explored animal mutilation mystery. Another film, *A Prairie Dawn*, focused on astronaut training in Denver. She has also produced documentaries in Ethiopia and Mexico for UNICEF about child survival efforts and for Turner Broadcasting in Atlanta about environmental challenges in the TV series *Earthbeat*.

In addition to television, Linda produces reports and edits the award-winning science, environment and earth mysteries news website, http://www.Earthfiles.com. Linda also reports monthly science, environment and earth mysteries news for Premiere Radio Networks *Coast to Coast AM with George Noory* and weekly news updates for *Dreamland Radio* at Unknowncountry.com.

Linda was an honored medical producer in Boston's WCVB Station Excellence Peabody Award. As Director of Special Projects at the KMGH-TV station in Denver, Colorado, she received the Aviation & Space Writers Association Award for Writing Excellence in Television, a Chicago Film Festival Golden Plaque for *A Radioactive Water*, Colorado's Florence Sabin Award for "outstanding contribution to public health" and several dozen other local and regional awards, including Emmys. She also worked as Director of International Programming for *Earthbeat*, an environmental series broadcast on Turner's WTBS Superstation, in Atlanta, Georgia.

Linda has traveled to Brazil, England, Norway, France, Switzerland, The Netherlands, Yugoslavia, Turkey, Ethiopia, Kenya, Egypt, Australia, Japan, Peru, Venezuela, Canada, Mexico, the Yucatan, and Puerto Rico for research and productions. Linda has written four books: *Mysterious Lights and Crop Circles*, 2nd edition, September 2002, about accounts and research regarding biophysical and biochemical changes in affected cereal crops by complex energy systems; *An Alien Harvest* about animal mutilation; *Glimpses of Other Realities, Volumes I and II* about U.S. military, intelligence and civilian testimonies concerning "unidentified phenomena interacting with earth life".[43]

Dr. Josef Allen Hynek (May 1, 1910 - April 27, 1986) was a United States astronomer, professor, and ufologist. He is perhaps best remembered for his UFO research. Hynek acted as scientific adviser to UFO studies undertaken by the U.S. Air Force under three consecutive names: Project Sign (1947-1949), Project Grudge (1949-1952), and Project Blue Book (1952 to 1969). For decades afterward, he conducted his own independent UFO research and is widely considered the father of the concept of scientific analysis of both reports and, especially, trace evidence purportedly left by UFOs.

Hynek was born in Chicago to Czech parents. In 1931, Hynek received a B.S. from the University of Chicago. In 1935, he completed his Ph.D. in astrophysics at Yerkes Observatory. He joined the Department of Physics and Astronomy at Ohio State University in 1936. He specialized in the study of stellar evolution and in the identification of spectroscopic binaries.

During World War II, Hynek was a civilian scientist at the Johns Hopkins Applied Physics Laboratory, where he helped to develop the United States Navy's radio proximity fuse.

336

After the war, Hynek returned to the Department of Physics and Astronomy at Ohio State, rising to full professor in 1950.

In 1956, he left to join Professor Fred Whipple, the Harvard astronomer, at the Smithsonian Astrophysical Observatory, which had combined with the Harvard Observatory at Harvard. Hynek had the assignment of directing the tracking of an American space satellite, a project for the International Geophysical Year in 1956 and thereafter.

After completing his work on the satellite program, Hynek went back to teaching, taking the position of professor and chairman of the astronomy department at Northwestern University in 1960.

In response to many "flying saucer" sightings (later unidentified flying objects), the United States Air Force established Project Sign in 1948; this later became Project Grudge, which in turn became Project Blue Book in 1952. Hynek was contacted by Project Sign to act as scientific consultant for their investigation of UFO reports. Hynek would study a UFO report and subsequently decide if its description of the UFO suggested a known astronomical object.

When Project Sign hired Hynek, he was initially skeptical of UFO reports. Hynek suspected that UFO reports were made by unreliable witnesses, or by persons who had misidentified man-made or natural objects. In 1948, Hynek said that the "the whole subject seems utterly ridiculous," and described it as a fad that would soon pass.

For the first few years of his UFO studies, Hynek could safely be described as a debunker. He thought that a great many UFOs could be explained as prosaic phenomena misidentified by an observer. But beyond such fairly obvious cases, Hynek often stretched logic to nearly the breaking point in an attempt to explain away as many UFO reports as possible. In his 1977 book, Hynek admitted that he enjoyed his role as a debunker for the Air Force. He also noted that debunking was what the Air Force expected of him.

Hynek's opinions about UFOs began a slow and gradual shift. After examining hundreds of UFO reports over the decades (including some made by credible witnesses, including astronomers, pilots, police officers, and military personnel), Hynek concluded that some reports represented genuine empirical observations. Another shift in Hynek's opinions came after conducting an informal poll of his astronomer colleagues in the early 1950s. Among those he queried was **Dr. Clyde Tombaugh**, who discovered the dwarf planet Pluto. Of 44 astronomers, five (over 11 percent) had seen aerial objects that they could not account for with established, mainstream science. Most of these astronomers had not widely shared their accounts for fear of ridicule or of damage to their reputations or careers (Tombaugh was an exception, having openly discussed his own UFO sightings). Hynek also noted that this 11% figure was, according to most polls, greater than those in the general public who claimed to have seen UFOs. Furthermore, the astronomers were presumably more knowledgeable about observing and evaluating the skies than the general public, so their observations were arguably more impressive. Hynek was also distressed by what he regarded as the dismissive or arrogant attitude of many mainstream scientists towards UFO reports and witnesses.

Early evidence of the shift in Hynek's opinions appeared in 1953 when Hynek wrote an article for the April 1953 issue of *The Journal of the Optical Society of America* titled "Unusual Aerial Phenomena," which contained what would become perhaps Hynek's best-known statement:

"Ridicule is not part of the scientific method, and people should not be taught that it is. The steady flow of reports, often made in concert by *reliable* observers, raises questions of scientific obligation and responsibility. Is there ... *any* residue that is worthy of scientific attention? Or, if the

re isn't, does not an obligation exist to say so to the public—not in words of open ridicule but seriously, to keep faith with the trust the public places in science and scientists?" (Emphasis in original)

The essay was very carefully worded: Hynek never states that UFOs are an extraordinary phenomenon. But it is clear that whatever his own views, Hynek was increasingly distressed by what he saw as the superficial manner most scientists looked at UFOs.

In 1953, Hynek was an associate member of the **Robertson Panel**, which concluded that there was nothing anomalous about UFOs and that a public relations campaign should be undertaken to debunk the subject and reduce public interest. Hynek would later come to lament that the Robertson Panel had helped make UFOs a disreputable field of study.

When the UFO reports continued at a steady pace, Hynek devoted some time to studying the reports and determined that some were deeply puzzling, even after considerable study. He once said, "As a scientist, I must be mindful of the past; all too often it has happened that matters of great value to science were overlooked because the new phenomenon did not fit the accepted scientific outlook of the time.

Hynek remained with Project Sign after it became Project Grudge (though with far less involvement than with Project Sign). Project Grudge was replaced with Project Blue Book in early 1952. Hynek continued as scientific consultant to Project Blue Book.

It was during the late stages of Blue Book in the 1960s that Hynek began speaking openly about his disagreements and disappointments with the Air Force. In late March 1966, in Michigan, two days of mass UFO sightings were reported and received significant publicity. After studying the reports, Hynek offered a provisional hypothesis for some of the sightings: a few of about 100 witnesses had mistaken swamp gas for something more spectacular. At the press conference where he made his announcement, Hynek repeatedly and strenuously made the qualification that swamp gas was a plausible explanation for only a portion of the Michigan UFO reports, and certainly, this conclusion did not apply to all UFO reports in general. But much to his chagrin, Hynek's qualifications were largely overlooked, and the words "swamp gas" were repeated ad infinitum in relation to UFO reports. The explanation was subject to national derision.

Late in his life, Hynek was critical of the popular extraterrestrial hypothesis and began expressing his doubts to theories that UFOs were physical spacecraft from other planets. As Hynek, himself has said in October 1976: "I have come to support less and less the idea that UFOs are 'nuts and bolts' spacecrafts from other worlds. There are just too many things going against this theory. To me, it seems ridiculous that super intelligences would travel great distances to do relatively stupid things like stop cars, collect soil samples, and frighten people. I think we must begin to re-examine the evidence. We must begin to look closer to home."

Hynek was the founder and head of the **Center for UFO Studies (CUFOS).** Founded in 1973 (originally in Evanston, Illinois but now based in Chicago), CUFOS is an organization stressing scientific analysis of UFO cases. CUFOS's extensive archives include valuable files from civilian research groups such as NICAP, one of the most popular and credible UFO research groups of the 1950s and 1960s.

In November 1978, a statement on UFOs was presented by Dr. Allen Hynek, in the name of himself, of Dr. Jacques Vallée, and of **Dr. Claude Poher**. This speech was prepared and approved by the three authors, before the United Nations General Assembly. The objective was to initiate a centralized United Nations UFO authority.

In 1973, at the MUFON annual symposium, held in Akron, Ohio, Hynek began to express his doubts regarding the extraterrestrial (formerly "interplanetary" or "intergalactic") hypothesis. His main point led him to the title of his speech: ***"The Embarrassment of the Riches."*** He was aware that the quantity of UFO sightings was much higher than the Project Blue Book statistics. Just this puzzled him. "A few good sightings a year, over the world, would bolster the extraterrestrial hypothesis—but many thousands every year? From remote regions of space? And to what purpose? To scare us by stopping cars, and disturbing animals, and puzzling us with theirseemingly pointless antics?"

In 1975, in a paper presented to the Joint Symposium of the American Institute of Aeronautics & Astronautics in Los Angeles, he wrote, "If you object, I ask you to explain – quantitatively, not qualitatively – the reported phenomena of materialization and dematerialization, of shape changes, of the noiseless hovering in the Earth's gravitational field, accelerations that – for an appreciable mass – require energy sources far beyond present capabilities – even theoretical capabilities, the well-known and often reported E-M (sc. electro-magnetic interference) effect, the psychic effects on percipients, including purported telepathic communications."

"There is sufficient evidence to defend both the ETI and the EDI hypothesis," Hynek continued. As evidence for the ETI (extraterrestrial intelligence) he mentioned, as examples, the radar cases as good evidence of something solid, and the physical trace cases. Then he turned to defending the EDI (extradimensional intelligence) hypothesis. Besides the aspect of materialization and dematerialization he cited the "poltergeist" phenomenon experienced by some people after a close encounter; the photographs of UFOs, sometimes on only one frame, not seen by the witnesses; the changing form right before the witnesses' eyes; the puzzling question of telepathic communication; or that in close encounters of the third kind the creatures seem to be at home in earth's gravity and atmosphere; the sudden stillness in the presence of the craft; levitation of cars or persons; the development by some of psychic abilities after an encounter. "Do we have two aspects of one phenomenon or two different sets of phenomena?" Hynek asked.

Finally, he introduced a third hypothesis. "I hold it entirely possible," he said, "that a technology exists, which encompasses both the physical and the psychic, the material and the mental.

I hypothesize a 'M&M' technology encompassing the mental and material realms. The psychic realms, so mysterious to us today, may be an ordinary part of an advanced technology

Hynek developed in his first book the close encounter scale to better catalogue various UFO reports. Dr. Hynek was also the consultant to Columbia Pictures and Steven Spielberg on the popular 1977 UFO movie, *Close Encounters of the Third Kind*, and made a brief, non-speaking appearance in the film (after the aliens disembark from the 'mother ship' at the end of the film, he can be seen, bearded and with pipe in mouth, stepping forward to view the spectacle).[69]

Jacobs, David Michael (1940 -), Historian and Professor of History at Temple University in Philadelphia. Teaches a curriculum course on the UFO phenomenon at Temple University. Jacobs main area of research is alien abductions and the hidden alien agenda towards humanity. [See the subsection of this book, *"Obtaining Your PhD. in Ufology and Returning your Doctorate When You've Done Bad Alien Abduction Research"*] [44]

Morris Ketchum Jessup (March 2, 1900 – April 20, 1959), had a Master of Science Degree in astronomy and, though employed for most of his life as an automobile-parts salesman and a photographer, is probably best remembered for his pioneering ufological writings and his role in "uncovering" the so-called "Philadelphia Experiment"

Born near Rockville, Indiana, Jessup grew up with an interest in astronomy. He earned a bachelor of science degree in astronomy from The University of Michigan in Ann Arbor, Michigan in 1925 and, while working at the Lamont-Hussey Observatory, received a master of science degree in 1926. Though he began work on his doctorate in astrophysics, he ended his dissertation work in 1931 and never earned the higher degree. He apparently dropped his career and studies in astronomy and worked for the rest of his life in a variety of jobs unrelated to science, although he is sometimes erroneously described as having been an instructor in astronomy and mathematics at the University of Michigan and Drake University.

Mr. Jessup has been referred to in ufological circles as "probably the most original extraterrestrial hypothesizer of the 1950s", and it has been said of him that he was "educated in astronomy and archeology and had working experience in both." Jessup took part in archeological expeditions to the Yucatan and Peru in the 1920s.

Jessup achieved some notoriety with his 1955 book *The Case for the UFO*, in which he argued that unidentified flying objects (UFOs) represented a mysterious subject worthy of further study. Jessup speculated that UFOs were "exploratory craft of "solid" and "nebulous" character." Jessup also "linked ancient monuments with prehistoric super science," years before similar claims were made by Erich von Däniken in *Chariots of the Gods?* and other books

Jessup wrote three further flying-saucer books, *UFOs and the Bible*, *The UFO Annual* (both 1956), and *The Expanding Case for the UFO* (1957). The latter suggested that transient lunar phenomenon were somehow related to UFOs in the earth's skies. Jessup's main flying-saucer scenario came to resemble that of the Shaver Hoax perpetrated by the science-fiction magazine editor Raymond A. Palmer—namely, that "good" and "bad" groups of space aliens were/are meddling with terrestrial affairs. Like most of the writers on flying saucers and the so-called contactees that emerged during the 1950s, Jessup displayed familiarity with the alternative mythology of human prehistory developed by Helena P. Blavatsky's cult of Theosophy.

Jessup also played a key role in the so-called "Philadelphia Experiment". In *The Case for the UFO*, Jessup theorized about the means of propulsion that flying saucer-style UFOs might use. Jessup speculated that antigravity and/or electromagnetism might have been responsible for the observed flight behavior of UFOs. On January 13, 1955, Jessup received a letter from a man identifying himself as one "**Carlos Miguel Allende**". In the letter, Allende informed Jessup of the so-called "**Philadelphia Experiment**", alluding to poorly sourced contemporary newspaper articles as proof. Allende also said that he had personally witnessed a U.S. Navy warship named the *USS Eldridge* disappear and reappear while he was serving aboard a merchant marine ship in her vicinity, the *SS Andrew Furuseth*.

In the spring of 1957, Jessup was contacted by the **Office of Naval Research (ONR)** in Washington, D.C., and asked to study the contents of a parcel that it had received. Upon arrival, the curious Jessup was astonished to find that a paperback copy of his book had been mailed to the ONR in a manila envelope marked "HAPPY EASTER". Furthermore, the book had been extensively annotated by hand in its margins, and an ONR officer asked Jessup if he had any idea as to who had done so.

The lengthy annotations were written in three different colors of ink and appeared to detail a correspondence between three individuals, only one of whom is given a name: "Jemi". The ONR labeled the other two "Mr. A." and "Mr. B." The annotators refer to each other as Gypsies and discuss two different types of "people" living in space. Their text contained nonstandard use of capitalization and punctuation and detailed a lengthy discussion of the merits of various suppositions that Jessup makes throughout his book, with oblique references to the **"Philadelphia Experiment"**, in a way that suggested prior or superior knowledge. (For example, "Mr. B." reassures his fellow annotators, who have highlighted a certain theory of Jessup's, "HE HAS NO KNOWLEDGE, HE COULD NOT HAVE. ONLY GUESSING. [sic]").

Based on the handwriting style and subject matter, and in comparison to the earlier letters that he had received, Jessup identified "Mr. A." as Carlos Allende/Carl Allen. Others have suggested that the three annotations are actually from the same person using three pens. Later the ONR contacted Jessup, claiming that the return address on Allende's letter to Jessup was an abandoned farmhouse. They also informed Jessup that the Varo Corporation, a research firm, was preparing a print copy of the annotated version of *The Case for the UFO*, complete with both letters he had received. Numbers vary, but it appears that around one hundred copies of the Varo Edition were printed and distributed within the U.S. Navy. Jessup was also sent three copies for his own use.

Jessup attempted to make a living writing on the subject of UFOs, but his follow-up books did not sell well and his publisher rejected several other manuscripts. In 1958 his wife left him, and his friends described him as being somewhat unstable when he traveled to New York. After returning to Florida, he was involved in a serious car accident and was slow to recover, apparently increasing his despondency. On April 19, 1959, Jessup contacted Doctor Manson Valentine and arranged to meet with him the next day, claiming to have made a breakthrough regarding an event known as the Philadelphia Experiment. However, on April 20, 1959, Jessup allegedly committed suicide in a Dade County, Florida, park, possibly the Matheson Hammock one, by inhaling automobile exhaust fumes by use of a hose connected to the exhaust pipe and a rear window of the vehicle. Some people believed that "The circumstances of Jessup's apparent suicide remain mysterious" and conspiracy theorists contended that it was connected to his knowledge of the "Philadelphia Experiment". Although some friends claimed that he possibly had been driven to suicide by the "Allende Case," other friends said that an extremely depressed Jessup had been discussing suicide with his friends for several months before his act. [45]

Donald Edward Keyhoe (June 20, 1897 - November 29, 1988) was an American Marine Corps naval aviator, writer of many aviation articles and stories in a variety of leading publications, and manager of the promotional tours of aviation pioneers, especially of Charles Lindbergh.

In the 1950s he became well-known as a UFO researcher and was widely regarded as the leader in the field" of Ufology in the 1950s and early-to-mid 1960s, arguing that the U.S. government should conduct appropriate research in UFO matters, and should release all its UFO files.

Keyhoe was born and raised in Ottumwa, Iowa. He earned a B.S. degree at the United States Naval Academy in 1919 and was commissioned a Marine Corps Lieutenant.

In 1922, he suffered an arm injury during an airplane crash in Guam during his long convalescence; Keyhoe began writing as a hobby. He eventually returned to active duty, but the injury gave Keyhoe persistent trouble, and, as a result, he retired from the Marines in 1923. He then worked for the National Geodetic Survey and U.S. Department of Commerce.

In 1927, Keyhoe managed a very popular coast-to-coast tour by Charles Lindbergh. This led to Keyhoe's first book, 1928's *Flying With Lindbergh*. The book was a quick success and led to a freelance writing career, with many of Keyhoe's articles and fictional stories (mostly related to aviation) appearing in a variety of leading publications.

Keyhoe returned to active duty during World War II in a Naval Aviation Training Division, retiring again a Major.

By the time his UFO books appeared, Keyhoe was already a well-established author with numerous appearances in the pulp magazines of the 1920s and 30s. Keyhoe wrote a number of air adventure stories, for *Flying Aces* and other magazines. He was also a freelancer for Saturday Evening Post, The Nation, and Reader's Digest.

Following Kenneth Arnold's report of odd, fast-moving aerial objects in the summer of 1947, interest in "flying disks" and "flying saucers" was widespread, and Keyhoe followed the subject with some interest, though he was initially skeptical of any extraordinary answer to the UFO question. Keyhoe had many friends and contacts in the military and the Pentagon and after some investigation, Keyhoe became convinced that the flying saucers were real.

As their forms, flight maneuvers, speeds, and light technology was apparently far ahead of any nation's developments, Keyhoe became convinced that they must be the products of unearthly intelligences, and that the U.S. government was trying to suppress the whole truth about the subject. This conclusion was based especially on the response Keyhoe found when he quizzed various officials about flying saucers. He was told there was nothing to the subject, yet was simultaneously denied access to saucer-related documents.

Keyhoe's article "Flying Saucers Are Real" appeared in the January 1950 issue of *True* (published December 26, 1949) and caused a sensation. Though such figures are always difficult to verify, Captain (U.S. Air Force), Edward J. Ruppelt, the first head of Project Blue Book, reported that "It is rumored among magazine publishers that Don Keyhoe's article in *True* (a men's magazine) was one of the most widely read and widely discussed magazine articles in history."

Capitalizing on the interest, Keyhoe expanded the article into a book, *The Flying Saucers Are Real* (1950); it sold over half a million copies in paperback. He argued that the Air Force knew that flying saucers were extraterrestrial, but downplayed the reports to avoid public panic. In Keyhoe's view, the aliens — wherever their origins or intentions — did not seem hostile, and had likely been surveilling the earth for two hundred years or more, though Keyhoe wrote that their "observation suddenly increased in 1947, following the series of A-bomb explosions in 1945."

Keyhoe wrote several more books about UFOs. *Flying Saucers From Outer Space* (1953) is perhaps the most impressive, being largely based on interviews and official reports vetted by the Air Force.

Carl Jung argued that Keyhoe's first two books were "based on official material and studiously avoid the wild speculations, *naiveté* or prejudice of other [UFO] publications."

Others have disagreed with Keyhoe's assessments. In his 1956 book, Edward J. Ruppelt wrote, "The Air Force wasn't trying to cover up", and declared that "The problem was tackled with organized confusion".

In 1956, Keyhoe cofounded the **National Investigations Committee On Aerial Phenomena (NICAP)**. He was one of several prominent professional, military or scientific figures on the board of directors, which lent the group a degree of legitimacy many of the other contemporary "flying saucer clubs" sorely lacked.

NICAP founder Thomas Townsend Brown was ousted as director in early 1957 after facing repeated charges of financial ineptitude. Keyhoe replaced him; he was only slightly better at managing NICAP's finances, and the group continued their efforts

With Keyhoe in the lead, NICAP pressed hard for Congressional hearings and investigation into UFOs. They scored some attention from the mass media, and the general public (NICAP's membership peaked at about 15,000 during the early and mid 1960s) but only very limited interest from government officials.

However, there was increasing criticism of the Air Force's Project Blue Book. Following a widely publicized wave of UFO reports in 1966, NICAP was among the chorus which called for an independent scientific investigation of UFOs. The Condon Committee was formed with this goal in mind, though it quickly became be mired in infighting and, later, controversy. Keyhoe publicized the so-called "Trick Memo", an embarrassing memorandum written by a Condon Committee coordinator which seemed to suggest that the ostensibly objective and neutral Committee had determined to pursue a debunking operation well before even beginning their studies.

On 22 January 1958, Keyhoe appeared on a CBS live television show the *Armstrong Circle Theatre* to speak on the topic of UFOs. Keyhoe charged that a U.S. Congressional committee was evaluating evidence that "will absolutely prove that the UFOs are machines under intelligent control". However, CBS stopped the audio portion of the live broadcast. Herbert A. Carlborg, CBS Director of Editing stated: *"this program had been carefully cleared for security reasons"*.

On 8 March 1958, Keyhoe appeared on *The Mike Wallace Interview* on ABC and spoke about flying saucers, contactees and the details of the *Armstrong Circle Theatre* censorship, which he blamed on the Air Force rather than CBS.

NICAP's membership plummeted in the late 1960s, and Keyhoe faced charges of incompetence and authoritarianism. By 1969 Keyhoe turned his focus away from the military and focused on the CIA as the source of the UFO cover-up. NICAP's board, headed by Colonel Joseph Bryan III, forced Keyhoe to retire as NICAP chief. Under Bryan's leadership, the NICAP disbanded its local and state affiliate groups, and by 1973 it had been completely closed.

In 1973, Keyhoe wrote his final book about UFO's, *Aliens from Space*. It promoted "Operation Lure", a plan to entice extraterrestrials to land on Earth, and described the problems Keyhoe had getting information from government agents.

Beyond this book, Keyhoe had little contact with Ufology as he settled into retirement. He did, however, speak at a few UFO conferences after his ouster from NICAP. In 1981 he joined MUFON's board of directors, but his membership was essentially in name only due to declining health, and he had little to do with the organization. He died in 1988 at the age of 91. He was buried in Green Hill Cemetery in Luray, Virginia. Jessup wrote three further flying-saucer books, *UFOs and the Bible*, *The UFO Annual* (both 1956), and *The Expanding Case for the UFO* (1957). The latter suggested that transient lunar phenomenon were somehow related to UFOs in the earth's skies. Jessup's main flying-saucer scenario came to resemble that of the Shaver Hoax perpetrated by the science-fiction magazine editor Raymond A. Palmer—namely, that "good" and "bad" groups of space aliens were/are meddling with terrestrial affairs. Like most of the writers on flying saucers and the so-called contactees that emerged during the 1950s, Jessup displayed familiarity with the alternative mythology of human prehistory developed by **Helena P. Blavatsky**'s cult of Theosophy. [46]

Dr. Lynne Kitei is an internationally acclaimed physician and health educator who pushed aside her successful medical career to pursue The Phoenix Lights book and internationally award-winning Documentary project. She was leading the cutting edge era of early disease detection and prevention as Chief Clinical Consultant of the Imaging-Prevention-Wellness Center at the world renowned Arizona Heart Institute in Phoenix, Arizona until coming forward, after seven years of anonymity, as a key witness to the historic and still unexplained mass sighting throughout Arizona on March 13, 1997

http://starworksusa.com/presenters_2012/lynne-d-kitei-m-d

Dr. Lynne graduated with a Bachelors of Science degree in Secondary Science Education, with minors in Communication and Voice from Temple University in Philadelphia, PA. 1970 and an MD Degree from Temple University School of Medicine 1974. She completed her postgraduate studies at the Medical College of Pennsylvania.

Before her medical training, Dr. Lynne appeared in over 30 featured and starring roles in professional musical comedies including Alice in Wonderland with Sherman Hemsley, Oklahoma starring Gordon MacRae, Guys & Dolls starring Betty Grable, and understudied for Barbara Eden in The Sound of Music. She also played the role of "Florence Arizona" in the 20th Century box office hit Raising Arizona starring Nicolas Cage, Holly Hunter, Frances McDormand, and John Goodman.

Dr. Lynne has been called the "woman pioneer" of medical communications in TV Guide after creating and producing innovative TV News health reports for NBC in Philadelphia in 1976. Since that time, Dr. Lynne has dedicated over 30 years to global public awareness, wellness, and health education. She has appeared as the resident health reporter for many of the major American news networks such as NBC has been featured on USA Cable, FOX TV News, FOX ABC, CNN, and CBS affiliates, etc.

As the producer, writer, and director of the "You Make It!" health video and workbook curriculum series. She worked under the auspices of the Arizona Community Foundation for 20 years to help fund and distribute her revolutionary prevention/education programs to schools, churches, youth groups and libraries throughout Arizona. Just Say KNOW to AIDS, Just Say KNOW To Drugs, Teen Pregnancy... Children Having Children and Drugs Don't Make IT! are currently being distributed worldwide by Discovery Education.

Dr. Lynne's innovative and renowned endeavors have been featured in publications including TV Guide Magazine, New Dawn Magazine, Runner Magazine, Physician's Management Magazine, Phoenix Magazine, UFO Magazine, National News Network, The Philadelphia Inquirer, Spirit Guide, L.A. Times Magazine, East Valley Tribune, What is Enlightenment Magazine, National UFO Magazine, Paradise Magazine, Brazil's Revista UFO Magazine, Open Minds Magazine, and numerous Arizona Republic Newspaper articles. She has also been recognized in the "Who's Who of American Women" and was chosen as Woman of the Year in Pennsylvania.

Dr. Lynne has been a keynote speaker at the University of Arizona Tucson Medical Center with Dr. John Mack, Bay Area UFO Conference, Hollywood National UFO Conference, Washington D.C. X-Conference, MUFON International Symposium in Denver, CO, Cosmic International Conference in Italy, Arizona State University Physical Science Auditorium, ASU Anthropology Museum Exhibit, Atlantic Coast UFO Conference, Trump Plaza in Atlantic City , N.J., International Institute of Human Sciences, Montreal, Canada, CSETI Conference Rio Rico, AZ.

Besides the hundreds of radio and TV appearances sharing the historic Arizona mass sighting, Dr. Lynne continues to tour the US and abroad, introducing screenings of the internationally award winning *"Phoenix Lights Documentary,"* as well as presentations for her best-selling book, *"The Phoenix Lights...A Skeptic's Discovery That We Are Not Alone"* and riveting power point lecture, *"Coincidence or Communication?"* [47] http://www.thephoenixlights.net/

, (April 18, 1952 -) was born in Woodbury, N.J., and raised mostly in Northern California. He graduated from high school in Stockton, Calif.; earned a bachelor's degree in communications from West Georgia College; and later earned a master's degree in communications from the University of the Pacific, where he also taught speech and debate and served as director of forensics. George also taught speech at California Polytechnic University, coached the debate team at the University of California at Berkeley and taught broadcast journalism at UNLV.

Along the way, George also worked as a hod carrier, farm laborer, carpenter's assistant and house painter. He moved to Las Vegas in 1979 and landed a job as a taxi driver. Later, he worked at KLVX-TV Channel 10 as a part-time studio cameraman and production assistant. KLAS-TV hired him in 1981 as a general assignment reporter. George also has co-anchored various newscasts for KLAS-TV.

Since 1995, George has been the chief reporter on Channel 8's I-Team investigative unit. In that capacity, he has earned *four* regional Edward R. Murrow awards and a national Edward R. Murrow award for his investigative stories on the voter registration fraud in the Clark County election of 2004.

Knapp has won a prestigious Peabody Award and *17* Emmy Awards. Nine times, he has won the Mark Twain Award for best news writing from the Associated Press. And in 1990, his series about UFOs was selected by United Press International as best in the nation for Individual Achievement by a Journalist.

His reporting on Nevada's infamous Area 51 military base was selected by UPI as Best Individual Achievement by a Reporter in 1989. He also writes an award-winning weekly column for a Las Vegas newspaper.

In the field of Ufology, the spotlight shone brightly on Knapp when he interviewed the Area 51 "insider" scientist, Robert (Bob) Lazar, whose story of working on captured alien flying and advanced energy propulsion technology brought Knapp, Lazar and Area 51 to the forefront of the public's attention worldwide. Much of the evidence has been hotly debated and controversial and resulted in video expose featuring the most comprehensive examination of the UFO phenomenon ever produced entitled *"UFOs: The Best Evidence."* [48]

Lorenzen, James Leslie "Jim" (birth name: Leslie James Lorenzen) (Jan. 2, 1922 - Aug. 28, 1986) in Grand Meadow, Minnesota USA and **Coral Elsie Lorenzen** (maiden name: Coral Elsie Lightner) (April 2, 1925 - April 12, 1988), founders and directors of **Aerial Phenomena Research Organization (APRO)**, one of the most prominent civilian UFO research groups of the 1960s and 1970s.

Jim Lorenzen was an electronics technician by trade. Between 1942 and 1945 he was a radio operator with the US Army Air Corps. Between 1945 and 1951 he played as a professional musician. From 1951 to 1954 he worked as an engineer at a radio station. Then from 1954 to 1960 he worked at as an electronics tech at Holloman AFB. His awards include an Air Medal with Cluster, Distinguished Flying Cross with Cluster and a Presidential Citation. From 1960 to 1967 Jim worked at Kitt Peak National Observatory as a technical associate. After that, he was the owner of Lorenzen Music Enterprises in Tuscon Arizona. He and his wife Coral authored a number of UFO books and co-founded the Aerial Phenomena Research Organization (APRO). Together they share the opinion that UFOs are best explained by the extraterrestrial hypothesis (ETH). [49]

See **APRO** under the subsection: **UFO Groups and Organizations** in this book for further information.

Maccabee, Dr. Bruce (1942 -) retired US Navy optical physicist, has analyzed numerous UFO videos and photos. Bruce spent his early year in Rutland, Vt. After high school, he studied physics at Worcester Polytechnic Institute in Worcester, Mass (B.S. in physics) and then at The American University, Washington, DC (M.S. and Ph. D. in physics). In 1972 commenced his long career at the Naval Surface Warfare Center, presently headquartered at Dahlgren, Virginia. He has worked on optical data processing, generation of underwater sound with lasers and various aspects of the Strategic Defense Initiative (SDI) and Ballistic Missile Defense (BMD) using high power lasers. Bruce Maccabee has been active in UFO research since the late 1960s when he joined the National Investigations Committee on Aerial Phenomena (NICAP) and was active in research and investigation for NICAP until its demise in 1980. He became a member of MUFON in 1975 and was subsequently appointed to the position of state Director for Maryland, a position he still holds. In 1979 he was instrumental in establishing the Fund for UFO Research and was the chairman for the about 13 years. He presently serves on the National Board of the Fund.

His UFO research and investigations (which are completely unrelated to his Navy work) have included the **Kenneth Arnold** sighting (Jun 24, 1947), the McMinnville, Oregon (Trent) photos of 1950, the Gemini 11 astronaut photos of September, 1966, the New Zealand sightings of December, 1978, the Japan Airlines (JAL1628) sighting of November 1986, the numerous sightings of Ed Walters and others in Gulf Breeze, Florida, 1987 - 1988, the "red bubba" sightings,1990-1992 (including his own sighting in September, 1991), the Mexico City video of August, 1997, the Phoenix lights sightings of March 13, 1997 at 10 PM, the Mexican Air Force case of March, 2004, the Iran Jet Case of Sept. 1976 and many others. He has also done historical research and was the first to obtain the "flying disc file" of the FBI (the REAL X-Files!).

http://www.openminds.tv/dr-bruce-maccabee-the-fbi-ufo-connection-march-8-2013/19875

Dr. Maccabee is the author or coauthor of about three dozen technical articles and more than a hundred UFO articles over the last 25 years, including many which appeared in the MUFON Journal and MUFON Symposium proceedings. He wrote the last chapter of "The Gulf Breeze Sightings" by Edward and Frances Walters (Morrow, 1990). He wrote the UFO history chapter of the book "UFOs: Zeugen und Zeichen," published in Germany in 1995. He is co-author with Edward Walters of "UFOs Are Real, Here's The Proof," (Avon Books, 1997), he is the author of "The UFO/FBI Connection" (Llewellyn Books, May 2000) and he is the author of novel, "Abduction in My Life," (November 2000). He is listed in Who's Who in Technology Today and American Men and Women of Science and he has appeared on numerous radio and TV shows and documentaries.

Bruce is an accomplished pianist and organist who performed at the 1997 and 1999 MUFON symposia. He lives in Frederick County, Maryland. [50]

Dr. James Edward McDonald (May 7, 1920 – June 13, 1971) was an American physicist and leading ufologist of the 1960s. He was born and raised in Duluth, Minnesota. He served as a cryptographer in the United States Navy during World War II, and afterward, married Betsy Hunt; they would have six children.

McDonald studied at the University of Omaha, the Massachusetts Institute of Technology, and earned his Ph.D. at Iowa State University. He taught at the University of Chicago for a year, and then in 1953, he was invited to help establish a meteorology and atmospherics program at the University of Arizona as a professor of meteorology. McDonald eventually became the head of the Institute of Atmospheric Physics but resigned as its administrator after about a year because he preferred to teach and research rather than oversee the department.

McDonald campaigned vigorously in support of expanding UFO studies during the mid and late 1960s, arguing that UFOs represented an intriguing, pressing and unsolved mystery which had not been adequately studied by science. He was one of the more prominent figures of his time who argued in favor of the extraterrestrial hypothesis as a plausible, but not completely proved, model of UFO phenomena.

He was one of the most outspoken scientists, who defended the need for legitimate scientific inquiry into the UFO phenomenon, a dedicated and tireless investigator who ranks as one of Ufology's premier researchers for his time. McDonald interviewed over 500 UFO witnesses, uncovered many important government UFO documents, and gave important presentations of UFO evidence.

In 1954, while driving through the Arizona desert with two meteorologists, McDonald spotted an unidentified flying object, a distant point of light, which none of the men could readily identify but, this sighting would spur McDonald's interest in UFOs. By the late 1950s, he was quietly investigating UFO reports in Arizona, and he had also joined NICAP, then the largest and most prominent civilian UFO research group in the nation. Given his training in atmospheric physics, McDonald was able to examine UFO reports in greater detail than most other scientists and was able to offer explanations for some previously unexplained reports. Using his security clearance with the US government, he also uncovered a number of well-documented UFO reports from the US Air Force's Project Blue Book, which he judged deeply puzzling even after stringent analysis. McDonald also lambasted the U.S. Air Force for what he saw as their inept handling of UFO studies.

In 1967 the **Office of Naval Research** granted McDonald a small budget in order to conduct his own UFO research, ostensibly to study the idea that some UFOs were misidentified clouds. He was able to peruse the files of Project Blue Book at Wright-Patterson Air Force Base, and eventually concluded that the Air Force was mishandling UFO evidence. Following the Robertson Panel's recommendations in 1953, the Air Force was following a debunking directive towards UFO reports, and only discussing UFO cases which were considered solved by a mundane explanation. All unexplained UFO cases were classified "secret" and not released to the public (see Robertson Panel for further information).

McDonald was particularly disturbed that astronomer J. Allen Hynek, had not alerted the scientific community to the fact that Project Blue Book was withholding some of the most anomalous and compelling UFO reports. Hynek argued that if he had exposed this, the Air Force would have dumped him as Blue Book's consultant. Their differences would never become reconciled in their lifetimes,

From the mid-1960s, McDonald devoted much of his time to trying to persuade journalists, politicians, and his colleagues that UFOs were the most pressing issue facing American science. He gave dozens of lectures, and wrote volumes of letters to newspapers, to his peers (especially at scientific journals) and to politicians. McDonald wrote to the Air Force Office of Scientific Research, arguing that they needed to radically shift what he saw as their superficial perspective towards UFOs. In response, the Air Force determined that they needed to "fireproof" themselves against McDonald's statements because of his unquestionable qualifications and credibility.

McDonald knew that promoting the extraterrestrial hypothesis could damage his credibility, but he was so convinced of its viability that he plowed ahead, regardless of consequences. He managed to secure limited support from a few prominent figures, such as **United Nations Secretary General U Thant,** who arranged for McDonald to speak to the UN's Outer Space Affairs Group on June 7, 1967. Additionally, in 1967, McDonald noted, "There is no sensible alternative to the utterly shocking hypothesis that UFOs are extraterrestrial probes".

In his Statement on Unidentified Objects to the House Committee on Science and Astronautics, McDonald made the following remarks regarding types of UFO accounts.

"The scope of the present statement precludes anything approaching an exhaustive listing of categories of UFO phenomena: much of what might be made clear at great length will have to be compressed into my remark that the scientific world at large is in for a shock when it becomes aware of the astonishing nature of the UFO phenomenon and its bewildering complexity. I make that terse comment well aware that it invites easy ridicule; but intellectual honesty demands that I make clear that my two years' study convinces me that in the UFO problem lie scientific and technological questions that will challenge the ability of the world's outstanding scientists to explain - as soon as they start examining the facts".

In the same statement, he said he had "become convinced that the scientific community ... has been casually ignoring as nonsense a matter of extraordinary scientific importance."

Following a widely publicized series of mass UFO sightings in southern Michigan in 1966, McDonald became one of several scientists to urge various authorities in the federal government and scientific community to undertake a formal study of UFOs. This public pressure, combined with pressure from some members of Congress (such as then-Congressman Gerald Ford), led the federal government to create the Condon Committee in late 1966. Based at the University of Colorado at Boulder, and named after Committee Chairman **Dr. Edward Condon**, a prominent physicist, the committee was advertised as an unbiased, objective, and thorough investigation into the UFO phenomenon.

However, McDonald and other UFO researchers soon became disillusioned with the committee, and in particular with its chairman, Dr. Condon, and his chief assistant, Dr. Robert Low. Condon's public comments to reporters ridiculing UFO eyewitnesses and his generally dismissive attitude towards the subject led many UFO researchers to doubt whether the investigation would be as neutral and unbiased as it proclaimed. McDonald formed alliances with those on the Condon Committee who disagreed with Condon's leadership and who wanted to undertake long-term UFO studies.

McDonald inadvertently played a major role in the controversy regarding the Condon Committee when one of the committee's investigators, who disagreed with Condon's attitudes privately, gave him a copy of the so-called **"Trick Memo"**. The memorandum, which was written by Condon's chief assistant **Dr. Robert Low**, outlined how the Committee could reach a predetermined conclusion that all UFO cases were explainable in mundane terms, while simultaneously appearing neutral during the actual investigation process. To many UFO investigators, including McDonald, the "Trick Memo" seemed to confirm their worst fears about the Condon Committee's bias regarding the UFO phenomenon. Following McDonald's release to the public of the now infamous "Trick Memo", Project Chair Edward Condon tried unsuccessfully to get McDonald fired from his tenured faculty position at the University of Arizona.

When the Condon Committee issued its final report in 1969, Dr. Condon wrote in the foreword to the report that, based on the committee's investigations, his conclusion was that there was nothing unusual about UFO reports; thus further scientific research into the UFO phenomenon was not worthwhile and should be discouraged. Condon's conclusions about UFOs were generally accepted by most scientists and the "mainstream" news media. McDonald, however, became one of a small number of prominent scientists and researchers who wrote detailed critiques and rebuttals of Condon's conclusions regarding UFOs. McDonald was particularly disturbed by the fact that, while Condon in his foreword had claimed that all UFO reports could be explained as hoaxes or misidentifications of manmade or natural objects or phenomena, the Report itself actually listed over 30% of the cases it investigated as "unexplained".

McDonald spoke before the United States Congress for a UFO hearing in 1968. In part, he stated his opinion that "UFOs are entirely real and we do not know what they are because we have laughed them out of court. The possibility that these are extraterrestrial devices; that we are dealing with surveillance from some advanced technology, is a possibility I take very seriously". McDonald emphasized that he accepted the extraterrestrial hypothesis as a possibility not due to any specific evidence in its favor, but because he judged competing hypotheses as inadequate.

James McDonald felt that the evidence provided in the final 30% of the cases which remained unexplained could have substantiated the opposite conclusion (that UFOs warranted much more scientific study) rather than the official conclusion, which was to recommend no further study.

In 1969, McDonald was a speaker at an American Association for the Advancement of Science UFO symposium. There he delivered a lecture, "Science in Default", which Jerome Clark calls "one of the most powerful scientific defenses of UFO reality ever mounted". McDonald discussed in detail a handful of well documented UFO cases which seemed, he thought, to defy interpretation by conventional science.

McDonald's tireless UFO efforts were exacting a toll: he was becoming professionally isolated, and his marriage was faltering. Beyond his verbal confrontations with Philip Klass and Edward Condon, McDonald butted heads with many other prominent figures, including **Donald Menzel** of Harvard University. Additionally, McDonald suffered a public humiliation when in 1970; he agreed to appear before a committee of the United States Congress to provide evidence against the development of the **supersonic transport (SST)** plane. Like many other atmospheric physicists who testified with him, McDonald was convinced that the plane could potentially harm the Earth's vital but fragile ozone layer. During his testimony Congressman Silvio Conte of Massachusetts - whose district contained factories that would help build the SST - tried to discredit McDonald's SST testimony by switching the hearing to a discussion of McDonald's UFO research. Although McDonald defended his UFO work and noted that his evidence regarding the SST had nothing to do with UFOs, Conte bluntly stated that anyone who "believes in little green men" was, in his opinion, not a credible witness. McDonald was deeply humiliated by Conte's mocking attitude, and by the open laughter of some committee members.

In March 1971, McDonald's wife Betsy told him she wanted a divorce. McDonald seems to have started planning his suicide not long afterward. He finished a few articles he was writing (UFO-related and otherwise) and made plans for the storage of his notes, papers, and research. In April 1971 he attempted suicide by shooting himself in the head. He survived, but was blinded and was wheelchair bound. For a short period, McDonald was committed to the psychiatric ward of a Tucson, Arizona hospital. He recovered a degree of peripheral vision and made plans to return to his teaching position. However, on June 13, 1971, a family, walking along a creek close to the bridge spanning the Canada Del Oro Wash near Tucson, found a body that was later identified as McDonald's. A .38 caliber revolver was found close to him, as well as a suicide note. [51].

354

Dr. John Edward Mack, (1929–2004), Harvard psychiatrist/professor, alien abduction researcher. [Please refer to the subsection; **"Obtaining Your PhD. in Ufology and Returning your Doctorate When You Have Done Bad Alien Abduction Research"** on Dr. John Mack for information]. [52]

William Leonard Moore (born October 31, 1943?) is an author and former UFO researcher. Prominent from the late 1970s to the late 1980s, he co-authored two books with **Charles Berlitz**, including The **Roswell Incident** - the first book written about the alleged Roswell UFO crash/retrieval. Bill Moore later became a controversial figure within Ufology due to his involvement with a group of intelligence contacts known as '**The Aviary'**. He played a role in The **Bennewitz Affair** and was a central figure in the release of the controversial Majestic 12 documents.

Bill Moore has a background in the liberal arts, history, teaching, and English. He speaks fluent French and holds a degree in Russian. Moore attended Thiel College, located in Greenville, Pennsylvania graduating in 1965. Before deciding to concentrate on his writing career he taught language and humanities at various high schools. Bill Moore was a subscriber to NICAP and during the mid, to late 1960s he was asked to investigate a couple of UFO cases for the group. He was later on the board of directors of APRO, based in Arizona.

Moore wrote *The Philadelphia Experiment* with Charles Berlitz. Published in 1979, it details the alleged 1943 US Navy cloaking experiment of the destroyer escort **USS *Eldridge*** at the Philadelphia Naval Yard.

Whilst Bill Moore was based in Minnesota and still publicizing *The Philadelphia Experiment* he met with fellow UFO researcher Stanton T. Friedman who he knew, to compare notes. Friedman stated to Moore that he had recently travelled to Louisiana and spoken to a retired Major Jesse Marcel, after following up on an earlier lead he received from a radio station manager whilst he was waiting for an interview. Jesse Marcel detailed to Friedman his involvement in an incident that had occurred almost 30 years previously - the handling of wreckage from an alleged crash of a craft during July 1947 in Roswell, New Mexico.

Moore also had some information pertaining to a UFO crash and when they thought it could be connected to the same story, they began to research the case further. Moore later raised the idea of a book with his writing partner Charles Berlitz who liked the idea. Within a relatively short period of time Moore conducted the field work, interviewed eyewitnesses, prepared reports and sent them to Berlitz in New York who wrote up the chapters of the book. *The Roswell Incident* was published in 1980 and was the first book written about the alleged Roswell UFO crash/retrieval.

In late 1980 after the release of *The Roswell Incident*, Moore received a phone call after appearing on a radio show where he was promoting the new book. In October 1988 on the nationally televised *UFO Cover-Up?: Live!* broadcast Bill Moore described:

"I got a phone call after appearing on a radio show from a man who said 'you're the only person we've heard talk about this subject, who seems to know what he's talking about'. He convinced me that he was a government intelligence agent and wanted to begin disseminating some information about UFOs to the public".

The original contact Moore later met was to be known as '**Falcon**'. In exchange for information received, Moore would keep tabs on sections of the UFO research community. This also included taking part in disinformation activities against Paul Bennewitz with AFOSI agent Richard C. Doty, as Moore would later admit at the 1989 Las Vegas MUFON conference. Throughout the 80s the group of intelligence contacts grew in number, each given a bird name - hence they became known as 'The Aviary'.

In June 1982 television producer Jaime Shandera joined forces with Moore to attend the meetings and collaborate on the research, including the discovery of the existence of MJ-12 - the purported high-level policy making group overseeing the UFO and extraterrestrial issue. In 1984 Moore and Shandera received the now infamous MJ-12 documents as a 35mm black and white film that was mailed to Jaime Shandera in a plain brown envelope. Their contacts with The Aviary continued until the 1990s when Moore and Shandera eventually left the UFO research field.

William Moore was also a central figure in Greg Bishop's 2005 book detailing the Bennewitz Affair *Project Beta: The Story of Paul Bennewitz, National Security, and the Creation of a Modern UFO Myth*. Some of Moore's activities and interactions with The Aviary have also been detailed in the 2006 book *Exempt from Disclosure: The Black World of UFOs* by Aviary members Robert M. Collins aka 'Condor' and Richard C. Doty - the surrogate 'Falcon' (see references below). [53] **http://en.wikipedia.org/wiki/Bill_Moore_(ufologist)**

Kevin D. Randle (1949 -), Captain in the US Air Force Reserves; also a leading investigator of the Roswell UFO Incident in 1947 has, for more than thirty years, studied the UFO phenomena in all its various incarnations. Training by the Army as a helicopter pilot, intelligence officer and military policeman, and by the Air Force as both an intelligence officer and a public affairs officer, provides Randle with a keen insight into the operations and protocols of the military, their investigations into UFOs, and into a phenomenon that has puzzled people for more than a century.

Randle's educational background is as diverse as his military experience. As an undergraduate at the University of Iowa, he studied anthropology. Graduate work included journalism, psychology and military science at the University of Iowa, California Coast University, and the American Military University. He has both a master and doctoral degree in psychology and a second master degree in the Art of Military Science.

During his investigations, Randle has traveled the United States to interview hundreds of witnesses who were involved in everything from the Roswell, New Mexico crash of 1947, to the repeated radar sightings of UFOs over Washington, D.C. in 1952, to the latest of the abduction cases. Randle was among the first writers to review the declassified Project Blue Book files, among the first to report on animal mutilations and among the first to report on alien abductions. He was the first to report the alien home invasions and among the first to suggest humans working with the aliens.

Randle has written extensively on UFOs beginning in 1973 with articles in various national magazines. He had published many books about UFOs starting with The UFO Casebook in 1989 and continuing to Roswell Revisited in 2007.

Randle was away from his UFO studies when recalled to active duty with the Army that included a tour in Iraq from 2003 to 2004. He recently retired from the Iowa National Guard as a lieutenant colonel.[54]

Nicholas "Nick" Redfern born 1964 in Pelsall, Walsall, Staffordshire is a British Ufologist and **Cryptozoologist** now living in Dallas, Texas, U.S.

Redfern is an active advocate of official disclosure, and has worked to uncover thousands of pages of previously-classified Royal Air Force, Air Ministry and Ministry of Defence files on UFOs dating from the Second World War from the Public Record Office and currently works as a feature writer and contributing editor for *Phenomena* magazine.

http://www.jerrypippin.com/UFO_Files_nick_redfern.htm

Redfern attended Pelsall Comprehensive School in Pelsall from 1976 to 1981.

Redfern joined a rock music and fashion magazine Zero in 1981; where he trained in journalism, writing, magazine production and photography, later going on to write freelance articles on UFOs during the mid-1980s.

From 1984 until 2001 he worked as a freelance feature writer for the Daily Express, People, Western Daily Press and Express & Star newspapers, as well as a full-time feature writer Planet on Sunday. Between 1996 and 2001 he worked as a freelance journalist for the following newsstand magazines in Britain: *The Weekender*, *Animals, Animals, Animals*, *Pet Reptile*, *Military Illustrated*, *Eye-Spy*, *The Unopened Files* and *The X-Factor*.

Between 1996 and 2000 Redfern signed a three-book publishing deal with Simon & Schuster of London for the publication of *A Covert Agenda: The British Government's UFO Top Secrets Exposed* (1997), *The FBI Files: The FBI's UFO Top Secrets Exposed* (1998) and *Cosmic Crashes: The Incredible Story of the UFOs That Fell to Earth* (2000). These books were published in the UK, Canada, Russia, Poland, Australia, New Zealand and Portugal.

Redfern has appeared on a variety of television programs in the UK, including The Big Breakfast; Channel 5 News; and GMTV and is in constant demand on the lecture circuit, both in the UK and overseas, and has appeared in internationally syndicated shows discussing the UFO phenomenon.

Redfern authored several best-selling books on UFOs including The FBI Files: The FBI's UFO Top Secrets Exposed; and Cosmic Crashes: The Incredible Story Of The UFOs That Fell To Earth.[55]

Edward J. Ruppelt (July 17, 1923, in Iowa – September 15, 1960) was a United States Air Force officer probably best-known for his involvement in Project Blue Book, a formal governmental study of unidentified flying objects. He is generally credited with coining the term "unidentified flying object", to replace the terms "flying saucer" and "flying disk", which had become widely known; Ruppelt thought the latter terms were both suggestive and inadequate.

Ruppelt was the director of Project Grudge from late 1951 until it became Project Blue Book in March 1952; he remained with Blue Book until late 1953. UFO researcher Jerome Clark writes, "Most observers of Blue Book agree that the Ruppelt years comprised the project's golden age when investigations were most capably directed and conducted. Ruppelt himself was open-minded about UFOs, and his investigators were not known, as Grudge's were, for force-fitting explanations on cases." (Clark, 517)

Ruppelt was born and raised in Iowa. He enlisted in the Army Air Corps during World War II, and served with distinction as a decorated bombardier: he was awarded "five battle stars, two theater combat ribbons, three Air Medals, and two Distinguished Flying Crosses." (Clark, 516)

After the war, Ruppelt was released into the Army reserves. He attended Iowa State College where, in 1951, he earned an aeronautical engineering degree. Shortly after finishing his education, Ruppelt was called back to active military duties after the Korean War began.

He was assigned to the Air Technical Intelligence Center headquartered at Wright-Patterson Air Force Base. Incidentally, the base had also headquartered two formal unidentified flying object investigations: Project Sign (1947–1948), which had come to favor the extraterrestrial hypothesis before being replaced with Project Grudge (1949–1951), which had a debunking mandate. Though not initially involved with Grudge, Ruppelt quickly learned that the project was facing troubles when high-ranking officers disapproved of the direction it had taken.

Eventually, Grudge was ordered dissolved, and Project Blue Book was planned to replace it. Lt. Col. N.R. Rosengarten asked Ruppelt to take over as the new project's leader, partly because Ruppelt "had a reputation as a good organizer" (Jacobs, 65), and had helped get other wayward projects back on track. though he was initially scheduled to stay with Blue Book for only a few months, when Project Grudge was upgraded in status in late 1951 and renamed Project Blue Book, Ruppelt (then a Captain) was kept on as director when normally, such an upgrade would require the appointment of at least a Colonel to oversee the project; this may well be a testament to Ruppelt's leadership and organizational skills.

Ruppelt quickly implemented a number of changes in the late stages of Project Grudge, which were carried over to most of his tenure with Blue Book. He streamlined the manner in which UFOs were reported to (and by) military officials, partly in hopes of alleviating the stigma and ridicule associated with UFO witnesses.

Knowing that factionalism had harmed the progress of Project Sign, Ruppelt did his best to recruit open-minded, but objective and neutral personnel to staff Blue Book. He tried to avoid the kinds of open-ended speculation that had led to Sign's personnel being split among advocates and critics of the extraterrestrial hypothesis. Ruppelt sought the advice of many scientists and experts and issued regular press releases (along with classified monthly reports for military intelligence).

Perhaps most importantly, Ruppelt also ordered the development of a standard questionnaire for UFO witnesses, hoping to uncover data which could be subject to statistical analysis. He commissioned the Battelle Memorial Institute to create the questionnaire and computerize the data. Using case reports and the computerized data, Battelle then did a massive scientific and statistical study of all Air Force UFO cases (completed in 1954 after Ruppelt had left Blue Book) and known as Project Blue Book Special Report No. 14. Battelle scientists found that even after stringent analysis, 22% of the cases remained classified as "unknown" and that these were different from the "knowns" at a very high level of statistical significance. The Battelle study also found that the best cases were twice as likely to be classified as unknowns as the worst cases.

During Ruppelt's tenure, Blue Book investigated a number of well-known UFO reports including the so-called **Lubbock Lights** and two highly-publicized radar-visual/jet-intercept cases which occurred over Washington DC in late July 1952 (see 1952 Washington D.C. UFO incident), which triggered the largest press conference since World War II to stop public panic. Also during

360

Ruppelt's tenure with Blue Book, most UFO cases were attributed to prosaic causes, but about twenty-five percent were deemed "unknown". As cases with little or no corroborative evidence were generally excluded from consideration during Ruppelt's tenure with Blue Book, the remaining unknowns arguably constitute some of the best-known, best studied, yet still perplexing UFO reports of the late 1940s and early 1950s.

The Air Force would be charged with a cover-up of UFO evidence. Ruppelt insisted, however, that at least during his tenure, conflict and confusion would be more accurately descriptive than to suggest that a deliberate cover-up was taking place. Ruppelt once wrote that the Air Force's approach to the UFO question "was tackled with organized confusion." (Ruppelt, 1956, p. 46) In defending General Samford's press conference on 29 July 1952, after the big UFO flap at Washington National Airport, Ruppelt wrote that "his [Samford's] people had fouled up in not fully investigating the sightings." (Ruppelt, 1956, p. 223) Astronomer and Blue Book consultant J. Allen Hynek thought that Ruppelt did his best, only to see his efforts stymied. Hynek wrote, "In my contacts with [Ruppelt] I found him to be honest and seriously puzzled about the whole phenomenon." (Hynek, 175)

Ruppelt requested reassignment from Blue Book in late 1953 shortly after the Robertson Panel issued its conclusions (based partly on the panel's official report, Ruppelt's Blue Book staff was reduced from more than ten personnel to three, including Ruppelt). He retired from the Air Force not long afterward and then worked in the aerospace industry.

Three years later, Ruppelt's book *The Report on Unidentified Flying Objects* (17 Chapters was published. The book is notable because it was, for several subsequent decades, the only account of Air Force UFO studies written by a participant. It remains arguably one of the most level-headed books about UFOs; Hynek suggested that Ruppelt's "book should be required reading for anyone seriously interested in the history of this subject." (Hynek, 175) In the book, Ruppelt detailed his time with Projects Grudge and Blue Book and offered his assessments of some UFO cases, including a portion he thought were puzzling and unexplained. Ruppelt also revealed much insider material and thinking, including the existence of previously unknown classified documents and studies, such as the Robertson Panel.

In 1956, Donald Keyhoe asked Ruppelt to join to serve as an adviser to NICAP. Ruppelt had recently suffered a heart attack and declined Keyhoe's offer. Ruppelt's book indicates that Ruppelt held some dim views of Keyhoe and his early writings; Ruppelt noted that while Keyhoe generally had his facts straight, his *interpretation* of the facts was another question entirely. He thought Keyhoe often sensationalized the material and accused Keyhoe of "mind reading" what he and other officers were thinking. Yet Keyhoe cites conversations with Ruppelt in later books, suggesting that Ruppelt may have occasionally advised Keyhoe.

In 1960 the expanded edition of Ruppelt's book (20 Chapters) was published by Doubleday & Co. The only change from earlier editions came in three more chapters which largely echoed the Air Force's position that there was nothing unusual about UFOs. Ruppelt seemed to have abandoned his early views that some UFO reports seemed mysterious and unexplained, and he declared UFOs a "space age myth". In an unusual manner, the date of the publication was omitted. The book, with the 1956 copyright note and the 1955 date of Ruppelt's Foreword, made the new edition appear to be the original edition. Only the dust jacket gives any hint that this is the second edition of the previous book.

Keyhoe and others would suggest that Ruppelt had caved into Air Force pressures to change his public statements about UFOs. Others argued against this, noting that Ruppelt had more than demonstrated his objectivity, and might have simply reached a conclusion after careful consideration of the evidence. Clark reports that Ruppelt's widow asserted that her husband's investigation of the contactee movement soured his opinion of UFO phenomena. Ruppelt's discussion of the contactees, particularly George Adamski, is arguably the most interesting portion of the revised book.

Ruppelt died of a heart attack on September 15, 1960, at the age of 37.[56]

Daniel Peter Sheehan was born in 1945 in Glen Falls, New York. He grew up in the tiny village of Warrensburg, right between Lake Champlain and Lake George at the southern edge of the Adirondack forest and by the border of the independent Six Nation Iroquois Confederacy. He was born into the Atomic Age, just 100 days before the detonation of the first atom bomb at the Trinity Test Site in New Mexico, in the first years of the post-World War II Baby Boom generation. At the age of 14, Daniel began attending Catholic Mass every Sunday, a tradition he continues to this day. During his teenage years, he developed a strong sense of justice and an inclusive view of humanity. These values motivated him in both his studies and pursuits.

Daniel P. Sheehan is a Constitutional and public interest lawyer. Over the last forty-five years, his work as an attorney, speaker, and educator has helped to expose the structural sources of injustice in the United States and abroad, protect the fundamental and inalienable rights of the world's citizens, and elucidate an inspiring and compelling vision for the direction of the human family. Daniel's dedication to his vision and work has put him at the center of some of the most important legal cases and social movements of our lifetimes. After meeting Sara Nelson during their work on the **Silkwood Case**, they started the **Christic Institute**, which would go on to uncover the **Iran-Contra Affair.**

In 1994, Daniel served as Legal Counsel to **John E. Mack**, Chair of the Department of Clinical Psychology at Harvard Medical School. Dr. Mack was a Pulitzer-Prize winning biographer who, utilizing the scientific methods of medical psychology, conducted extensive research on the phenomenon of alien abduction. In 1994 the Dean of Harvard Medical School called Dr. Mack before a special committee to defend the publication of his controversial book "Abduction: Human Encounters with Aliens". A clear violation of his free speech rights and position as a tenured professor—and highly respected clinician—Sheehan represented Dr. Mack before the committee, successfully protecting his right to academic freedom.

Daniel Sheehan's work with Dr. Mack was not the first time he had come into contact with the issue of life outside of Earth. In the late-1970s he acted as a Special Consultant to the **United States Library of Congress** investigation into the existence of an extraterrestrial intelligence that had been requested by **President Jimmy Carter**. Following this work, he presented a closed-door seminar on the Theological Implications of Contact to leading scientists at the **SETI Institute (Search for Extraterrestrial Intelligence)** at NASA's **Jet Propulsion Laboratory**.

Given these experiences, Sheehan was in a unique position to handle legal issues surrounding the extraterrestrial intelligence debate.

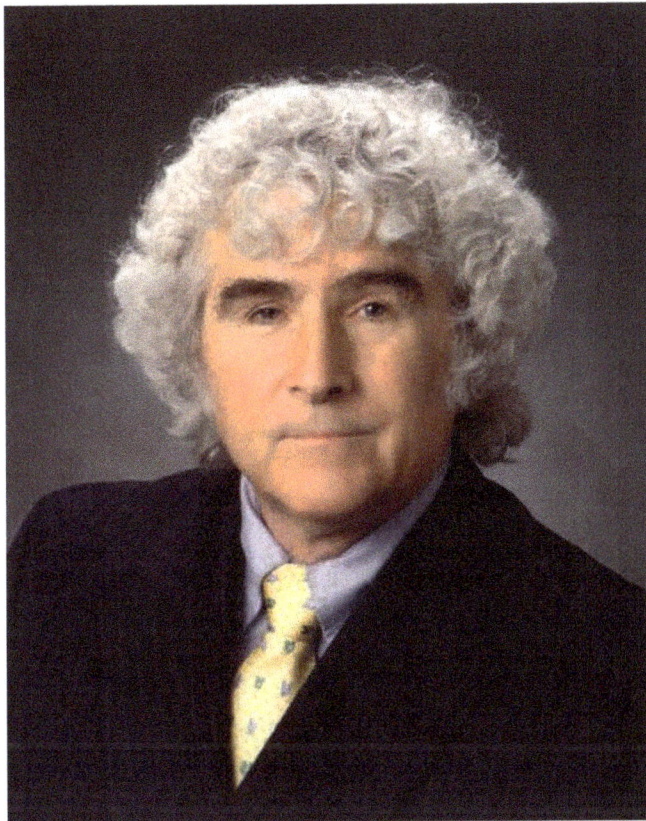

In 2001 he served as General Counsel to **The Disclosure Project**, which coordinated the sworn testimony—before members of the United States Congress—of former U.S. Military Officers, Federal Aviation Administration officials, and NASA employees attesting to direct personal knowledge of Government information concerning the UFO phenomenon and the potential existence of extraterrestrial intelligence. Sheehan also served as General Counsel to the Institute for **Cooperation in Space**, a U.S. citizens group dedicated to banning space-based weapons and the development of any weapons intended for offensive use against potential extraterrestrial civilizations. He has also presented, on multiple occasions, a talk on the Philosophical and Theological Implications of the Human Discovery of Extra-Terrestrial Intelligence at the **International UFO Congress**.

Daniel Sheehan's work has focused on identifying and cultivating non-dialectical, cooperative models for change. In 1999, he served as Director of the **New Paradigm Project at Mikhail Gorbachev's State of the World Forum**. There he joined world leaders in examining the obstacles to a new era of sustainable economic and social organization. Through the Romero Institute, he is currently working to establish the **New Paradigm Academy**, a year-long program that will give young leaders of the **Millennial Generation** the tools they need to effectively address global challenges. The Academy would house the **Worldview Institute**, a think-tank that incorporates an understanding of human worldviews into the resolution of global political and policy problems. http://en.wikipedia.org/wiki/Daniel_Sheehan_%28attorney%29

Brad Sparks is the leading expert on the CIA Robertson Panel and the history of the CIA investigation of UFO's, having interviewed every living CIA official involved with the Panel and having secured the release of the CIA's UFO files over the years since 1972. He was able to obtain the release of the complete, uncensored and declassified CIA Robertson Panel report and Minutes, as well as many other important CIA documents, some of which have never been released again to the public (because the CIA "lost" the documents). Brad discovered that the Air Force took known IFO cases falsely disguised as sensational unexplained UFO cases then planted them on the CIA and the Robertson Panel in a successful effort to sabotage the Panel and the CIA's UFO investigation. The AF's special file of 63 best UFO Unknowns was deliberately suppressed from the CIA.

https://www.youtube.com/watch?v=F-Aomm4O794

Brad also discovered that the CIA had concluded at that point, that UFO's were extraterrestrial (until the USAF deception), and this was confirmed by the CIA director and deputy director of its Office of Scientific Intelligence. Brad carried out a systematic investigation of the CIA's UFO activities, which included interviewing some 100 CIA Directors, Deputy Directors, Assistant Directors, and various intelligence officials of the CIA, NSA, DIA, Air Force and Naval Intelligence and other agencies, since 1975 in all of governmental history in UFO studies.

Brad was co-founder of **Citizens Against UFO Secrecy (CAUS)** and principal consultant in the **Freedom of Information Act (FOIA)** lawsuits managed by CAUS against the CIA and NSA.

Brad served as the assistant research director of the Aerial Phenomena Research Organization (APRO), founded by the late Coral and Jim Lorenzen, for APRO's last decade until 1988. Brad also is a co-founder of the BlueBookArchive.org which is putting online on the Web the entire U.S. Air Force Project Blue Book files, which he considers the best collection of publicly available UFO evidence in existence. He is a Research Associate of the National Aviation

Reporting Center on Anomalous Phenomena (NARCAP) founded by Dr. Richard Haines and is a MUFON member and consultant to the MUFON director.

Brad has worked with many scientists on UFO research projects such as astrophysicist **Dr. Peter Sturrock** and NASA scientist **Dr. Richard Haines**. He has worked with optical physicist **Dr. Bruce Maccabee**, on scientific investigations of photographic UFO cases, in particular, the McMinnville, Oregon, photos, since the 1970's, and presented his work on McMinnville to Sturrock's group in 1982. Based on his Lockheed training in radar systems, Brad has also worked with radar physicist Gordon Thayer and other scientists on analysis of the Lakenheath-Bentwaters AFB radar-visual sightings of UFOs, the RB-47 electronic intelligence UFO case, and other radar and electromagnetic cases.

Brad established this case for the first time in history that radar emissions from a UFO have been detected, measured and calibrated against a comparison ground radar detected at the same time as the moving airborne UFO radar signal, which proved the accuracy of the RB-47 ELINT measurements of the UFO radar beam. Brad has recently discovered a host of other similar UFO cases including somewhere the UFO's have broken our NSA codes in radio transmissions, which confounds conventional explanation.

Brad has investigated the startling case of the daylight UFO sighting by the world famed Lockheed aircraft designer **Clarence "Kelly" Johnson** and a separate team of his top Lockheed flight engineers, test pilots and chief aerodynamicist who simultaneously and independently sighted the hovering 200-foot UFO from widely separated locations. They watched it suddenly take off vertically at apparently earth escape velocity into space, as determined by Brad's triangulation of the two Lockheed groups' sightings.

Brad was the principal consultant for the Best Evidence: Top 10 UFO Sightings television a documentary produced by Redstar Films, Halifax, Nova Scotia, Canada, in 2005 (released worldwide 2007). Brad is listed in The UFO Encyclopaedia, edited by Margaret Sachs, and helped write and edit portions of The Encyclopaedia of UFO's, edited by Ronald Story. He has been interviewed on television about his UFO research in Best Evidence, on Hour Magazine with Gary Collins, by Omni and Fate magazines, and others on radio and in print. [57]

Wendelle Castyle Stevens is one of the most prominent ufologists in the United States. He was born (Jan. 18, 1923 - 2010) and raised in Minnesota, and enlisted in the Army shortly after high school. He graduated from the Lockheed Aircraft Maintenance & Repair School, Aviation Cadet Training and Fighter Pilot Advanced Training as a very young 2nd Lieutenant in the US Army Air Corps.

Wendelle C. Stevens is a retired U.S. Air Force pilot-turned-UFO investigator and researcher. Stevens' research began in 1946 when the Army Air Force assigned Stevens to Wright Field as part of a large technical review program of captured Nazi air technical documents. Stevens was then assigned to Alaska during the time the United States began extensive surveillance and mapping missions of Alaska and the north polar regions, as part of **Project Ptarmigan**. During that time, Stevens was in direct oversight of anomalous UFO sightings reports and film footage made by Air Force crews. This began a lifelong interest for Stevens in uncovering the truth regarding UFO activity.

After that, he attended the first Air Corps Flight Test Pilot School at Kelly Field, where he learned to fly all the aircraft of the Air Corps at the time, as well as a few US Navy aircraft. During his long career in the military, one of his assignments was the supervision of a highly classified team of technical specialists who were installing hi-tech data collecting equipment aboard the SAC B-29s of the Ptarmigan Project – a research project which was photographing and mapping every inch of the Arctic land and sea area. This equipment was designed to capture, record and analyze all EMF emissions in the Arctic, photograph all anomalous phenomena, and record all disturbances in the electrical and engine systems of the aircraft – looking for external influences caused by UFOs. The data was then couriered nightly to Washington.

Unable to possess any of this information for himself, Stevens, upon return, began his own research and collection effort, eventually amassing the largest private collection of UFO photographs in the world. He began to publish reports on the events and wrote many illustrated articles for many UFO publications. Disenchanted with the lack of detail on contact events reported in books and journals of the time, he began preparing detailed reports of his own investigations

Stevens has collected more than 4,000 actual UFO photographs and has written and co-authored over 30 books on extensive UFO contact cases. He has also been a Director of the International UFO Congress since its inception, as its original founding member. [58] Wendelle Stevens passed away in 2010. http://www.ufophotoarchives.org/

Leonard Stringfield (1920 - 1994) was an American Ufologist who took particular interest in crashed flying saucer stories. His contacts in the medical field gave him the first descriptions of the alien bodies allegedly recovered at Roswell or elsewhere.

Stringfield's interest in the subject began August 28, 1945, just three days before the end of the war, when he was an Army Air Force intelligence officer en route to Tokyo, Japan, along with twelve other specialists in the Fifth Air Force. As they approached Iwo Jima at about ten thousand feet in a sunlit sky, Stringfield related:

"I was shocked to see three teardrop-shaped objects from my starboard side window. They were brilliantly white, like burning magnesium, and closing in on a parallel course to our C-46. Suddenly our left engine feathered, and I was later to learn that the magnetic navigation-instrument needles went wild. As the C-46 lost altitude, with oil spurting from the troubled engine, the pilot sounded an alert; crew and passengers were told to prepare for a ditch! I do not recall my thoughts or actions during the next, horrifying moments, but my last glimpse of the three bogies placed them about 20 degrees above the level of our transport. Flying in the same, tight formation, they faded into a cloud bank. Instantly our craft's engine revved up, and we picked up altitude and flew a steady course to land safely at Iwo Jima."

Stringfield said his World War II encounter was so traumatic that he tried to forget about it. But he was drawn back into the UFO field in 1950 when two very sincere people related flying saucer sightings to him. Stringfield then wrote:

"This one experience near Iwo Jima was proof enough to me in 1950 that the 'foo fighter' of World War II--sometimes dubbed 'Kraut fireball' in the European Theater--and the flying saucer were one and the same kind of machine and from the same source: outer space."

Uneasy about the "rumored loss of Air Force interceptors chasing UFOs, the low-level green fireballs over Sweden and the southwestern United States" and his own experience, Stringfield related he was concerned about the "intent" behind the probes. In March 1954, he created

Civilian Research, Interplanetary Flying Objects (CRIFO), and published a monthly newsletter, *ORBIT*.

The newsletter caught the attention of radio newscaster Frank Edwards, who allowed Stringfield to announce it on his popular program in May. Instantly Stringfield was deluged with mail and newspapers, and radio stations from coast to coast called, wanting saucer news. Stringfield soon had 2500 paid subscribers to *ORBIT*. During the mid-1950s, CRIFO became the world's largest civilian UFO research group.

Then Stringfield wrote, "*Also taking note of CRIFO was the Air Force.*" Stringfield said the Air Defense Command in Columbus, Ohio called him September 9, 1955 and wanted his cooperation in obtaining immediate sighting reports using his large network of sources. To his surprise, he was also informed that the **Ground Observer Corps (GOC)** in southwestern Ohio had been instructed to report UFO activity directly to him for screening. (Stringfield lived in Cincinnati, Ohio.) He was then to call the ADC using a telephone code number ("Fox Trot Kilo 3-0 Blue") to report the better sightings. He was requested "not to ask any questions."

Later a member of the GOC informed him as to what happened to his screened reports. If the sighting was confirmed by radar, jets were then scrambled for intercept and the matter became classified. Stringfield recounted one such spectacular incident, on the night of August 23/24, 1955, when multiple UFOs were spotted on radar in the Columbus/Cincinnati region. Numerous jets were sent up for intercept over a wide region, but cloud cover prevented Stringfield from seeing what was happening, though he could hear the jets overhead.

To his surprise, the Air Force cleared his reporting of the incident in his newsletter. But when he tried to interest the local Cincinnati newspapers, the story was officially denied, as was his 377connection with the ADC.

368

Despite the official public denial of his work for the ADC, Stringfield wrote he received a letter in 1956 thanking him for his assistance from no less than Major General John A. Samford, director of Air Force Intelligence. He also received a letter in 1955 from Captain Edward J. Ruppelt, who had been director of the Air Force's public UFO investigation Project Blue Book from 1951-1953. Ruppelt was requesting information on CRIFO for the book he was writing at the time (*The Report on Unidentified Flying Objects*) and praised the report-collecting net Stringfield had established.

Stringfield said he continued his "cooperation" with the Air Force through 1956 until the GOC was disbanded and his screening duties for them ceased. Stringfield's relationship with the ADC during this period is recounted in his 1957 book *Inside Saucer Post, 3-0 Blue* and in his 1977 book *Situation Red.*

In 1957, Stringfield discontinued CRIFO and his monthly newsletter. The same year, he became public relations adviser for the newly formed civilian UFO group NICAP under the direction of Donald Keyhoe, a friend of his since 1953. He held the post until 1972, at which point he continued his private UFO research. It was during the 1970s that Stringfield began collecting witness accounts of UFO crash recoveries, including alien bodies. Many of these stories centered around activities at nearby Wright-Patterson AFB in Dayton, Ohio.

Stringfield first publicly reported his so-called "crash/retrieval" findings at a 1978 MUFON Symposium. He said he received two death threats beforehand but was never sure who was behind them or how serious they were. Thereafter, he self-published seven "Status Reports" on new crash-retrieval research until his death in 1994.

From 1967-1969, Stringfield served as an "Early Warning Coordinator" for the so-called Condon Committee, the government-sponsored scientific UFO investigation. His job, like his earlier one for the ADC, was to screen and report all UFO activity in southwestern Ohio.

Also when **Grenada Prime Minister Sir Eric Gairy** proposed the establishment of a UFO research agency within the **United Nations** in 1978, during the 32nd General Assembly of the UN, Stringfield served as his adviser.

Privately, Stringfield worked as Director of Public Relations and Marketing Services for DuBois Chemicals, a division of Chemed Corporation, Cincinnati. [59]
http://en.wikipedia.org/wiki/Leonard_H._Stringfield

Michael D. Swords is a Professor of Natural Science at Western Michigan University in Kalamazoo, Michigan. He graduated from the University of Notre Dame in 1962 with a B.S. degree in Chemistry, and also holds an M.S. degree from Iowa State University in Biochemistry (1965), and a Ph.D. from Case Western Reserve university (1972).

His major professional and involvements are teaching and writing in the areas of general sciences and anomalous phenomena. His teaching centers about human biology, the history, and philosophy of science, scientific methodology, and the "parasciences" of which Ufology is a member. His writings have concentrated mainly on topics in Ufology, parapsychology, and cryptozoology, and several have been published in the "MUFON UFO Journal". He has won University's Teaching Excellence Award (1978).

Photo by Thomas Gilpin (Yahoo images)

Swords is also interested in Ufology and is seen as an authority of the Condon Committee. He was a prominently featured talking head on the prime-time 2005 television special *Peter Jennings Reporting: UFOs — Seeing Is Believing*, discussing the early history of the U.S. Military's UFO investigations.

Dr. Swords is a member of several professional and parascientific societies including the American Association for the Advancement of Science, the American Association of University Professors, the Society for Scientific Exploration, and the Mutual UFO Network (MUFON). He serves as a MUFON consultant and was in the same capacity for the Aerial Phenomena Research Organization, the International Fortean Organization, and as a member of the advisory panel for the **Society for the Investigation of the Unexplained**. he is a Board member of the J. Allen Hynek **Center for UFO Studies**, and is the editor of Ufology's academic journal: "The Journal of UFO Studies".

Dr. Swords originally spoke at the MUFON 1986 UFO Symposium in East Lansing, Michigan. He has recently presented materials at the "Abductions" Conferences of TREAT II (VPI, 1990) and NCAE (Temple, 1991), as well as the National Science Teachers Association (Phoenix, 1989) [60]

Jacques Fabrice Vallée (born September 24, 1939, Pontoise, Val-d'Oise) is a French-born venture capitalist, computer scientist, author, ufologist and former astronomer. He currently resides in San Francisco, California in the United States. (He should not be confused with the Canadian astronomer Jacques Paul Vallée.)

370

In mainstream science, Vallée is notable for co-developing the first computerized mapping of Mars for NASA and for his work at SRI International in creating ARPANET, a precursor to the modern Internet. Vallée is also an important figure in the study of unidentified flying objects (UFOs), first noted for a defense of the scientific legitimacy of the extraterrestrial hypothesis and later for promoting the Interdimensional hypothesis similar partly to **Mac Tonnies'** crypto-terrestrial hypothesis.

Vallée was born in Pontoise, France. He received his Bachelor of Science degree in mathematics from the Sorbonne, followed by his Master of Science in astrophysics from the University of Lille. He began his professional life as an astronomer at the Paris Observatory in 1961. He was awarded the Jules Verne Prize for his first science-fiction novel in French.

He moved to the United States in 1962 and began working in astronomy at the University of Texas at Austin, at whose MacDonald Observatory he worked on NASA's first project making a detailed informational map of Mars.

In 1967, Vallée received a Ph.D. in computer science from Northwestern University. While at the Institute for the Future from 1972 to 1976 he was a principal investigator on the large NSF project for computer networking, which developed the first conferencing system, Planning Network (PLANET), on the ARPANET many years before the Internet was formed.

He has also served on the National Advisory Committee of the University of Michigan College of Engineering and was involved in early work on artificial intelligence.

Vallée has authored four books on high technology, including Computer Message Systems, Electronic Meetings, The Network Revolution, and The Heart of the Internet.

Along with his mentor, astronomer J. Allen Hynek, Vallée carefully studied the phenomenon of UFOs for many years and served as the real-life model for the character portrayed by **François Truffaut** in Steven Spielberg's film *Close Encounters of the Third Kind.*

His research has taken him to countries all over the world. Considered one of the leading experts in UFO phenomena, Vallée has written several scientific books on the subject.

His current endeavours include his involvement in SBV Ventures a venture capital fund as a general partner. He and the other general partner, Graham Burnette on SBV are also in the early stages of launching a second venture capital fund.

He is married and has two children.

In May 1955, Vallée first sighted an unidentified flying object over his Pontoise home. Six years later in 1961, while working on the staff of the French Space Committee, Vallée witnessed the destruction of the tracking tapes of an unknown object orbiting the earth. The particular object was a retrograde satellite - that is, a satellite orbiting the earth in the opposite direction. At the time he observed this, there were no rockets powerful enough to launch such a satellite, so the team was quite excited as they assumed that the Earth's gravity had captured a natural satellite (asteroid). A superior came and erased the tape. These events contributed to Vallée's long-standing interest in the UFO phenomenon.

In the mid-1960s, like many other UFO researchers, Vallée initially attempted to validate the popular Extraterrestrial Hypothesis (or ETH). Leading UFO researcher Jerome Clark [4] argues that Vallée's first two UFO books were among the most scientifically sophisticated defenses of the ETH ever mounted.

However, by 1969, Vallée's conclusions had changed, and he publicly stated that the ETH was too narrow and ignored too much data. Vallée began exploring the commonalities between UFOs, cults, religious movements, demons, angels, ghosts, cryptid sightings, and psychic phenomena. Speculation about these potential links was first detailed in Vallée's third UFO book, *Passport to Magonia: From Folklore to Flying Saucers.*

As an alternative to the extraterrestrial visitation hypothesis, Vallée has suggested a multidimensional visitation hypothesis. This hypothesis represents an extension of the ETH where the alleged extraterrestrials could be potentially from anywhere. The entities could be multidimensional beyond space-time, and thus could coexist with humans, yet remain undetected.

Vallée's opposition to the popular ETH hypothesis was not well received by prominent U.S. Ufologists; hence he was viewed as something of an outcast. Indeed, Vallée refers to himself as a "heretic among heretics".
Vallée's opposition to the ETH theory is summarized in his paper, "Five Arguments Against the Extraterrestrial Origin of Unidentified Flying Objects", Journal of Scientific Exploration, 1990.

Scientific opinion has generally followed public opinion in the belief that unidentified flying objects either do not exist (the "natural phenomena hypothesis") or, if they do, must represent evidence of a visitation by some advanced race of space travellers (the extraterrestrial hypothesis or "ETH"). It is the view of the author that research on UFOs need not be restricted to these two alternatives. On the contrary, the accumulated database exhibits several patterns tending to indicate that UFOs are real, represent a previously unrecognized phenomenon and that the facts do not support the common concept of "space visitors." Five specific arguments articulated here contradict the ETH:

1. unexplained close encounters are far more numerous than required for any physical survey of the earth;
2. the humanoid body structure of the alleged "aliens" is not likely to have originated on another planet and is not biologically adapted to space travel;
3. the reported behavior in thousands of abduction reports contradicts the hypothesis of genetic or scientific experimentation on humans by an advanced race;
4. the extension of the phenomenon throughout recorded human history demonstrates that UFOs are not a contemporary phenomenon; and
5. the apparent ability of UFOs to manipulate space and time suggests radically different and richer alternatives.

Vallée has contributed to the investigation of the Miracle at Fatima and Marian apparitions. His work has been used to support The Fatima UFO Hypothesis. Vallée is one of the first people to speculate about the possibility that the Miracle at Fatima was a UFO. This wasn't initially recognized as such due to lack of knowledge about UFOs at that time. It is believed by many that if they were aware of the UFO phenomenon that they would have initially assumed it was a UFO instead of a miracle. Vallée has also speculated about the possibility that other religious apparitions may have been the result of UFO activity including the Miracle of Lourdes and the revelations to Joseph Smith. Vallée and other researchers have advocated further study of unusual phenomena in the academic community. They don't believe that this should be handled solely by theologians

In the Steven Spielberg, film *Close Encounters of the Third Kind* Vallée served as the model for the French researcher character, Lacombe (François Truffaut).

In 1979, **Robert Emenegger** and **Alan Sandler** updated their 1974 *UFOs, Past, Present and Future* documentary with new 1979 footage narrated by Jacques Vallée. The updated version is entitled "UFOs: It Has Begun".

Jacques Vallée attempted to interest Spielberg in an alternative explanation for the phenomenon. In an interview on Conspire.com, Vallée said, "I argued with him that the subject was even more interesting if it wasn't extraterrestrials. If it was real, physical, but not ET. So he said, 'You're probably right, but that's not what the public is expecting -- this is Hollywood and I want to give people something that's close to what they expect.'"

Vallée proposes that there is a genuine UFO phenomenon, partly associated with a form of non-human consciousness that manipulates space and time. The phenomenon has been active throughout human history and seems to masquerade in various forms to different cultures. In his opinion, the intelligence behind the phenomenon attempts social manipulation by using deception on the humans with whom they interact. Vallée also proposes that a secondary aspect of the UFO phenomenon involves human manipulation by humans. Witnesses of UFO phenomena undergo a manipulative and staged spectacle, meant to alter their belief system, and eventually, influence human society by suggesting alien intervention from outer space. The ultimate motivation for this deception is probably a projected major change of human society, the breaking down of old belief systems and the implementation of new ones. Vallée states that the evidence if carefully analyzed, suggests an underlying plan for the deception of mankind by means of unknown, highly advanced methods (His colleague, Hynek, suggested that the UFO phenomenon was in fact demonic in nature).

Vallee states that it is highly unlikely that governments actually conceal alien evidence, as the popular myth suggests. Rather, it is much more likely that that is exactly what the manipulators want us to believe. Vallée feels the entire subject of UFOs is mystified by charlatans and science fiction. He advocates a stronger and more serious involvement of science in the UFO research and debate. Only this can reveal the true nature of the UFO phenomenon.

Vallée is often highly critical of UFO investigators overall, both believers and skeptics, asserting that what often passes for an acceptable level of investigation in a UFO context would be considered sloppy and seriously inadequate investigation in other fields. He has pointed out logical flaws and methodological flaws common in such research. Unlike many critics of UFO investigative efforts, his critiques are not condescending and dismissive and he indicates that he is simply interested in good science. Vallée expresses concern about the often authoritarian political and religious views expressed by many contactees. Amongst the groups profiled are the nascent Raelian movement and an early form of the Heaven's Gate suicide cult, against which Vallée prophetically warned potential converts, *"You only risk your life!"* He also argues that Scientology is another example of a UFO cult which has organized itself as a religious organization [61]

Alfred Lambremont Webre, J.D., M.Ed. (Canada) (1942 -), is an author, futurist, lawyer (member of the District of Columbia Bar), peace advocate, environmental activist, space activist who promotes the ban of space weapons, and is known as the founding father of exopolitics. He was a co-architect of the Space Preservation Treaty and the Space Preservation Act that was introduced to the U.S. Congress by Congressman Dennis Kucinich and is endorsed by more than 270 NGO's worldwide. He helped draft the Citizen Hearing in 2000 with Stephen Bassett and serves as a member of the Board of Advisors. Webre is also on the Board of Advisers at the Exopolitics Institute, is the congressional coordinator for The Disclosure Project, is a judge on the Kuala Lumpur War Crimes Tribunal, and is the International Director of the Institute for Cooperation in Space.

Webre entered Yale University in 1960 and graduated in 1964 with a Bachelor of Science in Industrial Administration Honors. His interest in law brought him to continue his education and to graduate from Yale Law School with a Juris Doctor in International law in 1967. In 1967-1968, Webre traveled to Montevideo, Uruguay and became a Fulbright Scholar in Economic Integration. In 1993, he entered the University of Texas at Brownsville and graduated with a Master of Education in Counselling in 1997.

Alfred Webre is a former Fulbright Scholar and graduate of Yale University, Yale Law School (Yale Law School National Scholar), and the University of Texas Counseling Program. Webre was General Counsel to the NYC Environmental Protection Administration and environmental consultant to the Ford Foundation, futurist at **Stanford Research Institute**, and author. He has taught Economics at Yale University (Economics Department) and Civil Liberties at the University of Texas (Government Department). Alfred Webre has been a delegate to the UNISPACE Outer Space Conference and NGO representative at the United Nations (Communications Coordination Committee for the UN; UN Second Special Session on Disarmament); elected Clinton-Gore Delegate to the 1996 Texas Democratic Convention; and a Member, Governor's Emergency Taskforce on Earthquake Preparedness, State of California (1980-82), appointed by Gov. Jerry Brown. Mr. Webre produced and hosted the Instant of Cooperation, the first live radio broadcast between the USA and the then Soviet Union, carried live by Gosteleradio and NPR satellite in 1987. He is a member of the District of Columbia Bar.

As Senior Policy Analyst in 1977 at the Center for the Study of Social Policy at Stanford Research Institute (now "SRI International", Menlo Park, California), Alfred Webre was Principal Investigator for a proposed civilian scientific Study of extraterrestrial communication, i.e. interactive communication between the terrestrial human culture and that of possible intelligent non-terrestrial civilizations. President Jimmy Carter reported his own 1969 UFO encounter and made a 1976 election campaign promise to create more transparency with the

'UFO' issue. This proposed Study was presented to and developed with interested White House staff of the Domestic Policy staff of President Jimmy Carter during the period from May 1977 until the fall of 1977, when it was abruptly terminated.

Today, Webre is the International Director of the Institute for Cooperation in Space (ICIS), promoting the vision of banning space weapons and transforming the permanent war economy into a peaceful and cooperative space exploration society focused on studying life in the universe.

Alfred Webre is a co-architect of the **Space Preservation Act and the Space Preservation Treaty** to ban space-based weapons.

Alfred Lambremont Webre's book, *Exopolitics: Politics, Government and Law In the Universe*, first defined the formal field of Exopolitics (the study of law, governance, and politics in the Universe).

Alfred Webre has also written "Earth Changes: A Spiritual Approach," "The Levesque Cases, "and "The Age of Cataclysm."

"The overall purpose of the proposed 1977 Carter White House Extraterrestrial Communication Study was to create, design and carry out an independent, civilian-led research compilation and evaluation of phenomena suggesting an Extraterrestrial and/or Inter-dimensional intelligent presence in the near-Earth environment.

"The scientific and public policy goal of the proposed 1977 Carter White House Extraterrestrial Communication Study was to fill a substantial gap in civilian scientific knowledge of the UFO (Unidentified Flying Object phenomenon), Extraterrestrial Biological Entities (EBEs), and related phenomena. This knowledge gap was created and maintained by excessive secrecy practices and regulations of U.S. Department of Defense and intelligence agencies in the various generations of its UFO-programs since the late 1940s, including but not limited to Project Grudge and Project Blue Book, as well as other alleged secret programs."

In 2000, he was a presenter at the Presidential Forum on Off-Planet Cultures Policy at the Santa Clara Convention Center in California. Since 2002, he is the host of "The Monday Brown Bagger", a public affairs radio talk show on Coop Radio CFRO 102.7 FM in Vancouver, British Columbia. He is a founding director of Canada's No Weapons in Space Campaign (NOWIS) established in 2002. In 2004, he created the Campaign for Cooperation in Space (CCIS), an international organization where he works with others to prevent the weaponization of space and promote the transformation of the war economy into a peaceful, cooperative space exploration industry.

Alfred Webre and Dr. Carol Rosin founded the **Institute for Cooperation in Space (ICIS)** in 2001. The ICIS mission is to educate decision-makers and the grassroots about why it is important to ban space weapons. Through the help of **Congressman Dennis Kucinich**, the **Space Preservation Act (HR 3657)** was introduced in the 108th Congress on January 23, 2002. ICIS continues to lobby for a **Space Preservation Treaty** conference where leaders of the world would gather to ban space weapons. Supported by former **Minister of National Defence Hon. Paul Hellyer**, who believes that this treaty would help put a cap on the war industry and open the door for international cooperation in outer space exploration. Thus, transforming the "war based" economy into a "peace based" economy.

The ICIS board is made up of various prominent individuals such as former astronauts Edgar Mitchell & **Dr. Brian O'Leary**, as well as **Arthur C. Clarke**, General Council **Daniel Sheehan** and **John McConnell** who is the founder of International Earth Day.

Webre believes that there is intelligent extraterrestrial life in our Universe. He is the author of the online e-book, "Towards a Decade of Contact" and the book *Exopolitics: Politics, Government, and Law in the Universe*. The exopolitics model functionally maps the operation of politics, government, and law in an intelligent Universe, and provides an operational bridge between models of terrestrial politics, government and law, and the larger models of politics, government and law in Universe society.

Webre believes that as exopolitics posts, the truest conception of our human circumstance may be that we are on an isolated planet in the midst of a populated, evolving, highly organized inter-planetary, intergalactic, multi-dimensional Universe society. He believes that we live on a planet that has been quarantined (**Zoo Hypothesis**) and that we are now being given an opportunity to join the rest of the spiritually evolved Universe Society in peace, thus an opportunity to avoid environmental global self-destruction or global self-destruction through war.

On March 10, 2007, Webre launched the Exopolitics Radio program, hosted by 1480 KPHX (which at the time was the Air America Radio affiliate, and Nova M Radio flagship station, in Phoenix) until the fall of 2008; the program remains in production and is distributed via podcast on its own website. Guests on the program have been generally advocates of views similar to that of Webre, and many are well known within the UFO research/enthusiast (and to a lesser extent the New Age) community. [62] [63] **http://exopolitics.blogs.com/about.html**

I had the opportunity to work with Alfred in the summer of 2001 to promote Dr. Steven Greer's *"Disclosure Project"* lecture tour in which we held Canada's first UFO Disclosure event at Simon Fraser University in Burnaby, B.C. (a municipality near Vancouver) on September 9, 2001. Dr. Steven Greer, Carol Rosin, and Dr. Alfred Webre were the guest speakers who presented talks on various aspects of the UFO and Extraterrestrial Intelligence phenomenon and the banning of space-based weapons. I emceed the event at that time.

OCEANIA - Australia

Michael Cohen, (1970-), Journalist and well-known psychic published numerous articles on UFO activity in Australia.[64]

Greg Taylor, editor of The Daily Grail, a daily news, commentary and information service for UFOs and the paranormal.[65]

SOUTH AMERICA - Chile

J. Antonio Huneeus, a Chilean-American journalist was born in New York in 1950, the son of a Chilean diplomat and United Nations official. He studied French at Sorbonne University in Paris in 1970 and Journalism at the University of Chile in Santiago in the early 1970's. Huneeus has lectured at dozens of conferences in the Americas, Europe, and Asia during the past 15 years.

He was science editor for the weekly magazine Que Pasa in Santiago, Chile, in the 1970s, and wrote a weekly science column in the 1980s for the New York City Tribune and the Hispanic NY daily Noticias del Mundo. He studied French language and civilization at the Sorbonne University in Paris in 1970, Journalism at the University of Chile in Santiago in 1972-73, and took a course on Communications Theory at the Catholic University in Santiago in 1974. He is fully fluent in Spanish, French, and English, and has some understanding of Portuguese and Italian.

He has appeared on many national and international publications, radio and TV programs, including Nippon-TV, Fox Encounter, the McLaughin Report on CNBC, the Sci-Fi Channel, the Russian documentary film UFO: Top Secret and other broadcasts in South America, Europe, and the Far East.

J. Antonio Huneeus is considered one of the world's top experts on UFOs. When Laurance S. Rockefeller commissioned Marie Galbraith and Sandra S. Wright to assemble a report on the best evidence for UFOs to send members of Congress and selected VIPs worldwide in the mid-90s, Huneeus was one of the co-authors (with Don Berliner and M. Galbraith). This report was recently published as a Dell paperback under the title of "UFO Briefing Document - The Best Available Evidence."

Huneeus has covered the UFO field from an international perspective as a science journalist, investigator, and lecturer during the past 24 years. His articles have appeared in a variety of newspapers, magazines, and journals in the Americas, Europe and Japan. These include Fate magazine, The Anomalist, and the New York City Tribune in the US; the magazines Mo Cero and Mundo Desconocido in Spain; Phenomena in France; Magazine 2000 in Germany; Planeta in Brazil; Cuarta Dimension in Argentina; Borderland and Super Science in Japan; the newspapers La Tercera and La Segunda in Chile and El Tiempo in Colombia; Anomalia in Russia; Notiziario UFO in Italy; UFO Magazine in Hungary, and other specialized publications around the world.

Huneeus edited the anthology 'A Study Guide to UFOs, Psychic and Paranormal Phenomena in the USSR' (NY: Abelard Publications, 1991); was consultant and producer for a Laser Active disc in Japan, 'UFO & ET,' produced for Pioneer Electronic by the Tokyo company Studio Garage in 1995; author of the 1998 and 1999 UFO Calendar (NY: Stewart, Tabori & Chang); one of several essayists of the book 'Of Heaven and Earth,' edited by Zecharia Sitchin (The Book Tree, 1996, published also in Germany); and commentator for the DVD 'Ultimate UFO! - The Complete Evidence' (Central Park Media, 2000).

Huneeus has lectured at dozens of Conferences in the Americas, Europe, and Asia during the past 18 years, including throughout the USA, Costa Rica, Chile, Argentina, Brazil, Spain, Italy, Hungary, Germany, Japan and South Korea. He received the "Ufologist of the Year" award at the National UFO Conference in Miami Beach in 1990; the 1991 annual award of the New York Fortean Society; and the In Memoriam Colman S. von Keviczky medal (international category) in 2000, awarded by UFO Magazine in Budapest, Hungary. He has served for many years as International Coordinator for the Mutual UFO Network (MUFON).
http://www.ufoevidence.org/researchers/detail59.htm [66]

Brazil

Ademar Jose Gevaerd (1962 -) is a Brazilian ufologist, editor of the Brazilian UFO Magazine. In 1983, he is the Founder and Director of the **Brazilian Center for Research of UFOs (CBPDV),** the largest organization of its kind in the world, based in Campo Grande (MS) and with more than 3,600 associates and he is the Brazilian Director for **Mutual UFO Network (MUFON).**

Gervaerd is a professor of chemistry until 1986 when he withdrew from the area to devote himself exclusively to UFO.

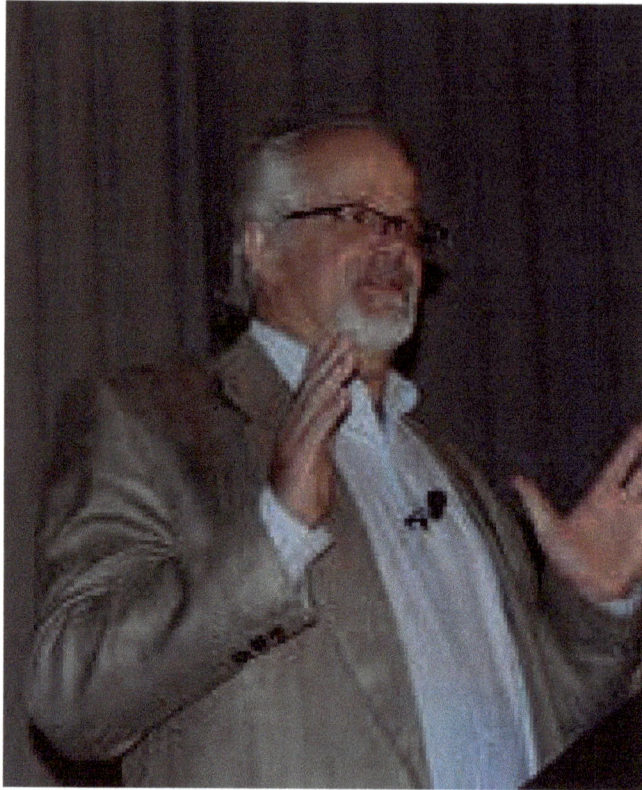

Gevaerd began his career in Ufology at the very young age of 11. Since 1989, he began to devote himself to the subject facilitating thousands of conferences in Brazil. He has researched and lectured in 54 other countries and is constantly in demand abroad.

In the 80s, Gevaerd was invited by Dr. J. Allen Hynek, a pioneer in the world of Ufology, to represent Brazil at the **Center for UFO Studies (CUFOS),** a position he holds to this day.

Gevaerd is also director for Latin America's Annual International UFO Congress, one of the busiest events in Ufology which is affiliated with the Las Vegas IUFO Congress and with numerous other organizations. Gevaerd was also, the mastermind of the campaign UFOs: Freedom of Information Now and coordinated the work of the **Brazilian Committee of Ufologists (CBU)** to demand from the **Brazilian Air Force (FAB)** to open its secret files and admits civilian participation in the process of investigating the UFO phenomenon in his country, which is now being carried out.

The campaign was based largely on statements by Brazilian Air Force Colonel Uyrange Holland to publicly admit the military's knowledge of the existence of UFOs, in an exclusive interview with Brazilian UFO Magazine. The move was successful and ufologists of CBU welcome the unprecedented announcement by the Brazilian military on May 20, 2005, to discuss the process of opening UFO files to the country.

Gevaerd has been active in almost all circles of the world where the UFO theme is taken seriously, participating in events, discussions, programs, campaigns etc. He is the coordinator for Latin America Program: Australian Hot Zone and had a decisive participation in programs like Rede Globo, the Discovery Channel, and the History Channel, among others. [67]
http://translate.google.com/translate?hl=en&sl=pt&u=http://www.aforteanosla.com.ar/Col aboraciones/autores/brasil/ademar%2520gevaerd.htm&ei=zMT9S7i2EJKyNqH6sN4H&sa =X&oi=translate&ct=result&resnum=3&ved=0C

Ecuador and USA

Dr. Carol Sue Rosin (born March 29, 1944; Wilmington, Delaware) is an award-winning educator, author, leading aerospace executive and space and missile defense consultant. She is a former spokesperson for **Wernher von Braun** and has consulted to a number of companies, organizations, government departments, and the intelligence community. She is the current President of the Institute for Cooperation in Space (ICIS) which she co-founded with Alfred Webre. Dr. Rosin has received the support of various prominent individuals such as U.S. Congressman **Dennis Kucinich**, and **Hon. Paul Hellyer**, a former Canadian Minister of National Defence. She is also a witness for The Disclosure Project.

http://starworksusa.com/presenters_2012/dr-carol-rosin

She received a Bachelor of Science from the University of Delaware and an Honorary Doctorate in Humanities from Archbishop Soloman Gbadebo, Nigeria.

Rosin was the first woman to hold the position of Corporate Manager at Fairchild Industries from 1974 through 1977, where she met the late Wernher von Braun. When they first met in early 1974, Von Braun was dying of cancer. According to Rosin's accounts, Von Braun spent the last years of his life explaining to her his position that space-based weapons are dangerous, destabilizing, too costly, unnecessary, and unworkable, and explaining the available alternatives. He asked Dr. Rosin to be his spokesperson and to appear on occasions when he was too ill to speak. According to Rosin, he also asked her to take on the challenge of promoting the ban of space weapons by educating decision-makers and the grassroots about transforming the military industrial complex into a peaceful space exploration industry.

According to Rosin, von Braun also spoke of the existence of **"Off Planet Cultures" (OPC)** (extraterrestrials) and government conspiracies to keep the existence of them a secret.

Rosin has been a consultant to corporations and organizations, including TRW, Disney, GE, IBM and the National Space Institute on Space and Defense. She has testified before the U.S. Congress, the U.S. Senate, and the President's Commission on Space and has met with people in over 100 countries about the feasibility of banning space-based weapons.

She is the founder and former director of the **Institute for Security and Cooperation in Outer Space (ISCOS)**, an NGO in consultative status with the **United Nations Economic and Social Counsel (UN-ECOSOC).**

She co-founded the **Institute for Cooperation in Space (ICIS)** in 2001 and is the current President. The ICIS board is made up of various prominent individuals such as former astronauts Edgar Mitchell & Dr. Brian O'Leary, as well as Arthur C. Clarke, General Council Daniel Sheehan and John McConnell who is the founder of International Earth Day.

She lived in Vilcabamba, Loja, Ecuador for a short period of time. She co-owns a hotel in Vilcabamba named Madre Tierra where she focuses on wildlife conservation, the study of organic foods and remedies as well as negative ions. Carol Rosin has since returned to the U.S. to live.

Rosin has received the Space Humanitarian Award in 2000 from the United Societies in Space for her 30 years of dedication on the subject, a Science Teachers Gold Medal Award, an Aviation Writers Award and an American Society of Engineering Educators Award for organizing the motivation program "It's Your Turn" for youth, women and minorities.

Argentina

Fabio Zerpa (born in Montevideo, Uruguay in 1928) is a parapsychologist and UFO researcher. Zerpa arrived in Argentina in 1951, and was already interested in extraterrestrial life and had already studied psychology. After some years of investigation, he started to give his first conferences at the beginning of the 1960s. In 1966 he created the radio program *Más allá de la* *cuarta dimensión* (Beyond the Fourth Dimension). Since then, Fabio Zerpa has reported on more than 3000 cases of UFO sightings and contact with extraterrestrials. Since 2001 he has been the director of the on-line magazine El Quinto Hombre (The Fifth Man). In December 2005 Zerpa was named cultural ambassador of the city of Colonia in his native country.[69]

Silvia Simondini (born in Buenos Aires, Argentina) she currently resides in Victoria, Entre Ríos Province, Argentina as a UFO researcher. She is founding member of Argentinean Cefora (Argentinean Republic Committee for UFO Phenomena).

Simondini had an incident with UFOs in 1968 and many times after. She arrived to live at Victoria city in 1990 and was already interested in extraterrestrial life. She went to Victoria knowing the strange events that were occurring in this location such as strange lights, and close sightings showed on TV. In 1992 she investigated the Cota Colman incident where a man was attacked by a craft in strange circumstances. After some years of investigation, she started to give her first conferences at her own museum and in the Argentinean UFO Congress in 2006.

She had created her own team of UFO investigators known as Visión Ovni (UFO vision) at the beginning of the 90s. In 1999, ufologist Nicolás Ojeda had donated all his material dating from the 60s to Visión Ovni. This is the most serious team of UFO research in Argentina. They have studied over 4000 cases of cattle mutilation in Argentina, abductions, and sightings.

In 2005 members of Visión Ovni started a radio emission where they speak about UFO phenomena weekly. [70] **http://en.wikipedia.org/wiki/Silvia_Simondini**

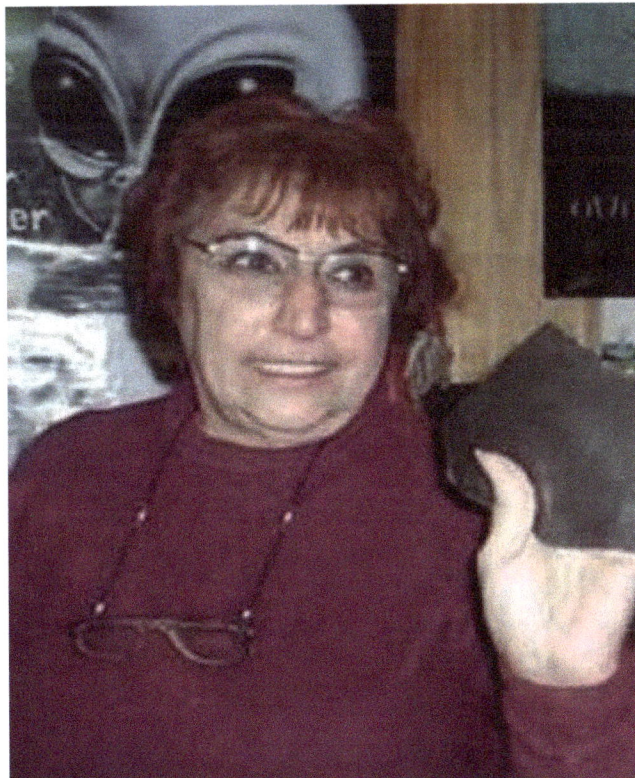

http://circuloesceptico.com.ar/tag/silvia-perez-simondini

383

References:

1. **"Skeptic Report/ UFO Bibliography: Salatun, J.".** http://www.skepticreport.com/ufo/ufo-s.htm. Retrieved 2010 -15-05.
2. **"BETA-UFO - Indonesian largest UFO Community, and** http://www.ufocasebook.com/2009/jakarta.html
3. **Pkcropcircles and UFOs, Karachi, Pakistan**
4. **"Ufo-Norway: MUFON USES SCANDINAVIAN CONTACTEE ON ITS ADVISORY BOARD OF CONSULTANTS: RAUNI-LEENA LUUKANEN."** http://www.kolumbus.fi/lauri.grohn/yk/skepsis/ufonorja.html. Retrieved 2010-15-05.
5. http://en.wikipedia.org/wiki/Marcel_Griaule. Retrieved 2010-15-05.
6. **"UFO EVIDENCE: Official French Gov't UFO study project to resume with new director Yves Sillard".** http://www.ufoevidence.org/documents/doc2008.htm. Retrieved 2008-01-01.
7. http://www.ufoevidence.org/researchers/detail93.htm. Retrieved 2010-15-05.
8. **"Ufology Library".** http://www.tarrdaniel.fw.hu/documents/ufology.html. Retrieved 2009-07-20.
9. **"RYUFOR Foundation, Hungarian Center for UFO Studies & Fund".** http://www.ryufor.hu/. Retrieved 2009-07-20.
10. **"Official site: Enrico Baccarini".** http://www.enricobaccarini.com/. Retrieved 2008-01-01.
11. **"Monsignor Corrado Balducci".** Paradigm Research Group. http://www.paradigmresearchgroup.org/Awards/Balducci_Corrado.htm. Retrieved 2008-01-01.
12. http://www.paolaharris.com/bio.htm. Retrieved 2010-15-05.
13. **"ufopsi.com Biography: Roberto Pinotti".** http://www.ufopsi.com/articles/robertopinotti.html. Retrieved 2008-01-01.

http://cezary_kwiatk.republika.pl/blania.htm. Retrieved 2010-15-05.

14. http://en.wikipedia.org/wiki/Doru_Davidovici. Retrieved 2010-15-05.

http://en.wikipedia.org/wiki/Juan_Jose_Benitez. Retrieved 2010-15-05.

15. **"Official Site: Erich von Däniken".** http://www.daniken.com/. Retrieved 2008-01-01.
16. **"Official website: Timothy Good- UFO Authority".** http://www.timothygood.co.uk/. Retrieved 2008-02-05.
17. **"Official site: Nick Pope".** http://www.nickpope.net/. Retrieved 2008-02-05.
18. **"UFO EVIDENCE: UFO Researchers & People- Jenny Randles.** http://www.ufoevidence.org/Researchers/Detail40.htm. Retrieved 2008-02-05.
19. **"(www.bibliotecapleyades.net) Brinsley Le Poer Trench: 'Legends and the case for Hollow Earth.'".** http://www.bibliotecapleyades.net/luna/esp_luna_19a.htm. Retrieved 2008-02-05.
20. **"Official site: Stanton T. Friedman".** http://www.v-j-enterprises.com/sfhome.html. Retrieved 2008-02-06.
21. http://orbwatch.com. Retrieved 2010-17-05.

22. "UFOBC.CA: Martin Jasek Bio". http://www.ufobc.ca/yukon/martin.htm. Retrieved 2008-01-01.
23. "UFO REVIEW #17: Interview with Kimball". http://alienworldsmag.com/uforeview_issues/UFO%20Review%20Issue%2017.pdf.
24. "Bio: Chris Rutkowski (Ufology Research of Manitoba)". http://www.ufon.org/uforom/mybio.html. Retrieved 2007-12-31.
25. "(www.alienresistance.org) David Sereda Ancient of Days 2004 Speaker Bio and Topic". http://www.alienresistance.org/aod04_david_sereda.htm. Retrieved 2008-02-10.
26. http://ufo-joe.tripod.com/gov/wbsebk.html Retrieved 2010-19- 05.
27. http://www.presidentialufo.com/grant-cameron-biography Retrieved 2010-19- 05.
28. http://www.shagharbourufo.com/ Retrieved 2010-27- 05
29. "(www.paradigmresearchgroup.org) Jaime Maussan: 2005 PRG Courage in Journalism". http://www.paradigmresearchgroup.org/Awards/Maussan_Jaime.htm. Retrieved 2008-02-05.
30. http://www.paradigmresearchgroup.org/stephenbassett.html Retrieved 2010-19- 05.
31. "Paradigm Research Group, Bassett bio". http://www.paradigmresearchgroup.org/stephenbassett.html.
32. Clark, Jerome (1998). *The UFOEncyclopedia: The Phenomenon from the Beginning, Volume 2, A-K Detroit: Omnigraphics*. Detroit, MI: Omnigraphics. ISBN 0-7808-0097-4.
33. "Own site (Robert Dean): Beyond Zebra". http://www.beyondzebra.com/bobdean.shtml. **Retrieved 2008-01-01.**
34. "Keyhole Publishing, Dolan bio". http://keyholepublishing.com/richard_dolan-bio.html.
35. http://www.nicap.org/ray.htm
36. http://www.cufos.org/books.html
37. Greer, Steven (1999). "About Steven M. Greer, M.D.". *Extraterrestrial Contact: The Evidence and Implications*. USA: Cataloging-in-Publication (Quality Books, Inc.). pp. 525–526. ISBN 978-0-96-732380-0. http://www.disclosureproject.org.
38. "Official site: Richard H. Hall". http://www.hallrichard.com/. Retrieved 2008-02-06.
39. "UFO EVIDENCE: UFO Researchers & People- James Harder, Ph.D.". http://www.ufoevidence.org/Researchers/Detail77.htm. Retrieved 2008-02-06.
40. "Intruders Foundation: Budd Hopkins UFO Abduction Research Foundation". http://www.intrudersfoundation.org/. Retrieved 2008-02-10.
41. "(www.nexusmagazine.com) Bio: Linda Moulton Howe http://www.nexusmagazine.com/Conf%202005/Linda%20Moulton%20Howe.html. Retrieved 2008-02-10.
42. "The J. Allen Hynek for UFO Studies". http://www.cufos.org/org.html. Retrieved 2008-02-10.
43. Ronald Story, ed., *The Encyclopedia of Extraterrestrial Encounters*, (New York: New American Library, 2001), s.v. "Morris K. Jessup," pp. 276.
44. Jerome Clark, *The UFO Book: Encyclopedia of the Extraterrestrial*. Visible Ink, 1998. ISBN 1-57859-029-9.
45. http://www.thephoenixlights.net/ Retrieved 2010-05-23

46. **"UFORC.com NEWS SERVICE: KLAS-TV Reporter - George Knapp to Give Keynote Banquet Speech at 4th UFO Crash Retrieval Conference"**. http://www.uforc.com/news/KLAS-TV_Knapp_UFOCRC_111006.htm. Retrieved 2008-01-01.

47. **"SCIFIPEDIA: APRO (the Aerial Phenomena Research Organization)"**. http://scifipedia.scifi.com/index.php?title=APRO&printable=yes. Retrieved 2008-02-10.

48. **"Dr. Bruce Maccabee Research Website"**. http://brumac.8k.com/. Retrieved 2008-08-24.

49. Berliner, Don. **"Congressional Hearings on UFOs"**. ufoevidence.org. http://www.ufoevidence.org/documents/doc1981.htm.

50. **"The John E Mack Institute (www.johnemackinstitute.org/)"**. http://www.johnemackinstitute.org/. Retrieved 2008-08-24

51. http://en.wikipedia.org/wiki/Bill_Moore_(ufologist) Retrieved 2010-05-28

52. *The Roswell Encyclopedia*, 2000, Collins Press, ISBN 0380798530

53. **"Nick Redfern Official site (www.nickredfern.com/)"**. http://www.nickredfern.com/. Retrieved 2008-08-24.

54. Diana Palmer Hoyt, "UFOCRITIQUE: UFOs, Social Intelligence and the Condon Committee"; Master's Thesis, Virginia Polytechnic Institute, 2000

55. **"MUFON 2007 Symposium, Brad Sparks bio"**. http://www.mufon.com/documents/2007MUFONSymposium-NewMJ12Revelations.pdf.

56. http://www.ufophotoarchives.org/ Retrieved 2010-05-23.

57. http://en.wikipedia.org/wiki/Leonard_H._Stringfield Retrieved 2010-05-26

58. **"The CUFON Interview of Michael D. Swords, Ph.D"**. http://www.cufon.org/cufon/cir/swords.htm. Retrieved 2008-08-24.

59. **Five Arguments Against the Extraterrestrial Origin of Unidentified Flying Objects** - Jacques Vallée, Ph.D.

60. **"Alfred Webre"**. Jerry Pippin. http://www.jerrypippin.com/UFO_Files_alfred_webre.htm.Retrieved 2008-01-01.

61. http://exopolitics.blogs.com/about.html Retrieved 2010-05-26

62. **"The UFO Casebook"**. http://ufocasebook.com/011209.html/. Retrieved 2009-01-14.

63. **"The Daily Grail"**. http://dailygrail.com/.

64. http://www.ufoevidence.org/researchers/detail59.htm Retrieved 2010-05-23.

65. http://translate.google.com/translate?hl=en&sl=pt&u=http://www.aforteanosla.com.ar/Colaboraciones/autores/brasil/ademar%2520gevaerd.htm&ei=zMT9S7i2EJKyNqH6sN4H&sa=X&oi=translate&ct=result&resnum=3&ved=0C Retrieved 2010-05-23.

66. Livingston, David. **"Dr. Carol Rosin**. The Space Show. http://www.thespaceshow.com/guest.asp?q=209. Retrieved 2008-01-01.

67. **"UFO INFO: "Argentina: Ufologist to Visit Policeman in Disappearance Case." (Fabio Zerpa)"**. http://www.ufoinfo.com/news/argentinapoliceman3.shtml. Retrieved 2008-02-05

68. http://en.wikipedia.org/wiki/Silvia_Simondini Retrieved 2010-05-26

CHAPTER 34

AN HISTORICAL LOOK AT UFO INVESTIGATIONS
BY GOVERNMENTS, MILITARIES AND
INTELLIGENCE COMMUNITIES

And the Intelligence *Community* When we think about who investigates the reports of strange UFO sightings, it comes as no surprise to many people that the government of their country would have a particular interest in the matter. The militaries acting on behalf of their nation's governments, who job it is to ensure the security their country and its people have a special interest in aerial phenomenon that intrude upon its country's airspace. In turn, **government intelligence agencies** like the **FBI, CIA, NSA** and many other alphabet soup intelligence agencies have also decided to tackle UFO phenomenon usually under the mantra of "National Security".

For several decades the United States government and especially the US Military and the US intelligence community took UFO reports very seriously. In fact, it was the Air Force that coined the term UFO in the first place, a term that acts as a metaphoric pall to the true nature and reality of the phenomenon. And although, the US government is tight-lipped as to the reality of the UFO phenomenon and has stated repeatedly when questioned that it no longer investigates cases of unidentified flying objects from the public, nevertheless, there are many governments around the world who still take UFO reports very seriously and continue to investigate them in an official capacity.

Ever since World War II many Allied pilots in the European campaign reported witnessing balls of light following their aircraft and eventually flying off at great speeds. They were dubbed **"Foo Fighters."** After the war, the militaries of the Allied Forces thinking these Foo Fighters may be a new secret weapon or aircraft of German design inquired of the Germans and Russians as to what these Foo Fighters were. In response, the Germans and Russians thought that the mysterious balls of light they had witnessed were assumed to be secret weapons of the United States. Thus, began over two decades of investigation as to what these unidentified flying objects were and whether or not they posed a threat. http://www.huffingtonpost.com/alejandro-rojas/ufo-cia-fbi-investigation_b_1856461.html

Reports of UFOs of varying shapes and sizes increased steadily into the 50s. After sightings over Washington, D.C., **President Truman** tasked the CIA to look into the matter. They convened the **Robertson Panel**, which determined that the phenomenon did not pose a direct threat; however, they did worry that it could pose a psychological threat that could be exploited by the Russians.

Eventually, due to the conclusions of an independent panel of investigators in 1969 with the University of Colorado, the **Condon Committee** commissioned by the Air Force, official UFO investigations in the United States ended. The panel concluded that there was no scientific benefit to the study of UFOs. However, the authors dedicate a chapter to UFO investigation that took place post-1969, demonstrating that there have been a number of important sightings that the Air Force could not ignore.

Although the United States ended its official investigations, other countries have continued to research the issue. News of the British Ministry of Defence's official UFO Desk closing has made headlines recently. However, the best example of a strong scientific effort has been France's **GEIPAN**, an organization that works with the **National Center of Space Sciences (CNES),** the French version of NASA. They post their findings on their website.
http://www.huffingtonpost.com/alejandro-rojas/ufo-cia-fbi-investigation_b_1856461.html

In this section, we will look in depth at these government investigation committees, panels and research departments such as **Project Blue Book** and the **Scientific Study of Unidentified Flying Objects** (aka. Condon Committee) sponsored by the US Air Force and many other projects and programs.

In contraposition to the government's investigations and research into the UFO phenomenon, many private UFO organizations sprang up because the public no longer trusted the USAF and the intelligence communities or accepted their explanations of public UFO sightings. Case in point has been in recent years the Wikileak's public releases of sensitive government documents that indicate that some politicians already know we are not alone in the universe. There are thousands upon thousands of UFO-related documents that are now being released into the public domain by many nations' governments.

Politicians, public officials and "heroes of the country" are coming forth and declaring the reality of the UFO phenomenon confirming what the general public has always suspected that we are not alone in the universe and that Et intelligences are coming to our planet routinely to monitor humanity.

"Decades ago, visitors from other planets warned us about the direction we were heading and offered to help. Instead, some of us interpreted their visits as a threat and decided to shoot first and ask questions after. Trillions of dollars have been spent on black projects which both congress and the commander in chief have deliberately been kept in the dark."– **Former Canadian Defense Minister - Paul Hellyer**

"In one of the cases during the cold war, 1961, there were about 50 UFOs in formation flying South from Russia across Europe. The supreme allied commander was very concerned and was about ready to press the panic button when they turned around and went back over the North Pole. They decided to do an investigation and they investigated for three years and they decided with absolute certainty that four different species, at least, have been visiting this planet for thousands of years. There's been a lot more activity in the past two decades, especially since we invented the atomic bomb." – **Paul Hellyer**

"Behind the scenes, high-ranking Air Force officers are soberly concerned about UFOs. But through official secrecy and ridicule, many citizens are led to believe the unknown flying objects are nonsense." - **Former head of CIA, Roscoe Hillenkoetter, 1960**

"There is a serious possibility that we are being visited and have been visited for many years by people from outer space, by other civilizations. Who they are, where they are from, and what they want should be the subject of rigorous scientific investigation and not be the subject of

'rubbishing' by tabloid newspapers." – **Lord Admiral Hill-Norton, Former Chief of Defence Staff, 5 Star Admiral of the Royal Navy, Chairman of the NATO Military Committee**

"This thing has gotten so highly-classified...it is just impossible to get anything on it. I have no idea who controls the flow of need-to-know because, frankly, I was told in such an emphatic way that it was none of my business that I've never tried to make it to be my business since. I have been interested in this subject for a long time and I do know that whatever the Air Force has on the subject is going to remain highly classified." – **Senator Barry Goldwater, Chairman of the Senate intelligence committee**

"There is abundant evidence that we are being contacted, that civilizations have been monitoring us for a very long time. That their appearance is bizarre from any type of traditional materialistic western point of view. That these visitors use the technologies of consciousness, they use toroids, they use co-rotating magnetic disks for their propulsion systems, that seems to be a common denominator of the UFO phenomenon"– **Dr. Brian O'Leary, Former NASA Astronaut, and Princeton Physics Professor**

Former Sen. Mike Gravel (D-Alaska), and 2008 presidential candidate said that there is *"an extraterrestrial influence that is investigating our planet. Something is monitoring the planet and they are monitoring it very cautiously."* http://www.collective-evolution.com/2013/06/29/wikileaks-cables-confirm-existence-of-extraterrestrial-life/

Did a Cloud Shape UFO Cause the Disappearance of the 5th Norfolk Regiment During WWI in Gallipoli, Turkey?

At the beginning of 1915, the French and British Governments decide to organize a common expedition against Turkey whose ports are open only to the German warships. The aim of this enterprise is to force through the Dardanelles Strait and take control of Constantinople (now Istanbul). The two Admiralties begin by sending a fleet which comes up against an altogether surprising Turkish defense. A French battleship, two English battleships and diverse cruisers and destroyers are sunk. It is then decided to undertake a landing on the Gallipoli peninsula.

In March, a French Expeditionary Corps embarks at Marseille alongside a British Army.

After many mishaps, these troops land on the Southern part of the peninsula, on 25 April. They would meet with violent resistance there. To the point that, three months later, despite furious combats led by General Gouraud, they had succeeded in penetrating only six kilometres towards the interior.

The Etats-Majors then decide to create a second Front by attacking the peninsula from the North-East. On 6 August, sixty thousand men land at Suvla. They too would come up against a solid Turkish Army.

After some terrible clashes at the foot of Mount Scimitar, the English head south to operate their junction with the Australians who have landed at Gafa Tepe. It is in the course of one of these marches that one of the most extraordinary events of the whole war takes place.

This occurs on 21 August, in the morning.

On this day, the 5th Norfolk Regiment, or rather what is left of it, that is to say, around four hundred men, receives the order to reinforce a Battalion of Australians and New Zealanders who are having trouble taking a certain Ridge 60, one of the key points in the region.

The 5th Norfolk Regiment (consisting of 266 British soldiers) therefore starts out. From the summit of a neighbouring hill, some New Zealand soldiers see it marching on a fairly steep slope, then entering a dip and climbing up a dried-up waterway.

The weather is splendid. However, the New Zealanders notice an anomaly in the scene. While the sky is clear, six or seven enormous clouds have been stationary since morning above Ridge 60. Clouds which a South wind of 6 or 7 kilometres an hour does not move from their position nor change their shape.

Further, another cloud comparable to a layer of very dense fog, which could be 250 metres long and 50 metres thick, seems to be clinging to the ground…
http://marilynkaydennis.wordpress.com/2012/04/25/a-regiment-disappears-inside-a-cloud/

The New Zealanders consider this phenomenon with surprise. One of them, a Sapper named Reichart, belonging to the 3rd Section of the 1st Company of Engineers, blurts out:

"They're strange, those clouds that aren't moving! I've been watching them since this morning, they look solid!"…

One of his mates says to him:

"Look at the one on the ground. It's reflecting the sunlight."

Meanwhile, the 5th Norfolk Regiment continues its climb amongst the stones of the dried-up waterway. The temperature is high in Turkey, in August, and the English soldiers are perspiring. After two hours of a difficult march, they finally arrive on a mound. There, they regroup and march in the direction of Ridge 60 which is partly covered by the strange layer of fog.

From the top of their hill, the New Zealanders observe the English. Sapper Reichart says to his companion:

"Look, the Pommies are getting to the cloud. We'll see if they're game enough to go in."

The other one says:

"Why wouldn't they be? It's not poisonous gas…"

Reichart replies:

"Maybe not; but I don't know why that fog doesn't look right!"

They soon see the 5th Norfolk Regiment reach the edge of the fog and plunge into it without hesitation. Reichart says:

"It's so thick that you can't see anyone in it."

In ranks of eight, the English Regiment is still penetrating the cloud.

When the last man has disappeared, the New Zealanders still watch the layer of fog. Sapper Reichart says:

"I wonder if they're all right."

The other smiles:

"It won't be long before we find out…"
http://marilynkaydennis.wordpress.com/2012/04/25/a-regiment-disappears-inside-a-cloud/

And they wait.

After five minutes, as no-one is reappearing, Reichart starts to worry:

"What can they be doing in there?"

Then he immediately cries out:

"Oh! Look!"

The strange cloud, inside which is the 5th Norfolk Regiment, has lifted from the ground and soon rises, not like ordinary layers of fog which disintegrate in the air but conserving its shape. Reichart hurls:

"But where are the Poms?"

On the ground, there is not one man, no weapon, nothing! The mound is absolutely empty.

These enormous lenticular clouds were photographed in Brazil. A few aviators imprudently penetrated them. Their aeroplanes disappeared.

The twenty-two men of the 1st New Zealand Company are rooted to the spot. While they are considering the place where four hundred English soldiers have just disappeared into thin air, the layer of fog continues to rise towards the clouds above it. When it reaches them, they all slowly move North and disappear into the sky.

No trace of the 5th Norfolk Regiment would ever be found again. Years pass by. And in 1918, after the capitulation of Turkey, England demands that the men of this Regiment, "Missing in Action", be returned to her.

The Turks search for them and reply that they have never heard of the 5th Norfolk Regiment. The English insist, furnish dates, precisions on the places, as well as the testimonies of the New Zealanders. The Turkish Etat-Major again hunts through its archives. Only to reply that no prisoners had been taken on 21 August 1915…

This story is authentic. It has been reported by numerous English magazines, by Returned Soldiers' newspapers which have published the New Zealanders' testimonies – notably that of Sapper Reichart – and it has been the subject of enquiries, searches, verifications, from both the British and Turkish authorities. No-one has ever been able to give an explanation…

This has become one of the true mysteries of WWI that still remains unexplained to this day. http://beforeitsnews.com/paranormal/2013/02/266-british-soldiers-disappear-in-1915-without-a-trace-2447930.html

Feu Fighters - ("Foo Fighters")

The term **"feu"** or **foo fighter** was used by Allied aircraft pilots in World War II to describe various UFOs or mysterious aerial phenomena often described as fireballs of light seen in the skies over both the European (Germany and France) and the Pacific (Indonesia and Japan) theaters of operations. The first sighting of objects similar to foo (feu) fighters was reported in 1941 in the Indian Ocean.

In **Andy Roberts'** article "Foo Fighters: The Story So Far" for the PROJECT 1947 he points out the problematic terminology of "Feu" or "foo fighter" initially used to describe a type of UFO often reported by air pilots from both the **Allied Forces** and **Axis Powers**. Its first usage may have originated from the U.S. 415th Night Fighter Squadron coining the term from the Smokey Stover comic strip. Smokey Stover (a fireman) always said in a somewhat nonsensical manner, *"Where there's foo, there's fire."* and the term was also commonly used to mean any UFO sighting from that period. Formally reported from November 1944 onwards, witnesses often assumed that the foo fighters were secret weapons employed by the enemy. http://midimagic.sgc-hosting.com/howifoof.htm and http://en.wikipedia.org/wiki/Foo_fighter

The term **Feu** comes from the French word for fire which more aptly describes the balls of fiery light seen by air pilots. The word "feu" is a typical bastardization of most foreign words that are often unpronounceable by most English speaking people from the U.K. and more so among Americans, into an anglicized pronunciation and thus, ended up being pronounced as "foo". It may also indicate that the objects were really feu-fighters, a type-1 fire balloons which many have been launched by the **French Underground** as a type of incendiary flak against German aircraft, particularly at nighttime. It may have also been serendipitous at the time that the US has a comic strip that used the term "Foo" with its reference to fire as the English equivalent to the French word for fire…"Feu". Bear in mind that it is more than likely that when bombing raids took placed over France that the word Feu was most probably already being used by the French first to describe some of the bombing devastation of their buildings long before it reached the ears of the English speaking part of the Allied Forces.

There were numerous sightings from airmen of the Allied Forces and the Axis Powers of "Feu" (Foo) Fighters over Europe and Japan during the Second World War
(Google Images)

The top brass (who knew what they were) probably used the name "feu-fighters," and the ranks (who didn't know what they were) heard "foo-fighters." One could also think of them as being *"faux-fighters"* (faux, pronounced "foe" means "fake"). But then they would have been heard as "foe-fighters." http://midimagic.sgc-hosting.com/howifoof.htm

The top brass as indicated above were **General Dwight D. Eisenhower** (head of US armed forces in Europe who later became the 34th US President) and **Winston Churchill** (head of the British armed forces) had already determined before the end of WWII that these strange aerial crafts were assessed as being extraterrestrial in origin. This fact became known through the release of secret UFO documents in from the British government files on this subject in the late '90s and the beginning of the 21st Century. http://www.dailymail.co.uk/sciencetech/article-1299994/Churchill-Eisenhower-agreed-cover-UFO-encounter-WWII.html

As **Andy Roberts** states, "This is pathetic really for an area of UFO activity which immediately proceeded the modern era and one which, if we are to believe the more *"enthusiastic"* ufologists, was the start of the so-called *"Government Cover-Up"*. The history of foo-fighters as represented within the subject of Ufology is riddled with problems which have put foo fighters in the historical niche they occupy today. These problems need stating and dealing with before the foo-fighter phenomenon can be seen in anything approaching a clear perspective."

Feu (foo)-fighters were described as brilliant fireballs of light that had the ability of:

1. Erratic flight
2. Fantastic speed (apparent - closer than they look)
3. Ability to hover (there were no helicopters yet)
4. Wobbling
5. Bright glow
6. No sound
7. Didn't appear on RADAR
8. Inability to be shot down
9. Don't drop bombs (maybe a spy device?)
10. Ability to fly in very close to aircraft (slip in between front and rear wings next to fuselage)
11. Sudden appearance and disappear
12. No crashed vehicles ever found

Now, after the war various explanations were offered as to the possibility that these objects were weather phenomena like **ball lightning** or **St. Elmo's Fire** the forerunner to the infamous **Chinese Lantern** explanation, that they were a type of incendiary paper lantern device. The most ridiculous explanation, of course, is that airmen were confused by or fired upon the planet Venus. Such an explanation makes absolutely no sense as anyone will tell you that the planet Venus or any other planet always maintains its distance from the observer which is tens of millions of miles away, something that no doubt most airmen would know when flying at night, it's basic astronomy 101, unless you lived most of your life in a cave and have never seen the stars or sky before!!!

The first sightings over Europe occurred in November 1944, when pilots flying over Germany by night reported seeing fast-moving round glowing objects following their aircraft. The objects were variously described as fiery, and glowing red, white, or orange. Some pilots described them as resembling Christmas tree lights and reported that they seemed to toy with the aircraft, making wild turns before simply vanishing. Pilots and aircrew reported that the objects flew formation with their aircraft and behaved as if under intelligent control, but never displayed hostile behavior. However, they could not be outmaneuvered or shot down. The phenomenon was so widespread that the lights earned a name – in the European Theater of Operations they were often called "kraut fireballs" but for the most part called "foo-fighters". The military took the sightings seriously, suspecting that the mysterious sightings might be secret German weapons, but further investigation revealed that German and Japanese pilots had reported similar sightings.

On 13 December 1944, the **Supreme Headquarters Allied Expeditionary Force** in Paris issued a press release, which was featured in the *New York Times* the next day, officially describing the phenomenon as a "new German weapon". Follow-up stories, using the term "Foo Fighters", appeared in the *New York Herald Tribune* and the British *Daily Telegraph*.
http://en.wikipedia.org/wiki/Foo_fighter

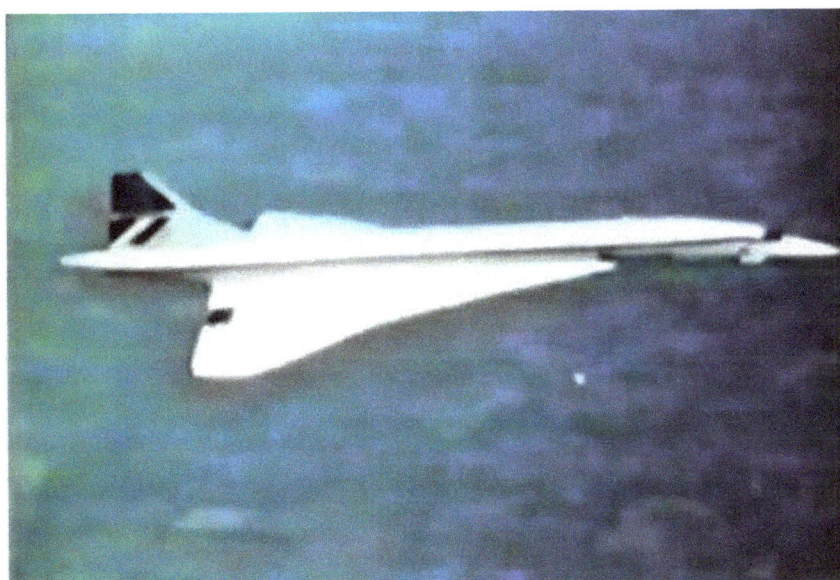

In the maiden flight of the British/French Concorde in 1976 a UFO similar to the Feu Fighters of WWII monitors and examines the aircraft before flying off into space
https://www.youtube.com/watch?v=SddSuyy-QUM

In its 15 January 1945 edition, *Time* magazine carried a story entitled "Foo-Fighter", in which it reported that the "balls of fire" had been following USAAF night fighters for over a month and that the pilots had named it the "foo-fighter". According to *Time*, descriptions of the phenomena varied, but the pilots agreed that the mysterious lights followed their aircraft closely at high speed. Some scientists at the time rationalized the sightings as an illusion probably caused by afterimages of dazzle caused by flak bursts, while others suggested **St. Elmo's Fire** as an explanation.

The "balls of fire" phenomenon reported from the **Pacific Theater of Operations** differed somewhat from the foo fighters reported from Europe; the "ball of fire" resembled a large burning sphere which "just hung in the sky", though it was reported to sometimes follow aircraft. On one occasion, the gunner of a B-29 aircraft managed to hit one with gunfire, causing it to break up into several large pieces which fell on buildings below and set them on fire. There was speculation that the phenomena could be related to the Japanese fire balloons' campaign. As with the European foo fighters, no aircraft was reported as having been attacked by a "ball of fire"

The postwar **Robertson Panel** cited foo fighter reports, noting that their behavior did not appear to be threatening, and mentioned possible explanations, for instance, that they were electrostatic phenomena similar to St. Elmo's fire, electromagnetic phenomena, or simply reflections of light from ice crystals. A type of electrical discharge from airplanes' wings similar to St. Elmo's Fire has been suggested as an explanation since it has been known to appear at the wingtips of aircraft. It has also been pointed out that some of the descriptions of foo fighters closely resemble those of ball lightning. The Panel's report suggested that "If the term "flying saucers" had been popular in 1943–1945, these objects would have been so labeled."
http://en.wikipedia.org/wiki/Foo_fighter

Foo fighters were reported on many occasions from around the world; a few examples are noted below:

- Sighting from September 1941 in the Indian Ocean was similar to some later foo fighter reports. From the deck of the S.S. *Pułaski* (a Polish merchant vessel transporting British troops), two sailors reported a "strange globe glowing with greenish light, about half the size of the full moon as it appears to us." They alerted a British officer, who watched the object's movements with them for over an hour.

- **Charles R. Bastien** of the Eighth Air Force reported one of the first encounters with foo fighters over the Belgium/Netherlands area; he described them as "two fog lights flying at high rates of speed that could change direction rapidly". During the debriefing, his intelligence officer told him that two RAF night fighters had reported the same thing, and it was later reported in British newspapers.

- Career U.S. Air Force pilot **Duane Adams** often related that he had witnessed two occurrences of a bright light which paced his aircraft for about half an hour and then rapidly ascended into the sky. Both incidents occurred at night, both over the South Pacific, and both were witnessed by the entire aircraft crew. The first sighting occurred shortly after the end of World War II while Adams piloted a B-25 bomber. The second sighting occurred in the early 1960s when Adams was piloting a KC-135 tanker.
http://en.wikipedia.org/wiki/Foo_fighter

Years after the war author **Renato Vesco** revived the wartime theory that the foo fighters were a Nazi secret weapon in his work *"Intercept UFO"*, reprinted in a revised English edition as *'Man-Made UFOs: 50 Years Of Suppression'* in 1994. Vesco claims that the foo fighters were, in fact, a form of ground-launched automatically guided jet-propelled flak mine called the *Feuerball* **(Fireball)**. The device, operated by special SS units, supposedly resembled a tortoise shell in shape and flew by means of gas jets that spun like a Catherine wheel around the fuselage. Miniature klystron tubes inside the device, in combination with the gas jets, created the foo fighters' characteristic glowing spheroid appearance. A crude form of collision avoidance radar ensured the craft would not crash into another airborne object, and an onboard sensor mechanism would even instruct the machine to depart swiftly if it was fired upon.

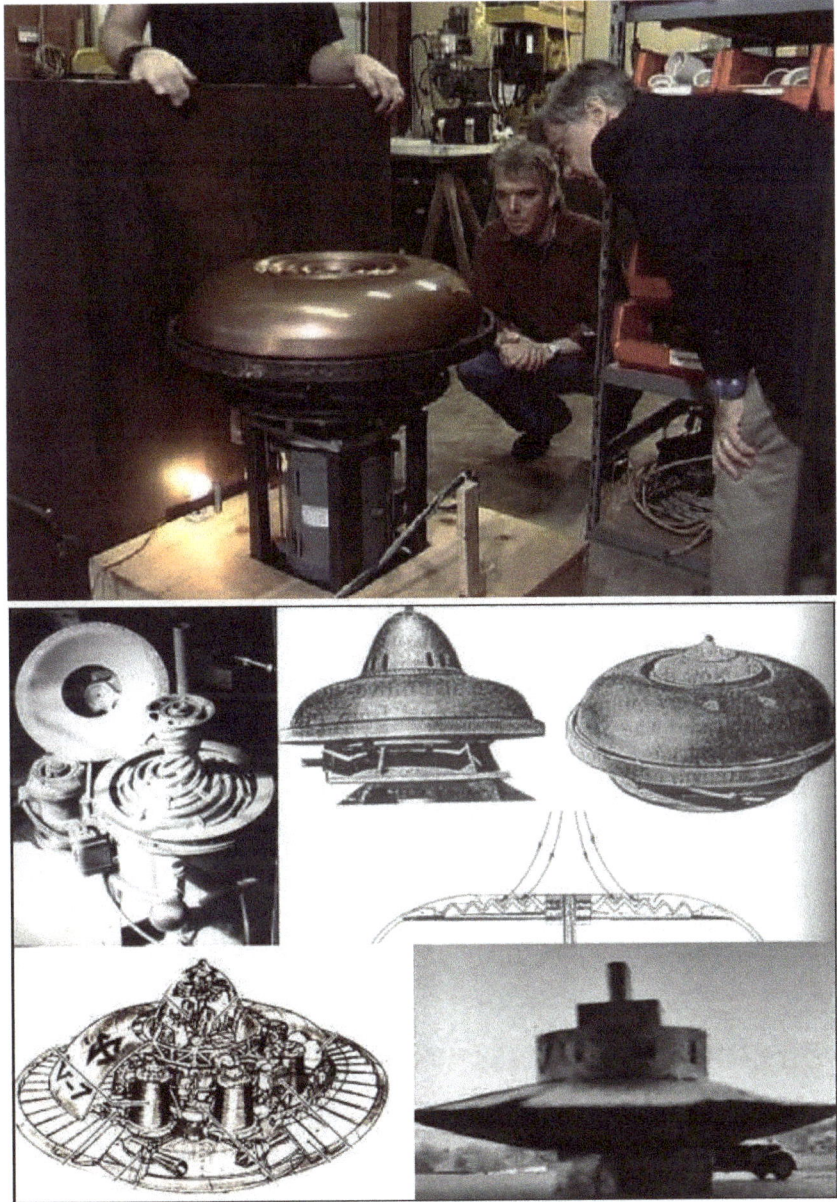

**The automatically guided jet-propelled flak mine *Feuerball*
(Fireball) operated by the German SS during WWII**
(Google Images)

The purpose of the *Feuerball*, according to Vesco, was two-fold. The appearance of this weird device inside a bomber stream would (and indeed did) have a distracting and disruptive effect on the bomber pilots, and Vesco alleges that the devices were also intended to have an *offensive* capability. Electrostatic discharges from the klystron tubes would, he states, interfere with the ignition systems of the bombers' engines, causing the planes to crash. Although there is no hard evidence to support the reality of the *Feuerball* drone, this theory has been taken up by other aviation/ufology authors and has even been cited as the most likely explanation for the phenomena in at least one recent television documentary on Nazi secret weapons. http://en.wikipedia.org/wiki/Foo_fighter

Unidentified Flying Objects (UFOs)

Let's begin with that most ambiguous term that seems to have caused so many people grief and puzzlement in trying to understand it… **Unidentified Flying Object** (**UFO** It became popularized in 1952 when the United States Air Force required an ambiguous term to describe any aerial phenomenon that could not be easily identified, even if it had a known conventional explanation or was of an unknown origin, as long as it remained unidentified to the casual observer or investigator from the public sector. This term was designed to conceal the true identity of an extraterrestrial spacecraft by not calling it by that name.

Today the term UFO is colloquially used to refer to any sighting of an unidentifiable flying object, regardless, of whether it had been investigated or not. In reality, the U.S. Air Force had already determined by the end of the Second World War that alien spacecraft were visiting our planet and thus, UFOs had been identified as Extraterrestrial spacecraft. According to the Air Force, the term UFO was allegedly for the benefit of the general public by way of ensuring that panic would not prevail, particularly so soon after a world war. The real reason was to allow the US Military the opportunity to investigate and potentially develop technologies as a spin-off from their research, while at the same time keeping the public confused on the matter with a generic term that would mislead them into false avenues of investigation. Unfortunately, the US Military underestimated the growing intelligence of the common people and their ability to deduce a rational conclusion from a set of unknown factors which inevitably came back to bite them in the form of persistent public inquiry.

It wasn't long before the public realized that strange craft traversing the countries skies were in fact not unidentified flying objects but were in reality Extraterrestrial spacecraft. It seemed that the "cat was out of the bag" and the problem was to how to get the cat back into the bag to the point that the public could be convinced that they were wrong in their assessment and that the answers lay with the higher educated, the professionals and the people in officialdom.

The problem of course to find the right people who were willing to do the US Air Force's bidding in an air that was above-board of any suspicion while acting in a quasi-scientific manner that would convince the public. And if necessary, ridicule those persisted in wanting to know the truth and threaten any who didn't back off from revealing what they had discovered about the subject matter.

"**Air Force Regulation 200-2** (text version pdf of the document), initially defined a UFO as "any airborne object that by performance, aerodynamic characteristics, or unusual features, does not conform to any presently known aircraft or missile type, or that cannot be positively identified as a familiar object." The Air Force added that *"Technical Analysis thus far has failed to provide a satisfactory explanation for a number of sightings reported."* A later version altered the definition to "Any aerial phenomena, airborne objects or objects that are unknown or appear out of the ordinary to the observer because of performance, aerodynamic characteristics, or unusual features," and added "Air Force activities must reduce the percentage of unidentifieds to a minimum. Analysis thus far has explained all but a few of the sightings reported. These unexplained sightings are carried statistically as unidentifieds."'

UFO reports increased precipitously after the first widely publicized U.S. sighting, reported by private pilot **Kenneth Arnold** on June 24, 1947, when he saw nine "flying saucers" moving at high speed near Mount Rainer, Washington, which gave rise to the popular terms **"flying saucer"** and **"flying disc."** His sighting gave rise to the modern UFO era as others soon began reporting sightings of similar UFOs, spawning a **"flap."**

Kenneth Arnold showing one of the objects he claims to have seen.
http://www.ufoevidence.org/cases/case511.htm

According to **Robert Carroll**, a professional skeptic states in his book, "**The Skeptic's Dictionary**": The phrase **"flying saucer,"** so familiar to Americans and UFO buffs, is the result of a reporter's error. After interviewing Arnold about his sighting, a reporter from the Eastern Oregonian newspaper reported that Arnold saw round, aerial objects (in fact he said they were "crescent shaped"). Arnold stated that the objects *"flew erratic, like a saucer if you skip it across the water"*—not that what he saw resembled an actual saucer. Yet that "saucer" interpretation stuck, prompting many eyewitnesses to repeat (and hoaxers to duplicate) Arnold's nonexistent description. This strongly shows the role of suggestion in UFO sightings; as skeptic, **Marty Kottmeyer** asks, *"Why would extraterrestrials redesign their craft to conform to [the reporter's] mistake?"* **http://www.skepdic.com/saucers.html**

Skeptics Carroll and Kottmeyer are correct up to a point in stating that **Kenneth Arnold** never coined the term flying saucer, however, nor did the reporter of Oregonian newspaper reporting on the Arnold sighting was the first person to coin the term flying saucer, he merely brought it back into the public awareness. The first usage of the term goes back to a newspaper report on a strange airship that was included in the **Denison Daily News** of Denison, Texas on January 25, 1878, when a Texas farmer, **John Martin** was credited with the first use of the term, **"flying saucer."**

Compare the above photo in Kenneth Arnold's hands to the actual drawing he made in his report to the Army Air Force Intelligence (see document below) which isn't the same depiction of the object he saw. In fact, his drawing looks very similar to the Canadian **Avro Y-2** aircraft and similar to the object photographed by William A. Rhodes that appeared in the July 9, 1947, edition of the Arizona Republic in Phoenix. (See photo images below).

As for Extraterrestrials redesigning their craft to conform to a reporter's mistaken perception, this statement distracts from the real issue that Arnold reported actual unidentified flying objects. The craft shape is really irrelevant as these ET spacecraft have many unusual design shapes, some are the typical saucer or disc shape, some are crescent shape, wedge shape, ball shape, pyramid shape, cone and diamond shape, etc. Perhaps, these debunkers need to spend more time doing their homework, instead of spending so much time spreading their false understanding of the subject.

Much later, the term ***UFO*** became synonymous for implying an alien spacecraft and generally most discussions of UFOs revolve around this presumption.

UFO enthusiasts and devotees have created organizations, religious cults have adopted extraterrestrial themes, and in general the UFO concept has evolved into a prominent mythos in modern culture. Some investigators now prefer to use the broader term **unidentified aerial phenomenon** (**UAP**), to avoid the confusion and speculative associations that have become attached to the term UFO. Another widely known acronym for UFO with its variant regional spellings can be found in Spanish, French, Portuguese and Italian as **OVNI (*Objeto Volador No Identificado*).**

The possibility that all UFO sightings are misidentifications of known natural phenomena inspired some debate in the scientific community about whether scientific investigation was warranted given the paucity of available empirical data. Very little peer-reviewed literature has been published in which scientists have proposed, studied or supported non-prosaic explanations for UFOs. Nevertheless, UFOs as a cultural phenomenon continues to be the subject of serious academic research and amateur investigators continue to advocate that UFOs represent real and unexplained events, usually associated with alien encounters. **Menzel, D. H.; Taves, E. H. (1977); The UFO enigma; Garden City (NY, USA); Doubleday** and **Sagan, Carl and Page, Thornton (1995); UFOs: A Scientific Debate; Barnes & Noble; pp. 310; ISBN 978076070916** also **McDonald, James. E.; (1968); Statement on Unidentified Flying Objects submitted to the House Committee on Science and Astronautics at July 29, 1968, Symposium on Unidentified Flying Objects; Rayburn Bldg., Washington, D.D.** and

COMETA Report http://www.ufoevidence.org/topics/Cometa.htm and **Politicking and Paradigm Shifting: James E. McDonald and the UFO Case Study** http://www.project1947.com/shg/mccarthy/shgintro.html

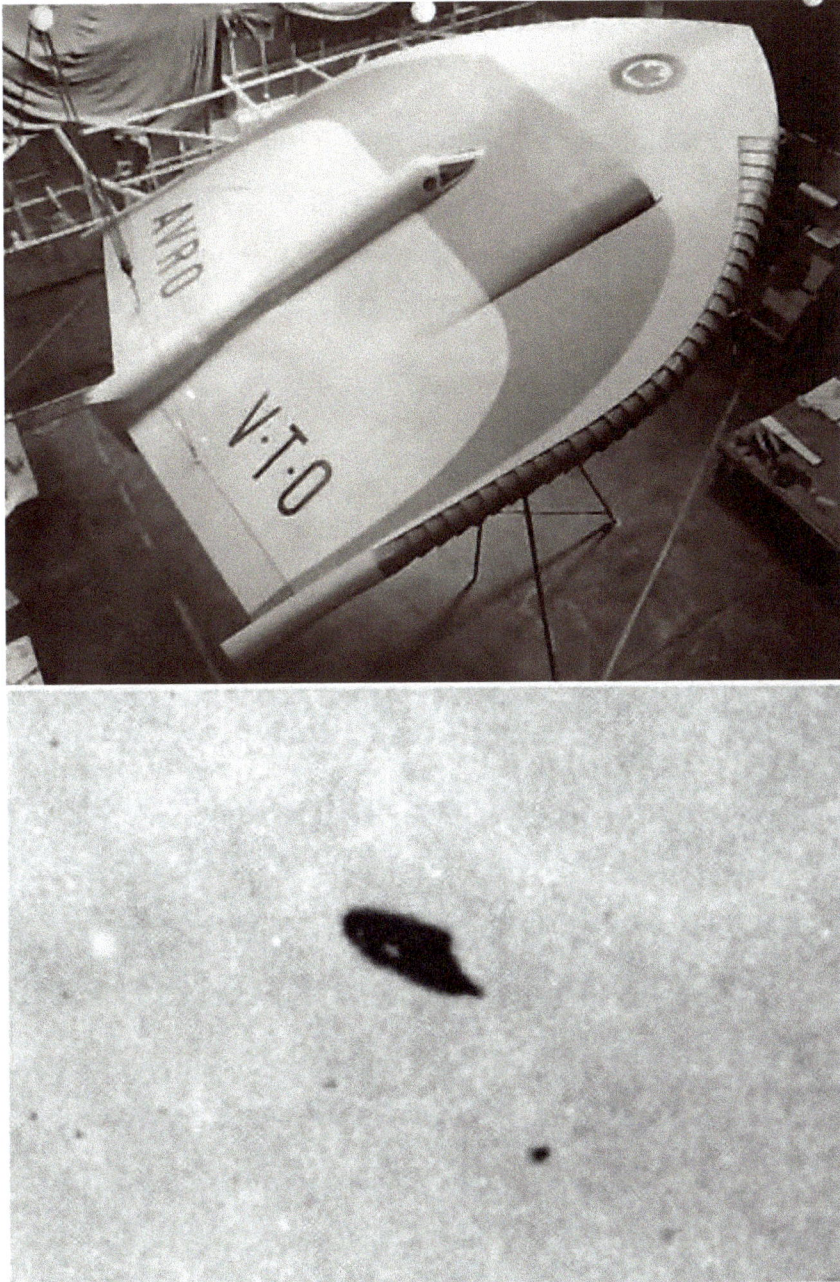

A comparison of the Avro Y-2 Canadian built prototype aircraft (top) and the alien spacecraft photographed by William A. Rhodes just before the Roswell, New Mexico saucer crash in 1947

http://discaircraft.greyfalcon.us/Rudolf%20Schriever.htm and http://www.ufocasebook.com/bestufopictures.html

Page 9

I have received lots of requests from people who told me to make a lot of wild guesses. I have based what I have written here in this article on positive facts and as far as guessing what it was I observed, it is just as much a mystery to me as it is to the rest of the world.

My pilot's license is 333487. I fly a Callair airplane; it is a three-place single engine land ship that is designed and manufactured at Afton, Wyoming as an extremely high performance, high altitude airplane that was made for mountain work. The national certificate of my plane is 33355.

Kenneth Arnold
Box 387
Boise, Idaho.

traveling this way →

Top

they seemed longer than wide their thickness was about 1/20 in of their width

mirror Bright

they did not appear to me to whirl or spin but seemed in fixed position traveling as I have made drawing.

Kenneth Arnold.

17

Kenneth Arnold's report to Army Air Forces (AAF) intelligence, dated July 12, 1947, includes annotated sketches of the typical craft in the chain of nine objects.
https://en.wikipedia.org/wiki/Kenneth_Arnold_UFO_sighting

1946: "Ghost Rockets" Over Scandinavia

Barely a year after the "foo fighter" episodes and the close of WWII, the second wave of UFO sightings began, this time in Scandinavia.

As reports started to come in from the general public from every corner of the globe, governments and their military began to investigate every reported case to seize any and all advantages that could be obtained from such knowledge before some other nation did.

On the night of June 9, 1946, a brilliant light streaked over Helsinki, Finland, with a smoke trail and the sound of thunder; its luminous trail persisted for ten minutes. Had this not been repeated the next night, it would have been written off as an unusually large meteor. The second one, according to news reports, turned and went back in the direction from which it had come.

On June 12, the Swedish Defense Staff asked military personnel to report their sightings through official channels, admitting that they had been aware of the phenomenon since May. On July 9 alone, more than 200 reports were received, many of them describing tubular or "spindle-shaped" objects, flying low and slowly with little or no sound.

A week after the establishment of a special **"ghost rocket"** committee by the Swedish Government, American Secretary of the Navy, **James Forrestal**, travelled to Stockholm to meet with the Swedish Secretary of War. According to a secret FBI memo of August 19, 1947, "the 'high brass' of the War Department exerted tremendous pressure on the (Army) Air Force's Intelligence to conduct research and collect information in an effort to identify the sightings."

On August 11, 1946, more than 300 reports of strange sightings were observed in just the Stockholm area. On August 20, General Jimmy Doolittle (in Stockholm on business for the Shell Oil Company) met with the head of the Swedish Air Force. This led to wide speculation in the Swedish press, as well as *The New York Times,* that "ghost rockets" were the subject of the meeting. In the 1980s, however, in an interview with UFO researchers, General Doolittle denied that his Swedish trip was officially connected with the "ghost rockets," although it is certainly likely that the subject came up in casual conversation.

Soon thereafter, Swedish newspapers began censoring most reports of "ghost rockets." However, reports appeared in other Scandinavian countries. According to a British Air Ministry Intelligence Report of September 1946:

"A large number of visual observations have been obtained from Scandinavia. Some of the best came from Norway. An analysis suggests the most notable characteristics of the projectiles to be: a) great speed; b) intense light frequently associated with a missile; c) lack of sound; d) approximate horizontal flight... Thus, if the phenomena now observed are of natural origin, they are unusual; sufficiently unusual to make possible the alternative explanation that at least some are missiles. If this is so, they must be of Russian origin."

There was a concerted effort on the part of the Swedish Government to blame many of the sightings on Soviet tests of captured German rockets. The Soviet Union had occupied

Peenemünde, the secret German test site across the Baltic Sea, where the V1 and V2 missiles were developed. Years later it was learned that the captured German equipment was immediately moved to Poland. There were no Soviet tests at **Peenemünde**, and thus the "official" explanation for the "ghost rockets" proved impossible.

This UFO was taken by Erik Reuterswärd on July 9th, 1946 near Guldsmedshyttan in mid-Sweden. Throughout the year, numerous "ghost rockets" were reported flying over Norway and Sweden and even crashing
https://commons.wikimedia.org/wiki/File:Ghostrocket_7-09-1946.jpg

As reports from Scandinavia began to taper off in September 1946, they were replaced by reports of similar sightings from Hungary, Greece, Morocco and Portugal. In 1984, when the Swedish Government finally opened its "ghost rocket" files, researchers found more than 1,500 reports had been secretly collected from 1946 on. One of the few official American reactions to the "ghost rockets" came in the January 9, 1947, issue of the Defense Department's *Intelligence Review* (classified "Secret" until 1978). This four-page summary of the "ghost rocket" events suggests that some of the sightings may have been of Soviet test missiles or jet airplanes (although no jets are known to have been in or near Scandinavia at the time).

A September, 1946 Top Secret Memorandum on Ghost Rockets
http://greyfalcon.us/restored/UFOs.htm

One sighting, detailed in the FBI report cited above, suggests there may have been more to it:

"On 14, August (1946) at 10 a.m. [a Swedish Air Force pilot]... was flying at 650 feet [200 m.]over central Sweden when he saw a dark, cigar-shaped object about 50 feet [15 m.] above and approximately 6,500 feet [2 km.] away from him travelling at an estimated 400 mph [650 km/hr].

The missile had no visible wings, rudder or other projecting part; and there was no indication of any fuel exhaust (flame or light), as had been reported in the majority of other sightings.

"The missile was maintaining a constant altitude over the ground and, consequently, was following the large features of the terrain. This statement casts doubt on the reliability of the entire report because a missile, without wings, is unable to maintain a constant altitude over hilly terrain."

Many years later, sophisticated cruise missiles, with tiny wings that would be invisible at such a distance, would be able to achieve "terrain-following" flight as a matter of routine. In1946, this was far beyond the capability of any existing technology.
http://www.bibliotecapleyades.net/ciencia/ufo_briefingdocument/1946.htm

Although the official opinion of the Swedish and U.S. military remains unclear, a Top Secret **USAFE (United States Air Force Europe)** document from 4 November 1948 indicates that at least some investigators believed the ghost rockets and later "flying saucers" had extraterrestrial origins. Declassified only in 1997, the document states:

"For some time we have been concerned by the recurring reports on flying saucers. They periodically continue to pop up; during the last week, one was observed hovering over Neubiberg Air Base for about thirty minutes. They have been reported by so many sources and from such a variety of places that we are convinced that they cannot be disregarded and must be explained on some basis which is perhaps slightly beyond the scope of our present intelligence thinking."

"When officers of this Directorate recently visited the Swedish Air Intelligence Service, this question was put to the Swedes. Their answer was that some reliable and fully technically qualified people have reached the conclusion that 'these phenomena are obviously the result of a high technical skill which cannot be credited to any presently known culture on earth.' They are therefore assuming that these objects originate from some previously unknown or unidentified technology, possibly outside the earth."

The document also mentioned a flying saucer crash search in a Swedish lake conducted by a Swedish naval salvage team, with the discovery of a previously unknown crater on the lake floor, believed caused by the object (possibly referencing the Lake Kölmjärv search for a ghost rocket discussed above, though the date is unclear). The document ends with the statement that "we are inclined not to discredit entirely this somewhat spectacular theory [extraterrestrial origins], meantime keeping an open mind on the subject." Jerome Clark, *The UFO Book: Encyclopedia of the Extraterrestrial,* 1998, Visible Ink   Press, ISBN 1-57859-029-9 and Timothy Good, *Above Top Secret,* 1988, William Morrow & Co., ISBN 0-688-09202-0 also, Timothy Good, *Need to Know: UFOs, the Military, and Intelligence,* 2007, Pegasus Books, ISBN 978-1-933648-38-5

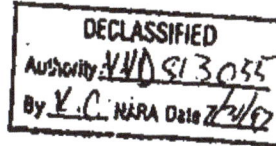

2-5317.

TOP SECRET

USAFE 14   TT 1524   TOP SECRET   4 Nov 1948

From OI OB

For some time we have been concerned by the recurring reports on flying saucers. They periodically continue to cop up; during the last week, one was observed hovering over Neubiberg Air Base for about thirty minutes. They have been reported by so many sources and from such a variety of places that we are convinced that they cannot be disregarded and must be explained on some basis which is perhaps slightly beyond the scope of our present intelligence thinking.

When officers of this Directorate recently visited the Swedish Air Intelligence Service. This question was put to the Swedes. Their answer was that some reliable and fully technically qualified people have reached the conclusion that "these phenomena are obviously the result of a high technical skill which cannot be credited to any presently known culture on earth." They are therefore assuming that these objects originate from some previously unknown or unidentified technology, possibly outside the earth.

One of these objects was observed by a Swedish technical expert near his home on the edge of a lake. The object crashed or landed in the lake and he carefully noted its azimuth from his point of observation. Swedish intelligence was sufficiently confident in his observation that a naval salvage team was sent to the lake. Operations were underway during the visit of USAFE officers. Divers had discovered a previously uncharted crater on the floor of the lake. No further information is available, but we have been promised knowledge of the results. In their opinion, the observation was reliable, and they believe that the depression on the floor of the lake, which did not appear on current Hydrographic charts, was in fact caused by a flying saucer.

Although accepting this theory of the origin of these objects poses a whole new group of questions and puts much of our thinking in a changed light, we are inclined not to discredit entirely this somewhat spectacular theory, meantime keeping an open mind on the subject. What are your reactions?

T O P   S E C R E T

(END OF USAFE ITEM 14)

November 1948 USAF Top Secret document citing extraterrestrial opinion
https://en.wikipedia.org/wiki/Extraterrestrial_hypothesis#/media/File:1948_Top_Secret_USAF_UFO_extraterrestrial_document.png

Green Fireballs

Green fireballs are another type of UFO which has been sighted in the sky since the late 1948 and 1949. Early sightings primarily occurred in the southwestern United States, particularly in New Mexico. They were once of notable concern to the US government because they were often clustered around sensitive research and military installations, such as Los Alamos and **Sandia National Laboratory**, then Sandia base. Furthermore, the strange green balls of light appeared suddenly and were reported many times per month near such New Mexico installations, but hardly anywhere else.

Green Fireball photographed over Texas, USA

Meteor expert **Dr. Lincoln LaPaz** headed much of the investigation into the fireballs on behalf of the military. LaPaz's conclusion was that the objects displayed too many anomalous characteristics to be a type of meteor and instead were artificial, perhaps secret Russian spy devices. The green fireballs were seen by so many people of high repute, including LaPaz and scientists at Los Alamos, that everybody agreed they were a real phenomenon. Secret conferences were convened at Los Alamos to study the phenomenon and in Washington by the Air Force Scientific Advisory Board.

In December 1949, **"Project Twinkle"**, a network of green fireball observation and photographic stations, was established but never fully implemented. It was discontinued two years later, with the official conclusion that the phenomenon was probably natural in origin. Green fireballs have been given natural, man-made, and Extraterrestrial origins and have become associated with both the Cold War and ufology. Because of the extensive government paper trail on the phenomenon, many ufologists consider the green fireballs to be among the best-documented examples of unidentified flying objects (UFOs).
http://en.wikipedia.org/wiki/Green_fireballs and
Jerome Clark, *The UFO Book: Encyclopedia of the Extraterrestrial*, Visible Ink Press, 1998.
Edward J. Ruppelt, *The Report on Unidentified Flying Objects*, 1956, Chap. 4
Brad Steiger, *Project Blue Book*, Ballantine Books, 1976 (Contains letter from Dr. J. Allen Hynek of Dr. LaPaz expressing final opinion on green fireballs)

Author's Rant: I must add a personal experience in this section, as it relates to the observation of *"green fireballs"*. I consider myself an experienced observer when it comes to identifying commercial and military aircraft, astronomical events and objects in space. Such observations throughout my life have enabled me to know the difference between meteorites showers and alleged meteorites that all seem to "fall" in the same geographical area.

I have seen the Lyrids (mid-April to late April), the Perseids (mid-July to late August), Leonids (mid-November), and the Geminids (mid-December) and I have also seen Halley's Comet with the naked eye, which was rather disappointing given all the hype about it, as it appeared more like a white blurry smudge in the night sky, I've seen Comet Bennett which was very close to the Earth and very spectacular in detail with just the unaided eye as well as through binoculars and telescope, and I have seen large and small meteors enter Earth's atmosphere with magnificent brilliance and with long fiery tails, and others that had long plumes of smoke stretching for miles. Most of these have been observed in different parts of Canada, but mostly in the province of British Columbia.

A meteorite is a natural object originating in outer space that often survives impact with the Earth's surface. Meteorites can be big or small. Most meteorites derive from small astronomical objects called meteoroids, but they are also sometimes produced by impacts of asteroids. When it enters the atmosphere, impact pressure causes the body to heat up and emit light, thus forming a fireball, also known as a meteor or shooting/falling star. The term bolide refers to either an extraterrestrial body that collides with the Earth or to an exceptionally bright, fireball-like meteor regardless of whether it ultimately impacts the surface.

When I was down in Crestone, Colorado in the summer of 1996, I witnessed what I thought was a rather unusual astronomical event that did not fall into the conventional observation of meteorites that enter the earth's atmosphere. Meteorites, for the most part, appear sporadically and travel randomly in the night sky on any giving day and especially so on

those predictable times of the years when they appear in great number. Some may appear to come in pairs and threes, but usually in singular fashion by nature and they tend to cover the horizon to horizon in their travel and trajectory.

The point I am making here is I know what meteorites look like and I have no difficulty in identifying them in the night sky.

What I saw in Crestone with other witnesses while "UFO vectoring" did not appear in this traditional and natural manner. These "so-called meteorites" came into the atmosphere from different directions (this part is typically conventional in observation) but their trajectory was at a single point toward the ground and they appeared "green" in coloured without smoky trails when observed through binoculars. The next day, again at night time, the same occurrence took place, but this time, a180 degrees in the opposite direction. On both occasions, the "green fireballs" appeared to "land" as if to converge on a single area. This would be an unusual observation for any trained astronomer or observer. My assessment given the nature of my activity at the time was that these "Green fireballs" *were artificial in nature and were being controlled in a plan re-entry trajectory*. [Bold Italics added for emphasis].

Are there actual green coloured meteorites? Of course, there are but, not all meteorites are green and no massive cluster of meteorites ever descends on one point on the ground over a repeated time span!

Here is a very brief sampling of some interesting accounts of unidentified flying objects seen and reported throughout history that sparked serious investigations by world governments:

Shen Kuo (1031–1095), a statesman of the Song dynasty Chinese government scholar-official and prolific polymath inventor and scholar, wrote a vivid passage in his *Dream Pool Essays* (1088) about an unidentified flying object. He recorded the testimony of eyewitnesses in 11th-century Anhui and Jiangsu (especially in the city of Yangzhou), who stated that a flying object with opening doors would shine a blinding light from its interior (from an object shaped like a pearl) that would cast shadows from trees for ten miles in radius, and was able to take off at tremendous speeds. http://en.wikipedia.org/wiki/Unidentified_flying_object and http://www.openminds.tv/remarkable-pearl-ufo-reported-by-famous-chinese-medieval-scientist-shen-kuo-1079/22837

Flying Saucers

In a newspaper report on a strange airship included in the **Denison Daily News** of Denison, Texas on January 25, 1878, a Texas farmer, **John Martin**, was credited with one of the first uses of the term **"flying saucer."** Martin had actually seen a "balloon-shaped" UFO but used the saucer term to describe the size of the object from his perspective. Martin's sighting was on January 2. What he saw was a dark object high in the sky. The object was moving closer to him all the while. Because the object maintained a dark color, there was speculation that the object was solid and backlit.

The headlines of the 25th would read, "A STRANGE PHENOMENON." Some of the reports is listed here:

"From Mr. John Martin, a farmer who lives some six miles south of this city, we learn the following strange story: Tuesday morning while out hunting, his attention was directed to a dark object high up in the southern sky. The peculiar shape and velocity with which the object seemed to approach riveted his attention and he strained his eyes to discover its character."

"When first noticed, it appeared to be about the size of an orange, which continued to grow in size. After gazing at it for some time, Mr. Martin became blind from long looking and left off viewing it for a time in order to rest his eyes. On resuming his view, the object was almost overhead and had increased considerably in size and appeared to be going through space at wonderful speed." http://www.americanchronicle.com/articles/view/17732

Other UFO sightings in the state of Texas since the Denison sighting by John Martin have been the UFO sightings of **1883 in Marfa, Texas, Aurora, Texas of 1897, Laredo, Texas of 1948, Lubbock, Texas of 1951, Levelland, Texas of 1957, Seguin, Texas of , 1975, and the UFO sighting of 1980 in Dayton, Texas**. There is obviously something about Texas that attracts the attention of Extraterrestrial visitors.

On February 28, 1904, there was a sighting by three crew members on the *USS Supply* 300 miles west of San Francisco, reported by Lt. Frank Schofield, later to become Commander-in-Chief of the Pacific Battle Fleet. Schofield wrote of three bright red egg-shaped and circular objects flying in echelon formation that approached beneath the cloud layer, then changed course and "soared" above the clouds, departing directly away from the earth after two to three minutes. The largest had an apparent size of about six suns.[17]

1916 and 1926: The three oldest known pilot UFO sightings, of 1305 catalogued by **National Aviation Reporting Center on Anomalous Phenomena (NARCAP)**. On January 31, 1916, a UK pilot near Rochford reported a row of lights, like lighted windows on a railway carriage that rose and disappeared. In January 1926, a pilot reported six "flying manhole covers" between Wichita, Kansas and Colorado Springs, Colorado. In late September 1926, an airmail pilot over Nevada was forced to land by a huge, wingless cylindrical object.

On August 5, 1926, while traveling in the Humboldt Mountains of Tibet's Kokonor region, **Nicholas Roerich** reported that members of his expedition saw "something big and shiny reflecting the sun, like a huge oval moving at great speed. Crossing our camp the thing changed in its direction from south to southwest. And we saw how it disappeared in the intense blue sky. We even had time to take our field glasses and saw quite distinctly an oval form with shiny surface, one side of which was brilliant from the sun." Another description by Roerich was, "...A shiny body flying from north to south. Field glasses are at hand. It is a huge body. One side glows in the sun. It is oval in shape. Then it somehow turns in another direction and disappears in the southwest."

In the Pacific and European theatres during World War II, "Foo-fighters" (metallic spheres, balls of light and other shapes that followed aircraft) were reported and on occasion photographed by

Allied and Axis pilots. Some proposed Allied explanations at the time included **"St. Elmo's Fire"**, the planet Venus, hallucinations from oxygen deprivation, or German secret weapon.

On February 25, 1942, U.S. Army observers reported unidentified aircraft both visually and on radar over the Los Angeles, California region. Antiaircraft artillery was fired at what was presumed to be Japanese planes. No readily apparent explanation was offered, though some officials dismissed the reports of aircraft as being triggered by anxieties over expected Japanese air attacks on California. However, **Army Chief of Staff Gen. George C. Marshall** and **Secretary of War Henry Stimson** insisted real aircraft were involved. The incident later became known as the **Battle of Los Angeles**, or the West coast air raid.

In 1946, (this has already been discussed above and merely repeated here for the purpose of a chronological timeline) there were over 2000 reports, collected primarily by the Swedish military, of unidentified aerial objects in the Scandinavian nations, along with isolated reports from France, Portugal, Italy and Greece, then referred to as "Russian hail", and later as "ghost rockets", because it was thought that these mysterious objects were possibly Russian tests of captured German V1 or V2 rockets. Although most were thought to be natural phenomena like meteors, over 200 were tracked on radar and deemed to be "real physical objects" by the Swedish military. In a 1948 top secret document, the Swedish military told the USAF Europe in 1948 that some of their investigators believed them to be extraterrestrial in origin.
http://en.wikipedia.org/wiki/Unidentified_flying_object

There are of course hundreds of thousands of UFO case reports worldwide in just this century alone, enough to ignite a fuse of investigation under most government and military officials.

Perhaps, nowhere else has any one country taken such measures to investigate, assess, formulated action response plans, controlled news media release on the enigma, involved public scientists to convey the appearance of a "sincere, honest and open" investigation into the UFO phenomenon for the sake of public inquiry, while secretly covering up the real evidence of their own top secret research, than has the United States of America.

Before we look at the history of government and military involvement in the U.S. we will look first at what other governments have done in their research efforts into the phenomenon.

Government and Military Investigations of UFOs in Australia

The study of unidentified flying objects in Australia appears to be one of public interest for the subject both within the country and from outside, chiefly from the U.S.A. and the Philippines. It is a history that begins around 1947 with news of the Kenneth Arnold sighting and the Roswell saucer crash. From that point, onward many UFO organizations spring up with personality issues, powers struggles, and renaming of branch groups from their original parent organization.

The first gathering of UFO enthusiasts occurred in Melbourne Victoria (Australia) in March 1949 at the *Aeronautical Research Laboratory* – Melbourne, Victoria (ARL)· The non-government meeting minutes show an attendance of 23 enthusiasts from various state and local groups such as the **British Interplanetary Society, Royal Aeronautical Society (RAS - Melb)**,

412

Commonwealth Aircraft Corp - Melbourne (CAC), RAAF and the Army Research Establishment. By May 1949 this early gathering became known as the *Aeronautical & Meteorological Phenomena Research* (AMPR) under Mr. Brian Boyle and Jack Seers. AMPR started to list and research "Flying Saucer" stories and produced a small limited quarterly publication called "Interplanetary Saucer".

Here we see the first involvement of the Royal Australian Air Force and Army establishments into a civilian organization. AMPR workload increased as the world ran into its next major sightings flap. In 1951 AMPR changed their structure and name to the **Aeronautics & Phenomena Research Victoria (APRV)**. In May 1952 Mr. R. M. Seymour, Superintendent of the *Federal Civil Aviation Department*, Air Traffic Control Branch Melbourne, reported that Australian Intelligence officers had refused his Department permission to investigate flying saucer reports on the grounds that UFO's were "security matters".

AMPR decided on 6 February 1953, to form an auxiliary group called the **Australian Flying Saucer Investigating Committee (AFSIC)**, in partnership with the *Astronomical Society of Victoria*.

On the 20 November 1953, a question was asked in the House of Representatives 'Question time' by Mr. Downer (MP), about numerous sighting of UFO's over Australia. The then Minister for Air, Hon. William McMahon (MP) (later Prime Minister) replied that the "*saucers*" were a problem more for the psychologist than the defense authorities.

Government Minister Richard Gardiner Casey (later Baron Casey and Governor-General of Australia - 1965 to 1969) wrote a 'Letter to the Editor' in an Australian newspaper dated 5 February 1954 stating his own interest in reported cases of UFOs and a possible correlation between certain meteorite showers.

Baron Casey was a member of the Victorian research group APRV. His time as Minister in charge of the CSIRO and as Minister for External Affairs enabled him to act as a conduit between governmental research, the public, and public enthusiast clubs. Casey's curiosity on the subject related to UFO remained with him until his death in June 1976.

Then on March 1954 Jarrold, who started his own one man UFO organization and considered a "loose cannon "in the Ufology field in Australia was contacted by South Australian, *Fred Stone*. Stone agreed to form a branch of Jarrold's group and so the AFSB (SA) was formed. But this relationship was fragile and as Jarrold's behavior changed due to stress, the relationship deteriorated. The RAAF also noted problems with Jarrold, whereupon Sq Leader A.H. Birch, AFC, Air Force Head-Quarters, Victoria Barracks, Victoria, formerly noted this in a letter dated concerning the possibility of his Society causing embarrassment to the Royal Australian Air Force. The RAAF then changed their policy and shifted their help to other organizations within Australia.

Queensland members call for an independent structure and formed the **Queensland Flying Saucer Research Bureau (QFSRB)** - now known as **UFO Research Queensland (UFORQld)**, and went their own separate way. These new groups are infiltrated and members come under the watchful eye of the **Australian Security Intelligence Organization (ASIO)**, as seen in this report dated 4 August 1959.

More in-house fights and squabbles have caused UFO groups within the country to divide and reorganize into new entities. Is it possible that the Australian military and intelligence communities are using the same tactics that their cousins the Americans employ in their own country? To infiltrate, cause dissention and dissolution of civilian UFO groups, so as to let the military get on with the business of studying the phenomenon without the public looking over its shoulder.

On 27 February 1965, in the country town of Ballarat, Victoria, Australia held its first UFO group congregation. The RAAF was represented by Mr. B.G. Roberts, a senior research scientist with the **Operational Research Office (ORO),** Department of Air, Canberra, and two RAAF officers to look after a display. Air Marshal Sir George Jones (who had an interest in UFOs) also attended. Keynote speakers, including the Rev. William Gill and Charles Brew, gave accounts on their very public UFO experience. At the conference, a new public group structure named *Commonwealth Aerial Phenomena Investigation Organisation* **(CAPIO)** was voted into existence, out of proceedings lawyer Peter E. Norris LL.B (Melb) became CAPIO's first President, under the patronage of Air Marshal Sir George Jones.

On the 12 October 1966, the Department of Air invited the CSIRO to comment on specific UFO cases which the RAAF would send to them. CSIRO agreed to the DOA's request.

On the 26 June 1968, in a letter from the Australian Governments Department of External Affairs to the Secretary Prime Minister's Department, the following policy observations are recorded; *"...the history of this subject reveals that the more time and effort that is spent by experienced scientists in investigating, the smaller becomes the residue of unexplained phenomena...in spite of these difficulties, the Australian Government continues to keep records of "sighting" and associated phenomena reported within Australian and associated territories."*

Victorian groups VUFORS and PRA invite Dr. J. Allen Hynek to visit Australia. Hynek accepts the invitation and in 1973 Hynek arrives in Australia, spending 4 days in Melbourne, followed by shortstops in Sydney and Brisbane, then finally to Papua New Guinea.

In 1974, Harry Griesberg and David Seargent established the *Australian Co-ordination Section* (ACOS) of the US based *Center for UFO Studies* (CUFOS). This followed the 1973 visit to Australia of Dr. J. Allen Hynek, who requested that Australian ufologists forward copies of interesting Australian sighting reports to CUFOS in the USA.

Although the UFO files were available to Defence personnel and Civilians Defence personnel for years, these files remained closed to outside researchers. Then in Oct 1979, extensively through the efforts of VUFORS researchers and Fl/Lt Brett Biddington, (Later Group Captain - RAAF) , the RAAF invited Victorian researchers to visit the Intelligence Cell at RAAF Headquarters Support Command, Victoria Barracks, Victoria, to view the majority files collection, and copy what they wished. This was a significant 'turn around' by the Defence Force, but it was the start of a period of remarkable co-operation between Government Departments and Civilian groups. It

414

was not until 1981 that another change in policy was instigated. A selection of general UFO files could be sent to Defence HQ, Russell Office, Canberra, when requested. The limited selection became open to other civilians for research after being vetted by a second controlling officer, but after their inspection, they were returned to the Victorian collection.

The structure of Australian Government is in three tiers. The lowest level is *Local Council*, above that, is the *State Government* and over these two levels sits the *Federal Government*. Throughout the modern era of Australian Ufology, departments of the Federal government have played the major role of official inquisitor. The State Government does play a minor role, however, archival evidence reveals that when a State body gets involved their primary object is to pass the case up to the Federal sphere.

RAAF dealt with more than 400 reports between 1950 to1959 and over 1,300 reports between 1960 to1980. The Department of Air produced Unusual Aerial Sightings (UAS) lists for the public.

Three such examples are Summary No1 January 60 - Dec 68, Summary No2 January 69 - Dec 69, Summary No3 January 71 - Dec 71.

These above samples lists are from the West Australian, National Archives of Australia (NAA) files. However, it is known that there are over 10,000 files in over 130 folders on the topic of UFO or Flying Saucers located within the NAA 60 million file collection. After 1980, Government UFO reports that were in external departments, were culled and then scattered throughout Australia to NAA storage areas, away from their original central point in Victoria.

From 1930 to 1959 the majority collection of Defence UFO files were held at HQ Southern Air, G Block, Albert Part Barracks, Melbourne with Army Intelligence. After 1959 the files were moved to RAAF Headquarters, Support Command, Victoria Barracks, Victoria Intelligence Cell, within the main building. These case files were controlled by the Australian Army through the Australian Army Intelligence Corps staff within the Australian Intelligence system, and they played a principle role in the field investigation of any UFO phenomenon in Australia.

However, the Army kept a low public profile but filtered reports for action. All Departments' Central Offices were located in Melbourne, Victoria, between 1930 and 1969, then due to Government restructure, a majority of departments moved to Canberra. The main collection of UFO files stayed in Victoria until 1989 later going to Canberra or the National Archives of Australia

January 4 1994, RAAF Wing Commander Brett Biddington, on behalf of the Chief of Air Staff, informed every civilian UFO groups around Australia that "*The number of reports made to the RAAF in the past decade had declined significantly, which may indicate that organizations such as yours are better known and are meeting the community's requirements.*" Therefore, the RAAF was not going to investigate or collect any more public reports from that date on.

Although the RAAF have officially removed themselves from the public report collection phase, there is evidence that internal investigations, by other Government departments, still continue under the new title of **Unusual Aerial Phenomena (UAP)** or **Unusual Aerial Sightings (UAS).** Called the *"1996 - Defense Instructions (General) ADMIN 55-1"*, this 1996 Department of Defence (DOD) policy document better known as 'ADMIN 55-1' concerns UAS Policy is still in used. http://en.wikipedia.org/wiki/Australian_ufology#Government

Government and Military Investigations of UFOs in Greece

The "ghost rocket" reports were not confined to Scandinavian countries. Similar objects were soon reported early the following month by British Army units in Greece, especially around Thessaloniki. In an interview on September 5, 1946, the Greek Prime Minister, Konstantinos Tsaldaris, likewise reported a number of projectiles had been seen over Macedonia and Salonika on September 1. In mid-September, they were also seen in Portugal, and then in Belgium and Italy.

The Greek government conducted their own investigation, with their leading scientist, physicist Dr. Paul Santorini, in charge. Santorini had been a developer of the proximity fuse on the first A-bomb and held patents on guidance systems for Nike missiles and radar systems. Santorini was supplied by the Greek Army with a team of engineers to investigate what again were believed to be Russian missiles flying over Greece.

In a 1967 lecture to the Greek Astronomical Society, broadcast on Athens Radio, he first publicly revealed what had been found in his 1947 investigation. "We soon established that they were not missiles. But, before we could do anymore, the Army, after conferring with foreign officials *[U.S. Defense Dept.],* ordered the investigation stopped. Foreign scientists *[from Washington]* flew to Greece for secret talks with me." Later Santorini told UFO researchers such as Raymond Fowler that secrecy was invoked because officials were afraid to admit of a superior technology against which we have "no possibility of defense."
http://en.wikipedia.org/wiki/Ghost_rockets#Greek_government_investigation and
Timothy Good, *Above Top Secret*, 1988, William Morrow & Co., ISBN 0-688-09202-0
Donald Keyhoe, *Aliens From Space*, 1973, Doubleday & Co., ISBN 0-385-06751-8

Government and Military Investigations of UFOs in Sweden

Although the official opinion of the Swedish and U.S. military remains unclear, a Top Secret **USAFE (United States Air Force Europe)** document from 4 November 1948 indicates that at least some investigators believed the ghost rockets and later "flying saucers" had extraterrestrial origins. Declassified only in 1997, the document states:

"For some time we have been concerned by the recurring reports on flying saucers. They periodically continue to pop up; during the last week, one was observed hovering over Neubiberg Air Base for about thirty minutes. They have been reported by so many sources and from such a variety of places that we are convinced that they cannot be disregarded and must be explained on some basis which is perhaps slightly beyond the scope of our present intelligence thinking.

416

"When officers of this Directorate recently visited the Swedish Air Intelligence Service, this question was put to the Swedes. Their answer was that some reliable and fully technically qualified people have reached the conclusion that 'these phenomena are obviously the result of a high technical skill which cannot be credited to any presently known culture on earth.' They are therefore assuming that these objects originate from some previously unknown or unidentified technology, possibly outside the earth."

The document also mentioned a flying saucer crash search in a Swedish lake conducted by a Swedish naval salvage team, with the discovery of a previously unknown crater on the lake floor, believed caused by the object (possibly referencing the Lake Kölmjärv search for a ghost rocket discussed above, though the date is unclear). The document ends with the statement that "we are inclined not to discredit entirely this somewhat spectacular theory [extraterrestrial origins], meantime keeping an open mind on the subject." Timothy Good, *Need to Know: UFOs, the Military, and Intelligence*, 2007, Pegasus Books, ISBN 978-1-933648-38-5

Government and Military Investigations of UFOs in United Kingdom

What follows is considered as the official history of Ufology in Britain as conveyed by **Georgina Bruni** and **Nick Pope** to **Graham Birdsall** in his magazine, "UFO Magazine".

The mysterious wave of airship sightings that took place over America in 1896 and 1897 was mirrored by a series of sightings that took place in Britain, starting in 1909. One of the first of these so-called 'scareship' sightings occurred in the early hours of 23 March 1909, when PC Kettle from Peterborough heard a strange buzzing sound from above. When he looked up, he saw a bright light attached to an immense, oblong-shaped craft, which moved at a fairly high speed across the sky. Numerous further sightings were reported.

Another report of a landed airship concerned an event that took place on 18 May 1909, on Caerphilly Mountain in South Wales. The witness reported having seen two strangely-dressed occupants who he heard talking to each other in a strange language that he was unable to identify. A subsequent examination of the alleged landing site revealed some damage to the ground.

The public perception was that these were sightings of German airships carrying out reconnaissance missions. But there is no indication that **Count Ferdinand von Zeppelin's** airship program was sufficiently advanced in 1909 to conduct such operations over the UK. In any case, German airships of the period could manage nothing remotely close to the sorts of speeds and manoeuvre that were being reported.

The sighting of airship type crafts marks the beginning of official interest in unexplained aerial phenomena. The 1909 wave was followed by further reports in 1912 of airships and with tension between Britain and Germany being so high, it was suggested that a Zeppelin was involved. On 27 November 1912 **William Johnson-Hicks MP** raised the matter in Parliament and quizzed the First Lord of the Admiralty, who was Winston Churchill at this time about the events. The latter confirmed that reports had been received, but said that subsequent investigation had not produced any explanation for what had been seen.

Sightings continued throughout 1913 and one consequence of this was the strengthening of the Aerial Navigation Act of 1911. A Bill was duly passed which set up prohibited areas. If these were violated or if an airship failed to respond to signals from the ground, it could then be shot down and to enable this to be carried out, the War Office stepped up efforts to produce a gun capable of bringing down an airship. The War Office continued to investigate the 1913 sightings but drew a blank. http://www.nickpope.net/official_history.htm

Most UFO researchers are familiar with the Foo Fighter mystery, which involved strange balls of light and small, metallic objects seen by both Allied and Axis pilots during the Second World War. File AIR 14/2800 at the Public Record Office contains one of the few surviving official British reports of these objects, detailing how aircrew from Bomber Command's 115 Squadron saw some of these strange objects on bombing raids in December 1943. Foo Fighter sightings, however, were viewed by the British Government as *"... the imaginations of men under strain interpreting fearfully observations which had a natural explanation."* so it seems, were dismissed out of hand by officialdom. Or were they?

R. V. Jones features prominently in this history of officialdom's involvement with the UFO issue. He was a protégé of Churchill's key scientific advisor Frederick Lindemann (later Lord Cherwell) and Sir Henry Tizard. He played a key role in anticipating and countering German technical advances in fields such as radar, radio-beam navigation, V-1 and V-2 weapons and the embryonic German nuclear program. He was appointed as Assistant Director of Intelligence (Science) in 1941 and promoted to Director of Intelligence in 1946. During his government service, Jones forged very close links with the Americans, especially the CIA, who in 1993 honoured him with a perpetual intelligence medal in his name. When he died in 1997 a press release was issued containing eulogies from **CIA Director George Tenet** and former **CIA Director James Woolsey.**

In the immediate aftermath of the Second World War, a new mystery was to emerge, which again involved R. V. Jones. This was the so-called **"Ghost Rocket"** wave of sightings that occurred in Scandinavia in 1946. Jones is as dismissive of these sightings as he had been about Foo Fighters, believing them to be either "imaginary" or meteors. But many personnel working in Air Technical Intelligence believed these were sightings of Russian "flying bombs" and investigated the matter thoroughly. Jones reveals how the Swedish authorities recovered what they believed were pieces that had fallen from a Ghost Rocket which turn out to be fragments of coke.

We should never underestimate the power of the media, or its capability to set the political agenda, even to the extent that it can drive government policy. This is as true today as it was in the post-war years. The year that Ufology first really hit the headlines in the UK was 1950. Prior to that coverage had been largely concerned US sightings and the reporting was often dismissive. But it was not just the media who were clamouring for answers and pressing the Government for action. Some very senior Establishment figures felt that something should be done and lobbied on the subject, sometimes openly and sometimes behind the scenes. Some of these figures were quite prepared to express openly the view that some UFO sightings might well be extraterrestrial in origin.

Another senior Establishment figure whose interest and belief in UFOs is widely known and documented is the wartime Commander-in-Chief of Fighter Command, **Air Chief Marshal Sir Hugh Dowding**. He was as outspoken as Mountbatten on the issue. Writing in the *Sunday Dispatch* on 11 July 1954 he said:

"I am convinced that these objects do exist and that they are not manufactured by any nation on Earth. I can, therefore, see no alternative to accepting the theory that they come from some extraterrestrial source."

Sir Peter Horsley, who died on 20 December 2001, was a former Air Marshal whose distinguished RAF career saw him retire as Deputy Commander-in-Chief at HQ Strike Command. A chapter of his 1997 autobiography *Sounds from another Room* relates to his interest in UFOs and the interest of friends and colleagues such as Air Chief Marshal Sir Arthur Barratt, General Sir Frederick Browning and General Martin.
http://www.nickpope.net/official_history.htm

While serving as a Royal equerry in 1952, Horsley began a study into the UFO phenomenon, with the full knowledge of the **Duke of Edinburgh**, who was briefed on Horsley's findings. Horsley has said that the Duke of Edinburgh was interested and open-minded on the subject, though keen that Horsley's inquiry should be low-key.

An Establishment figure whose interest in UFOs is less well known is **Sir Henry Tizard**. Tizard is best known for his pioneering work on the development of radar technology prior to the Second World War and his various wartime posts included Scientific Adviser to the Air Staff. He returned to the Ministry of Defence in 1948 as Chief Scientific Adviser, a post that he held until 1952. Tizard had followed the official debate about ghost rockets with interest and was intrigued by the increasing media coverage of UFO sightings in the UK, America and other parts of the world. Using his authority as Chief Scientific Adviser at the MOD he decided that the subject should not be dismissed without some proper, official investigation. Accordingly, he agreed that a small Directorate of Scientific Intelligence/Joint Technical Intelligence Committee (DSI/JTIC) working party should be set up to investigate the phenomenon.

The **Flying Saucer Working Party** was set up in October 1950 but operated under such secrecy that its existence was known to very few. Nevertheless, there were two clues that such a study had been carried out. One of these clues was obvious, but the other was more obscure.

The first clue was in the Secretary of State for Air's response to Prime Minister Winston Churchill's famous 28 July 1952 memo in which he enquired "What does all this stuff about flying saucers amount to? What can it mean? What is the truth? Let me have a report at your convenience". The response, dated 9 August 1952, began "The various reports about unidentified flying objects, described by the Press as "flying saucers", were the subject of a full Intelligence study in 1951".

The second clue was in a minute dated 29 May 1959, written by an official in S6 (a now defunct MOD division whose responsibilities for researching and investigating UFOs were latterly taken on by DS8, Sec(AS) and now DAS). This minute contained a sentence which read "The subject was reviewed by the J.I.C. some years ago and their views agree with a more extensive review carried out by the Americans". http://www.nickpope.net/official_history.htm

There was some considerable discussion and debate about the terms of reference of the Flying Saucer Working Party. The final version read as follows:

1. To review the available evidence in reports of "Flying Saucers".
2. To examine from now on the evidence on which reports of British origin of phenomena attributed to "Flying Saucers" are based.
3. To report to DSI/JTIC as necessary.
4. To keep in touch with American occurrences and evaluation of such.

The five-man working party was chaired by Mr. G. L. Turney from one of the MOD's scientific intelligence branches. All the members were specialists in the field of scientific and technical intelligence. One member, Wing Commander M. Formby, Assistant Director of Intelligence (Technical) at the Air Ministry, also chaired the Guided Missiles Working Party.

The working party's conclusions were set out in a document dated June 1951 and bearing the designation DSI/JTIC Report No. 7. It was entitled "Unidentified Flying Objects" and classified "Secret Discreet". http://www.nickpope.net/official_history.htm

The report begins with a history of the UFO phenomenon, covering the Scandinavian "Ghost Rocket" wave of 1946, Kenneth Arnold's sighting, the death of **Captain Thomas Mantell** and the work of Projects Sign and Grudge. Curiously, Foo Fighters were not mentioned at all. Through our study of various DSI/JTIC minutes it seems that this oversight occurred because while Fighter Command was invited to submit views to the Flying Saucer Working Party, Bomber Command was not.

Roswell is not mentioned, although there is a reference to a report of a "crashed flying saucer full of the remains of very small beings". But the Report states that the author of these claims had admitted that it had been a fabrication and it is clear that this is a reference not to Roswell but to Frank Scully's claims about the recovery of a UFO at Aztec, New Mexico, in 1948.

The report then details some British UFO sightings, concentrating on three cases involving military witnesses. But in each case, the sightings are dismissed as either optical illusions or misidentifications of ordinary aircraft or meteorological balloons. One visual sighting from a pilot had apparently been correlated by radar, but this was attributed to interference from another radar system.

The report concludes that all UFO sightings could be explained as misidentifications of ordinary objects or phenomena, optical illusions, psychological delusions or hoaxes. The main body of the report ends with the following statement:

"We accordingly recommend very strongly that no further investigation of reported mysterious aerial phenomena be undertaken, unless and until some material evidence becomes available".

In looking at the activities of the Flying Saucer Working Party, one cannot overstate the influence of the Americans. The phrase "following the lead given by the Americans on this subject" which we quote in the previous paragraph is extremely revealing and it is clear from the report itself that much of the material comes from liaison with those involved with Projects Sign and Grudge. There are other clues. As we have said, R. V. Jones forged extremely close links with the Americans on a range of intelligence issues and it is interesting to note that the fourth item of the Flying Saucer Working Party's terms of reference (requiring them to liaise with US authorities) was a late - though undoubtedly sensible - addition to the original remit.

Once the terms of reference included a requirement to get alongside the Americans on the UFO question, active liaison began. A member of the Flying Saucer Working Party duly travelled to America to meet with US authorities. It is also known that H. Marshall Chadwell was consulted and sat in on at least one of the Flying Saucer Working Party's meetings.
http://www.nickpope.net/official_history.htm

Chadwell was Assistant Director of the CIA's Office of Scientific Intelligence and in 1952 and

1953 was one of the key figures in the Scientific Panel on UFOs, better known as the **Robertson Panel**, after its chairman **H. P. Robertson**, an eminent physicist from the California Institute of Technology.

Robertson had been President Eisenhower's Scientific Adviser during the war, holding the rank of a four-star General. He had worked closely with R. V. Jones on various scientific intelligence matters and moved seamlessly between government service and academia. His post-war appointments included a post as theoretical physicist in Pasadena, associated with the Mount Wilson and Mount Palomar Observatories, and a spell as head of the Weapons Systems Evaluation Group at the Pentagon.

The Robertson Panel's skeptical report concluded that further study of the UFO phenomenon was not warranted, though as CIA Chief Historian Gerald Haines has confirmed, the CIA did not abandon their interest in the phenomenon.

It is also interesting to note what **Edward Ruppelt** (former head of the USAF's **Project Blue Book**) says about the British UFO research effort. Writing in his 1956 book *The Report on Unidentified Flying Objects* he makes a number of specific references to the UK.

In chapter 3 he states that the 1948 document *Estimate of the Situation* (prepared by staff on the USAF's Project Sign, initially classified Top Secret and concluding that some UFOs were extraterrestrial) mentioned that "**Ghost Aeroplanes**" had been detected on British radar early in 1947.

In chapter 10 there is a sentence that reads as follows:

"Two RAF intelligence officers who were in the US on a classified mission brought six single-spaced typed pages of questions they and their friends wanted answered".

Chapter 14 mentions the September 1952 UFO sightings during Operation Mainbrace (including the sightings at RAF Topcliffe). Ruppelt comments:

"It was these sightings, I was told by an RAF exchange intelligence officer in the Pentagon that caused the RAF to officially recognize the UFO".

In chapter 17 Ruppelt reveals that even after he had left Project Blue Book and the USAF, friends in RAF intelligence kept him informed about latest developments, on a private basis.

Another indication of the strong US influence on the Flying Saucer Working Party is the fact that their June 1951 final report was entitled *Unidentified Flying Objects*. This term had been devised by Ruppelt himself, early in 1951, but was not at the time in use outside US Government circles. http://www.nickpope.net/official_history.htm

To put the above remarks about US influence into context, it is worth noting the extent to which Britain was in thrall to America more generally by the early Fifties. This process had started

during the Second World War with the **Lend-Lease Bill**, the terms of which had contributed to the decline of British power and influence. By the end of the war, it was clear that in a very real sense the British Empire had been supplanted by an American one. In intelligence matters too, the historic position had been reversed and in post-war years Britain was very much the junior partner to the US.

The **Flying Saucer Working Party** had been dissolved in 1951 amidst a frenzy of skepticism that had clearly been fuelled by the Americans. The response that Churchill received to his 1952 inquiry showed that the skeptics still had the upper hand within the MOD. But this was soon to change. During the period 1952 to 1957, there were a series of UFO sightings involving the military, which forced the MOD to rethink and then reverse its policy. These included sightings during **Operation Mainbrace** in September 1952 (including those at RAF Topcliffe), the West Malling incident on 3 November 1953, **Flight Lieutenant Salandin's** near-collision with a UFO on 14 October 1954, the Lakenheath/Bentwaters radar/visual sightings on 13 and 14 August 1956 and the RAF West Freugh incident on 4 April 1957.

High-profile sightings such as these, together with the increasing number of reports from the general public, pushed the skeptics within MOD onto the defensive. The Flying Saucer Working Party's recommendation that UFO sightings should not be investigated was overturned and by the mid-Fifties two Air Ministry Divisions were actively involved in investigating UFO sightings. The divisions concerned were S6, a civilian secretariat division on the air staff, and DDI (Tech), a technical intelligence division. Their brief was to research and investigate the UFO phenomenon looking for evidence of any threat to the UK. http://www.nickpope.net/official_history.htm

The British Ministry of Defence (MoD) published in 2006 the "Scientific & Technical Memorandum 55/2/00a" of a four-volume, a 460-page report titled "Unidentified Aerial Phenomena in the UK Air Defence Region ", based on a study on behalf of the Defence Intelligence Staff, section DI55, codenamed **Project Condign**. The report is dated December

2000 and was previously classified as top secret. It discusses the British UFO reports (the MoD used the term Unidentified Aerial Phenomena, UAP) received between 1959 and 1997.

The report affirms that UFOs are a real phenomenon, but points out that they present no threat to national defense. The report further states that there is no evidence that UFO sightings are caused by incursions of intelligent origin, or that any UFOs are solid objects which might cause a collision hazard. Although the study admits of being unable to explain all analyzed UFO sightings with certainty, it recommends that section DI55 ceases monitoring UFO reports, as they do not provide information useful for Defence Intelligence. **Ministry of Defense (December 2000). "Unidentified Aerial Phenomena in the UK Air Defence Region: Executive Summary" (PDF). pp. 4, 10, 11. http://www.mod.uk/NR/rdonlyres/7D2B11E0-EA9F-45EA-8883-A3C00546E752/0/uap_exec_summary_dec00.pdf." Retrieved May 5, 2010.**

"That UAP exist is indisputable … [they] clearly can exhibit aerodynamic characteristics well beyond those of any known aircraft or missile – either manned or unmanned."

Government and Military Investigations of UFOs in Canada

The study of Ufology in Canada has evolved over the last sixty years in much the same way as in many other countries, namely in a disunified, disorganized and in an erratic fashion. Given the diversity of those attracted to the study of the phenomenon spanning a wide spectrum of backgrounds, interests, and motivations. It's little wonder that disagreement has existed among them, the general public and those in authorities, being comprised of those to whom the media seem to turn for comments whenever someone reports a UFO sighting.

At BUFORA's 2nd London **International UFO Congress** in London, England, May 25, 1981**David Haisell** stated in his lecture that these problems that unfortunately faced Canada were also, compounded by two additional factors, namely its immense size and sparsely distributed population. The area of coverage for each investigator/researcher adds up to many thousands of square miles. Also, basic extrapolation would imply that most UFO related events in Canada would go unnoticed and hence unreported because of the large areas of the nation with little or no population. To compound the problem even further there are only a few active UFO groups in Canada, covering the immensity of this great land and its population. http://www.ufoevidence.org/documents/doc682.htm

Author's Rant: However, there is a basic flaw with this type of argument in which a sparsely populated country would tend to miss most of the "UFO action". This assumption presumes that many UFOs are off enjoying the countryside or possibly sampling the flora and fauna of our native country. The fact is that situation is far different as the evidence supports that many sightings occur in around populated areas as people seem to be the main interest of ETI. Perhaps, this conclusion may have been true back in the early 80s, but, it is certainly not true in 2014.]

Nevertheless, Ufology in Canada has had its share of the world's attention focused on it during in the early years of its history. Serious investigation into the flying saucer phenomenon began about three years after the Kenneth Arnold sighting in the U.S. and the accompanying UFO flap of 1947. People like **Wilbert B. Smith** and his involvement in the government's **Project Magnet** and **Second Story** investigations and his familiarity with some of the board members of the secret U.S. based "**Majestic 12**" **(MJ-12)** has certainly placed Canada in a key position of UFO investigative research. http://www.ufoevidence.org/documents/doc682.htm

There were also the astounding accounts of the **Steven Michalak/Falcon Lake UFO Close Encounter,** the **Shag Harbour Incident,** and the **Carp, Ontario UFO Landing/Guardian event**. Then, there was the UFO over flights of the Parliament Buildings on **Parliament Hill** as well as over the Prime Minister's residence and the Governor General's house, all witnessed by many RCMP officers at the time of the event and duly noted in their logged reports.

More recently, Canada held its first public UFO/ETI disclosure event on September, 2001in Vancouver, B.C. in conjunction with **Dr. Steven Greer's Disclosure Project** lecture tour, sponsored and organized by the **Vancouver CSETI Working Group** (a branch of CSETI based in the U.S.) which was well received by the British Columbia major news media but, it appears not to have registered as a blip on the national news media's radar.

Former Liberal Party Defence Minister, **Paul Hellyer** recently came out in support of a government initiative to ban space-based weapons and particularly in demanding that world governments disclose the reality of the ET presence and their associated alien technology. Canadian politics, when it comes to parliamentarian discussions on the subject of UFOs if somewhat incompetent, ignorant, apathetic or possibly covert is anything, but boring.

We must go back to the beginning of Canada's initial involvement in UFO or flying saucer investigation, which came about three years after the **Kenneth Arnold** incident in the U.S., and the accompanying **"flaps"** of UFO sightings of 1947.

As is the case in most situations of this kind whenever something truly unique and world-shaking in its implications comes to the public's attention and then to the attention of politicians, it is often treated in an inauspicious manner, plagued by the usual amount of intrigue and double talk which seems to be characteristic of most government sponsored projects surrounding controversial material. The initiator of the UFO investigation was one **Wilbert B. Smith**, a senior radio engineer with the **Canadian Department of Transport (DOT).** http://www.ufoevidence.org/documents/doc682.htm

Smith's area of research was in radio wave propagation, a study which eventually led him into fields such as auroras, cosmic radiation, atmospheric radioactivity and geomagnetism. It was the latter of these fields which really attracted his attention and caused him to speculate that the potential energy of the Earth's magnetic field might be extracted and used*. (Was this a rediscovery of Nicola Tesla's work in the early 1900s of radiant energy or a new discovery of the Zero Point Energy field that is all around us?)* [My italics added].

He had already progressed to the stage of developing a crude experimental model to demonstrate his theory, and successfully tested the unit which, in his words, *"demonstrated the soundness of the basic principles in a qualitative manner and provided useful data for the design of a better unit."* http://www.ufoevidence.org/documents/doc682.htm

He believed, *"that we are on the track of something which may well prove to be the introduction to a new technology."* This **"is borne out by the investigations which are being carried on at the present time in relation to flying saucers."**

The investigations he was referring to were those being carried out by the US Government at that time. In late 1950 Smith had attended a North American Radio Broadcasting conference in Washington, DC, and while there, made enquiries through the Canadian Embassy staff who were able to inform him that:

a) the matter of UFOs was the most highly classified subject in the US, rating higher than the H-bomb,
b) UFOs exist,
c) their modus operandi is unknown, but concentrated effort is being made by a small group headed by Dr. Vannevar Bush, (Of **"MJ-12"** fame)
d) the entire matter is considered by US authorities to be of tremendous significance.
http://www.ufoevidence.org/documents/doc682.htm

Such explosive confirmations from the Canadian Embassy became the foundation upon which the secret **Majestic Twelve** documents leaked years later; eventually came into the hands of UFO researchers and sparked decades of UFO controversy among Ufologists as to government involvement and cover-up. People like **Stanton Freidman, Don Berliner, Dr. Bob** and **Ryan Wood** have been its chief supporters whereas **Kevin Randle** as been one of its vocal detractors.

A concerted effort is now being made to discover the identities of these Canadian Embassy officials in the hope of identifying their sources for these claims.

Smith felt the preliminary result of his work in geomagnetism correlated with the available data on reported UFO behavior, and that they were fairly close to some of the answers. And this was thirty-one years ago! The Defence Research Board liaison officer at the Canadian Embassy in Washington evidently agreed with Smith for he was extremely anxious for him to get in touch with **Dr. Solandt**, Chairman of the **Defence Research Board (DRB)** upon Smith's return to Ottawa, to discuss with him future investigations along the line of geomagnetism energy release.

Consequently, upon his return to Canada, Smith met with Solandt on November 20 and obtained his support. Solandt agreed that work on geo-magnetic energy should proceed as rapidly as possible and offered DRB's cooperation in providing laboratory facilities, acquisition of equipment, and specialized personnel.

On November 21 he outlined his proposal in writing to the Controller of Telecommunications, indicating DRB's support and requesting that a project be set up and carried out on a part-time basis "until such time as sufficient results would warrant more definitive action". Smith proposed a classified program that would explore seven avenues of investigation since he believed that the little-explored aspects of magnetism might hold the key to a new and significant technology, with profound impact on our civilization. http://www.ufoevidence.org/documents/doc682.htm

It is curious that the avenues of investigation **Smith** proposed made no reference to UFOs. Could it have been that Smith willfully omitted such reference in order to ensure a greater probability that the project would be approved? Or was he only interested in UFOs because they seemed to be demonstrating that some of his concepts were apparently being applied, whereas his main interest was indeed in the new technology which he felt he was on the verge of initiating?

There seems to be evidence for both points of view, but since I don't want to analyze Smith's career at this time I shall not pursue that issue any further. It is enough at this moment to recognize that Smith's curiosity was responsible for **Project Magnet's** initiation in November 1950 and for its relatively secret progress for a few years at least. It is significant, though, that the official 'Magnet' report, when eventually released many years later, dealt only with UFO sighting analysis, and made no mention of Smith's geo-magnetics research.

Curiously, the Canadian government in all its wisdom saw the need for still another project to analyze UFO reports, less classified than Project Magnet, but still confidential. During the early months of 1952, there was a noticeable increase in the number of UFO incidents covered by the Canadian Press. (8, 9, 10) Several of these involved reports of disc-shaped craft over **Royal Canadian Air Force** bases, many reported by service personnel themselves.

The **Defence Research Board (DRB)** noted this increase and DRB chairman **Solandt** asked staff member **Harold Oatway** to get a committee together "to see if we can make anything out of these flying saucer reports".

On April 22, 1952, potential committee members gathered by Harold Oatway held its first meeting, with **Peter Millman**, head of the Dominion Observatory, as its chairman. Smith, **Commander C.P. Edwards** and Solandt were also among those present. it was felt that a formal investigation of UFOs was needed and a new committee be established under the name of **Project Theta** to "standardize procedures, instructions for observers, examine interrogation and to establish a standard method of recording and indexing for subsequent analysis."

By May 19, 1952, a change in project name was deemed required to reflect the valid name protocols and thus, the new name of '**Project Second Storey**' replace Theta.

On June 25, 1952 Smith submitted an interim report on Project Magnet in which he stated that it appeared evident that flying saucers are emissaries from other civilizations and actually operate on magnetic principles, magnetic principles which we have failed to grasp due to our not paying enough attention to the structure of fields in our study of physics. http://www.ufoevidence.org/documents/doc682.htm

It seemed that Smith's conclusions were not shared by the project committee in any of their meetings, it was probably felt irrelevant since Chairman Millman noted *"that evidence to date did not seem to warrant an all-out investigation by the Canadian Services but it might be well to continue to collect at some central location all forms which may be submitted to the Services."*

426

And as it turned out, Millman's conclusion was based on activities in the U.S. in the wake of the Robertson Panel, which is now known to have been a CIA whitewash. So here is further evidence of top-level U.S. Canadian inter-relationship in the UFO field, and if we assume this inter-relationship continued after the **Robertson Panel**, it is safe to assume that investigation of UFOs in Canada was placed under the control of some branch of Canadian Intelligence. We have more evidence for this from none other than **Prime Minister Trudeau**. We can only guess that Smith's interim 'Magnet' report probably played a significant part in this assumed decision.

During the summer of 1953, **Wilbur Smith** obtained approval from the **Department of Transport (DOT)** to set up some UFO detection equipment at Shirley's Bay, near Ottawa, and by the end of October, the installation was complete. The instruments included a gamma-ray counter, a magnetometer, a radio receiver (to detect the presence of radio noise, and a recording gravimeter.

News of Magnet finally leaked to the media presumably because of the conspicuous nature of the Shirley's Bay installation. As expected, denials were attempted; on the very day the station went into operation **Dr. Solands** was quoted as saying reports of the station's establishment were completely untrue. However, he was forced to back down on this, and later claimed he actually had only said that such a station was not being operated by his department and that he personally had no knowledge of its existence. Even this was difficult to accept since the installation was located next to DOT's Ionosphere Station on **Defence Research Board (DRB)** property, and in fact, the building was loaned to Smith by DRB.
http://www.ufoevidence.org/documents/doc682.htm

Public awareness of this project was to be a source of frustration, annoyance, and embarrassment to DRB and DOT, and it put Smith in an awkward position since he was still officially a member of the **Second Storey Committee**. This was probably a contributing factor to the contents of **Millman's** November 21st summary report of Project Second Storey. He announced that Project Second Storey's forms and instructions for filing of sighting information were available for any government department seriously interested in pursuing the matter further, but the committee felt that, owing to the impossibility of checking independently the details of the majority of sightings, most of the material did not lend itself to a scientific method of investigation.

Could it be that they were not aware of the scientific study being conducted in the U.S. at that time by the Air Force on UFO reports collect from June 1, 1947, to December 31, 1952? The study was initiated in 1952 and continued through 1954, and proved beyond a doubt the existence of an unidentified phenomenon, even though the conclusions were worded in such a way as to divert attention from the evidence. The study being referred to was known as **'Project Blue Book Special Report #14'** which is probably the most constantly referred to in the literature of **Stanton Friedman**, and more recently was the subject of a paper by **Bruce Maccabee** in the Journal of UFO Studies, Vol. 1, No. 1, entitled the *'Scientific Investigation of Unidentified Flying Objects'*.

It is difficult to believe Millman's statement and feel more inclined to think his remarks were designed to appease 'somebody' in the event Smith's association with Second Storey eventually became public knowledge, and also to save face in view of the Adamski & contactee activity now in the public eye. http://www.ufoevidence.org/documents/doc682.htm

Just after 3 PM in the afternoon of August 8, 1954, the instrumentation at the Shirley's Bay installation registered an unusual disturbance. In Smith's words *"the gravimeter went wild"*, as a much greater deflection was registered than could be explained by conventional interference such as passing aircraft. **Smith** and his colleagues rushed outside only to find a heavy overcast. Whatever was up there was hidden in the clouds. The only evidence they had was the deflection registered on the chart recorder paper.
Two days later the controller of Telecommunications issued a form letter, which was also authorised as a press release, admitting that the DOT had been engaged in the study of UFOs for three and a half years, that considerable data had been collected and analysed, but it had not been possible to reach any definite conclusion, and since new data simply confirmed existing data, there seemed little point in carrying the project any further on an official level.

Project Magnet was to be dropped, but Smith would continue to receive and catalogue data on an unofficial basis. In Smith's words, Magnet went *"underground"* probably joining Second Storey. By the way, isn't the fact that *"new data... confirmed existing data."* supposed to be what you would expect of a positive scientific experiment?

The detection of "whatever it was" two days before had evidently inspired rapid action. Does it seem likely that a project, which had finally apparently detected what it was looking for, would be terminated?

That doesn't make sense; instead, it seems pretty good justification for changing to a 'Top Secret' classification. It is apparent that pressure was applied to Smith to downplay or even deny the Shirley's Bay incident. http://www.ufoevidence.org/documents/doc682.htm

Researcher **Greg Kanon** writes: *"In an abrupt about-face, Smith announced, before the May 17th, 1955 session of the Commons' Special Committee on Broadcasting, that no UFOs had ever been detected at the Shirley's Bay Station. At about this same time, **Captain Edward J. Ruppelt** (who earlier served as chief UFO investigator for the U.S. Air Force) was reportedly told by RCAF Intelligence officers that only **'officially'** had the Shirley's Bay Station produced negative results. These developments led some UFO researchers to conclude that Smith had been successfully silenced by officialdom."*

Whatever the case, Smith kept busy over the next few years, and we get a glimmer of what he was up to from a presentation he gave about five years later to the Illuminating Engineering Society's Canadian Regional Conference during which he stated:

"We know that gravity is not all Newton visualised. Far from being a basic force in nature, it is really a derived function and is the consequence of a dynamic condition, not a static one. We know what goes into its makeup; we know its formula and we have a pretty good idea of how to go about bringing it under control. We have conducted experiments that show that it is possible to create artificial gravity (not Centrifugal force) and to alter the gravitational field of the Earth. This we have done. It is Fact. The next step is to learn the rules and do the engineering

necessary to convert the principle into workable hardware." That statement was made 22 years ago. The question is, what has been achieved since then?

Smith died of cancer on December 27, 1962. Smith was one of the foremost thinkers of his time - a well-respected ufologist - one of the first of our breed.
http://www.ufoevidence.org/documents/doc682.htm and
https://www.youtube.com/watch?v=1sjpKvxu0Bk

Government and Military Investigations of UFOs in Uruguay

The Uruguayan Air Force has had an ongoing UFO investigations since 1989 and analyzed 2100 cases, of which they consider only 40 (about 2%) definitely lacking any conventional explanation. All files have recently been declassified. The unexplained cases include military jet interceptions, abductions, cattle mutilations, and physical landing trace evidence. **Colonel Ariel Sanchez**, who currently heads the investigation, summarized their findings as follows: "The commission managed to determine modifications to the chemical composition of the soil where landings are reported. The phenomenon exists. It could be a phenomenon that occurs in the lower sectors of the atmosphere, the landing of aircraft from a foreign air force, up to the extraterrestrial hypothesis. It could be a monitoring probe from outer space, much in the same way that we send probes to explore distant worlds. The UFO phenomenon exists in the country. I must stress that the Air Force does not dismiss an extraterrestrial hypothesis based on our scientific analysis". http://www.inexplicata.blogspot.com/

USAF Current Official Statement on UFOs

Former Senator **Barry Goldwater** was one of the more prominent American politicians to openly show an interest in UFOs.

On March 28, 1975, Goldwater wrote to Shlomo Arnon: "The subject of UFOs has interested me for some long time. About ten or twelve years ago I made an effort to find out what was in the building at Wright-Patterson Air Force Base where the information has been stored that has been collected by the Air Force, and I was understandably denied this request. It is still classified above Top Secret." Goldwater further wrote that there were rumors the evidence would be released and that he was "just as anxious to see this material as you are, and I hope we will not have to wait much longer."

The April 25, 1988 issue of *The New Yorker* carried an interview where Goldwater said he repeatedly asked his friend, **Gen. Curtis LeMay**, if there was any truth to the rumors that UFO evidence was stored in a secret room at Wright-Patterson Air Force Base, and if he (Goldwater) might have access to the room. According to Goldwater, an angry LeMay gave him "holy hell" and said, "Not only can't you get into it but don't you ever mention it to me again."

In a 1988 interview on **Larry King's** radio show, Goldwater was asked if he thought the U.S. Government was withholding UFO evidence; he replied: *"Yes, I do."* He added:

"I certainly believe in aliens in space. They may not look like us, but I have very strong feelings that they have advanced beyond our mental capabilities...I think some highly secret government UFO investigations are going on that we don't know about – and probably never will unless the Air Force discloses them." http://en.wikipedia.org/wiki/Barry_Goldwater#UFOs

Below is the United States Air Force's official statement regarding UFOs, as noted in USAF Fact Sheet 95-03:

"From 1947 to 1969, the Air Force investigated Unidentified Flying Objects under Project Blue Book. The project, headquartered at Wright-Patterson Air Force Base, Ohio, was terminated December 17, 1969. Of a total of 12,618 sightings reported to Project Blue Book, 701 remained "unidentified."

The decision to discontinue UFO investigations was based on a number of factors: an evaluation of a report prepared by the University of Colorado entitled, "Scientific Study of Unidentified Flying Objects;" a review of the University of Colorado's report by the National Academy of Sciences; previous UFO studies and the Air Force's own experience of investigating UFO reports during 1940 to 1969.

As a result of these investigations, studies, and experience gained from investigating UFO reports since 1948, the conclusions of Project Blue Book were:

1. **No UFO reported, investigated, and evaluated by the Air Force has ever given any indication of threat to our national security.**
2. **There has been no evidence submitted to or discovered by the Air Force that sightings categorized as "unidentified" represent technological developments or principles beyond the range of present-day scientific knowledge.**
3. **There has been no evidence indicating the sightings categorized as "unidentified" are extraterrestrial vehicles.**

With the termination of Project Blue Book, the Air Force regulation establishing and controlling the program for investigating and analyzing UFOs was rescinded. Documentation regarding the former Blue Book investigation was permanently transferred to the Modern Military Branch, National Archives, and Records Service, and is available for public review and analysis.

Since the termination of Project Blue Book, nothing has occurred that would support a resumption of UFO investigations by the Air Force.

There are a number of universities and professional scientific organizations that have considered UFO phenomena during periodic meetings and seminars. A list of private organizations interested in aerial phenomena may be found in "Encyclopaedia of Associations", published by Gale Research. Interest in and timely review of UFO reports by private groups ensures that sound evidence is not overlooked by the scientific community. Persons wishing to report UFO sightings should be advised to contact local law enforcement agencies.
http://en.wikipedia.org/wiki/Project_Blue_Book#USAF_current_official_statement_on_UFOs

Identifying **Unidentified Flying Objects** is a difficult task due to the normally poor quality of the evidence provided by those who report sighting the objects. Nevertheless, most officially investigated UFO sightings, such as from the U.S. Air Force's **Project Blue Book**, have been identified as being due to honest misidentifications of natural phenomena, aircraft, or other prosaic explanations. In early U.S. Air Force attempts to explain UFO sightings, unexplained sightings routinely numbered over one in five reports. However, in early 1953, right after the CIA's **Robertson Panel**, percentages of unexplained sightings dropped precipitously, usually being only a few percent in any given year. When Project Blue Book closed down in 1970, only 6% of all cases were classified as being truly unidentified.

UFOs that can be explained are sometimes termed **"IFOs" (Identified Flying Objects)**.

CHAPTER 35

GOVERNMENT UFO STUDIES IN THE U.S.A.

The following are some major studies undertaken during the past 50 years that reported on identification of UFOs:

- **Project Blue Book Special Report No. 14** (referred to further below as **BBSR**) was a massive statistical study the Battelle Memorial Institute did for the USAF of 3,200 UFO cases between 1952 and 1954. Of these, 22% remained were classified as unidentified ("true UFOs"). Another 69% were deemed identified (IFOs). There was insufficient information to make a determination in the remaining 9%.

- The official French government UFO investigation **(GEPAN/SEPRA)**, run within the **French space agency CNES** between 1977 and 2004, scientifically investigated about 6000 cases and found that 13% defied any rational explanation (UFOs), while 46% were deemed readily identifiable and 41%, lacked sufficient information for classification.

- The USAF-sponsored **Condon Committee** study reported that all 117 cases studies were or could probably be explained. A 1971 review of the results by the American Institute of Aeronautics and Astronautics concluded that 30% of the 117 cases remained unexplained.

- Of about 5,000 cases submitted to and studied by the civilian UFO organization **NICAP**, 16% were judged unknowns.

In contrast, much more conservative numbers for the percentage of UFOs were arrived at individually by astronomer Allan Hendry, who was the chief investigator for the **Center for UFO Studies (CUFOS)**. CUFOS was founded by astronomer Dr. Allen Hynek (who had been a consultant for the Air Force's Project Blue Book) to provide a serious scientific investigation into UFOs. Hendry spent 15 months personally investigating 1,307 UFO reports. In 1979, Hendry published his conclusions in *The UFO Handbook: A Guide to Investigating, Evaluating, and Reporting UFO Sightings*. Hendry admitted that he would like to find evidence for extraterrestrials but noted that the vast majority of cases had prosaic explanations. He found 89% of reports definitely or probably identifiable and only 9% unidentified. "Hardcore" cases—well-documented events which defied any conceivable conventional explanation—made up only 1.5% of the reports. http://en.wikipedia.org/wiki/Identification_studies_of_UFOs

The first official response by the US Military to the UFO mystery occurred during the Second World War with "Foo Fighters" seen over war-torn Europe followed by the alleged "Attack over LA" by what was thought to be a Japanese precursor invasion of America. Then, the statement made by **General Douglas MacArthur** at the end of the War that "the next great war may come from outer space from extraterrestrials," and with the enigmatic **Roswell Saucer Crash** in July 1947 in New Mexico, it's little wonder that when UFOs flew over Washington, D.C. on July 19–20 and again on July 26–27 in 1952, the U.S. Military's UFO balloon had finally burst. With the constant stream of reports coming into Project Blue Book and clogging the channels of communication with irrelevant reports finally made the Military and in particular the U.S. Air force to sit up and take notice of the increasing attention and interest being paid by the public to the UFO phenomenon and the possible ensuing hysteria that might be generated from these more recent events.

UFO Fleet Flies over Washington, D.C. on July 19 – 20 and July 26 – 27, 1952
http://www.sheeplety.com/1952-washington-dc-ufos-over-white-house/

Project Sign (1947 - 1948)

Project Sign was an official U.S. government study of unidentified flying objects (UFOs) undertaken by the United States Air Force in late 1947 and dissolved in late 1948.

Formally, Project Sign came to no conclusion about UFOs with their final report stating that the existence of "flying saucers" could neither, be confirmed or denied. However, prior to this, Sign officially argued that UFOs were likely of extraterrestrial origin, and most of the project's personnel came to favor the extraterrestrial hypothesis before this opinion was rejected and Sign was dissolved. Blum, Howard, *Out There: The Government's Secret Quest for Extraterrestrials,* Simon and Schuster, 1990

433

Sign was instigated following a recommendation from Lt. General Nathan F. Twining, then the head of Air Materiel Command. Just before this, **Brig. Gen. George Schulgen**, of the Army Air Forces air intelligence division, had completed a preliminary review of the many UFO reports—then called "**flying discs**" by military authorities—which had received considerable publicity following the Kenneth Arnold sighting of June 24, 1947. Schulgen's study, completed in late July 1947, concluded that the flying discs were real craft. Schulgen then asked Twining and his command, which included the intelligence and engineering divisions located at Wright-Patterson Air Force Base (then Wright Field), to carry out a more exhaustive review of the data.

In his formal letter to Schulgen on September 23, 1947, in part, Twining wrote:

- The phenomenon reported is something real and not visionary or fictitious.
- There are objects probably approximately the shape of a disc, of such appreciable size as to appear to be as large as a man-made aircraft.
- There is the possibility that some of the incidents may be caused by natural phenomena, such as meteors.
- The reported operating characteristics such as extreme rates of climb, maneuverability (particularly in roll), and action which must be considered evasive when sighted or contacted by friendly aircraft and radar, lend belief to the possibility that some of the objects are controlled either manually, automatically or remotely.
- The apparent common description of the objects is as follows: ...
- It is possible within the present U.S. knowledge... to construct a piloted aircraft which has the general description ...
- Any development in this country along the lines indicated would be extremely expensive...
- Due consideration must be given to the following:-

The possibility that these objects are of domestic origin - the product of some high security project not known to AC/AS-2 or this command.

The lack of physical evidence in the shape of crash recovered exhibits which would undeniably prove the existence of these objects.

The possibility that some foreign nation has a form of propulsion, possibly nuclear, which is outside of our domestic knowledge.

He recommended that "... Army Air Forces issue a directive assigning a priority, security classification and code name for detailed study of this matter." **(Clark, 489)** Though conducted by the Army Air Force, the study's information and conclusions would be made Twining's recommendation was approved on December 30 by Major General Laurence C. Craigie, Director of Research and Development under the Deputy Chief of staff for Materiel at Headquarters U.S. Air Force. According to Craigie's directive, it would be the role of Sign to: "...collect, collate, evaluate and distribute to interested government agencies and contractors all information concerning sightings and phenomena in the atmosphere which can be construed to be of concern to the national security." Jerome Clark, *The UFO Book: Encyclopedia of the Extraterrestrial,* Visible Ink, 1998; ISBN 1-57859-029-9

On January 22, 1948, Project Sign formally began its work as a branch of **Air Technical Intelligence Center (ATIC)** at Wright-Patterson Air Force Base, under the direction of **Captain Robert R. Sneider**.

Sign was seen as a very important undertaking: Ruppelt wrote that Sign "was given a 2A priority, 1A being the highest priority an Air Force project could have." Though it was classified "restricted", the study's existence was eventually known to the general public, and was often called "Project Saucer". However, UFO historian Wendy Connors established, through an interview with a surviving Sign secretary, that "Project Saucer" was the project's original informal name and had actually begun in late 1946. If this was the case, then the Army Air Force had already begun an investigation of UFOs well before the Kenneth Arnold sighting that launched the first flood of UFO reports of June-July 1947 in the United States. (See, e.g., WWII foo fighter UFOs and the post-war ghost rockets)
http://ufoinfo.com/ufobooks/projectbluebook.shtml

Studies were undertaken by Air Intelligence at the Air Force base nearest to any particular UFO report, though some cases were studied directly by Air Materiel Command. Allen Hynek, then teaching astronomy at Ohio State University, was hired as a consultant to help weed out UFO reports which could be misidentified meteors, stars and the like. Hynek was mentioned by name in Project Sign memos as early as May 1948 and was formally involved as a consultant a few months later. In 1985, Hynek reported, "I was quite negative in most of my evaluations. I stretched far to give something a natural explanation, sometimes when it may not have really had it." CLOSE ENCOUNTER WITH DR. J. ALLEN HYNEK By Dennis Stacy; An Interview With The Dean, 1985; Re-Edited for CUFON by Dale Goudie 1991

Early in the modern UFO age, it was taken for granted by the U.S. Air Force that the flying saucers existed. As Ruppelt wrote,

> *"ATIC's intelligence specialists were confident that within a few months or a year they would have the answer to the question, "What are UFO's?" The question, "Do UFO's exist?" was never mentioned. The only problem that confronted the people at ATIC was, "Were the UFO's of Russian or interplanetary origin?" Either case called for a serious, secrecy-shrouded project. Only top people at ATIC were assigned to Project Sign."*

Sign's first major undertaking was the study of a widely publicized UFO encounter known as the Mantell Incident. On 7 January 1948, Air Force pilot Thomas Mantell—in pursuit of an aerial artifact Mantell reportedly described as "a metallic object ... it is of tremendous size." (Clark, 352) —died when his aircraft crashed near Franklin, Kentucky. Project Sign investigators were officially inconclusive about the Mantell case, but Hynek determined that Mantell had been chasing the planet Venus—a conclusion that eventually met with widespread incredulity. The Venus explanation was later formally withdrawn, and the Mantell incident remains listed as an "unknown."

According to later Project Blue Book director Edward J. Ruppelt, Project Sign investigators were less skeptical about the **Chiles-Whitted UFO Encounte**r over Montgomery, Alabama on 24 July 1948. In this case, two airline pilots reported that a rocket-shaped UFO, glowing blue and

seeming to emit reddish flames, approached them on a near-collision course. Pilots Chiles and Whitted reported the object appeared to show a double row of ports or windows emitting an intense bluish-white light. (A similar object with a double row of windows was also seen over The Hague, Netherlands a few days earlier and independently reported to Project Sign.) Some Sign researchers were deeply impressed by the close UFO sighting from two highly credible pilot-witnesses. The reports of "windows" also suggested the objects were possibly occupied.
Jerome Clark, *The UFO Book: Encyclopedia of the Extraterrestrial*, Visible Ink, 1998; ISBN 1-57859-029-9
Edward J. Ruppelt, *The Report on Unidentified Flying Objects*, Doubleday & Co., 1956

The Estimate of the Situation (1948)

The **Estimate of the Situation** (the term "estimate of the situation" is generic, often used in military intelligence to describe a type of early report on an important subject) was a document supposedly written in 1948 by the personnel of United States Air Force's Project Sign -including the project's director, Captain Robert R. Sneider - which explained their reasons for concluding that the extraterrestrial hypothesis was the best explanation for unidentified flying objects.

As late as 1960, U.S.A.F. personnel claimed that the document never existed. However, several Air Force officers and one consultant describe the report as being a real document that was suppressed. The Estimate document has been described by some Ufologists as the "Holy Grail of Ufology" and noted that Freedom of Information Act requests for the document have been fruitless. (Randles and Hough, 85) Jerome Clark; *The UFO Book: Encyclopedia of the Extraterrestrial*; Visible Ink, 1998; ISBN 1-57859-029-9

Project Sign investigated earlier UFO reports, but the highly-publicized **Chiles-Whitted UFO Encounter** of July 24, 1948, had made an incredible impact on Sign". In that encounter, two experienced airline pilots claimed a torpedo-shaped object nearly collided with their commercial airplane. Sign personnel judged the report convincing and compelling, partly because the alleged object also closely matched the description of an independent sighting from The Hague a few days earlier.

The project members had several dozen aerial observations that they could not explain, many of them by military pilots and scientists. The objects seemed to act like real technology, but their sources said they were not ours. The flying fuselage encounter (**Chiles-Whitted**) intrigued them. The Prandtl theory of lift indicated that such an odd shape *can* fly, but the technology of an advanced power plant was still beyond earth-based nuclear technology. Sign personnel generally accepted that the more reliable witnesses were describing accurately what they'd seen. Given that there was no evidence that either the U.S. or the U.S.S.R. had anything remotely like the UFOs reported, Sign personnel gradually began considering extraterrestrial origins for the objects. The result was the legendary **Estimate of the Situation.**

Michael Swords argued that this consideration of non-earthly origin was "not as incredible in intelligence circles as one might think." Because many in the military were "pilots, engineers, and technical people" they had a "'can do' attitude" and tended to regard unavailable technologies not as impossibilities, but as challenges to be overcome. Rather than dismissing UFO reports out of hand, they considered how such objects might function. This perspective, argues Swords, "contrasted markedly with many scientists' characterizations of such concepts as impossible, unthinkable or absurd." http://en.wikipedia.org/wiki/Project_Sign

At about the same time the Estimate was working its way up the ranks, another group was arguing against any extraterrestrial origins for the saucers. Informally led by Major Aaron J. Boggs, this group was profoundly skeptical of the saucers' reality; Boggs was described by his peers as "the Pentagon's saucer killer".

Under Boggs' guidance, a competing document prepared by the anti-extraterrestrial group in the Directorate of Intelligence was also making the rounds in military intelligence. With input from the **Office of Naval Intelligence**, this study argued that the flying saucers were probably real, though it suspected that they were craft made by the Soviet Union. There was concern in U.S. military intelligence circles that the Soviets could make aeronautical advances on the work of Nazi scientists, especially the **Horten Brothers**, a pair of brilliant aeronautical engineers far in advance of their U.S. counterparts. If the saucers were of Soviet origin, then they were operated so openly in U.S airspace by the Soviets probably as a method of psychological warfare to negate U.S. confidence in atom bomb as the most advanced and decisive weapon. However, Sign researchers could find no hard evidence supporting this hypothesis. With the emergence of cases like the Chiles-Whitted sighting, a rift developed within Sign's staff between those who thought UFOs might be extraterrestrial (see the extraterrestrial hypothesis or ETH) and those who rejected this idea in favor of a more prosaic explanation. Air Intelligence Report AIR 100-203-79: ANALYSIS OF FLYING OBJECT INCIDENTS IN THE U.S. URL accessed February 23, 2007. http://en.wikipedia.org/wiki/Project_Sign

According to Swords, the Estimate was probably completed in September 1948. The Estimate also argued that UFO reports might closely coincide with the approach of the planets Mercury, Venus or Mars to Earth, that the UFOs might be using the planets as launching bases, and predicted a wave of UFO reports in mid-October.

In late September or early October 1948, the Estimate was approved by Colonels William Clingerman and Howard McCoy (Sneider's superiors), who then submitted it to the office of General Charles Cabell, the chief of Air Force intelligence.

According to Swords, The Pentagon went into an "uproar" over the Estimate, which generated "intense" debate. Cabell was newly-appointed, and found himself in charge of a "split house:" some were sympathetic and intrigued, if not entirely convinced of the Estimate's accuracy, while others rejected the very idea of interplanetary saucers as impossible. Unsure of how to proceed, Cabell eventually submitted the Estimate to his superior, **General Hoyt Vandenberg**, Chief-of-Staff of the U.S. Air Force. http://en.wikipedia.org/wiki/Project_Sign

According to Ruppelt, the Estimate was rejected by Vandenberg primarily due to lack of supporting physical evidence, and was "batted back down" the chain of command. In a letter dated November 3, 1948, Cabell wrote to Sign, via McCoy, describing flying saucers as real, but rejecting the interplanetary hypothesis and asking for another Estimate. Cabell wrote:

"The conclusion appears inescapable that some type of flying object has been observed. Identification and the origin of these objects are not discernible to this Headquarters. It is imperative, therefore, that efforts to determine whether these objects are of domestic or foreign origin must be increased until conclusive evidence is obtained. The needs of national defence require such evidence in order that appropriate countermeasures may be taken."
http://www.project1947.com/fig/1948b.htm

 As Swords writes, "Despite Vandenberg's negative reception of the Estimate, Project Sign personnel did not back off." Sign continued their investigations of UFO reports and continued favoring the ETH. Swords speculates that this refusal to change their approach was due to strong minority support for the ETH within the Pentagon and/or a rather mild rejection of the Estimate. In one case, seeking evidence of an advanced propulsion system, Sign personnel tested the radiation levels of George Gorman's National Guard aircraft which was said to have had a "dogfight" with a flying saucer over Fargo, North Dakota.
http://en.wikipedia.org/wiki/Project_Sign

General Charles P. Cabell asked Sign for a second, non-extraterrestrial opinion of the flying saucers. Sign submitted a brief response which did not explicitly mention extraterrestrial ideas but strongly hinted at them, even citing the works of Charles Fort to argue that unusual objects had been flying in the earth's skies for decades before the Arnold encounter. A document signed by Sneider, dated December 20, 1948, but written earlier, has been called the "Ghost of the Estimate" since it echoes many of the same ideas explained in the lost Estimate of the Situation. McCoy wrote,

"...there remains a certain number of reports for which no reasonable everyday explanation is available. So far, no physical evidence of the existence of the unidentified sightings has been obtained...
The possibility that the reported objects are vehicles from another planet has not been ignored. However, tangible evidence to support conclusions about such a possibility are completely lacking..."
McCoy Memo - 1948; and http://www.project1947.com/fig/1948c.htm

Nonetheless, the rejection of the Estimate quickly took its toll on Project Sign and supporters of the ETH. As Ruppelt wrote,

"By the end of 1948, Project Sign had received several hundred UFO reports. Of these, 167 had been saved as good reports. About three dozen were "Unknown." Even though the UFO reports were getting better and more numerous, the enthusiasm over the interplanetary idea was cooling off ... More and more work was being pushed off onto the other investigative organization that was [sic] helping ATIC. The kickback on the Top Secret Estimate of the Situation was beginning to dampen a lot of enthusiasms. It was definitely a bear market for UFO's."

The first public report of the Estimate was in Captain Edward J. Ruppelt's 1956 book, *The Report on Unidentified Flying Objects*. He wrote:

"In intelligence, if you have something to say about some vital problem you write a report that is known as an **"Estimate of the Situation."** A few days after the DC-3 was buzzed [the Chiles-Whitted UFO report], the people at **ATIC (Air Technical Intelligence Centre)** decided that the time had arrived to make an Estimate of the Situation. **The situation was the UFO's; the estimate was that they were interplanetary!** [Bold font added for emphasis]

> *"It was a rather thick document with a black cover and it was printed on legal sized paper. Stamped across the front were the words TOP SECRET.*
>
> *It contained the Air Force's analysis of many of the [UFO] incidents I have told you about plus many similar ones. All of them had come from scientists, pilots, and other equally credible observers, and each one was an unknown ("unknowns" defined by Ruppelt as well-documented UFO reports from reliable observers which defied analysis).*
>
> *... When the estimate was completed, typed, and approved, it started up through channels to higher command echelons. It drew considerable comment but no one stopped it on its way up."*
> **Edward J. Ruppelt; *The Report on Unidentified Flying Objects*; 1956; Doubleday & Company**

Ruppelt's 1956 book, which first publicly disclosed the Estimate, was cleared by the Air Force. Clark writes that as late as 1960, Air Force officials denied that the Estimate was real, despite the fact that censors had approved Ruppelt's book a few years before. According to Clark the U.S. Air Force later formally admitted the Estimate was real, but Clark's bibliography does not make clear what statement or document confirmed the Estimate's reality.

Additionally, according to Clark, the Estimate's existence was confirmed by U.S. Air Force Major Dewey J. Fournet, who as an Air Force major in the Pentagon served as liaison with official UFO project headquartered at Wright-Patterson Air Force Base in Dayton, Ohio." (Clark, 178) Fournet has been described as being "unimpressed" with the Estimate and was furthermore quoted as describing the ET conclusion as an "extreme extrapolation" based on scant evidence. **http://www.project1947.com/bg/ufogov.htm** and **Jerome Clark; *The UFO Book: Encyclopedia of the Extraterrestrial*; Visible Ink, 1998; ISBN 1-57859-029-9**

An Air Force consultant, astronomer Dr. Allen Hynek, also verified the Estimate's existence. **J. Allen Hynek; *The UFO Experience: A Scientific Inquiry*; 1972; Henry Regnery Company**

In the early 1980s, researcher Kevin D. Randle (Randle, 1989) said he spoke with an unnamed colonel who claimed to have helped write the Estimate when he was a lieutenant. According to the colonel, when Vandenberg was sent a working draft of the report, he allegedly ordered the paragraphs giving physical evidence (metal recovered in New Mexico) removed from the report. After doing so, Vandenberg then rejected the final version as lacking physical evidence. Randle claimed that he realized the significance of this anecdote only a few years later while investigating the 1947 Roswell, New Mexico crash. According to Randle, the colonel had died by that point, and a follow-up interview was not possible. **Kevin Randle; *UFO\ Casebook*; Warner Books; 1989; ISBN 0-446-35715-4**

The McCoy letter of November 8, 1948, which mentioned that there was no physical evidence of extraterrestrial origins for flying saucers, has sometimes been cited as evidence against the Roswell UFO incident of July 1947, where a UFO allegedly crashed in the New Mexico desert. Swords argues, that the McCoy letter should not be interpreted this way because the U.S. Military usually operates in a highly compartmentalized, need to know basis. If something as extraordinary as an alien spacecraft had crashed in the summer of 1947, Swords contends that fact would have almost certainly been quickly suppressed and that Sign would not necessarily have been informed of it. http://en.wikipedia.org/wiki/Estimate_of_the_Situation

By late 1948, Project Sign was discontinued in name and replaced by a much more negatively oriented **Project Grudge**. Ruppelt reported that the choice of the word "Grudge" to describe the new project was deliberate.

Ruppelt wrote,

"On February 11, 1949, an order was written that changed the name of the UFO project from Project Sign to Project Grudge. The order was supposedly written because the classified name, Project Sign, had been compromised. This was always my official answer to any questions about the name change. I'd go further and say that the names of the projects, first Sign, then Grudge, had no significance. This wasn't true; they did have significance, a lot of it."

Ruppelt referred to the Project Grudge era as the **"Dark Ages"** of official Air Force UFO investigations. Still, by late 1949, some 20 percent of UFO sightings remained classified as "unknown" by Grudge. By late 1951, according to Ruppelt, some highly influential Pentagon generals had become so disenchanted with Grudge's debunking that Grudge itself was dismantled and replaced by Project Blue Book, with Ruppelt in charge. **Blum, Howard, *Out There: The Government's Secret Quest for Extraterrestrials*, Simon and Schuster, 1990**

Project Grudge (1949 - 1951)

Project Grudge was a short-lived project by the U.S. Air Force to investigate unidentified flying objects (UFOs). Grudge succeeded Project Sign in February 1949, and was then followed by Project Blue Book. The project formally ended in December 1949, but actually continued on in a very minimal capacity until late 1951.

Project Sign had been active from 1947 to 1949. Some of Sign's personnel including director Robert Sneider favored the extraterrestrial hypothesis as the best explanation for UFO reports. They prepared the Estimate of the Situation arguing their case. This theory was ultimately rejected by high-ranking officers, and Project Sign was dissolved and replaced by Project Grudge.

It was announced that Grudge would take over where Sign had left off, still investigating UFO reports. But as Air Force Captain Edward J. Ruppelt would write, "In doing this, standard intelligence procedures would be used. This normally means an *unbiased evaluation* of intelligence data. But it doesn't take a great deal of study of the old UFO files to see that standard intelligence procedures were not being followed by Project Grudge. Everything was being evaluated on the premise that UFOs couldn't exist. No matter what you see or hear, don't believe it." (Ruppelt, 59-60, emphasis his) *Edward J. Ruppelt, The Report on Unidentified Flying Objects*

Ruppelt noted that some of "ATIC's top intelligence specialists who had been so eager to work on Project Sign were no longer working on Project Grudge. Some of them had drastically and hurriedly changed their minds about UFOs when they learned the Pentagon was no longer sympathetic to the UFO cause." (Ruppelt, 60) *Edward J. Ruppelt, The Report on Unidentified Flying Objects*

As **Dr. Michael D. Swords** writes, "Inside the military, [Maj. Aaron J.] Boggs in the Pentagon and [Col. Harold] Watson at AMC [Air Material Command] were openly giving the impression that the whole flying saucer business was ridiculous. Project Grudge became an exercise of derision and sloppy filing. Boggs was so enthusiastically anti-saucer that General Cabell ordered General Moore to create a more proper atmosphere of skeptical respect for the reports and their observers." **(Swords, 98)** and <u>http://en.wikipedia.org/wiki/Project_Grudge</u>

Critics charged that, from its formation, Project Grudge was operating under a debunking directive: *all* UFO reports were judged to have prosaic explanations, though little research was conducted, and some of Grudge's "explanations" were strained or even logically untenable. In his 1956 book, Edward J. Ruppelt would describe Grudge as the "Dark Ages" of USAF UFO investigation. Grudge's personnel were, in fact, conducting little or no investigation, while simultaneously relating that all UFO reports were being thoroughly reviewed. Ruppelt additionally reported that the word "Grudge" was chosen deliberately by the anti-saucer elements in the Air Force.

Like Project Sign, Grudge thought that the vast bulk of UFO reports could be explained as misidentified clouds, stars, sun dogs, conventional aircraft or the like. However, unlike Sign which thought some UFOs might have an extraordinary answer, Grudge's personnel thought the remaining minority of reports could be explained away as normal phenomena. Grudge began a public relations campaign to explain their conclusions to the general public.

The first salvo in the PR campaign came via Sidney Shallet of the *Saturday Evening Post*, one of the more popular magazines of the era. Shallet's article appeared in two consecutive issues of the *Post* (April 30 and May 7, 1949) and generally echoed the Grudge line: Most UFO reports could be easily explained as mundane phenomena misidentified by an eyewitness, the subject was blown out of proportion by the mass media. Shallet suggested that hoaxes and crackpots played a prominent role in popularizing UFOs, and the opinions of many high-ranking military personnel were featured.

The article also included a few misrepresentations of the facts. Shallet asserted that the Air Force thought the subject was nonsense and was more or less forced to investigate flying saucers due to public interest--this was manifestly false, as the Air Force took the UFO subject seriously nearly from the beginning. Shallet, of course, did not have access to some secret information, such as the 1947 memorandum by **Gen. Nathan Twining** that had declared flying saucers a "real and not visionary" phenomenon and had kick-started Project Sign, and did not mention Sign's secret Estimate of the Situation that had argued in favor of an extraterrestrial origin for UFOs.

Shallet's article was perhaps the first detailed public discussion of UFOs, let alone with the endorsement of such prominent military men. Grudge had hoped the article would reduce public interest in flying saucers, but the effect was just the opposite: Shallet had mentioned in passing that a small minority of UFO reports seemed to defy analysis, and these statements were seized upon by the press and the curious. Ruppelt wrote that rather than squelching interest, Shallet had "planted the seed of doubt" in the general public.

Project Grudge issued its only formal report in August 1949. Though over 600 pages long, the report's conclusions stated:

> *A. There is no evidence that objects reported upon are the result of an advanced scientific foreign development; and, therefore they constitute no direct threat to the national security. In view of this, it is recommended that the investigation and study of reports of unidentified flying objects be reduced in scope. Headquarters AMC [Air Materials Command] will continue to investigate reports in which realistic technical applications are clearly indicated.*
> *NOTE: It is apparent that further study along present lines would only confirm the findings presented herein. It is further recommended that pertinent collection directives be revised to reflect the contemplated change in policy.*
>
> *B. All evidence and analyses indicate that reports of unidentified flying objects are the result of:*
> *1. Misinterpretation of various conventional objects.*
> *2. A mild form of mass hysteria and war nerves.*
> *3. Individuals who fabricate such reports to perpetrate a hoax or to seek publicity.*
> *4. Psychopathological persons.*

Not long after this report was released, it was reported that Grudge would soon be dissolved. Despite this announcement, Grudge was not quite finished. A few personnel were still assigned to the project, and they aided the authors of a few more debunking mass media articles.

In April 1951, Bob Ginna of *Life* magazine visited Wright-Patterson Air Force Base. Investigating Grudge, he uncovered what Clark describes as "the project's manifest shortcomings". (Clark, 239) In response (at least "for appearances sake" according to Clark (ibid.) some of the more obviously anti-UFO personnel at Wright-Patterson were reassigned. By mid-1951, Grudge consisted only of Lt. Gerry Cummings. According to Ruppelt, Cummings took his job seriously but found little help or success in his efforts to reverse several years of apathy and dubious research.

On September 10, 1951, there was a radar/visual UFO encounter near Fort Monmouth in New Jersey. Pilots and radar operators reported encounters with a number of fast-moving, highly maneuverable disc-shaped aircraft. High-ranking personnel ordered an investigation, and

Cummings and Lt. Colonel N.R. Rosegarten spent most of 13 September interviewing witnesses and gathering documentation at Ft. Monmouth.

The duo were then ordered to relate the results of their investigation directly to Major General Charles P. Cabell, then the head of Air Force intelligence at the Pentagon. Cummings and Rosegarten arrived at a meeting already in progress and found the atmosphere thick with tension. Cabell, in particular, was distressed by what he saw as the sloppy debunking and lackadaisical attitude Project Grudge brought to bear on a subject he thought deserved serious scrutiny. Cummings and Rosegarten related their conclusions of the Fort Monmouth incident: they agreed with Monmouth personnel who judged the fast moving objects sighted there as being "intelligently controlled." (Clark, 240)

When given permission to speak freely to Cabell and the others, Cummings (as Ruppelt wrote) "cut loose. He told how every UFO report [submitted to Grudge] was taken as a huge joke" and Grudge had become all but moribund. (Clark, 240) Jerome Clark; *The UFO Book: Encyclopedia of the Extraterrestrial*; Visible Ink, 1998; ISBN 1-57859-029-9

When **General Charles P. Cabell** learned that Grudge had essentially ignored UFO reports, he became furious. The Fort Monmouth case had highlighted what critics saw as Air Material Command's sloppy debunking, and at a meeting, a frustrated Cabell was reported to have said, "I want an open mind; in fact, I *order* an open mind! Anyone who doesn't keep an open mind can get out now! ... Why do *I* have to stir up the action? Anyone can see that we do not have a satisfactory answer to the saucer question." (Swords, p. 103) At another meeting--this one of high-ranking military Colonels--Cabell said, "I've been lied to, and lied to, and lied to. I want it to stop. I want the answer to the saucers and I want a good answer." (Swords, p. 103) Cabell also characterized the 1949 Grudge report as "tripe". http://en.wikipedia.org/wiki/Project_Grudge

Lt. Col. N.R. Rosegarten asked, Ruppelt to take over as the new project's leader in late 1951, partly because Ruppelt "had a reputation as a good organizer." (Jacobs, 65) While Cabell wanted Grudge reactivated, he did not want the general public to know that he and some others in the military took UFOs seriously, and ordered the project to keep a low profile. This, he hoped, would protect the military's reputation on both fronts: if the saucer phenomenon was groundless, they could not be accused of sensationalism, but if the phenomenon proved to have some basis in fact, the military could produce serious studies of the subject. Cabell especially did not want the military to be perceived as belittling civilians who had reported UFOs.

Grudge became Project Blue Book. Its first era--directed by Ruppelt--is generally seen as its most open-minded and productive era. Jerome Clark; *The UFO Book: Encyclopedia of the Extraterrestrial*; Visible Ink, 1998; ISBN 1-57859-029-9
Edward J. Ruppelt, *The Report on Unidentified Flying Objects*
http://en.wikipedia.org/wiki/Project_Grudge

Robertson Panel Report (1953)

The **Robertson Panel** was a committee commissioned by the Central Intelligence Agency in 1952 in response to widespread reports of unidentified flying objects, especially in the

Washington, D.C. area. The panel was briefed on U.S. military activities and intelligence; hence the report was originally classified Secret.

Before the final Battelle report was published, the **Central Intelligence Agency (CIA)** had developed an interest in UFOs as a national security (not scientific) issue, and set up a committee to examine existing UFO data. The panel, headed by mathematician and physicist **Howard Percy Robertson** met from 14 to 17 January 1953. It concluded unanimously that the UFO sightings posed no direct threat to national security, but did find that a continued emphasis on UFO reporting might threaten government functions by causing the channels of communication to clog with irrelevant reports and by inducing mass hysteria. Also, the panel worried that nations hostile to the US might use the UFO phenomena to disrupt air defenses. To meet these problems, the Robertson panel stated that a policy of public education on the lack of evidence behind UFOs was needed, to be done through the mass media, schools etc. in order to "debunk" UFOs, and reduce public interest in the subject. There is evidence this was carried out more than two decades after the Panel's conclusion. It also recommended monitoring private UFO groups for subversive activities. Haines, Gerald K. (April 14, 2007). "CIA's Role in the Study of UFOs, 1947-90". https://www.cia.gov/library/center-for-the-study-of-intelligence/csi-publications/csi-studies/studies/97unclass/ufo.html. Retrieved June 28, 2010.

Most UFO reports, they concluded, could be explained as misidentification of mundane aerial objects, and the remaining minority could, in all likelihood, be similarly explained with further study.

The Panel concluded that a public relations campaign should be undertaken in which the national security agencies take immediate steps to strip the Unidentified Flying Objects of the special status they have been given and the aura of mystery they have unfortunately acquired."

Critics (including a few panel members) would later lament the Robertson Panel's role in making UFOs a somewhat disreputable field of study.

In 1952, a wave of UFO reports in the United States centered around Washington DC. So many civilians contacted various government agencies regarding UFO's that daily governmental duties were impacted; the New York Times reported on August 1, 1952, "regular intelligence work has been affected." Various newspapers, such as the Baltimore Sun, Washington Star, Denver Post, and Los Angeles Times, reported on July 31 that Air Force Chief of Staff Hoyt S. Vandenberg thought that the recent spate of UFO sightings and reports had generated "mass hysteria". There was a general concern among the military that the hysteria and confusion generated by UFO reports could be utilized by the United States' enemies, primarily the Soviet Union.

Documents indicate that the CIA became involved at the request of the **National Security Council** after President Truman personally expressed concern over UFOs at a July 28, 1952, NSC meeting. (However, there was no formal NSC meeting on that date). The CIA's study was largely conducted by the CIA's **Office of Scientific Intelligence (OSI).** The CIA thought the question so pressing that they authorized an ad hoc committee in late 1952. http://en.wikipedia.org/wiki/Robertson_Panel

The Robertson Panel first met formally on January 14, 1953, under the direction of Howard Percy Robertson. He was a physicist, a CIA employee, and the director of the **Defense Department Weapons Evaluation Group**. He was instructed by OSI to assemble a group of prominent scientists to review the Air Force's UFO files.

In preparation for this, Robertson first personally reviewed Air Force files and procedures. The Air Force had recently commissioned the **Battelle Memorial Institute** to scientifically study all of the UFO reports collected by Project Sign, Project Grudge, and Project Blue Book. Robertson hoped to draw on their statistical results, but Battelle insisted that they needed much more time to conduct a proper study. (For more, see below).

Other panel members were respected scientists and military personnel who had worked on other classified military projects or studies. All were then skeptical of UFO reports, though to varying degrees. Apart from Robertson, the panel included:

- **Luis Alvarez**, physicist, radar expert (and later, a Nobel Prize recipient);
- **Frederick C. Durant**, missile expert;
- **Samuel A. Goudsmit**, Brookhaven National Laboratories nuclear physicist
- **Thornton Page**, astrophysicist, radar expert, deputy director of Johns Hopkins Operations Research Office;
- **Lloyd Berkner**, physicist, and J. Allen Hynek, astronomer, were associate panel members.

Most of what is known about the actual proceedings of the meetings comes from sketchy minutes kept by Durant, later submitted as a memo to the NSC. It is the only declassified document to date that details the panel's discussions. In addition, various participants would later comment on what transpired from their perspective. Ed Ruppelt, then head of Project Blue Book, first revealed the existence of the secret panel in his 1956 insider book, but without revealing names of panel members. http://en.wikipedia.org/wiki/Robertson_Panel

The Panel had four consecutive days of formal meetings. In total, they met for only 12 hours. Only 23 cases out of 2,331 Air Force UFO cases on record (or about 1%) were reviewed. Although Ruppelt wrote that the Panel studied their best cases, Hynek would opine that the panel, in fact, seemed to have neither the time nor the desire to study the more puzzling ones. For example, the radar experts on the panel (Alvarez and Page) seemed to show little interest in reports of radar UFO cases, which they dismissed as "anxiety over fast radar tracks" by the Air Defense Command. http://en.wikipedia.org/wiki/Robertson_Panel

Of the Panel members, Ruppelt would write in his private papers that Goudsmit was exceptionally hostile to the subject: "Goudsmit was probably the most violent anti-saucer man at the panel meeting. Everything was a big joke to him which brought down the wrath of the other panel members on numerous occasions." Goudsmit even stated later that reporters of UFOs were as dangerous to society as drug addicts.

Alvarez was also extremely skeptical but more professional in his conduct. Page at the time was likewise hostile, later recalling that he made a statement during the meeting that UFOs were "nonsense", bringing about a reprimand from Robertson, despite their good friendship. Ruppelt, however, felt that Page was more open-minded, and although he obviously did not know much about UFOs, he tended to line up with Hynek against Alvarez and Goudsmit in their adamancy that UFOs could not exist.

In contrast, Robertson, Berkner, and Durant seemed to have a personal interest in the subject. It was noted, for example, in a CIA memo that although Berkner was not keen to participate, he "felt strongly that the saucer problem should be thoroughly investigated from a scientific point of view." Another CIA memo produced after the completion of the panel's work indicates that Durant, despite the panel's negative conclusions, thought that materials on flying saucers should continue to be maintained by a major division of OSI, such as Physics and Electronics.
http://en.wikipedia.org/wiki/Robertson_Panel

The first day, the panel viewed two amateur motion pictures of UFOs: the **Mariana UFO film footage** and 1952 Utah UFO Film (the latter was taken by Navy **Chief Petty Officer Delbert C. Newhouse**, who had extensive experience with aerial photography). Two Navy photograph and film analysts (Lieutenants R.S. Neasham and Harry Woo) then reported their conclusions: based on more than 1,000 man hours of detailed analysis, the two films depicted objects that were not any known aircraft, creature or weather phenomena. Air Force Captain Edward J. Ruppelt then began a summary of Air Force efforts regarding UFO studies.

The second day, Ruppelt finished his presentation. Hynek then discussed the Battelle study, and the panel discussed with Air Force personnel the problems inherent in monitoring UFO sightings.

The third day, Air Force **Major Dewey J. Fournet** spoke to the panel. For over a year he had coordinated UFO affairs for the Pentagon. Fournett supported the extraterrestrial hypothesis as the best explanation for some puzzling UFO reports. For the remainder of the third day, the panel discussed their conclusions, and Robertson agreed to draft a preliminary report.

The fourth and final day, the panel rewrote and finalized their report.

The Robertson Panel's official report concluded that 90 percent of UFO sightings could be readily identified with meteorological, astronomical, or natural phenomena and that the remaining 10 percent of UFO reports could, in all likelihood, be similarly explained with detailed study. It was suggested that witnesses had misidentified bright stars and planets, meteors, auroras, mirages, atmospheric temperature inversions, and lenticular clouds; other sightings were judged as likely misinterpretation of conventional aircraft, weather balloons, birds, searchlights, kites, and other phenomena.

None of the Panel's members was formally trained in motion picture or photographic analysis and only one had any experience with photography (astronomic still photography and not motion picture film). Nonetheless, after screening the films only a few times, they dismissed the idea that either the **1950 Montana UFO Film** or the **1952 Utah UFO Film** showed "genuine" UFOs. The Panel's members instead argued that the "UFOs" in the Montana film were actually the reflections of two jet fighters alleged to be in the area at the time and that those in the Utah film

446

were actually seagulls flying near the Great Salt Lake. However, the Panel's conclusions contradicted U.S. Air Force photo analysts who had earlier specifically ruled out birds as an explanation for the Utah film and had thought that jets were a highly unlikely, but remotely plausible, explanation for the Montana film (Clark, 1998). The Panel's conclusions also seemingly ignored eyewitness testimony in both film cases that the objects, while closer to the camera operators, were clearly-defined metallic flying saucers, not the rather indistinct lights seen on the films. Jerome Clark, *The UFO Book: Encyclopedia of the Extraterrestrial*, ISBN 1-57859-029-9

Furthermore, the Panel suggested the Air Force should begin a "debunking" effort to reduce "public gullibility" and demystify UFO reports, partly via a public relations campaign, using psychiatrists, astronomers, and assorted celebrities to significantly reduce public interest in UFOs. It was also recommended that the mass media be used for the debunking, including influential media giants like the **Walt Disney Corporation**. The primary reasoning for this recommendation lay in the belief that the Soviets might try to "mask" an actual invasion of the USA by causing a wave of false "UFO" reports to swamp the Pentagon and other military agencies, thus temporarily blinding the US government to the impending Communist invasion.

Their formal recommendation stated, "That the national security agencies take immediate steps to strip the Unidentified Flying Objects of the special status they have been given and the aura of mystery they have unfortunately acquired." Report of Scientific Advisory Panel on Unidentified Flying Objects Convened by Office of Scientific Intelligence, CIA January 14 - 18, 1953

Also recommended was government monitoring of civilian groups studying or researching UFOs "because of their potentially great influence on mass thinking... the apparent irresponsibility and possible use of such groups for subversive purposes should be kept in mind." Two UFO groups, in particular, were singled out: **APRO** and **Civilian Saucer Investigations (CSI).**

Joint-Army-Navy-Air Force Publication 147 (JANAP 147) and Air Force Regulation 200-2 (AFR 200-2) 1953, 1954, 1958

The recommendations of the Robertson Panel were implemented by a series of special military regulations. **Joint-Army-Navy-Air Force Publication 147 (JANAP 147)** of December 1953 made reprinting of any UFO sighting to the public a crime under the Espionage Act, with fines of up to ten thousand dollars and imprisonment ranging from one to ten years. This act was considered binding on all who knew of the act's existence, including commercial airline pilots. A 1954 revision of **Air Force Regulation 200-2 (AFR 200-2)** made all sighting reports submitted to the air force classified material and prohibited the release of any information about UFO sightings *unless* the sighting was able to be positively identified. In February 1958 a revision of AFR 200-2 allowed the military to give the FBI the names of people who were **"illegally or deceptively bringing the subject [of UFOs] to public attention".** Because of the Robertson Panel the Air Force's Project Blue Book's procedures of investigating UFOs also changed, attempting to find a quick explanation and then file them away. Project Blue Book was a successor of Project Grudge.

In 1956 retired Marine Major Donald Keyhoe founded the **National Investigations Committee on Aerial Phenomena (NICAP)** a UFO investigations organization. By 1969 Keyhoe turned his focus on the CIA as the source of the UFO cover-up. NICAP's board, headed by **Colonel Joseph Bryan III**, forced Keyhoe to retire as NICAP chief. Bryan was actually a former covert CIA agent who had served the agency as founder and head of its psychological warfare division. Under Bryan's leadership, the NICAP disbanded its local and state affiliate groups, and by 1973 it had been completely closed. http://en.wikipedia.org/wiki/Robertson_Panel

Ruppelt's 1956 book *The Report On unidentified Flying Objects* contained the first publicly-released information about the Robertson Panel, with a summary of their proceedings and conclusions. Ruppelt's book did not include the names of the Panel members, nor any institutional or governmental affiliations.

In 1958, the National Investigations Committee on Aerial Phenomena (NICAP), a civilian UFO research group, requested that the Air Force release the panel's report. The Air Force released three summary paragraphs and the names of the panel's members. In 1966 a nearly full-length version of the report was printed in the science column of the *Saturday Review*. See Clark, 1998

Panel member Thornton Page would later change some of his more stridently skeptical conclusions regarding the Panel's report, and regarding UFOs in general. In his 1969 critique of the Condon Report, Page would lament the "excessive levity" he brought to the Panel's proceeding, detailing how he later thought the UFO subject deserved serious scrutiny. Dr. Thornton Page's Review of the Condon Report

Hynek's opinions changed in later years as well, so much that he became, to many, the scientifically respectable voice of Ufology. He would lament that the Robertson Panel had "made the subject of UFOs scientifically unrespectable, and for nearly 20 years, not enough attention was paid to the subject to acquire the kind of data needed even to decide the nature of the UFO phenomenon." See Clark, 1998

According to Swords, the Robertson Panel's report had an "enormous" impact throughout the U.S. Government: the CIA abandoned a "major high level [UFO] investigation" planned in conjunction with the **National Security Council**; UFO research projects by personnel in The Pentagon were quashed; and Project Blue Book's hopes to establish a scientific advisory board were dashed. Blue Book was also downgraded in status and stripped of most responsibility for investigating serious, well-attested UFO cases, which were instead secretly turned over to a newly-formed division of the Air Defense Command. Directives were also issued not to discuss the unexplainable cases with the public and to reduce the percentage of "unknowns". http://en.wikipedia.org/wiki/Robertson_Panel

Though the CIA's official history suggests that the Robertson Panel's conclusions were never carried out, there is evidence that contradicts this. Perhaps the most unambiguous evidence for the Robertson Panel's covert impact on news media reporting about UFOs is a personal letter by Dr. Thornton Page, discovered in the Smithsonian archives by biochemist Michael D. Swords. The 1966 letter, addressed to former Robertson Panel Secretary Frederick C. Durant, confides that Page "helped organize the CBS TV show around the Robertson Panel conclusions."

Thornton Page was no doubt referring to the CBS Reports TV broadcast of the same year, "UFOs: Friend, Foe, or Fantasy?" narrated by **Walter Cronkite**. (Incidentally, this program was criticized for inaccurate and misleading presentations. Page's letter indicates that the Robertson Panel was still putting a negative spin on UFO news at least 13 years after the panel met. **Jerome Clark, *The UFO Book: Encyclopedia of the Extraterrestrial*, ISBN 1-57859-029-9**

Furthermore, according to Swords, there is ample evidence to prove that CSI was pressured to disband by the U.S. Government. FBI documents indicate that noted engineer Walther Riedel was pressured to resign from CSI, and not long afterwards, the group disbanded; in response, Robertson wrote to Marshall Chadwell, stating "that ought to fix the Forteans." (Robertson was referring to the devotees of American writer Charles Fort (1874-1932), whose books argued in favor of the reality of extraterrestrials on Earth.) APRO was active through the late 1980s. There has also been speculation that UFO group NICAP was infiltrated by CIA operatives. http://en.wikipedia.org/wiki/Robertson_Panel

Even later, Randles and Hough note that there was a "CIA memo from 1976" which "tells how the agency is still having to 'keep in touch with reporting channels' in ufology (in other words, to spy on UFO groups." http://en.wikipedia.org/wiki/Robertson_Panel

Some scholars and investigators have suggested that the Robertson Panel's true objective was to justify a CIA domestic propaganda-and-surveillance campaign, rather than to investigate UFOs. For example, journalist Howard Blum writes that it is difficult to accept any argument that the Robertson Panel was ever intended as a serious scientific analysis: Blum argues that the Panel's perfunctory rejection of the U.S. Navy's detailed examination of the UFO films is all but impossible to justify on scientific grounds. Similarly, Swords has argued that the Panel seems to have been designed as an elaborate theater exercise instead of a serious attempt to get to the bottom of the UFO issue. Although the Panel put on a show of evaluating some UFO evidence, its scientific analysis was cursory and its conclusions mostly likely pre-ordained. Also, the Panel looked only at the evidence in the public domain, not higher-quality classified military evidence. Psychologist David R. Saunders, a member of the University of Colorado's UFO study (the Condon Committee), had earlier expressed similar conclusions. Given that Robertson had worked as a high-level scientific intelligence officer during World War II, he would have been familiar with the use of such tactics to hide a sensitive national-security problem from scrutiny by outsiders. **Blum, Howard, Out There: The Government's Secret Quest for Extraterrestrials, Simon and Schuster, 1990.** and
David R. Saunders and R. Roger Harkins, *UFOs? Yes! Where the Condon Committee Went Wrong*, New York: Signet, 1968, p. 105

It is a widely-held conclusion amongst UFO investigators that the Robertson Panel's conclusions and recommendations had a great influence on official United States policy regarding UFOs for many decades. See **Clark, 1998, Blum, 1990,** http://en.wikipedia.org/wiki/Robertson_Panel

When the Battelle Memorial Institute finally finished their massive review of Air Force UFO cases in 1954 (called **"Project Blue Book Special Report No. 14"),** their results were markedly different from those of the Robertson Panel. Whereas the Robertson Panel spent only twelve hours reviewing a limited number of cases, the Battelle Institute had four full-time scientific analysts working for over two years analyzing 3201 reports. Classifying a case as "unknown" required agreement among all four analysts, whereas a "known" or conventional classification required agreement by only two analysts. Still, they concluded 22% of the cases remained unsolvable. The percentage climbed to 35% when considering only the best cases and fell to 18% for the worst cases. Not only are the percentages of unknowns much higher than those for the Robertson Panel, but the higher percentages for the better cases are directly opposite one conclusion of the panel that their remaining 10% of unknowns would disappear if further investigated and more information was available. Furthermore, the Battelle study had already thrown out cases they deemed to have insufficient information to make a determination (9% of all cases). Thus, the fact that a case was classified as "unknown" had nothing to do with lack of information or investigation.

The study also looked at six characteristics of the sightings: duration, speed, number, brightness, color, and shape. For all characteristics, the knowns and unknowns differed at a highly statistically significant level, further indicating that the knowns and unknowns were distinctly different classes of phenomena.

Despite this, the summary section of the final report declared it was "highly improbable that any of the reports of unidentified aerial objects... represent observations of technological developments outside the range of present-day knowledge." A number of researchers have noted that the conclusions of the analysts were usually at odds with their own statistical results, displayed in 240 charts, tables, graphs and maps. Possibly the analysts simply had trouble accepting their own results. Others conjecture this was another result of the Robertson Panel, the conclusions being written to satisfy the new political climate within Project Blue Book following the panel.

Project Blue Book (1952 - 1969)

Project Blue Book was one of a series of systematic studies of unidentified flying objects (UFOs) conducted by the United States Air Force (U.S.A.F.). Started in 1952, it was the second revival of such a study (the first two of its kind being **Projects Sign** and **Grudge**). A termination order was given for the study in December 1969, and all activity under its auspices ceased in January 1970.

Project Blue Book had two goals:

1. to determine if UFOs were a threat to national security, and
2. to scientifically analyze UFO-related data.

Thousands of UFO reports were collected, analyzed and filed. As the result of the Condon Report, which concluded there was nothing anomalous about UFOs, Project Blue Book was ordered shut down in December 1969 and the Air Force continues to provide the following summary of its investigations:

1. No UFO reported, investigated and evaluated by the Air Force was ever an indication of threat to our national security;

450

2. There was no evidence submitted to or discovered by the Air Force that sightings categorized as "unidentified" represented technological developments or principles beyond the range of modern scientific knowledge; and
3. There was no evidence indicating that sightings categorized as "unidentified" were extraterrestrial vehicles.[1]

By the time Project Blue Book ended, it had collected 12,618 UFO reports and concluded that most of them were misidentifications of natural phenomena (clouds, stars, etc.) or conventional aircraft. The UFO reports were archived and are available under the Freedom of Information Act, but names and other personal information of all witnesses have been changed.

Public USAF UFO studies were first initiated under Project Sign at the end of 1947, following many widely publicized UFO reports (see Kenneth Arnold). Project Sign was initiated specifically at the request of General Nathan Twining, chief of the Air Force Materiel Command at Wright-Patterson Air Force Base. Wright-Patterson was also to be the home of Project Sign and all subsequent official USAF public investigations.

Sign was officially inconclusive regarding the cause of the sightings. However, according to US Air Force **Captain Edward J. Ruppelt** (the first director of Project Blue Book), Sign's initial intelligence estimate (the so-called **Estimate of the Situation**) written in the late summer of 1948, *"concluded that the flying saucers were real craft, were not made by either the Russians or US, and were likely extraterrestrial in origin."* (See also **Extraterrestrial Hypothesis**.) This Estimate was forwarded to the Pentagon, but subsequently ordered destroyed by Gen. Hoyt Vandenberg, USAF Chief of Staff, citing a lack of physical proof. Vandenberg subsequently dismantled Project Sign.

Project Sign was succeeded at the end of 1948 by Project Grudge, which had a debunking mandate. Ruppelt referred to the era of Project Grudge as the "dark ages" of early USAF UFO investigation. Grudge concluded that all UFOs were natural phenomena or other misinterpretations, although it also stated that 23 percent of the reports could not be explained.

According to Captain Edward J. Ruppelt, by the end of 1951, several high-ranking, very influential USAF generals were so dissatisfied with the state of Air Force UFO investigations that they dismantled Project Grudge and replaced it with Project Blue Book in early 1952. One of these men was Gen. Charles P. Cabell. Another important change came when General William Garland joined Cabell's staff; Garland thought the UFO question deserved serious scrutiny because he had witnessed a UFO [2]. (*Garland was probably not in the "loop" on the UFO matter and did not have a need to know*). **[My italics added]**

The new name, Project Blue Book, was selected to refer to the blue booklets used for testing at some colleges and universities. The name was inspired, said Ruppelt, by the close attention that high-ranking officers were giving the new project; it felt as if the study of UFOs was as important as a college final exam. Blue Book was also upgraded in status from Project Grudge, with the creation of the **Aerial Phenomenon Branch**.[3]

Ruppelt was the first head of the project. He was an experienced airman, having been decorated for his efforts with the Army Air Corps during World War II, and having afterwards earned an aeronautics degree. ***He officially coined the term "Unidentified Flying Object", to replace the many terms* ("flying saucer", "flying disk"** and so on) the military had previously used; Ruppelt thought that "unidentified flying object" was a more neutral and accurate term. Ruppelt resigned from the Air Force some years later, and wrote the book ***The Report on Unidentified Flying Objects,*** which described the study of UFOs by United States Air Force from 1947 to 1955. Swords writes that "Ruppelt would lead the last genuine effort to analyze UFOs"[4].

Ruppelt implemented a number of changes: He streamlined the manner in which UFOs were reported to (and by) military officials, partly in hopes of alleviating the stigma and ridicule associated with UFO witnesses. Ruppelt also ordered the development of a standard questionnaire for UFO witnesses, hoping to uncover data which could be subject to statistical analysis. He commissioned the Battelle Memorial Institute to create the questionnaire and computerize the data. Using case reports and the computerized data, Battelle then did a massive scientific and statistical study of all Air Force UFO cases, completed in 1954 and known as **"Project Blue Book Special Report No. 14" (see summary below).**

Knowing that factionalism had harmed the progress of Project Sign, Ruppelt did his best to avoid the kinds of open-ended speculation that had led to Sign's personnel being split among advocates and critics of the extraterrestrial hypothesis. As **Michael Hall** writes, "Ruppelt not only took the job seriously but expected his staff to do so as well. If anyone under him either became too skeptical or too convinced of one particular theory, they soon found themselves off the project."[5] In his book, Ruppelt reported that he fired three personnel very early in the project because they were either "too pro" or "too con" one hypothesis or another. Ruppelt sought the advice of many scientists and experts and issued regular press releases (along with classified monthly reports for military intelligence).

Each U.S. Air Force Base had a Blue Book officer to collect UFO reports and forward them to Ruppelt [6]. During most of Ruppelt's tenure, he and his team were authorized to interview any and all military personnel who witnessed UFOs and were not required to follow the chain of command. This unprecedented authority underlined the seriousness of Blue Book's investigation.

Under Ruppelt's direction, Blue Book investigated a number of well-known UFO cases, including the so-called Lubbock Lights, and a widely publicized 1952 radar/visual case over Washington D.C. According to Jacques Vallee[7], Ruppelt started the trend, largely followed by later Blue Book investigations, of not giving serious consideration to numerous reports of UFO landings and/or interaction with purported UFO occupants.

Astronomer Dr. J. Allen Hynek was the scientific consultant of the project, as he had been with Projects Sign and Grudge. He worked for the project up to its termination and initially created the categorization which has been extended and is known today as Close Encounters. He was a pronounced skeptic when he started, but said that his feelings changed to a more wavering skepticism during the research, after encountering a few UFO reports he thought were unexplainable.

452

Ruppelt left Blue Book in February 1953 for a temporary reassignment. He returned a few months later to find his staff reduced from more than ten to two subordinates. Frustrated, Ruppelt suggested that an Air Defense Command unit (the 4602nd Air Intelligence Service Squadron) be charged with UFO investigations.

In July 1952, after a build-up of hundreds of sightings over the previous few months, a series of radar detections coincident with visual sightings were observed near the National Airport in Washington, D.C. (see 1952 Washington D.C. UFO incident). Future Arizona Senator and 2008 presidential nominee **John McCain** is alleged to be one of these witnesses.

After much publicity, these sightings led the Central Intelligence Agency to establish a panel of scientists headed by **Dr. H. P. Robertson,** a physicist of the California Institute of Technology, which included various physicists, meteorologists, and engineers, and one astronomer (Hynek). The Robertson Panel first met on January 14, 1953, in order to formulate a response to the overwhelming public interest in UFOs.

Ruppelt, Hynek, and others presented the best evidence, including movie footage that had been collected by Blue Book. After spending 12 hours reviewing 6 years of data, the Robertson Panel concluded that most UFO reports had prosaic explanations and that all could be explained with further investigation, which they deemed not worth the effort.

In their final report, they stressed that low-grade, unverifiable UFO reports were overloading intelligence channels, with the risk of missing a genuine conventional threat to the U.S. Therefore, they recommended the Air Force de-emphasize the subject of UFOs and embark on a debunking campaign to lessen public interest. They suggested debunkery through the mass media, including The **Walt Disney Company**, and using psychologists, astronomers, and celebrities to ridicule the phenomenon and put forward prosaic explanations. Furthermore, civilian UFO groups "should be watched because of their potentially great influence on mass thinking. The apparent irresponsibility and the possible use of such groups for subversive purposes should be kept in mind."

It is the conclusion of many researchers[6][8] that the Robertson Panel was recommending controlling public opinion through a program of official propaganda and spying. They also believe these recommendations helped shape Air Force policy regarding UFO study not only immediately afterwards, but also into the present day. There is evidence that the Panel's recommendations were being carried out at least two decades after its conclusions were issued.

In December 1953, **Joint Army-Navy-Air Force Regulation number 146** made it a crime for military personnel to discuss classified UFO reports with unauthorized persons. Violators faced up to two years in prison and/or fines of up to $10,000.

In his book (see external links) Ruppelt described the demoralization of the Blue Book staff and the stripping of their investigative duties following the Robertson Panel. As an immediate consequence of the Robertson Panel recommendations, in February 1953, the Air Force issued Regulation 200-2, ordering air base officers to publicly discuss UFO incidents only if they were judged to have been solved, and to classify all the unsolved cases to keep them out of the public eye.

The same month, investigative duties started to be taken on by the newly formed 4602nd Air Intelligence Squadron (AISS) of the Air Defense Command. The 4602nd AISS was tasked with investigating only the most important UFO cases with intelligence or national security implications. These were deliberately siphoned away from Blue Book, leaving Blue Book to deal with the more trivial reports.

General Nathan Twining, who got Project Sign started back in 1947, was now Air Force Chief of Staff. In August 1954, he was to further codify the responsibilities of the 4602nd AISS by issuing an updated Air Force Regulation 200-2. In addition, UFOs (called "**UFOBs**") were defined as "any airborne object which by performance, aerodynamic characteristics, or unusual features, does not conform to any presently known aircraft or missile type, or which cannot be positively identified as a familiar object." Investigation of UFOs was stated to be for the purposes of national security and to ascertain "technical aspects." AFR 200-2 again stated that Blue Book could discuss UFO cases with the media only if they were regarded as having a conventional explanation. If they were unidentified, the media was to be told only that the situation was being analyzed. Blue Book was also ordered to reduce the number of unidentified to a minimum.

All this was done secretly. The public face of Blue Book continued to be the official Air Force investigation of UFOs, but the reality was it had essentially been reduced to doing very little serious investigation and had become almost solely a public relations outfit with a debunking mandate. To cite one example, by the end of 1956, the number of cases listed as unsolved had dipped to barely 0.4 percent, from the 20 to 30% it had been only a few years earlier.

Eventually, Ruppelt requested reassignment; at his departure in August 1953, his staff had been reduced from more than ten (precise numbers of personnel varied) to just two subordinates and himself. His temporary replacement was a noncommissioned officer. Most who succeeded him as Blue Book director exhibited either apathy or outright hostility to the subject of UFOs or were hampered by a lack of funding and official support.

UFO investigators often regard Ruppelt's brief tenure at Blue Book as the high-water mark of public Air Force investigations of UFOs, when UFO investigations were treated seriously and had support at high levels[8]. Thereafter, Project Blue Book descended into a new "Dark Ages" from which many UFO investigators argue it never emerged[8]. However, Ruppelt later came to embrace the Blue Book perspective that there was nothing extraordinary about UFOs; he even labeled the subject a "Space Age Myth." *"One possible reason for the about turn in attitude was that Ruppelt may have been let in on the UFO secret toward the end of his tenure as a reward for the hard work he dutifully carried out on behalf of the Air force but, the information revealed to him may have been of limited importance and not necessarily the full picture, as, like all things that were designated at the **"above top secret"** level, it is based on a need to know thus, his "Space Age Myth" statement regarding UFOs. He finally knew, they were real!"*[**My Italics added**].

In March 1954, **Captain Charles Hardin** was appointed the head of Blue Book. However, most UFO investigations were conducted by the 4602nd, and Hardin had no objection. Ruppelt wrote that Hardin "thinks that anyone who is even interested [in UFOs] is crazy. They bore him."[9]

In 1955, the Air Force decided that the goal of Blue Book should be not to investigate UFO reports, but rather to reduce the number of unidentified UFO reports to a minimum. By late 1956, the number of unidentified sightings had dropped from the 20-25% of the Ruppelt era to less than 1%.

Captain George T. Gregory took over as Blue Book's director in 1956. Clark writes that Gregory led Blue Book "in an even firmer anti-UFO direction than the apathetic Hardin."[9] The 4602nd was dissolved, and the 1066th Air Intelligence Service Squadron was charged with UFO investigations.

In fact, there was actually little or no investigation of UFO reports; a revised AFR 200-2 issued during Gregory's tenure emphasized that unexplained UFO reports must be reduced to a minimum.

One way that Gregory reduced the number of unexplained UFOs was by simple reclassification. "Possible cases" became "probable", and "probable" cases were upgraded to certainties. By this logic, a *possible* comet became a *probable* comet, while a probable comet was flatly declared to have been a misidentified comet. Similarly, if a witness reported an observation of an unusual balloon-*like* object, Blue Book usually classified it as a balloon, with no research and qualification. These procedures became standard for most of Blue Book's later investigations; **(see Hynek's comments below).**

Major Robert J. Friend was appointed the head of Blue Book in 1958. Friend made some attempts to reverse the direction Blue Book had taken since 1954. Clark writes that "Friend's efforts to upgrade the files and catalog sightings according to various observed statistics were frustrated by a lack of funding and assistance."[9]

Heartened by Friend's efforts, Hynek organized the first of several meetings between Blue Book staffers and ATIC personnel in 1959. Hynek suggested that some older UFO reports should be reevaluated, with the ostensible aim of moving them from the "unknown" to the "identified" category. Hynek's plans came to naught.

During Friend's tenure, ATIC contemplated passing oversight Blue Book to another Air Force agency, but neither the Air Research and Development Center, nor the Office of Information for the Secretary of the Air Force was interested.

In 1960, there were U.S. Congressional hearings regarding UFOs. Civilian UFO research group NICAP had publicly charged Blue Book with covering up UFO evidence and had also acquired a few allies in the U.S. Congress. Blue Book was investigated by the Congress and the CIA, with critics—most notably the civilian UFO group NICAP[8] asserting that Blue Book was lacking as a scientific study. In response, ATIC added personnel (increasing the total personnel to three military personnel, plus civilian secretaries) and increased Blue Book's budget. This seemed to mollify some of Blue Book's critics,[8] that but it was only temporary. A few years later **(see below)**, the criticism would be even louder.

By the time he was transferred from Blue Book in 1963, Friend thought that Blue Book was effectively useless and ought to be dissolved, even if it caused an outcry amongst the public.

Major Hector Quintanilla took over as Blue Book's leader in August 1963. He largely continued the debunking efforts, and it was under his direction that Blue Book received some of its sharpest criticism. UFO researcher **Jerome Clark** goes so far as to write that, by this time, **Blue Book** had "lost all credibility."[10]

Physicist and UFO researcher **Dr. James E. McDonald** once flatly declared that Quintanilla was "not competent" from either a scientific or an investigative perspective.[11] However, McDonald also stressed that Quintanilla "shouldn't be held accountable for it", as he was chosen for his position by a superior officer, and was following orders in directing Blue Book.[11]

Blue Book's explanations of UFO reports were not universally accepted, however, and critics — including some scientists — suggested that **Project Blue Book** was engaged in questionable research or, worse, perpetrating cover up.[8] This criticism grew especially strong and widespread in the 1960s.

Take, for example, the many mostly nighttime UFO reports from the midwestern and the southeastern United States in the summer of 1965: Witnesses in Texas reported "multicolored lights" and large aerial objects shaped like eggs or diamonds.[8] The Oklahoma Highway Patrol reported that Tinker Air Force Base (near Oklahoma City) had tracked up to four UFO's simultaneously and that several of them had descended very rapidly: from about 22000 feet to about 4000 feet in just a few seconds,[8] an action well beyond the capabilities of conventional aircraft of the era. John Shockley, a meteorologist from Wichita, Kansas, reported that using the state Weather Bureau radar, he tracked a number of odd aerial objects flying at altitudes between about 6000 and 9000 feet.[8] These and other reports received wide publicity.

Project Blue Book officially determined[8] the witnesses had mistaken Jupiter or bright stars (such as Rigel or Betelgeuse) for something else.

Blue Book's explanation was widely criticized as inaccurate. Robert Riser, director of the Oklahoma Science and Art Foundation Planetarium offered a strongly-worded rebuke of Project Blue Book that was widely circulated: "That is as far from the truth as you can get. These stars and planets are on the opposite side of the earth from Oklahoma City at this time of year. The Air Force must have had its star finder upside-down during August."[8]

A newspaper editorial from the *Richmond News Leader* opined that "Attempts to dismiss the reported sightings under the rationale as exhibited by Project Bluebook (sic) won't solve the mystery … and serve only to heighten the suspicion that there's something out there that the air force doesn't want us to know about",[8] while a Wichita-based UPI reporter noted that "Ordinary radar does not pick up planets and stars."[8]

456

Another case that Blue Book's critics seized upon was the so-called Portage County UFO Chase, which began at about 5.00am, near Ravenna, Ohio on April 17, 1966. Police officers Dale Spaur and Wilbur Neff spotted what they described as a disc-shaped, silvery object with a bright light emanating from its underside, at about 1000 feet in altitude.[8][12] They began following the object (which they reported sometimes descended as low as 50 feet), and police from several other jurisdictions were involved in the pursuit. The chase ended about 30 minutes later near Freedom, Pennsylvania, some 85 miles away. *This famous police chase of UFOs was used in one of the early scenes from Steven Spielsberg's movie, "Close Encounters of the Third Kind".*

The UFO chase made national news, and the police submitted detailed reports to Blue Book. Five days later, following brief interviews with only one of the police officers (but none of the other ground witnesses), Blue Book's director, Major Hector Quintanilla, announced their conclusions: The police (one of them an Air Force gunner during the Korean War) had first chased a communications satellite, then the planet Venus.

This conclusion was widely derided,[8] and was strenuously rejected by the police officers. In his dissenting conclusion, Hynek described Blue Book's conclusions as absurd: in their reports, several of the police had unknowingly described the moon, Venus, *and* the UFO, though they unknowingly described Venus as a bright "star" very near the moon. **Ohio Congressman William Stanton said that** *"The Air Force has suffered a great loss of prestige in this community … Once people entrusted with the public welfare no longer think the people can handle the truth, then the people, in return, will no longer trust the government."*

In September 1968, Hynek received a letter from **Colonel Raymond Sleeper** of the Foreign Technology Division. Sleeper noted that Hynek had publicly accused Blue Book of shoddy science, and further asked Hynek to offer advice on how Blue Book could improve its scientific methodology. Hynek was to later declare that Sleeper's letter was "the first time in my 20 year association with the air force as scientific consultant that I had been officially asked for criticism and advice [regarding] … the UFO problem."[13]

Hynek wrote a detailed response, dated October 7, 1968, suggesting several areas where Blue Book could improve. In part, he wrote:

A. ... neither of the two missions of Blue Book [determining if UFOs are a threat to national security and using scientific data gathered by Blue Book] is being adequately executed.
B. The staff of Blue Book, both in numbers and in scientific training, is grossly inadequate...
C. Blue Book suffers … in that it is a closed system ... there is virtually no scientific dialogue between Blue Book and the outside scientific world...
D. The statistical methods employed by Blue Book are nothing less than a travesty.
E. There has been a lack of attention to significant UFO cases ... and too much time spent on routine cases ... and on peripheral public relations tasks. Concentration could be on two or three potentially scientific significant cases per month [instead of being] spread thin over 40 to 70 cases per month.
F. The information input to Blue Book is grossly inadequate. An impossible load is placed on Blue Book by the almost consistent failure of UFO officers at local air bases to transmit adequate information...

G. The basic attitude and approach within Blue Book is illogical and unscientific...

H. Inadequate use had been made of the Project scientific consultant [Hynek himself]. Only cases that the *project monitor* deems worthwhile are brought to his attention. His scope of operation ... has been consistently thwarted ... He often learns of interesting cases only a month or two after the receipt of the report at Blue Book.[14]

Despite Sleeper's request for criticism, none of Hynek's commentary resulted in any substantial changes in Blue Book.

Quintanilla's own perspective on the project is documented in his manuscript, "UFOs, An Air Force Dilemma." **Lt. Col Quintanilla** wrote the manuscript in 1975, but it was not published until after his death. Quintanilla states in the text that he personally believed it arrogant to think human beings were the only intelligent life in the universe. Yet, while he found it highly likely that intelligent life existed beyond earth, he had no hard evidence of any extraterrestrial visitation.[15]

Criticism of Blue Book continued to grow through the mid-1960s. NICAP's membership ballooned to about 15,000, and the group charged the U.S. Government with a cover-up of UFO evidence.

Following U.S. Congressional hearings, the **Condon Committee** was established in 1966, ostensibly as a neutral scientific research body. However, the Committee became mired in controversy, with some members charging director Edward U. Condon with bias, and critics would question the validity and the scientific rigor of the Condon Report.

In the end, the Condon Committee suggested that there was nothing extraordinary about UFOs, and while it left one case unexplained, further research would not be likely to yield very significant results.

In response to the Condon Committee's conclusions, **Secretary of the Air Force Robert C. Seamans, Jr.** announced that Blue Book would soon be closed, because further funding "cannot be justified either on the grounds of national security or in the interest of science."[16] The last publicly acknowledged day of Blue Book operations was December 17, 1969. However, researcher Brad Sparks,[17] citing research from the May 1970 issue of NICAP's *UFO Investigator*, reports that the last day of Blue Book activity was actually January 30, 1970. According to Sparks, Air Force officials wanted to keep the Air Force's reaction to the UFO problem from overlapping into a fourth decade, and thus altered the date of Blue Book's closure in official files.

Blue Book's files were sent to the Air Force Archives at Maxwell Air Force Base in Alabama. Major David Shea was to later claim that Maxwell was chosen because it was "accessible yet not too inviting."[16]

Ultimately, Project Blue Book stated that UFOs sightings were generated as a result of:

- A mild form of mass hysteria.

458

- Individuals who fabricate such reports to perpetrate a hoax or seek publicity.
- Psychopathological persons.
- Misidentification of various conventional objects.

These official conclusions were directly contradicted by the USAF's own commissioned *Blue Book Special Report #14*. Psychological factors and hoaxes actually constituted less than 10% of all cases and 22% of all sightings, particularly the better-documented cases, remained unsolved. (See section below for details and Identified flying object.)

As of April 2003, the USAF has publicly indicated that there are no immediate plans to re-establish any official government UFO study programs.[18]

Below is the United States Air Force's official statement regarding UFOs, as noted in USAF Fact Sheet 95-03: [18]

From 1947 to 1969, the Air Force investigated Unidentified Flying Objects under Project Blue Book. The project, headquartered at Wright-Patterson Air Force Base, Ohio, was terminated December 17, 1969. Of a total of 12,618 sightings reported to Project Blue Book, 701 remained "unidentified."

The decision to discontinue UFO investigations was based on an evaluation of a report prepared by the University of Colorado entitled, **"Scientific Study of Unidentified Flying Objects;"** a review of the University of Colorado's report by the National Academy of Sciences; previous UFO studies and Air Force experience investigating UFO reports during 1940 to 1969.

As a result of these investigations, studies, and experience gained from investigating UFO reports since 1948, the conclusions of Project Blue Book were:

1. **No UFO reported, investigated, and evaluated by the Air Force has ever given any indication of threat to our national security.**
2. **There has been no evidence submitted to or discovered by the Air Force that sightings categorized as "unidentified" represent technological developments or principles beyond the range of present-day scientific knowledge.**
3. **There has been no evidence indicating the sightings categorized as "unidentified" are extraterrestrial vehicles.**

With the termination of Project Blue Book, the Air Force regulation establishing and controlling the program for investigating and analyzing UFOs was rescinded. Documentation regarding the former Blue Book investigation was permanently transferred to the Modern Military Branch, National Archives and Records Service, and is available for public review and analysis.

Since the termination of Project Blue Book, nothing has occurred that would support a resumption of UFO investigations by the Air Force.
http://en.wikipedia.org/wiki/Project_Blue_Book#USAF_current_official_statement_on_UFOs

Identifying **Unidentified Flying Objects** is a difficult task due to the normally poor quality of the evidence provided by those who report sighting the objects. Nevertheless, most officially investigated UFO sightings, such as from the U.S. Air Force's **Project Blue Book**, have been identified as being due to honest misidentifications of natural phenomena, aircraft, or other prosaic explanations. In early U.S. Air Force attempts to explain UFO sightings, unexplained sightings routinely numbered over one in five reports. However, in early 1953, right after the CIA's **Robertson Panel**, percentages of unexplained sightings dropped precipitously, usually being only a few percent in any given year. When Project Blue Book closed down in 1970, only 6% of all cases were classified as being truly unidentified.

UFOs that can be explained are sometimes termed **Identified Flying Objects (IFOs).**

Post-Blue Book U.S.A.F. UFO Activities (1969)

An Air Force memorandum (released via the **Freedom of Information Act**) dated October 20, 1969, and signed by Brigadier General C.H. Bolander states that even after Blue Book was dissolved, that "reports of UFOs" would still "continue to be handled through the standard Air Force procedure designed for this purpose." Furthermore, wrote Bolander, *"Reports of unidentified flying objects which could affect national security ... are not part of the Blue Book system."*[19] To date, these other investigation channels, agencies or groups are unknown.

Additionally, Blum reports[6] that Freedom of Information Act requests show that the U.S. Air Force has continued to catalog and track UFO sightings, particularly a series of dozens of UFO encounters from the late 1960s to the mid-1970s that occurred at U.S. military facilities with nuclear weapons. Blum writes that some of these official documents depart drastically from the normally dry and bureaucratic wording of government paperwork, making obvious the sense of "terror" that these UFO incidents inspired in many U.S.A.F. personnel.

Project Blue Book Special Report No. 14 (1952 - 1954)

In late December 1951, Ruppelt met with members of the **Battelle Memorial Institute,** a think tank based in Columbus, Ohio. Ruppelt wanted their experts to assist them in making the Air Force UFO study more scientific. It was the Battelle Institute that devised the standardized reporting form. Starting in late March 1952, the Institute started analyzing existing sighting reports and encoding about 30 report characteristics onto IBM punched cards for computer analysis.

Project Blue Book Special Report No. 14 was their massive statistical analysis of Blue Book cases to date, some 3200 by the time the report was completed in 1954 after Ruppelt had left Blue Book. Even today, it represents the largest such study ever undertaken. Battelle employed four scientific analysts, who sought to divide cases into *"knowns", "unknowns",* and a third category of *"insufficient information."* They also broke down knowns and unknowns into four categories of quality, from excellent to poor. E.g., cases deemed excellent might typically involve experienced witnesses such as airline pilots or trained military personnel, multiple witnesses, corroborating evidence such as radar contact or photographs, etc. In order for a case to be deemed a "known", only two analysts had to independently agree on a solution. However, for a case to be called an "unknown", all four analysts had to agree. Thus the criterion for an "unknown" was quite stringent.

In addition, sightings were broken down into six different characteristics — color, number, duration of observation, brightness, shape, and speed — and then these characteristics were compared between knowns and unknowns to see if there was a statistically significant difference.

The main results of the statistical analysis were:

- About 69% of the cases were judged known or identified (38% were considered conclusively identified while 31% were still "doubtfully" explained); about 9% fell into insufficient information. About 22% were deemed "unknown", down from the earlier 28% value of the Air Force studies.
- In the known category, 86% of the knowns were aircraft, balloons, or had astronomical explanations. **Only 1.5% of all cases were judged to be psychological or "crackpot" cases.** A "miscellaneous" category comprised 8% of all cases and included possible hoaxes.
- The higher the quality of the case, the more likely it was to be classified unknown. 35% of the excellent cases were deemed unknowns, compared to only 18% of the poorest cases being unknown. *This was the exact opposite result predicted by skeptics, who usually argued unknowns were poorer quality cases involving unreliable witnesses that could be solved if only better information were available.* [Italics added]
- In all six studied sighting characteristics, the unknowns were different from the knowns at a highly statistically significant level: in five of the six measures, the odds of knowns differing from unknowns by chance was only 1% or less. When all six characteristics were considered together, the probability of a match between knowns and unknowns was less than 1 in a billion. **(More detailed statistics can be found at Identified Flying Objects.)**

Despite this, the summary section of the Battelle Institute's final report declared it was "highly improbable that any of the reports of unidentified aerial objects... represent observations of technological developments outside the range of present-day knowledge." A number of researchers, including Dr. Bruce Maccabee, who extensively reviewed the data, have noted that the conclusions of the analysts were usually at odds with their own statistical results, displayed in 240 charts, tables, graphs and maps. Some conjecture that the analysts may simply have had trouble accepting their own results or may have written the conclusions to satisfy the new political climate within Blue Book following the Robertson Panel.

When the Air Force finally made Special Report #14 public in October 1955, it was claimed that the report scientifically proved that UFOs did not exist. Critics of this claim note that the report actually proved that the "unknowns" were distinctly different from the "knowns" at a very high statistical significance level. The Air Force also incorrectly claimed that only 3% of the cases studied were unknowns, instead of the actual 22%. They further claimed that the residual 3% would probably disappear if more complete data were available. Critics counter that this ignored the fact that the analysts had already thrown such cases into the category of "insufficient information", whereas both "knowns" and "unknowns" were deemed to have sufficient information to make a determination. Also, the "unknowns" tended to represent the higher quality cases, i.e. reports that already had better information and witnesses.

The result of the monumental BMI study was echoed by a 1979 French GEPAN report which stated that about a quarter of over 1,600 closely studied UFO cases defied explanation, stating, in part, "These cases … pose a real question."[20] When GEPAN's successor SEPRA closed in 2004, 5800 cases had been analyzed, and the percentage of inexplicable unknowns had dropped to about 14%. The head of SEPRA, Dr. Jean-Jacques Velasco, found the evidence of extraterrestrial origins so convincing in these remaining unknowns, that he wrote a book about it in 2005.[1]

Hynek's Criticism

Hynek was an associate member of the Robertson Panel, which recommended that UFOs needed debunking. A few years later, however, Hynek's opinions about UFOs changed, and he thought they represented an unsolved mystery deserving scientific scrutiny. As the only scientist involved with US Government UFO studies from the beginning to the end, he could offer a unique perspective on Projects Sign, Grudge, and Blue Book.

After what he described as a promising beginning with a potential for scientific research, Hynek grew increasingly disenchanted with Blue Book during his tenure with the project, leveling accusations of indifference, incompetence, and of shoddy research on the part of Air Force personnel. Hynek notes that during its existence, critics dubbed Blue Book "The Society for the Explanation of the Uninvestigated."[21]

Blue Book was headed by Ruppelt, then Captain Hardin, Captain Gregory, Major Friend, and finally Major Hector Quintanilla. Hynek had kind words only for Ruppelt and Friend. Of Ruppelt, he wrote, "In my contacts with him I found him to be honest and seriously puzzled about the whole phenomenon."[22] Of Friend, he wrote "Of all the officers I worked with in Blue Book, Colonel Friend earned my respect. Whatever private views he may have held, he was a total and practical realist, and sitting where he could see the scoreboard, he recognized the limitations of his office but conducted himself with dignity and a total lack of the bombast that characterized several of the other Blue Book heads."[23]

He held Quintanilla in especially low regard: "Quintanilla's method was simple: disregard any evidence that was counter to his hypothesis."[24] Hynek wrote that during Air Force Major Hector Quintanilla's tenure as Blue Book's director, "the flag of the utter nonsense school was flying at its highest on the mast." Hynek reported that Sergeant David Moody, one of Quintanilla's subordinates, "epitomized the conviction-before-trial method. Anything that he didn't understand or didn't like was immediately put into the psychological category, which meant 'crackpot'.

Hynek reported bitter exchanges with Moody when the latter refused to research UFO sightings thoroughly, describing Moody as "the master of the possible: possible balloon, possible aircraft, possible birds, which then became, by his own hand (and I argued with him violently at times) the probable."

"UNIDENTIFIED FLYING OBJECTS AND AIR FORCE PROJECT BLUE BOOK"
URL accessed February 21, 2010

1. Dr. Michael D. Swords; "UFOs, the Military, and the Early Cold War Era", pages 82-121 in "UFOs and Abductions: Challenging the Borders of Knowledge" David M. Jacobs, editor; 2000, University Press of Kansas, ISBN 0-7006-1032-4; p103.
2. see Clark, 1998
3. Dr. Michael D. Swords; "UFOs, the Military, and the Early Cold War Era", p102
4. http://www.nicap.dabsol.co.uk/51-69.htm
5. Blum, Howard, *Out There: The Government's Secret Quest for Extraterrestrials*, Simon and Schuster, 1990
6. Jacques Vallee, Passport to Magonia: On UFOs, Folklore and Parallel Worlds (1969)
7. Jerome Clark, *The UFO Book: Encyclopedia of the Extraterrestrial*, 1998; Detroit: Visible Ink Press, ISBN 1-57859-029-9
8. Jerome Clark, *The UFO Book: Encyclopedia of the Extraterrestrial*, p468
9. Jerome Clark, *The UFO Book: Encyclopedia of the Extraterrestrial*, p592
10. Ann Druffel; *Firestorm: Dr. James E. McDonald's Fight for UFO Science*; 2003, Wild Flower Press; ISBN 0-926524-58-5, p63
11. see also Portage County UFO chase page for further details
12. Jerome Clark *The UFO Book: Encyclopedia of the Extraterrestrial*, p477
13. Jerome Clark, *The UFO Book: Encyclopedia of the Extraterrestrial*, 1998; Detroit: Visible Ink Press, p478–479; emphasis as in original
14. http://www.nidsci.org/pdf/quintanilla.pdf | Quintanilla, H. (1974). UFOs, An Air Force Dilemma.
15. Jerome Clark, *The UFO Book: Encyclopedia of the Extraterrestrial*, p480
16. Proceedings of the Sign Historical Group UFO History Workshop
17. USAF Fact Sheet 95-03
18. Jenny Randles and Peter Houghe; *The Complete Book of UFOs: An Investigation into Alien Contact and Encounters*; Sterling Publishing Co, Inc, 1994; ISBN 0-8069-8132-6, p179
19. Jenny Randles and Peter Houghe; *The Complete Book of UFOs: An Investigation into Alien Contact and Encounters*; Sterling Publishing Co, Inc, 1994; ISBN 0-8069-8132-6, p202
20. J. Allen Hynek; *The UFO Experience: A Scientific Inquiry*; 1972; Henry Regenery Company, p180
21. J. Allen Hynek; *The UFO Experience: A Scientific Inquiry*; 1972; Henry Regenery Company, p175
22. J. Allen Hynek; *The UFO Experience: A Scientific Inquiry*; 1972; Henry Regenery Company, p187
23. J. Allen Hynek; *The UFOExperience: A Scientific Inquiry*; 1972; Henry Regenery Company, p103

- ABC's Peter Jennings reports on Blue Book and history of UFOs in America
- Project Blue Book Archive Online version of USAF Project Blue Book microfilm
- The Project Bluebook Story and Facts UFO.Whipnet.org
- *The Report on Unidentified Flying Objects – Edward J. Ruppelt* The book on-line

- **Project Blue Book in FBI's FOIA reading room** A summary of the project
- **Footnote Project Blue Book Archive** View and annotate original Project Blue Book report documents from the National Archives (NARA) microfilm collection.
- **Official U.S. Air Force factsheet**
- **Informal Dept of Defense collection of references on the web**

CHAPTER 36

UNIVERSITY OF COLORADO UFO PROJECT
(CONDON REPORT) (1966 - 1968)

The **Condon Committee** was the informal name of the **University of Colorado UFO Project**, a study of unidentified flying objects, undertaken at the University of Colorado from 1966 to 1968 under the direction of physicist **Edward Condon**.

From the outset, the US Air Force wanted out of all investigations into unidentified flying objects as they were getting increased sighting reports yearly and felt that there wasn't anything to it and that it was taking up too much of the Air Force's time in investigations. In order to appear above board on the matter and to placate the public fears and concerns while downplaying the interest in the subject matter, a scientific committee was struck with a biased agenda to discredit all and any existing and future UFO sightings as being real!
https://www.youtube.com/watch?v=H8jlNttLUOQ

The Condon Committee was instigated at the behest of the United States Air Force, which had studied UFOs since the 1940s. After examining many hundreds of UFO files from the Air Force's **Project Blue Book** and from civilian UFO groups **NICAP** and **APRO**, the Committee selected 56 to analyze in detail for the purpose of deciding whether "analysis of new sightings may provide some additions to scientific knowledge of value to the Air Force"[1] and "to learn from UFO reports anything that could be considered as adding to scientific knowledge".[2]

This final report (normally titled **Scientific Study of Unidentified Flying Objects** but commonly called the **Condon Report**) was published in 1968. Arguing that the study of UFOs was unlikely to yield major scientific discoveries, the report also suggested that "persons with good ideas for specific studies in this field should be supported" by Federal government agencies on a case by case basis. In particular, the Committee noted that there were gaps in scientific knowledge in the fields of "atmospheric optics, including radio wave propagation, and of atmospheric electricity" that might benefit from further research in the UFO field.[2]

The Report was reviewed by a panel of the National Academy of Sciences, which endorsed its scope, conclusions, and recommendations.[3] The Report's conclusions were generally welcomed by the scientific community and have been cited as a decisive factor in the generally low levels of interest regarding UFOs among academics in subsequent years. **Peter Sturrock** writes that the report is "the most influential public document concerning the scientific status of this [UFO] problem. Hence, all current scientific work on the UFO problem must make reference to the Condon Report."[4] However, the report has faced much criticism as to its methodology and bias, from both investigators who worked on the project and others.

Beginning in 1947 with **Project Sign** (which then became **Project Grudge** and finally Project Blue Book), the U.S. Air Force had undertaken a formal study of UFOs, which had become a subject of considerable public (and some governmental) interest. Yet Blue Book had come under increasing criticism in the 1960's. Growing numbers of critics—including U.S. politicians, newspaper writers, UFO researchers, scientists and some of the general public—were suggesting that Blue Book was conducting shoddy, unsupported research, or worse, perpetrating a cover up. UFO researcher Jerome Clark goes so far as to write that Blue Book had "lost all credibility." (p. 592) [5] The Air Force wished to stop studying UFOs, yet they found themselves in a bind: If they simply ended Blue Book, they'd risk inflaming cover-up accusation, but UFOs had become such a controversial issue that no other governmental agency was willing to take responsibility for further UFO studies.

Following a wave of UFO reports in 1965, astronomer and Blue Book consultant J. Allen Hynek wrote a letter to the **Air Force Scientific Advisory Board (AFSAB)** suggesting that a panel convene to re-examine Blue Book, and offer some new ideas as to goals and directions. The AFSAB agreed and asked Brian O'Brien to chair a committee. The Ad Hoc Committee to Review Project Blue Book (also called the O'Brien Committee) convened for one day in February 1966. All Committee members but astronomer Carl Sagan had formal ties to the AFSAB. While none of the O'Brien committee accepted as viable anything so radical as the **Extraterrestrial Hypothesis (ETH),** they did suggest that previous UFO studies had been lacking, and could be undertaken "in more detail and depth than had been possible to date" and that the U.S. Air Force should work "with a few selected universities to provide scientific teams" to study UFOs. [5] The O'Brien Committee suggested that, ideally, about 100 well-documented UFO sightings should be studied annually, with about 10 man-days devoted to each case. **(Saunders and Harkins, 25)**

In late March 1966, two days of mass UFO sightings were reported in Michigan. After studying the reports, Hynek offered a provisional hypothesis for some of the sightings: a few of about 100 witnesses had mistaken swamp gas for something more spectacular. At the press conference where he made his announcement, Hynek made repeated, strenuous qualification that swamp gas was a plausible explanation for only a portion of the Michigan UFO reports and certainly not for UFO reports in general. His qualifications were largely overlooked, and the words "swamp gas" were repeated ad infinitum in relation to UFO reports, and the explanation was subject to national derision. Soon, a UFO hearing was scheduled for April 5, 1966, before the United States Congress, directed by L. Mendel Rivers.

At the hearing, Air Force secretary Harold Brown defended the Air Force's UFO studies, but he also echoed the O'Brien Committee in stating that there was room for "even stronger emphasis on the scientific aspects". [5] At the same hearing, Hynek suggested that "a civilian panel of physical and social scientists ... examine the UFO problem critically for the express purpose of determining whether a major problem exists." [6] Shortly after the congressional hearing, the Air Force announced it was seeking one or more universities to undertake a study of UFOs. The Air Force wanted a respected figure with no publicly declared opinions on UFOs to direct the study, hoping such an effort might reduce or eliminate Air Force critic. **(Saunders, 25)** The Air Force ideally wanted to have several groups active at several universities, but it took some time to find even a single school willing to accept the Air Force's offer. Both Hynek and James E. McDonald suggested their own campuses (Northwestern University and the University of Arizona, respectively), but they were not accepted, because both men had become lighting rods for UFO-related controversy, though for very different reasons: to some, Hynek was tainted by his Air Force association, while McDonald was publicly discussing the extraterrestrial hypothesis as a viable explanation for UFOs. Astronomer **Donald Menzel** was suggested to lead the project, but he was rejected because many saw him as a biased debunker.

Harvard University, the University of California, Berkeley, the Massachusetts Institute of Technology and the University of North Carolina at Chapel Hill were all asked to consider the UFO project, but all declined. Some schools were afraid of attracting controversy if they mishandled the study, but more often, UFOs were seen as a somewhat suspect field of study.

After the National Center for Atmospheric Research declined to undertake the Air Force's UFO study, its director Walter Orr Roberts suggested that the Air Force ask physicist Edward Condon of the University of Colorado to take the project.

In the summer of 1966, Condon agreed to consider the Air Force's offer. Condon was among the best known and most distinguished scientists of his time, but he required some persuading to accept the Air Force's project. Condon would later report that Air Force Col. Ratchford had appealed to his vanity and sense of civic responsibility, telling him that the UFO project was "a dirty chore", but a man of Condon's reputation would produce results readily accepted by the scientific community. **(Jacobs, 208)**

Dr. Edward U. Condon

467

Despite his reticence, Condon was an ideal project director, from the Air Force's perspective. Perhaps most impressively, Condon had years earlier bucked the House Un-American Activities Committee when it investigated him due in part to his wife's Czechoslovakian background; Saunders characterized Condon's tenacious encounters with the HUAC as "almost legendary" among fellow scientists **(Saunders and Harkins, 33)**. Saunders writes that this and other occasions had created an impression that Condon "was a scientist who spoke the public language" and who was willing to point out governmental abuses where he saw them. **(Saunders and Harkins, 33)** Hynek noted that Condon *"was noted not only for his scientific record but also for his courage in speaking out on controversial issues."* [6]

Condon asked **Robert J. Low** -- an assistant dean of the university's graduate school program— his opinion of the University of Colorado undertaking the study. Low approved of the idea and presented it to several professors and deans. Reactions were mixed. Some thought the UFO project could be worthwhile, but others rejected it as too controversial or too disreputable.

The Trick Memo

On August 9, 1966, Low wrote a memorandum intended to persuade the more reluctant faculty to accept the UFO project. This so-called "Trick Memo" explained how the University could perform the project without risking their reputation, and how the University UFO research project could arrive at a predetermined conclusion while appearing objective. In part, Low wrote:

> "Our study would be conducted almost entirely by non-believers who, though they couldn't possibly *prove* a negative result, could and probably would add an impressive body of thick evidence that there is no reality to the observations. The trick would be, I think, to describe the project so that, to the public, it would appear a totally objective study but, to the scientific community, would present the image of a group of non-believers trying their best to be objective but having an almost zero expectation of finding a saucer." Low also suggested that if the study focused less on "the physical reality of the saucer", and more on the "psychology and sociology of person and groups who report seeing UFOs", then "the scientific community would get the message"[5] emphasis as in original).

In the same article cited above, Klass suggests that the word "trick," as used by Low, did not have the "devious" connotation perceived by Americans, but rather that the Oxford University-educated Low might have absorbed the British usage, meaning, "the art or knack of doing something skillfully." Low insisted he'd meant the word "trick" this way in the memo, and one of Low's colleagues reported that Low had sometimes used the word "trick" more in line with British usage, but Klass's interpretation of this memo, however, seems to be in the minority, with critics suggesting that Low's meaning is obviously deceptive, by its context. This Trick Memo would later come to public attention, and would generate considerable controversy. Though Committee member David Saunders had some sharp criticism for Low, he also wrote that even considering the Trick Memo, "to present Low as a plotter or conspirator is unfair and hardly accurate" (Saunders and Harkins, 128), though Saunders does suggest that it was "hasty and foolish to express such ideas on paper--especially foolish if Low really believed what he was saying." **(Saunders and Harkins, 129)** Similarly, Hynek wrote that "I believe Low has been unduly criticized for this memo. I can appreciate the dilemma Low faced. He wanted his university to get the contract (for whatever worldly reason) and to convince the university administration that they should take it ... He wanted to invoke a path of respectability. But the path he chose was unfortunate." [6]

Afterwards, Low approached several members of the school's psychology staff, notably **William A. Scott** and **David R. Saunders**, who agreed to aid for the project, though they were initially unaware that the prime focus of the study would be psychological. Saunders would become a co-principal investigator, would play a major role in the project, and also in the subsequent controversies and publicity. Saunders was a NICAP member and the only committee member with more than a passing interest in UFOs.

Critics (including Jerome Clark) have suggested that finances were a factor in persuading the school to accept the Air Force's project: The University of Colorado had recently seen substantial budget cuts, while the Air Force offered $313,000 for the study (the total funding would later rise to over $500,000). Condon dismissed this suggestion, noting that $313,000 was a rather modest budget for an undertaking scheduled to last more than a year with a staff of over a dozen. **(Saunders and Harkins, 29)**

The Study Begins

On October 6, 1966, the University of Colorado formally agreed to undertake the UFO study. Condon would be the director, while Low was coordinator and Saunders a co-principal investigator, along with astronomer Franklin Roach. The other primary Committee members were astronomer William K. Hartmann; psychologists Michael Wertheimer, Dan Culbertson and James Wadsworth (a graduate student); chemist Roy Craig; electrical engineer Norman Levine; physicist Frederick Ayer; and administrative assistant Mary Louise Armstrong. Several other scientists or experts would serve in part-time and temporary roles, or as consultants.

Two days after the Committee had formally accepted the project, the Denver Post quoted Low as saying that the project had met the University's acceptance threshold by the narrowest of margins, and furthermore that the project was accepted largely because it was difficult to say no to the Air Force.

Public response to the Committee's announcement was generally positive; historian Jacobs argues that there was "optimism on all sides." (Jacobs, 225). Hynek characterized Condon's perspective towards UFOs as "basically negative", but he also assumed the Condon's opinions would change once he familiarized himself with evidence in some of the more puzzling UFO cases. NICAP's Donald Keyhoe was publicly supportive, but privately expressed fears that the Air Force would be controlling things from behind the scenes. That a scientist of Condon's standing would involve himself with UFO research marked something of a sea change, and heartened some academics who had long expressed interest in the subject, such as atmospheric physicist James E. McDonald. Many other scientists who'd earlier been hesitant to speak out on the subject now offered their opinions, whether skeptical, supportive or somewhere in between.

One of the **Condon Committee's** first formal duties was a briefing by **Hynek** and astrophysicist/mathematician **Jacques Vallee**. Both men stressed the importance of implementing a fast, consistent, statistical rating system to sort UFO reports and focus attention towards the best documented and most puzzling cases. The Committee also met with Major **Hector Quintanilla** (then head of **Project Blue Book**), and with Col. Robert Hippler of the Air Force Office of Scientific Research.

The Committee secured the help of civilian UFO research group **APRO**, though they would play a relatively minor role in the project when compared to **NICAP's** involvement. In November 1966, **Keyhoe** and **Richard Hall** (both of NICAP), briefed the panel. They agreed to share NICAP's considerable research files, and also to implement an **Early Warning System** to better collect UFO reports. Eventually, Hall and **Saunders** would form a "close working relationship" after Hall worked for the Committee as a paid consultant for two weeks.[5]

The remainder of 1966 was devoted primarily to assembling a library, and determining how to best collect on-site investigations of UFO reports as quickly as possible. Despite these advances, the Committee was somewhat adrift for a few months, due in no small part to disagreements among the Committee's members as use of funding **(Saunders and Harkins, 77)** and methodology debates.[6] A particular problem was that by seeking out people with no position on UFOs, the Committee was staffed by persons with no experience regarding (or knowledge of) previous UFO studies. One Committee member suggested filming UFOs using stereo cameras mounted with diffraction gratings in order to study the spectrum of light emitted by UFOs. This had been attempted some fifteen years earlier following a specific suggestion regarding UFOs made by Dr. Joseph Kaplan in 1954 but was quickly judged impractical after a number of such cameras were distributed to Air Force bases.[6]

When the Air Force asked for a progress report in January 1967, Committee members scrambled to prepare a presentation. During the meeting with Air Force officials, Committee members noted that they had decided to focus more on alleged UFO eyewitnesses than on UFO reports themselves, and hoped to stage false UFO reports to test witness perception and memory. This plan was quashed by Air Force Colonel Hippler, who feared what Michael D. Swords described as "a public relations catastrophe for the Air Force." The meeting was unproductive until Low asked specifically what the Air Force expected from the Committee. The Air Force representatives had no ready reply, but a few days later, Col. Hippler wrote to Low. Though he wrote on Air Force letterhead, Hippler stressed that he was writing not in any official capacity, and suggested that ultimately, the UFO project "ought to be able to come to an anti-ETH conclusion", as Clark writes. Hippler went on to write that the Air Force wanted to cease its UFO studies and that an official study reporting that there was nothing unusual about UFO reports would be the best way to accomplish this goal.[5] Low replied to the letter, thanking Hippler for clarifying the Air Force's expectations. With this sequence of events, Swords argued some years later, "the fix was in," and the Committee had formally abandoned any pretense of objectivity.[5]

Internal Tensions Begin

In late January 1967, Keyhoe and Hall gave Saunders a clipping of *The Elmira Star-Gazette*, dated January 26. Condon was quoted as saying that he thought the government should not study

UFOs because the subject was nonsense, adding, "but I'm not supposed to reach that conclusion for another year."[5] Saunders was stunned. He asked if Condon could have been misquoted, but Keyhoe reported that several NICAP members had been present when Condon delivered his lecture; one of them had resigned from NICAP in protest, arguing that the Condon Committee was nothing more than pretense. The next day Saunders confronted Condon about the press clipping. Saunders feared that NICAP would end their association with the Committee (thus eliminating a valuable source of case files), and furthermore that the negative publicity following a split from NICAP could harm public perception of the committee.

In the meantime, Condon had taken no part in the field investigations; he would ultimately investigate at most four or five UFO cases, mostly contactees, of several hundred cases which the Committee examined. Furthermore, the committee's members found it difficult to speak with Condon: they usually had to speak to coordinator Low with questions or problems but were often unsatisfied with Low's efforts. On at least one occasion, Condon fell asleep while a consultant was offering a presentation. Consultant James E. McDonald had initially been hopeful for the Committee, but after making a few presentations and feeling as though Condon completely ignored his contributions, McDonald grew increasingly vocal in his criticism. He would soon begin to detail his view of the committee's problems in letters to Frederick Seitz, president of the National Academy of Science.

Despite the growing internal tension, the Committee's members continued to collect, study and analyze UFO reports, including nearly 40 field investigations around the United States. They investigated a few well-known reports, including an early cattle mutilation report. There was, however, an increasing suspicion among the Committee's members that their research would be used to support a foregone conclusion. Most of the Committee's regular members objected to the manner in which Condon and Low were directing the Committee, and several members were considering writing a dissenting minority report if Condon overruled their conclusions that some UFO reports seemed anomalous and deserving of closer scrutiny. The Committee was disturbed that Condon and Low tried to insulate them from Hynek, Vallee, McDonald and others who thought UFOs deserved study, while simultaneously openly consulting with avowed UFO debunkers. That Condon focused most of his interest towards the lunatic fringe of UFO reports disturbed much of the Committee as well. Another particular irritation was that while NICAP and Blue Book had promised to share new UFO reports as quickly as possible, only NICAP had done so. Even Condon—so often criticized for bias and ambivalence—formally complained to the Air Force about their lack of cooperation.

The Committee's members usually worked solo, and rarely (if ever) met as a group to discuss their progress, to critique one another's work, or to reach a consensus on disagreements. Because of this, individuals embraced a number of approaches, sometimes resulting in conflict or disagreements. Notably, the Committee's members differed in their opinions regarding the extraterrestrial hypothesis. Some (especially Saunders) thought the ETH should be included as one of a range of hypotheses to explain UFOs; others (notably Low and Wertheimer) rejected any consideration of the ETH. Low wrote a position paper characterizing the ETH as "nonsense"; Wertheimer adamantly argued that the ETH could neither be proved nor disproved, and he afterwards had little to do with the Committee. **(Jacobs, 228)** This ETH dispute developed into an ideological and methodological schism among the Committee's members: One group, championed by Low, thought that, as Jacobs writes "the solution to the UFO mystery was to be found in the psychological makeup of witnesses"; the other group, championed by Saunders, "wanted to look at as much as the data as possible." **(Jacobs, 230)**

In September 1967, another collision with NICAP was narrowly averted. Keyhoe learned that Condon had given a lecture to the National Bureau of Standards, a group Condon had once chaired. In his lecture, Condon had discussed three UFO reports made by obviously unstable kooks and had intimated that many or most UFO reports came from such persons. An irritated Keyhoe asked Saunders why NICAP's time and money should be used in collecting and forwarding UFO reports to the Committee when Condon's bias was obvious. Keyhoe threatened to sever NICAP's association with the Condon Committee. In spite of his own growing doubts, Saunders convinced Keyhoe that Condon could separate his own opinions from his work, and had simply forgotten to state where his personal opinions began. Keyhoe accepted this but also warned that if the committee could not demonstrate a more objective manner, NICAP would cease their involvement and publicize their complaints.

After Keyhoe was mollified (at least temporarily), Saunders told Condon of the development. Condon was nonplussed; if NICAP chose to sever their association he had no objection. After some thirty minutes of discussion, Saunders persuaded Condon to write Keyhoe and report that the quotes from the National Bureau of Standards speech were taken out of context. Shortly after this, both Low and Condon were quoted in the *Rocky Mountain News* as expressing their approval of an article in *Science* arguing against the ETH. Privately and publicly—including during Committee proceedings—Low and Condon were repeatedly arguing that UFO studies were a waste of time. Clark writes that "By now all that was keeping the staff from open revolt was one hold-out: Roy Craig, who insisted that Condon still had his full confidence."[5]

Fearing the worst from NICAP following the *Rocky Mountain News* story, Low flew from Colorado to Washington, DC for a meeting with Keyhoe. Keyhoe asked Low if the Committee was "on the level". According to Keyhoe, Low replied, "I see no reason why you have to determine whether the Colorado Project is on the level or not," and furthermore admitted that Condon had a very negative opinion of the Project and of UFO studies in general. Low noted that much of the Committee held opinions very different from Condon's, but Keyhoe countered that as director, Condon could override any dissenting opinions when the final report was written.

Despite these problems, Low urged Keyhoe to continue sending case files and reports to the Committee. When Keyhoe asked why NICAP should continue supporting a project which had effectively reached its findings, Keyhoe reported Low's reply as, "If you don't, the project could be accused of reaching a conclusion without all of NICAP's evidence."[5]

Cracks in the Dam

Due to several developments in 1966 and 1967, the internal conflicts in the Condon Committee were about to burst into public awareness.

On November 14, 1966, Keyhoe wrote a long letter to Condon (c.c. to Low), detailing his concerns and questions regarding the project. Were Condon and Low's biases tainting the

472

project? Were the Air Force's orders directing the project? Had Condon himself read any of the NICAP case studies? Why had Condon himself done so little field research? Condon and Low replied by telling Keyhoe that they were under no obligation to answer his queries. With their non-answer, Keyhoe had nearly reached his breaking point; NICAP was no longer sending UFO case files to the Committee.

The Trick Memo Exposed

In July 1967, Committee Member Roy Craig was scheduled to speak before a Portland, Oregon audience regarding the Condon Committee. When Craig asked Low for some documentation regarding the Committee's origins, Low gave him a stack of papers, unaware that a copy of the Trick Memo was included. After giving the speech, Craig—previously Condon's staunchest ally on the Committee, other than Low—showed the Trick Memo to Committee member Norman Levine, saying, "See if this doesn't give you a funny feeling in the stomach."[5]

Levine showed the memo to Saunders, who was saddened but not surprised; the memo seemed to explain the attitudes Low and Condon had demonstrated from the project's beginning. Copies of the Trick Memo were circulated to the entire Committee, barring Low and Condon. Public disclosure of the memo was considered, but decided against: there was still hope that the final report might recommend further study of the UFO phenomenon. Eventually, however, Saunders gave a copy of the memo to Keyhoe. In turn, Keyhoe told James E. McDonald of the memo's contents, but, citing confidentiality promises, did not give him a copy of the memo. Eventually, McDonald located a copy of the memo in the project's open files.

The Trick Memo confirmed McDonald's worst suspicions about the Committee. In response, he wrote a seven-page letter to Condon, explaining point by point, his problems, frustration, and disappointment with the Committee's shortcomings. Apparently unaware that the Trick Memo was never intended for public circulation, McDonald quoted a few lines from it (the same "...the trick would be..." portion cited above), then added, "I am rather puzzled by the viewpoints expressed there ... but I gather that they seem straightforward to you, else this part of the record would, presumably, not be available for inspection in the open Project files."[5]

When Condon read McDonald's letter on February 5, 1968, he became furious. Low read the letter, and Armstrong reported that he "exploded," suggesting that whoever was responsible for McDonald's having the memo should be fired, before calming down and discussing the affair with Condon. **(Saunders and Harkins, 188)** The next day, Condon called a meeting of the Committee to uncover the chain of events that had led to McDonald's receiving the Trick Memo. Saunders characterized Condon's manner as imperious, behaving as though he were "the Grand Inquisitor." **(Saunders and Harkins, 190)** Condon asked the Committee to read McDonald's letter. When they did, the Committee was initially occupied with the substance McDonald's incisive, pointed critique and all but ignored the few lines quoted from the Trick Memo. When Condon wanted to know how McDonald had received a copy of a project memo, Saunders admitted that he'd forwarded the Trick Memo to Keyhoe. Condon reportedly called Saunders "disloyal" and said, "For an act like that you deserve to be ruined professionally." **(Saunders and Harkins, 189)** Saunders responded, he said, by stating he was loyal to the American public, while Condon seemed beholden to the Air Force.

The next day, in brief letters, Saunders and Levine were fired "for cause", and Condon issued a press release reporting that the men had been fired "for incompetence." The Colorado Daily asked Condon to elaborate on the nature of the incompetence, and he declined. Fearing libel charges from Saunders and Levine if the paper ran unqualified accusations of incompetence, the *Colorado Daily* omitted the reason for Saunders and Levine's termination, thus angering Condon. (Saunders and Harkins, 193) Though the Trick Memo had never formally been declared confidential or personal, and though McDonald had located the memo in the project's open files, Condon repeatedly insisted in subsequent months that McDonald had "stolen" it from Low's personal files. **(Saunders and Harkins, 201)**

Condon telephoned the president of the University of Arizona to report that McDonald had stolen the trick memo from the Project's files, and also wrote a letter to the Air Force to deprecate Levine in an attempt to harm his security clearance. These were not the only instances in which Condon tried to damage someone's career after they'd dissatisfied him regarding the UFO Project. Condon had earlier tried to get Committee consultant Robert M. Wood fired from his McDonnell Douglas position after Wood had written "a critical but polite letter listing his concerns about project shortcomings"; and Condon would later consider blocking Carl Sagan's entry into the distinguished Cosmos Club because Sagan--though quite skeptical of UFOs--had argued the subject deserved serious scrutiny.[5]

On February 24, 1968, administrative assistant Mary Lou Armstrong resigned from the Condon Committee. In her letter she wrote staff morale had reached a deplorable depth, and that "there is an almost universal 'lack of confidence'" in coordinator Low, arguing that much of the Committee's troubles was Low's fault. "Had *you* [Condon] handled the direction of our activities, there would not have been such a serious conflict."[6]

Publicity

On April 30, 1968, Keyhoe held a press conference to announce that NICAP had severed all ties with the Committee. He circulated copies of the Trick Memo, which received wide publicity.

By now, the Condon Committee's conflicts were being covered in the mass media, including a John G. Fuller article, "Flying Saucer Fiasco" in the May 1968 issue of the popular magazine *Look*. Including interviews with Saunders and Levine, Fuller detailed the controversy and accusations leveled against the Condon Committee and described the project as a "$500,000 trick." [5] Condon responded by writing to Look, declaring that Fuller's article contained unspecified "falsehoods and misrepresentations". **(Jacobs, 231)**

The press had earlier occasionally mention of the Committee's troubles, but Fuller's article brought a much higher level of attention, especially from scientific and technical journals, many of which began discussing the Committee in their editorial and letters pages. *Industrial Research* reprinted the Trick Memo, while *Scientific Research* interviewed Saunders and Levine, who reported that they were considering a libel suit against Condon for terminating them for alleged "incompetence"; they furthermore said that Condon had used an "unscientific approach" in directing the Committee. **(Jacobs, 231)** Condon said that calling his methods "unscientific" was itself libelous, and in turn threatened to sue Saunders and Levine.

When the American Association for the Advancement of Science covered the ongoing Committee controversy in an issue of its official journal *Science*, Condon first promised to grant an interview apparently in the hopes of offering his side of the conflict. Shortly thereafter, however, *Science* editor Daniel S. Greenberg reported that Condon announced a change of opinion and refused to cooperate. When Greenberg pressed Condon for help, Condon refused to communicate further and resigned from the AAAS in protest when the article was published without his input. **(Jacobs, 233)**

The Fuller article also helped inspire Congressional hearings. Representative J. Edward Roush spoke on the House floor, arguing that Fuller's article brought up "grave doubts about as to the scientific profundity and objectivity of the project"; in a Denver Post interview, Roush suggested that the Trick Memo proved that the Air Force had been dictating the Project's direction and conclusions, despite denials. **(Jacobs, 233)**

Even before the Condon Report was released, astronomer **Frank Drake** wrote to the National Academy of Sciences, suggesting that the Condon Committee's final report was tainted, and should thus be discredited. The General Accounting Office announced that they were considering an investigation of the Committee's finances.

The Condon Report

In spite of the ongoing controversy, the Committee's members largely continued their work. By late 1968, they'd completed their reports and handed them over to Condon, who wrote summaries of each case study and then offered the manuscript to the NAS, then headed by Condon's longtime friend and former student, Frederick Seitz. A panel of 11 NAS members claimed they reviewed the report (the nature of their review has been debated), then issued a statement that supported the manuscript's conclusions. In response to the report's findings, Project Blue Book formally closed down in January 1970.

The Report ran to 1,485 pages in hardcover and 965 pages in the Bantam paperback edition. It divided UFO cases into five categories: old UFO reports (from before the Committee convened), new reports, photographic cases, radar/visual cases, and UFOs reported by astronauts (some UFO cases fell into multiple categories). The entire Condon Report is available online; see External Links section below.

In the second paragraph of his introductory "Conclusions and Recommendations", Condon wrote:

"Our general conclusion is that nothing has come from the study of UFOs in the past 21 years that has added to scientific knowledge. Careful consideration of the record as it is available to us leads us to conclude that further extensive study of UFOs probably cannot be justified in the expectation that science will be advanced thereby." **(Condon, 1)**

This was the core of Condon's position of UFOs, and these are his words which received wide attention in the mass media. Many reviews of the book and newspaper editorials supported Condon's position that the UFO question was answered and the case was closed.

The report earned a mixed reception from scientists and academic journals. Astrophysicist Peter A. Sturrock notes that, in general, "critical reviews [of the Committee's report] came from scientists who had actually carried out research in the UFO area, while the laudatory reviews came from scientists who had not carried out such research." **(Sturrock, 46)** Sturrock also writes that "most of the scientific community paid little attention when the report was published, and none later." **(Sturrock, 49)** Furthermore, Sturrock writes that while the Condon report received "almost universal praise from the news media", responses from "scientific journals were mixed." The esteemed journal *Nature*, printed *A Sledgehammer for Nuts*, a largely positive review, while *Icarus* (then edited by Carl Sagan) published both an approving review by Dr. Hong-Yee Chiu and a negative appraisal by **Dr. James E. McDonald**.

Sturrock summarized his analysis of the report as follows:

> In sum, it is my opinion that weaknesses of the Condon Report are an understandable but regrettable consequence of a misapprehension concerning the nature and subtlety of the phenomenon. It is also my opinion that there is much in the Condon Report that could be used in support of the proposition that an analysis of the totality of UFO reports would show that a signal emerges from the noise and that the signal is not readily comprehensible in terms of phenomena now well known to science. If this is so, then the Report makes a case for the further scientific study of UFO reports. It appears that this opinion is, in fact, shared by certain members of the Colorado Project staff. For instance, Professor David R. Saunders, who left the project in unfortunate circumstances, has published a book **(Saunders & Hawkins 1968)** challenging the findings of the Condon Report. Gordon D. Thayer also has continued his interest in the phenomenon, as is evident from his report on the Lakenheath case for the journal Astronautics and Aeronautics **(Thayer, 1971)**. [1]

To no one's surprise, however, a number of critics—several of whom had already attacked the Committee—argued that the Report was profoundly flawed, or even unscientific. Journalist **C.D.B. Bryan** writes that the final report "left nearly everyone dissatisfied." **(Bryan, 189)**

Positive Responses

Science and *Time* were among the many newspapers, magazines, and journals which published approving reviews or editorials related to the Condon Report. Some compared any continued belief in UFOs as an unusual phenomenon to those who insisted the earth was flat; others predicted that interest in UFOs would wane and in a few generations be only dimly remembered, like relics of spiritualism such as ectoplasm or table-raising.

The March 8, 1969, issue of *Nature* offered a generally positive review for the Condon Report but seemed to suggest that UFO studies were a wasteful, futile indulgence, writing, "The Colorado project is a monumental achievement, but one of perhaps misapplied ingenuity. It would doubtless be inapt to compare it with earlier centuries' attempts to calculate how many angels could balance on the point of a pin; it is more like taking a sledgehammer to crack a nut, except that the nuts will be quite immune to its impact."

On January 8, 1969, the New York Times headline reported, "U.F.O. Finding: No Visits From Afar." The article (by **Walter Sullivan**) declared that due to the report's finding, the ETH could finally be dismissed and all UFO reports had prosaic explanations. Sullivan noted that the report had its critics, but characterized them as "U.F.O. enthusiasts", a term which would subsequently reappear (often with the same dismissive tone) in later descriptions of UFO researchers. Clark argues that Sullivan had a conflict of interest by failing to disclose his as-yet-unpublished introduction to the report's paperback edition for Bantam Books[5] Furthermore, Clark characterizes Sullivan's introduction as *"a revisionist history of the project."* [5]

Negative Responses

Several observers have criticized the report as a sloppy work: Jacobs describes the report as "a rather unorganized compilation of independent articles on disparate subjects, a minority of which dealt with UFOs." **(Jacobs, 240)** Hynek agrees with this characterization, he argues that the report is "a voluminous, rambling, poorly organized report ... considerably less than half of which was addressed to the investigation of UFO reports." [6] Hynek also contended that beyond Condon's introduction, "the rest of the lengthy report defies succinct description. It is a loose compilation of partly related subjects, each by a different author."[6] Swords contends that the report's daunting structure "indicates very clearly that the organization's chaos and personnel dislocations ... made the creation of a smooth document impossible."

In the April 14, 1969, issue of *Scientific Research*, Robert L. M. Baker, Jr. wrote that rather than settling the UFO issue, the Condon Committee's report "seems to justify scientific investigation along many general and specialized frontiers." [7]

In the December 1969 issue of *Physics Today*, Condon Committee consultant Gerald Rothberg wrote that he had thoroughly investigated about 100 UFO cases, three of four of which left him puzzled. He thought that this "residue of unexplained reports [indicated a] legitimate scientific controversy." [5]

In the November 1970 issue of *Astronautics and Aeronautics*, The **American Institute of Aeronautics and Astronautics (AIAA)** published their review of the Condon Report. The AIAA subcommittee appreciated the difficulty of the undertaking, and generally agreed with Condon's suggestion that little of value had been uncovered by scientific UFO studies, but had some criticism for the Condon Report, stating that the AIAA "did not find a basis in the report for [Condon's] prediction that nothing of scientific value will come of further studies." [8]

30 of the report's 56 UFO cases are classified as unknown, though some were regarded as possibly hoaxes or misidentifications. In a review published in *Bulletin of the Atomic Scientists*, Hynek noted that the percentage of unknowns in the Condon report was well above the unknowns in **Project Sign, Project Grudge and Project Blue Book**[5].

Critics charge that Condon's case summaries are inaccurate or misleading. For example, **Gordon David Thayer** [5] was the Committee's consultant on the 35 radar-visual cases. Thayer concluded that 19 of 35 cases were almost certainly due to "anomalous propagation": so-called "radar ghosts" which appear to be a solid object, but are actually generated by fog, clouds, birds, insect swarms, or temperature inversions. Though Thayer offered anomalous propagation as a likely explanation for just over 50% of the cases he studied, Condon suggested that anomalous propagation was responsible for *all* the radar cases.

Committee members regarded a few of the UFO reports as genuinely anomalous, yet in his summaries, Condon makes no mention of these conclusions. Jacobs argues that these enigmatic reports were "buried" among the confirmed cases. **(Jacobs, 241)** In his analysis of a 1965 Lakenheath, England radar-visual case, Thayer wrote, "The apparently rational, intelligent behavior of the UFO suggests a mechanical device of unknown origin as the most probable explanation of this sighting ... The probability of at least one UFO involved appears to be fairly high", yet Condon completely ignores this conclusion. Another instance is Case Number 46, a series of photographs taken in 1950 in McMinnville, Oregon. After inspecting original photo negatives, Committee investigator **William K. Hartmann** wrote that, "This is one of the few UFO reports in which all factors investigated, geometric, psychological, and physical appear to be consistent with the assertion that an extraordinary flying object, silvery, metallic, disk-shaped, tens of meters in diameter, and evidently artificial, flew within sight of two witnesses."[9] Condon made no mention of this conclusion.

In the section devoted to UFO reports made by astronauts, **Franklin Roach** declared that three accounts related by astronauts **Frank Borman** aboard Gemini 7 and **James McDivitt** aboard Gemini 4 were "a challenge to the analyst" and "puzzling". Roach writes that if **NORAD**'s list of space objects near the Gemini 4 spacecraft was accurate (as he concluded) than the objects McDivitt reported remained unidentified. **(Condon, 312)** Again, Condon's summary doesn't mention Roach's conclusion.

In 1969, as part of his lecture "Science in Default", physicist **James E. McDonald** said, "The Condon Report, released in January 1968, after about two years of Air Force-supported study is, in my opinion, quite inadequate. The sheer bulk of the Report, and the inclusion of much that can only be viewed as 'scientific padding', cannot conceal from anyone who studies it closely the salient point that it represents an examination of only a tiny fraction of the most puzzling UFO reports of the past two decades, and that its level of scientific argumentation is wholly unsatisfactory. Furthermore, of the roughly 90 cases that it specifically confronts, over 30 are conceded to be unexplained. With so large a fraction of unexplained cases (out of a sample that is by no means limited only to the truly puzzling cases, but includes an objectionably large number of obviously trivial cases), it is far from clear how Dr. Condon felt justified in concluding that the study indicated 'that further extensive study of UFOs probably cannot be justified in the expectation that science will be advanced thereby.'"[10]

In a 1969 issue of *The American Journal of Physics,* Thornton Page reviewed the Condon Report and wrote, *"Intelligent laymen can (and do) point out the logical flaw in Condon's conclusion based on a statistically small (and selected) sample. Even in this sample, a consistent pattern can be recognized; it is ignored by the 'authorities,' who then compound their 'felony' by recommending that no further observational data be collected."*[11] Ironically, Page had been a member of the Robertson Panel which suggested UFOs should be debunked to reduce public interest, though his opinions regarding UFOs changed in his later years.

J. Allen Hynek's criticism

In his 1972 book *The UFO Experience: A Scientific Inquiry*, astronomer J. Allen Hynek discussed the Condon Report at length in a chapter titled "Science Is Not Always What Scientists Do." He argues that the report is so flawed as to be nearly worthless as a scientific study. In brief, Hynek argued, "The Condon Report settled nothing."[6] He also suggested that people should essentially read the Condon Report backwards: the case studies first, then Condon's summaries.

Hynek described Condon's introduction as "singularly slanted," but also notes that it "avoided mentioning that there was embedded within the bowels of the report a remaining mystery; that the committee had been unable to furnish adequate explanations for more than a quarter of the cases examined."[6] Hynek argues that "Unimpeachable evidence shows that Condon did not understand the nature and scope of the problem" he was charged with studying.[6]

Like many other critics, Hynek notes that some of the unsolved cases were judged most puzzling. Particularly bothersome to Hynek was the overriding notion that UFOs were tied inexorably to the idea of extraterrestrial life. By focusing on this one hypothesis, the report "did not try to establish whether UFOs really constituted a problem for the scientist, whether physical or social."[6]

Furthermore, Hynek notes that the report relies on so few UFO reports that overarching trends may have been ignored.

Hynek also argues that the Condon Report was not scientific. They chose to hinge on the ETH, but, Hynek insists, the data are simply lacking to analyze that hypothesis and reach an informed conclusion. By not being able to demonstrate that a hypothesis was falsifiable, they violated one of the fundamental rules of the **"scientific method".** The only hypothesis the Committee could have tested, Hynek wrote, was "There exists a phenomenon, described by the content of the UFO reports, which presently is not physically explainable."[6]

Peter A. Sturrock's Criticism

Astrophysicist **Peter A. Sturrock** has offered a number of detailed critiques of the Condon Report.

A review of Sturrock's critique notes that "This report has clouded all attempts at legitimate UFO research since its release. Much of the public, including the scientific community and the press, erroneously assumes that this project represents a serious, in-depth look into the issues. Sturrock assiduously dissects the Condon Report and makes it clear that the study is scientifically flawed. In fact, anyone who actually reads the report carefully will be surprised to find that Edward Condon, who personally wrote the Summary and Conclusions, did not investigate any of the cases. Rather it was his staff that did the legwork. That is why the report is internally inconsistent with the body of the document supporting some UFO cases, while the summary does not."[12]

In his own detailed critiques of the Condon Committee, Sturrock writes that "Another important point of scientific methodology is that, if one is evaluating a hypothesis (such as ETH), it is beneficial to regard this hypothesis as one member of a complete and mutually exclusive set of hypotheses. This point seems to have been recognized by Thayer ... but it was apparently ignored by Condon and other members of the project staff. It is of little use to argue that the evidence does not support one hypothesis unless one knew what the surviving hypotheses are." **(Sturrock, 40)**

Sturrock also criticizes the Condon Committee for heavy reliance on what he calls "'theory dependent' arguments. This requirement, above all, makes the appraisal of the UFO phenomenon very difficult: if we entertain the hypothesis that the phenomenon may be due to an extremely advanced civilization, we must face the possibility that many ideas we accept as simple truths may, in a wider and more sophisticated context, not be as simple, and may not even be truths." (Sturrock, 40) As a specific example of "theory dependent" analysis in the Condon Report, Sturrock notes a case where an allegedly supersonic UFO did not produce a sonic boom. He notes that "we should *not* assume that a more advanced civilization could not find some way of traveling at supersonic speeds without producing a sonic boom." Furthermore, Sturrock notes that J.P. Petit "has proposed a procedure involving magneto-hydrodynamic processes whereby the shock wave of a supersonic object would be suppressed." **(Sturrock, 40)**

References

1. ^ Condon Report: Section II, Summary of the Study, Edward Condon
2. ^ *a b* Condon Report: Section I Conclusions and Recommendations, Edward Condon
3. ^ Review of the University of Colorado Report on Unidentified Flying Objects by a Panel of the National Academy of Sciences, 1969
4. ^ "An Analysis of the Condon Report on the Colorado UFO Project" by Peter A. Sturrock, 1987, J. Scientific Exploration, Vol. 1, No. 1, p. 75.
5. ^ *a b c d e f g h i j k l m n o p q r s* Clark, Jerome, *The UFO Book: Encyclopedia of the Extraterrestrial*, Visible Ink, 1998, pp. 593-604, ISBN 1-57859-029-9.
6. ^ *a b c d e f g h i j k l m* Hynek, J. Allen, *The UFO Experience: A Scientific Inquiry*, Henry Regnery Company, 1972, pp. 192-244.
7. ^ The UFO Report: Condon Study Falls Short, Robert L. M. Baker, Jr., Scientific Research, April 14, 1969, p. 41.
8. ^ UFO - An Appraisal of the Problem, 1968 Statement of the American Institute of Aeronautics and Astronautics (AIAA) Subcommittee on UFOs
9. ^ Condon Report, Photographic Case Studies: Cases 46 - 59
10. ^ Science in Default - Twenty-Two Years of Inadequate UFO Investigations - James E. McDonald -1969
11. ^ Dr. Thornton Page's Review of the Condon Report
12. ^ Review of Peter Sturrock's 'The UFO Enigma' - UFO Evidence

- "A Sledgehammer for Nuts"; *Nature*, Volume 221, March 8, 1969; pages 899-900
- C.D.B. Bryan; Close Encounters of the Fourth Kind: Alien Abduction, UFOs and the Conference at M.I.T.; Alfred A. Knopf, 1995; ISBN 0-679-42975-1

- Edward W. Condon, Director, and Daniel S Gillmor, Editor; Final Report of the Scientific Study of Unidentified Flying Objects; Bantam Books, 1968
- David Michael Jacobs; The UFO Controversy In America; Indiana University Press, 1975; ISBN 0-253-19006-1
- David R. Saunders and R. Roger Harkins; UFO's? Yes! Where the Condon Committee Went Wrong; World Publishing, 1969
- Peter A. Sturrock; The UFO Enigma: A New Review of the Physical Evidence; Warner Books, 1999; ISBN 0-446-52565-0

External links and Sources

- Scientific Study of Unidentified Flying Objects, Dr. Edward U. Condon, Scientific Director, Daniel Gilmor, Editor
- UFO: An Appraisal of the Problem (from the AIAA)
- Robert M. L. Baker Jr., The UFO Report: Condon Study Falls Short
- Flying Saucer Fiasco, by John G. Fuller
- The Condon UFO Study: A Trick or a Conspiracy? by Philip J. Klass. 1986
- Science in Default: Twenty-Two Years of Inadequate UFO Investigations, by James E. McDonald
- UFOs And The Condon Report - A Scientist's Critique (James MacDonald Critique of the Condon Report)
- THE TRUTH ABOUT THE CONDON REPORT (from NICAP)
- Dr. Thornton Page's Review of The Condon Report
- An Analysis of the Condon Report on the Colorado UFO Project. Dr. Peter. A. Sturrock
- USAF-Sponsored Colorado Project for the Scientific Study of UFOs, by Michael D. Swords, Ph.D.
- Links to Related Online Documents regarding Scientific Study of Unidentified Flying Objects, Dr. Edward U. Condon, Scientific Director, Daniel Gilmor, Editor

CHAPTER 37

GLOBAL CIVILIAN STUDIES OF THE UFO AND ETI PHENOMENON

The Sturrock Panel and Report

Sturrock is a prominent British scientist who with initial healthy skepticism came to the conclusion that the subject of **unidentified flying objects (UFOs)** required further serious scientific study.

Sturrock's interest traces back to the early 1970s when, seeking someone experienced with both computers and astrophysics, he hired **Dr. Jacques Vallee** for a research project. Upon learning that Vallee had written several books about UFOs, Sturrock—previously uninterested in UFOs—felt a professional obligation to at least peruse Vallee's books. Though still largely skeptical, Sturrock's interest was piqued by Vallee's books. Sturrock then turned to the Condon Report (1969), the result of a two-year UFO research project that had been touted as the answer to the UFO question. Sturrock commented that "The upshot of this was that, far from supporting Condon's conclusions [that there was nothing extraordinary about UFOs], I thought the evidence presented in the report suggested that something was going on that needed study." *The UFO Enigma: A New Review of the Physical Evidence*; 2000; Edited by Peter A. Sturrock; Aspect Books; ISBN 0446677094

At about the same time that the Condon Committee was conducting its investigation, the **American Institute of Aeronautics and Astronautics (AIAA)** in 1967 had set up a subcommittee to bring the UFO phenomenon to the attention of serious scientists. In 1970 this subcommittee published a position paper also highly critical of how the Condon Committee had conducted its investigation and how Condon's written conclusions often didn't match the cases detailed in the final report. Overall, the AIAA deemed about a third of the cases still unsolved. Unlike Condon, they felt these unsolved cases represented the essential core of the UFO problem and deserving of further scientific scrutiny.

Sturrock was curious what the general attitudes of the members of the AIAA might be and in 1973 surveyed the San Francisco branch of the AIAA, with 423 out of 1175 members responding. Opinions were widespread as to whether UFOs were a scientifically significant problem. Most seemed unsure or neutral on the question. Sturrock was also curious as to whether fellow scientists like the AIAA members ever reported seeing UFOs, i.e., anomalous aerial phenomena that they couldn't identify.

The survey indicated that about 5% had, which is typical for what is usually reported for the general population as a whole. *The UFO Enigma: A New Review of the Physical Evidence. 2000. Edited by Peter A. Sturrock. Aspect Books. ISBN 0446677094*

In 1975, Sturrock did a more comprehensive survey of members of the **American Astronomical Society**. Of some 2600 questionnaires, over 1300 were returned. Only two members offered to waive anonymity, and Sturrock noted that the UFO subject was obviously a very sensitive one for most colleagues. Nonetheless, Sturrock found a strong majority favoured continued scientific studies, and over 80% offered to help if they could. Sturrock commented that the AAS members seemed more open to the question than the AIAA members in his previous survey. As in the AIAA survey, about 5% reported puzzling sightings, but skepticism against the **Extraterrestrial Hypothesis (ETH)** ran high. Most thought that UFO reports could ultimately be explained conventionally. Sturrock also found that skepticism and opposition to further study were correlated with lack of knowledge and study: only 29% of those who had spent less than an hour reading about the subject favored further study versus 68% who had spent over 300 hours. ttp://www.ufoevidence.org/documents/doc604.htm

Peter Sturrock Astronomer and Emeritus Professor of Applied Physics
http://www.anomalyarchives.org/public-hall/collections/books/the-ufo-enigma/

Noting that many scientists wished to see UFOs discussed in scientific journals (and at the same time, an almost complete absence of such articles in journals) Sturrock helped establish the **Society for Scientific Exploration** in 1982 to give a scientific forum to subjects that are neglected by the mainstream. Their publication, the *Journal of Scientific Exploration* has been published since 1987.

In 1998, Sturrock organized a scientific panel to review various types of physical evidence associated with UFOs. The panel felt that existing physical evidence that might support the ETH was inconclusive, but also deemed extremely puzzling UFO cases worthy of further scientific study. Sturrock subsequently wrote up the work of the panel in a book *The UFO Enigma: A New Review of the Physical Evidence*.

RAND Corporation Paper (USA, 1968)

The **RAND Corporation** produced a short internal document titled "UFOs: What to Do?" published in November 1968. The paper gives a historical summary of the UFO phenomenon, talks briefly about issues concerning extraterrestrial life and interstellar travel, presents a few case studies and discusses the phenomenological content of a UFO sighting, reviews hypotheses and finally conclude by recommending organizing a central UFO report receiving agency and conducting more research on the phenomenon. Kocher, George (November 1968). "UFOs: What to Do?" RAND Corporation. http://www.theblackvault.com/documents/ufoswhattodo.pdf. Retrieved June 23, 2010.

Project Identification (USA, 1973-1980)

In 1973, a wave of UFO sightings in southeast Missouri prompted Harley D. Rutledge, physics professor at the University of Missouri to conduct an extensive field investigation of the phenomenon. The findings were published in the book *Project Identification: the first scientific field study of UFO phenomena*. Although taking a specific interest in describing unidentified aerial phenomena and not identifying them, the book goes in later chapters to reference the presumed intelligence of the sighted objects.

Rutledge's study results were not published in any peer-reviewed journal or other scientific venue or format. http://en.wikipedia.org/wiki/Ufology

Studies by GEPAN, SERPA & GEIPAN (France, 1977-present)

In 1977, the **French Space Agency CNES** Director General set up a unit to record UFO sighting reports. The unit was initially known as *Groupe d'Etudes des Phénomènes Aérospatiaux Non identifiés* (GEPAN), changed in 1988 to *Service d'expertise de rentrée atmosphérique Phenom* (SERPA) and in 2005 to *Groupe d'études et d'informations sur les phénomènes aérospatiaux non identifiés* (GEIPAN).

GEIPAN has found a mundane explanation for the vast majority of recorded cases, but in 2007, after 30 years of investigation, 1600 cases (circa 28 % of total cases) remain unexplained "despite precise witness accounts and good-quality evidence recovered from the scene." and are categorized as "Type D" In April 2010, GEIPAN statistics state that 23% of all cases are of Type D. Jean-Jacques Velasco, the head of SEPRA from 1983 to 2004, wrote a book in 2004 saying that 13.5% of the 5800 cases studied by SEPRA were without any rational explanation and stated that UFOs are extraterrestrial in origin. CNES (March 26, 2007). "GEIPAN UAP investigation unit opens its files". http://www.cnes.fr/web/CNES-en/5866-geipan-uap-investigation-unit-opens-its-files.php. Retrieved June 23, 2010.

GEIPAN, CNES (April 8, 2010). "GEIPAN statistics" (in French). http://www.cnes-geipan.fr/geipan/statistiques.html. Retrieved June 23, 2010.
La Dépêche du Midi (April 18, 2004). "'Yes, UFOs exist': Position statement by SEPRA head, Jean-Jacques Velasco". http://www.ufoevidence.org/documents/doc1627.htm. Retrieved June 23, 2010.

United Nations (1977-1979)

Because of the lobbying of Eric Gairy, the Prime Minister of Grenada, the United Nations General Assembly addressed the UFO issue in the late 1970s. On July 14, 1978, a panel, with Gordon Cooper, J. Allen Hynek, and Jacques Vallée among its members, held a hearing to inform the **UN Secretary General Kurt Waldheim** about the matter.[84] As a consequence of this meeting, the UN adopted decisions A/DEC/32/424 and A/DEC/33/426, which called for "establishment of an agency or a department of the United Nations for undertaking, coordinating and disseminating the results of research into unidentified flying objects and related phenomena".

Meeting at the United Nations on 14 July 1978 to discuss the need for UN support for UFO studies. Left to right: Gordon, Jacques Vallee, Claude Poher, J. Allen Hynek, Prime Minister of Grenada Sir Eric Gairy, UN Secretary General Kurt Waldheim and (near right) David Saunders. (Description credit: Good)
http://www.ufoevidence.org/documents/doc2006.htm

A/DEC/32/424 UNBISnet- United Nations Bibliographic Information System, Dag Hammarskjöld Library (Retrieved June 23, 2010)
A/DEC/33/426, UNBISnet (Retrieved June 23 , 2010)
UN (December 18, 1978). "Recommendation to Establish UN Agency for UFO Research - UN General Assembly decision 33/426".
http://www.ufoevidence.org/documents/doc902.htm. Retrieved June 23, 2010.

"The General Assembly invites interested Member States to take appropriate steps to coordinate on a national level scientific research and investigation into extraterrestrial life, including unidentified flying objects, and to inform the Secretary-General of the observations, research, and evaluation of such activities."

COMETA Report (France, 1999)

COMETA (*COMité d'ÉTudes Approfondies*, "Committee for in-depth studies") is a private French group, which is mainly composed of high-ranking individuals of the French Ministry of Defence. In 1999 the group dropped a bombshell on Ufology, particularly in America and in the world at large when they published a 90-page report titled "*Les Ovni Et La Defense: A quoi doit-on se Préparer?*" ("UFOs and Defense: What Should We Prepare For?"). It became the first of many unofficial government and non-government disclosures with the release of thousands of UFO reports. The report analyzed various UFO cases and concluded that UFOs are real, complex flying objects and that the extraterrestrial hypothesis (see below) has a high probability of being the correct explanation for the UFO phenomenon. The study recommended that the French government should adjust to the reality of the phenomenon and conduct further research.

Skeptic Claude Maugé criticized COMETA for research incompetency and claimed that the report tried to present itself as an official French document when in fact, it was published by a private group.

COMETA was a high-level French UFO study organization from the late 1990s, composed of high-ranking officers and officials, some having held command posts in the armed forces and aerospace industry. The name "COMETA" in English stands for "Committee for in-depth studies." The study was carried out over several years by an independent group of mostly former "auditors" at the Institute of Advanced Studies for National Defence, or IHEDN, a high-level French military think-tank, and by various other experts.

The group was responsible for the 'COMETA Report' (1999) on UFOs and their possible implications for defense in France. The report concluded that about 5% of the UFO cases they studied were utterly inexplicable and the best hypothesis to explain them was the extraterrestrial hypothesis (ETH). The authors also accused the United States government of engaging in a massive cover-up of the evidence.

The 'COMETA Report' was not solicited by the French government, although before its public release, it was first sent to French President Jacques Chirac and to Prime Minister Lionel Jospin. Immediately afterward, a French weekly news and leisure magazine called VSD referred to it as an "official report", though technically this wasn't the case since COMETA was the work of a private, non-profit, ufological study group. Skeptic Claude Maugé wrote about this: "By letter dated 23 February General Bastien, of the Special Staff of the President of the Republic, wrote: 'To answer your question, this 'report' compiled by members of an association organized under the law of 1901 (ruling most non-commercial private associations in France) did not respond to any official request and does not have any special status'."

The report drew largely on the research of GEPAN / SEPRA, a section of the French space agency CNES, unique in being the only official French government-sponsored organization to investigate UFOs. The birth of GEPAN / SEPRA in the mid-1970s was in large part due to the intense wave of high strangeness UFO sightings in France in 1954.

COMETA Report Membership

The COMETA Report was prefaced by General Bernard Norlain of the Air Force, former Director of IHEDN. The preamble was by **André Lebeau**, former President of the **National Center for Space Studies (Centre National D'études Spatiales), or CNES**. The authors of the report were an association of experts, many of whom were or had been auditors (defense and intelligence analysts) of IHEDN. The group was presided over by General Denis Letty of the Air Force, another former auditor of IHEDN.

Other members included:

- General Bruno Lemoine, Air Force (former auditor of IHEDN)
- Admiral Marc Merlo, (former auditor of IHEDN)
- Michel Algrin, Doctor in Political Sciences, attorney at law (former auditor of IHEDN)
- General Pierre Bescond, engineer for armaments (former auditor of IHEDN)
- Denis Blancher, Chief National Police superintendent at the Ministry of the Interior
- Christian Marchal, chief engineer of the National Corps des Mines and Research Director at the National Office of Aeronautical Research (ONERA)
- General Alain Orszag, Ph.D. in physics, armaments engineer

Outside contributors included **Jean-Jacques Velasco**, head of SEPRA at CNES, **François Louange**, President of Fleximage, specializing in photo analysis, and **General Joseph Domange**, of the Air Force, general delegate of the Association of Auditors at IHEDN.

Although members of COMETA consisted mostly of ex-members of IHEDN, IHEDN made it clear that it had nothing to do with this report. As Claude Maugé wrote in his article: "According to Lieutenant-Colonel Pierre Bayle, head of the Communication Service of IHEDN, 'The Institute for Advanced Studies in National Defence wishes to make it clear that statements made by these individuals only engage them, and them alone, and are in no way a reflection of the thoughts of IHEDN, which has no special element of information on this topic.'"

COMETA Report, part 1 (July 1999). "UFOs and Defense: What Should We Prepare For?" (PDF). ufoevidence.org.
http://www.ufoevidence.org/newsite/files/COMETA_part1.pdf. Retrieved June 23, 2010.

"UFOs and Defense: What Should We Prepare For?" (PDF). ufoevidence.org. pp. 38. http://www.ufoevidence.org/newsite/files/COMETA_part2.pdf. Retrieved June 23, 2010.
"[…] almost certain physical reality of completely unknown flying objects […], apparently operated by intelligent [beings]." A single hypothesis sufficiently takes into account the facts […] . It is the hypothesis of extraterrestrial visitors.
COMETA Report, part 2 (July 1999). http://en.wikipedia.org/wiki/Ufology

Project Condign (UK, 1996-2000)

The British Ministry of Defence (MoD) published in 2006 the "Scientific & Technical Memorandum 55/2/00a" of a four-volume, 460-page report titled "Unidentified Aerial Phenomena in the UK Air Defence Region ", based on a study on behalf of the Defence Intelligence Staff, section DI55, codenamed **Project Condign**. The report is dated December 2000 and was previously classified as top secret. It discusses the British UFO reports (the MoD used the term Unidentified Aerial Phenomena, UAP) received between 1959 and 1997. Wired (5-10-2006). *"It's Official: UFOs Are Just UAPs"*. http://www.wired.com/science/space/news/2006/05/70862. Retrieved June 23, 2010.

The report affirms that UFOs are a real phenomenon, but points out that they present no threat to national defence. The report further states that there is no evidence that UFO sightings are caused by incursions of intelligent origin, or that any UFOs are solid objects which might cause a collision hazard. Although the study admits of being unable to explain all analyzed UFO sightings with certainty, it recommends that section DI55 ceases monitoring UFO reports, as they do not provide information useful for Defence Intelligence. Ministry of Defense (December 2000). *"Unidentified Aerial Phenomena in the UK Air Defence Region: Executive Summary"* (PDF). pp. 4. http://www.mod.uk/NR/rdonlyres/7D2B11E0-EA9F-45EA-8883-A3C00546E752/0/uap_exec_summary_dec00.pdf. Retrieved June 23, 2010. "That UAP exist is indisputable … [they] clearly can exhibit aerodynamic characteristics well beyond those of any known aircraft or missile – either manned or unmanned."
Telegraph (September 20, 2009). "Britain's X-Files: RAF suspected aliens of "tourist" visits to Earth". http://www.telegraph.co.uk/news/newstopics/howaboutthat/ufo/6209684/Britains-X-Files-RAF-suspected-aliens-of-tourist-visits-to-Earth.html. Retrieved June 23, 2010.
"Unidentified Aerial Phenomena in the UK Air Defence Region: Executive Summary" (PDF). pp. 10. http://www.mod.uk/NR/rdonlyres/7D2B11E0-EA9F-45EA-8883-A3C00546E752/0/uap_exec_summary_dec00.pdf.
Ministry of Defense (December 2000). *"Unidentified Aerial Phenomena in the UK Air Defence Region: Executive Summary"* (PDF). pp. 11. http://www.mod.uk/NR/rdonlyres/7D2B11E0-EA9F-45EA-8883-A3C00546E752/0/uap_exec_summary_dec00.pdf. Retrieved June 23, 2010.

Project Hessdalen / Project EMBLA (Norway, 1983–present / Italy 1999-2004)

Since 1981, in an area near the mountain valley Hessdalen in Norway unidentified flying, glowing objects have been commonly observed. This so-called Hessdalen phenomenon has been two times the subject of scientific field studies: the "**Project Hessdalen**" (1983–1985 and 1995 to date) secured technical assistance from the Norwegian Defense Research Establishment, the University of Oslo and the University of Bergen, while the **"Project EMBLA"** (1999–2004) was an Italian team of scientists led by Massimo Teodorani (Istituto di Radioastronomia di Bologna).

Both studies confirmed the presence of the phenomenon and were able to record it with cameras and various technical equipment such as radar, laser, and infrared. The origin and nature of the lights remain unclear. **Teodorani, Massimo (2004). "A Long-Term Scientific Survey of the Hessdalen Phenomenon".** *Journal of Scientific Exploration* 18 (12): 222–224. http://www.scientificexploration.org/journal/jse_18_2_teodorani.pdf. **Erling Strand. "Project Hessdalen 1984 - Final Technical Report".** http://www.hessdalen.org/reports/hpreport84.shtml. Retrieved June 23, 2010. "Beside the light measurements, it can be "measured" by radar and laser. Perhaps the measurements we did on the magnetograph and spectrum analyzer are due to this phenomenon as well. We have to do more measurements with these instruments before we can be sure of that." **Erling Strand. "Project Hessdalen 1984 - Final Technical Report".** http://www.hessdalen.org/reports/hpreport84.shtml. Retrieved June 23, 2010. "We have not found out what this phenomenon is. That could hardly be expected either. But we know that the phenomenon, whatever it is, can be measured." **Teodorani, Massimo (2004). "A Long-Term Scientific Survey of the Hessdalen Phenomenon".** *Journal of Scientific Exploration* 18 (12): 217–251. http://www.scientificexploration.org/journal/jse_18_2_teodorani.pdf. "A self-consistent definitive theory of the phenomenon's nature and origin in all its aspects cannot be constructed yet quantitatively".

The researchers of Project EMBLA speculated about the possibility of atmospheric plasma as the origin of the phenomenon, and also considered the possibility of extraterrestrial intelligence being the cause of the lights.

Massimo Teodorani, Gloria Nobili (2002). **"EMBLA 2002 - An Optical and Ground Survey in Hessdalen"** (PDF). pp. 16 http://www.hessdalen.org/reports/EMBLA_2002_2.pdf. Retrieved June 23, 2010. "Whatever these things are, if some "alien intelligence" is behind the Hessdalen phenomenon, that hypothetical intelligence has shown no interest in searching a direct, continuative and structurally evolved communication with mankind and went on behaving in such a way that the light phenomenon itself appears to be totally elusive."

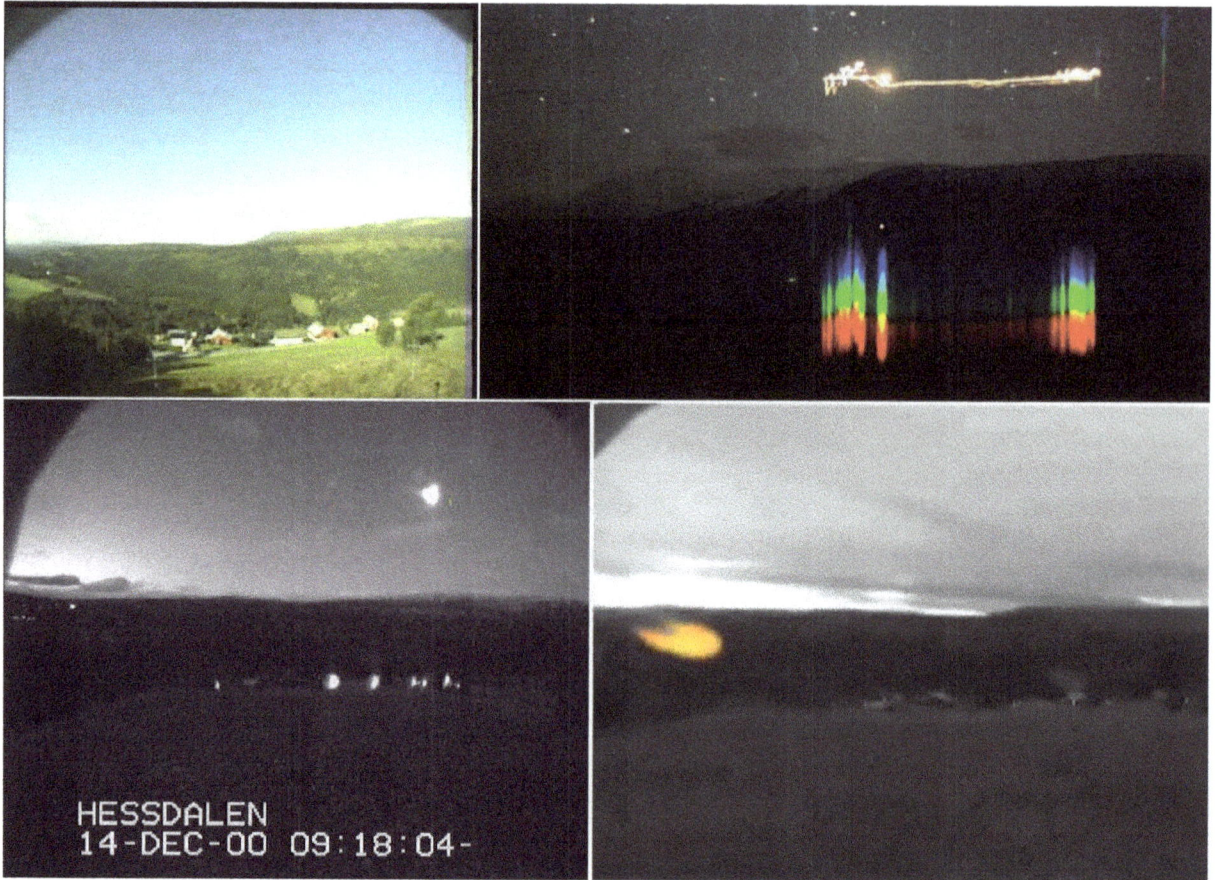

The Hessdalen Valley in Norway (top left), a spectrographic image of an anomalous nocturnal light (top right) in the same valley, (bottom left) light taken at late dusk and (bottom right) a bright orange coloured light photographed at early dawn
(Google Images)

Disclosure Project Press Conference (USA, 2001)

On May 9, 2001, twenty government workers from military and civilian organizations spoke about their experiences regarding UFOs and UFO confidentiality at the **National Press Club** in Washington DC. The press conference was initiated by **Steven M. Greer**, founder of the **Disclosure Project**, which has the goal of disclosing alleged government UFO secrecy.

The purpose of the press conference was to build public pressure through the media to obtain a hearing before the United States Congress on the issue. The event was televised and went out over the internet digital airwaves where an estimated 250,000 people tuned in over the internet to hear the Disclosure witnesses' testimonies which broke the NPC record of 25,000 listeners by a factor of ten! It was also estimated that 2 billion people either saw on television or through related new stories Disclosure Project event later that day.

Dr. Steven Greer, International Director of the Disclosure Project
delivers a speech at the National Press Club

Although major American media outlets reported on the conference, the interest quickly died down again, and no hearing came forth. This is very strange given the extremely high public interest within the US and from other countries around the globe for this type of earth-shattering news. It is highly suspected that intelligence operatives (shills) or agencies (like the CIA or **NSA**) embedded within the major news media order the news media to "kill the story" after one or two days of airtime or public press exposure.

The fact that major media news sources in America had no desire to run with a history-making event did not stop the public interest from waning away, as was evident by the public turnouts to smaller Disclosure Witness forums held around the country including across Canada beginning in Vancouver, BC and ending in Montreal, Que. and over into the U.K. Katelynn Raymer (May 10, 2001). "Group Calls for Disclosure of UFO Info". ABC News. http://abcnews.go.com/Technology/story?id=98572. Retrieved June 23, 2010. Rob Watson (May 10, 2001). "UFO spotters slam 'US cover-up'". BBC News. http://news.bbc.co.uk/2/hi/americas/1322432.stm. Retrieved June 23, 2010. Watson (2001) Sharon Kehnemui (May 10, 2001). "Men in Suits See Aliens as Part of Solution, Not Problem". Fox News. http://www.foxnews.com/story/0,2933,24364,00.html. Retrieved June 23, 2010

Fife Symington Press Conference (USA, 2007)

On November 12, 2007, a press conference, moderated by former Governor of Arizona Fife Symington was held at the National Press Club in Washington DC. Nineteen former pilots and military and civilian officials spoke about their experiences with UFOs and demanded that the U.S. government engage in a new investigation of the phenomenon. Bonnie Malkin (November 14, 2007). "Pilots call for new UFO investigation". Telegraph. http://www.telegraph.co.uk/news/worldnews/1569371/Pilots-call-for-new-UFO-investigation.html. Retrieved June 23 , 2010. "I touched a UFO: ex-air force pilot". The Sydney Morning Herald. November 13, 2007. http://www.smh.com.au/news/unusual-tales/i-touched-a-ufo-exair-forcepilot/2007/11/13/1194766648633.html?page=fullpagecontentSwap1. Retrieved May 5, 2010. Retrieved June 23, 2010.

Citizens Hearing on UFO Disclosure (USA, 2014)

It wasn't until 2013 that another hearing on UFO disclosure was held once again, at the **National Press Club** and organized this time by lobbyist, **Stephen Bassett** of the **Paradigm Research Project**.

From April 29 to May 3, 2013, researchers, activists, and military/agency/political witnesses representing ten countries gave testimony in Washington, DC to six former members of the United States Congress about events and evidence indicating an extraterrestrial presence engaging the human race.

The **Citizen Hearing on Disclosure** was an unprecedented event in terms of size, scope and the involvement of former members of the U. S. Congress. With over 30 hours of testimony from 40 witnesses over five days. The event was the most concentrated body of evidence regarding the extraterrestrial subject ever presented to the press and the general public at one time.

Once again, the major news network failed to do their due diligence by providing this event with full media coverage and attention to what can only be viewed as another history-making news story. Instead, it was marginalized after a few days of press coverage and left to the smaller tabloids and internet blogs run by UFO skeptics with ridiculing accounts about the whole proceedings as just the ranting of an eclectic bunch of "UFO believers" seeking publicity for their cause!

Former Minister of Defence for Canada (yellow arrow), the Honourable Paul T. Hellyer gives testimony before a mock congressional hearing on UFO Disclosure
http://www.jessemarceljr.com/the-citizen-hearing-on-disclosure.html

However, with the overall success of the event, Stephen Bassett is now pushing for an actual congressional hearing instead of the mock hearing heard by former congress members. The testimony of all forty witnesses was recorded on DVD and sent to every congressman and senator in the White House and a swelling citizens email, twitter and fax campaign is being generated to bring public pressure to bear upon congress and the White House to finally deal with the UFO and ETI matter, once and for all without further delays or excuses.

BIBLIOGRAPHY, WEBLIOGRAPHY AND VIDEOGRAPHY

The following list includes all books and major journals, newspapers and web based material, including other reference ebooks and materials such as web links and video links found on the internet in researching this book. Not all chapters use reference material and therefore, these chapter are not listed.

Bibliography listed refers to books marked in RED,
Webliography refers to websites marked in BLUE,
Videography refer to video websites marked in GREEN,
News Service websites are marked in LIGHT BLUE, and
Newspaper websites, magazines and professional papers are marked in PURPLE.

It does not include specific government documents, archival repositories, or various journals or other web based material.

VOLUME TWO (BOOK TWO)
UFO DISCLOSURE AND COVERT PROGRAMS OF DECEPTION

CHAPTER 17
THE GHETTOIZATION OF THE UFO COMMUNITY

"Scientific Study of Unidentified flying Objects" Conducted by the University of Colorado under Research Contract Number F44620-67-C-0035 with the U.S. Air Force. Dr. Edward U. Condon – Project Director

Clark, Jerome. The Emergence of a Phenomenon: UFOs from the Beginning through 1959; The UFO Encyclopedia. Vol. 2. Detroit: Omnigraphics, 1992.

Lorenzen, Coral E. The Great Flying Saucer Hoax: The UFO Facts and Their Interpretation. New York: William Frederick Press, 1962. Rev.:

Flying Saucers: The Startling Evidence of Invasion from Outer Space. New York: New American Library, 1966.

Lorenzen, Coral, and Jim Lorenzen. *Abducted! Confrontations with Beings from Outer Space.* New York: Berkley, 1977.

——. Encounters with UFO Occupants. New York: Berkley, 1976.

——. *UFOs: The Whole Story.* New York: New American Library, 1969.

——. *UFOs Over the Americas.* New York: New American Library, 1968.

The UFO Book – Encyclopedia of the Extraterrestrial" 1998 by Jerome Clark

"The UFO Encyclopedia" by Margaret Sachs, 1980

http://www.cufos.org/org.html

http://en.wikipedia.org/wiki/Mutual_UFO_Network

http://en.wikipedia.org/wiki/Raelism

CHAPTER 18
THE COMMON PEOPLE SPEAK UP, SOME ONE LISTENS, A REPUTATION TO BE MADE, PLEASE TAKE ME SERIOUSLY

http://en.wikipedia.org/wiki/Joe_Firmage

CHAPTER 19
HOAXERS, NAY SAYERS, PROFESSIONAL SKEPTICS AND DEBUNKERS

http://www.hyper.net/ufo/skeptics.html

http://en.wikipedia.org/wiki/UFO

http://en.wikipedia.org/wiki/Donald_Howard_Menzel

(Jerome Clark, The UFO Book: Encyclopedia of the Extraterrestrial, 1998)

http://en.wikipedia.org/wiki/Philip_Klass

http://answers.yahoo.com/question/index?qid=20070704110934AA7xodJ

http://en.wikipedia.org/wiki/James_Oberg

http://www.bibleufo.com/debunking.htm_Patrick Cooke

http://www.stantonfriedman.com/index.php?ptp=ufo_challenge

http://www.hyper.net/ufo/skeptics.html

http://ezinearticles.com/?Anatomy-of-the-UFO-Debunkers---Part-Four&id=4272238

http://en.wikipedia.org/wiki/Critical_thinking

CHAPTER 20
THE UFO INVESTIGATOR - A HEALTHY SKEPTIC

http://theethicalskeptic.com/category/ethical-skepticism/what-is-ethical-skepticism/

http://en.wikipedia.org/wiki/Jacques_Vall%C3%A9e

CHAPTER 21
WHY IS THE SEARCH FOR TRUTH ALWAYS LANDMINED WITH MISINFORMATION, DISINFORMATION, EXCORIATION, DENIALS, LIES, AND OBFUSCATION?

Gleanings, LXX, p.136 and The Power of the Covenant – Part One, pp.18

https://www.youtube.com/watch?v=7AAJ34_NMcI

CHAPTER 23
MISINFORMATION, MISINTERPRETATION AND MISPERCEPTIONS OF UFOS AND ETI

http://en.wikipedia.org/wiki/George_Adamski

Williamson, George Hunt, Other Tongues—Other Flesh. Amherst, Wisconsin: Amherst Press, 1953.

http://en.wikipedia.org/wiki/George_Adamski

Travis Walton, The Walton Experience, 1978;

Travis Walton, The Fire in the Sky, 1996;

Timothy Good, Alien Base – Earth's Encounters with Extraterrestrials 1998;

Jerome Clark, The UFO Book: Encyclopedia of the Extraterrestrial, 1998;

C.D.B. Bryan, Close Encounters of the Fourth Kind 1995

http://en.wikipedia.org/wiki/Truman_Bethurum

http://ufocasebook.com/saucerstory.html

http://en.wikipedia.org/wiki/Daniel_Fry

http://en.wikipedia.org/wiki/George_Van_Tassel

CHAPTER 24
GIVE ME THAT NEW TIME SPACE RELIGION BECAUSE
THE RELIGIONS HERE ON PLANET EARTH
ARE JUST NOT WACKY ENOUGH!

Jerome Clark, The UFO Book: Encyclopedia of the Extraterrestrial, 1998)

Jacques Vallee; Messengers of Deception; 1979

Balch, Robert W. *"Waiting for the ships: disillusionment and revitalization of faith in Bo and Peep's UFO cult."* In James R. Lewis, ed. The Gods have Landed: New Religions from Other Worlds; Albany: SUNY. 1995.

http://en.wikipedia.org/wiki/Heaven%27s_Gate_%28religious_group%29

Balch, Robert W. The Gods have Landed: New Religions from Other Worlds. Albany: SUNY. 1995

http://synchromysticismforum.com/viewtopic.php?f=4&t=2257

Robert W. Balch; 1995; *"Waiting for the ships: disillusionment and revitalization of faith in Bo and Peep's UFO cult."* In James R. Lewis, ed. The Gods have Landed: New Religions from Other Worlds; Albany: SUNY.

http://en.wikipedia.org/wiki/Raelian_movement

http://en.wikipedia.org/wiki/Ra%C3%ABlism.

http://en.wikipedia.org/wiki/File:Raelian_symbols.svg

http://en.wikipedia.org/wiki/The_Urantia_Book

http://en.wikipedia.org/wiki/Unarius

Diana Tumminia and R. George Kirkpatrick *"Unarius: Emergent Aspects of an American Flying Saucer Group"* In James R. Lewis, ed. The Gods have Landed: New Religions from Other Worlds; Albany: SUNY. 1995

John A. Saliba, *"Religious Dimensions of UFO phenomena"*, In James R. Lewis, ed. The Gods have Landed: New Religions from Other Worlds. Albany; SUNY. 1995

http://en.wikipedia.org/wiki/Aetherius_Society

http://en.wikipedia.org/wiki/Scientology

https://www.youtube.com/watch?v=JKORjwXqh70

CHAPTER 25
CATTLE MUTILATIONS, CHUPACABRA AND ABDUCTIONS

Saunders David R, Harkins R Roger (1969), UFO's? Yes! Where the Condon Committee Went Wrong; World Publishing; ASIN B00005X1J1

Christopher O'Brien (1996); The Mysterious Valley; St. Martin's Press; ISBN D-312-95883-8

https://www.youtube.com/watch?v=xfkbv_a8yc4

Good Timothy, (1993), Alien contact: top-secret UFO files revealed, William Morrow & Co., ISBN 0-688-12223-X; Linda Moulton Howe (1980 1989), A Strange Harvest (Documentary)

http://www.crystalinks.com/animal_mutilation.html;

Fawcett Lawrence, Greenwood Barry (1993), UFO Coverup, Fireside, ISBN 0671765558

http://www.daily-times.com/four_corners-news/ci_24351581/new-book-says-government-not-aliens-were-behind

http://newsok.com/medical-tool-of-trade-researcher-helps-develop-portable-laser/article/2469190

CHAPTER - 26
CHUPACABRAS, SASQUATCHES, AND THE MOTHMAN, OH, MY!

http://www.monstropedia.org/index.php?title=Chupacabra

http://www.ornl.gov/sci/techresources/Human_Genome/home.shtml

http://en.wikipedia.org/wiki/Vacanti_mouse;

BBC News, Sci/Tech, Monday, April 13, 1998, "Girl may be first to grow artificial ear"

http://www.sciencechannel.com/video-topics/sci-fi-supernatural/kapow-superhero-science-spider-silk-gene-goats/

http://en.wikipedia.org/wiki/Phosphorescence

BBC News, Thursday, 12 January 2006, "Taiwan breeds green-glowing pigs"

http://www.mnn.com/green-tech/research-innovations/photos/12-bizarre-examples-of-genetic-engineering/glow-in-the-dark

http://www.smithsonianmag.com/science-nature/the-glow-in-the-dark-kitty-77372763/?no-ist

http://library.thinkquest.org/20830/Frameless/Manipulating/Experimentation/Cloning/long doc.htm

Dolly the Sheep, 1996-2003 from the Science Museum, London

"Dolly the sheep clone dies young", BBC News, Friday, 14 February 2003

http://en.wikipedia.org/wiki/Dolly_the_sheep

"Researchers clone monkey by splitting embryo" CNN - January 13, 2000

https://www.youtube.com/watch?v=Vvj78a7CP2k

https://www.youtube.com/watch?v=D5iSm6auHi4
https://www.youtube.com/watch?v=Tz8HxNfIG8Q

CHAPTER 27
PROGRAMMABLE LIFE FORMS (PLFs)

https://www.youtube.com/watch?v=aYk_nPoIu8E
https://www.youtube.com/watch?v=p7JF0aGdaI0

and https://www.youtube.com/watch?v=8Wv-Igj8E7U

https://www.youtube.com/watch?v=vx5F0r84wwg

https://www.youtube.com/watch?v=eUF8NJIYrJY

https://www.youtube.com/watch?v=okacAlE2DQ8

http://en.wikipedia.org/wiki/Alien_abductions

https://www.youtube.com/watch?v=DZtil_ZsrOE

CHAPTER 28
THE HUMAN PRESERVATION PROGRAM HYPOTHESIS –
ETI SAMPLING BIOLOGICAL SPECIMENS INCLUDING HUMAN DNA

Michel, Aime. "The Valensole Affair"; Flying Saucer Review 11, 6.

Confrontations: A Scientist's Search for Alien Contact by Vallee, Jacques; 1990; New York; Ballantine Books

http://nawewtech.angelfire.com/soil.html

http://www.ufohastings.com/

http://en.wikipedia.org/wiki/Alien_abductions

http://en.wikipedia.org/wiki/Perspectives_on_the_abduction_phenomenon

The Controllers: a New Hypothesis of Alien Abductions; Martin Cannon; 1990

CHAPTER 29
FAMOUS ALIEN ABDUCTION CASES

Jerome Clark; The UFO Book: Encyclopedia of the Extraterrestrial; Visible Ink; 1998; ISBN 1-57859-029-9

Jenny Randles and Peter Houghe; The Complete Book of UFOs: An Investigation into Alien Contact and Encounters; Sterling Publishing Co, Inc, 1994; ISBN 0-8069-8132-6

Travis Walton; Fire in the Sky; Marlowe & Company Publishing, 1996; ISBN 1-56924-840-0

Travis Walton; The Walton Experience; 1978; Berkley Publishing Corp; ISBN 425-03675-8

http://en.wikipedia.org/wiki/Travis_Walton

Jacques Vallee; Confrontations: A Scientist's Search for Alien Contact; Random House, Inc., 1990; ISBN 0-345-36501-1

https://www.youtube.com/watch?v=Nwy-A0OtQIA

Ralph Blum and Judy Blum; Beyond Earth, Man's Contact With UFOs; Bantam Books, Inc; 1974

Unidentified Flying Object (the Schirmer Abduction case) Source: James Spears, Vancouver Province (BC, Canada), Mar. 20, 1976
http://www.ufoevidence.org/Cases/CaseSubarticle.asp?ID=661

Jacques Vallee; Dimensions; 1988; Ballantine Books; ISBN 0-345-36002-8

The U.S. Air Force; Dr. Edward U. Condon – Project Director; Scientific study of Unidentified Flying Objects; Bantam books, Inc; 1969

http://www.nicap.org/hicksontape.htm

Ralph Blum and Judy Blum; Beyond Earth, Man's Contact With UFOs; Bantam Books, Inc; 1974

Clark, Jerome, The UFO Book: Encyclopedia of the Extraterrestrial; Visible Ink Press, 1998

http://ufos.about.com/od/aliensalienabduction/p/pascagoula.htm

http://www.nicap.dabsol.co.uk/newwitness.htm

http://www.ufocasebook.com/pascagoulanewwitness.html

"The Allagash Abductions"; Raymond E. Fowler; 1993; Wild Flower Press; ISBN 0-926524-22-4

https://www.youtube.com/watch?v=Da9UKjvdvJk

CHAPTER 30
MIND CONTROL AND ABDUCTIONS

The Controllers: a New Hypothesis of Alien Abductions; Martin Cannon; 1990

UFO CONTACT CENTER INTERNATIONAL GROUP MEETING; 1988; Speaker: Martin Cannon; Subject: UFOs and Mind Control

"The Power of the Covenant, Part One and Two"; National Spiritual Assembly of the Baha'is of Canada; 2000; Baha'I Canada Publications; ISBN 0-88867-101-6

Ralph Blum and Judy Blum; Beyond Earth, Man's Contact With UFOs; Bantam Books, Inc; 1974

The Controllers: a New Hypothesis of Alien Abductions; Martin Cannon; 1990

Kathleen McAuliffe, "The Mind Fields," OMNI Magazine, February 1985.

http://www.wikidoc.org/index.php/Rauni-Leena_Luukanen-Kilde

http://www.examiner.com/article/secretly-forced-brain-implants-a-brief-history

Microchip Implants, Mind control, Cybernetics

http://www.wikidoc.org/index.php/Rauni-Leena_Luukanen-Kilde

http://www.israelect.com/reference/WillieMartin/NEWS-35.htm

http://www.haarp.alaska.edu/haarp/factSheet.html. Retrieved 2009-09-27 https://www.youtube.com/watch?v=VkMhDT3P7mc

Clark, Jerome (1998);The UFO Book: Encyclopedia of the Extraterrestrial; Visible Ink; ISBN 1578590299

Paul Norman, (1996), "The Frederick Valentich Disappearance", Victorian U.F.O. Research Society Inc. (2007-04-27)

CHAPTER 32
DO YOU RETURN YOUR PHD OR DOCTORATE IN ALIEN ABDUCTION RESEARCH WHEN YOUR WORK IS FAULTY?

Bryan, C.D.B.; (1995); Close Encounters of the Fourth Kind: Alien Abduction, UFOs, and the Conference at M.I.T., New York; Alfred A. Knopf; ISBN 0-679-42975-1

http://www.ufoabduction.com/biography.htm. Retrieved 2009-11-14.
http://www.ufoabduction.com/

CHAPTER 33
UFOLOGISTS FROM AROUND THE WORLD

http://www.ufocasebook.com/2009/jakarta.html

http://en.wikipedia.org/wiki/Marcel_Griaule

http://www.ufoevidence.org/researchers/detail93.htm

http://www.paolaharris.com/bio.htm

http://cezary_kwiatk.republika.pl/blania.htm

http://en.wikipedia.org/wiki/Doru_Davidovici

http://en.wikipedia.org/wiki/Juan_Jose_Benitez

http://orbwatch.com

http://ufo-joe.tripod.com/gov/wbsebk

http://ufo-joe.tripod.com/gov/wbsebk.html

www.presidentialufo.com.

http://www.presidentialufo.com/grant-cameron-biography

http://www.shagharbourufo.com/

http://www.paradigmresearchgroup.org/stephenbassett.html

http://www.nicap.org/bios/fowler.htm

http://www.thephoenixlights.net/

http://en.wikipedia.org/wiki/Bill_Moore_(ufologist)

http://en.wikipedia.org/wiki/Daniel_Sheehan_%28attorney%29

http://www.ufophotoarchives.org/

http://en.wikipedia.org/wiki/Leonard_H._Stringfield

http://exopolitics.blogs.com/about.html

http://www.ufoevidence.org/researchers/detail59.htm

http://translate.google.com/translate?hl=en&sl=pt&u=http://www.aforteanosla.com.ar/Colaboraciones/autores/brasil/ademar%2520gevaerd.htm&ei=zMT9S7i2EJKvNqH6sN4H&sa=X&oi=translate&ct=result&resnum=3&ved=0C

http://en.wikipedia.org/wiki/Silvia_Simondini

References:

69. "Skeptic Report/ UFO Bibliography: Salatun, J.". http://www.skepticreport.com/ufo/ufo-s.htm. Retrieved 2010 -15-05.

70. "BETA-UFO - Indonesian largest UFO Community, and http://www.ufocasebook.com/2009/jakarta.html

71. Pkcropcircles and UFOs, Karachi, Pakistan

72. "Ufo-Norway: MUFON USES SCANDINAVIAN CONTACTEE ON ITS ADVISORY BOARD OF CONSULTANTS: RAUNI-LEENA LUUKANEN." http://www.kolumbus.fi/lauri.grohn/yk/skepsis/ufonorja.html. Retrieved 2010-15-05.

73. http://en.wikipedia.org/wiki/Marcel_Griaule. Retrieved 2010-15-05.

74. "UFO EVIDENCE: Official French Gov't UFO study project to resume with new director Yves Sillard". http://www.ufoevidence.org/documents/doc2008.htm. Retrieved 2008-01-01.

75. http://www.ufoevidence.org/researchers/detail93.htm. Retrieved 2010-15-05.

76. "Ufology Library". http://www.tarrdaniel.fw.hu/documents/ufology.html. Retrieved 2009-07-20.

77. "RYUFOR Foundation, Hungarian Center for UFO Studies & Fund". http://www.ryufor.hu/. Retrieved 2009-07-20.

78. "Official site: Enrico Baccarini". http://www.enricobaccarini.com/. Retrieved 2008-01-01.

503

79. "Monsignor Corrado Balducci". **Paradigm Research Group.**
http://www.paradigmresearchgroup.org/Awards/Balducci_Corrado.htm. **Retrieved 2008-01-01.**

80. http://www.paolaharris.com/bio.htm. **Retrieved 2010-15-05.**

81. "ufopsi.com Biography: Roberto Pinotti".
http://www.ufopsi.com/articles/robertopinotti.html. **Retrieved 2008-01-01.**

http://cezary_kwiatk.republika.pl/blania.htm. Retrieved 2010-15-05.

82. http://en.wikipedia.org/wiki/Doru_Davidovici. **Retrieved 2010-15-05.**

http://en.wikipedia.org/wiki/Juan_Jose_Benitez. Retrieved 2010-15-05.

83. "Official Site: Erich von Däniken". http://www.daniken.com/. **Retrieved 2008-01-01.**

84. "Official web site: Timothy Good- UFO Authority".
http://www.timothygood.co.uk/. **Retrieved 2008-02-05.**

85. "Official site: Nick Pope". http://www.nickpope.net/. **Retrieved 2008-02-05.**

86. "UFO EVIDENCE: UFO Researchers & People- Jenny Randles".
http://www.ufoevidence.org/Researchers/Detail40.htm. **Retrieved 2008-02-05.**

87. "(www.bibliotecapleyades.net) Brinsley Le Poer Trench: "Legends and the case for Hollow Earth."". http://www.bibliotecapleyades.net/luna/esp_luna_19a.htm.
Retrieved 2008-02-05.

88. "Official site: Stanton T. Friedman". http://www.v-j-enterprises.com/sfhome.html.
Retrieved 2008-02-06.

89. http://orbwatch.com. **Retrieved 2010-17-05.**

90. "UFOBC.CA: Martin Jasek Bio". http://www.ufobc.ca/yukon/martin.htm.
Retrieved 2008-01-01.

91. "UFO REVIEW #17: Interview with Kimball".
http://alienworldsmag.com/uforeview_issues/UFO%20Review%20Issue%2017.pdf.

92. "Bio: Chris Rutkowski (Ufology Research of Manitoba)".
http://www.ufon.org/uforom/mybio.html. **Retrieved 2007-12-31.**

93. "(www.alienresistance.org) David Sereda Ancient of Days 2004 Speaker Bio and Topic". http://www.alienresistance.org/aod04_david_sereda.htm. **Retrieved 2008-02-10.**

94. http://ufo-joe.tripod.com/gov/wbsebk.html **Retrieved 2010-19- 05.**

95. http://www.presidentialufo.com/grant-cameron-biography **Retrieved 2010-19- 05.**

96. http://www.shagharbourufo.com/ **Retrieved 2010-27- 05**

97. "(www.paradigmresearchgroup.org) Jaime Maussan: 2005 PRG Courage in Journalism".
http://www.paradigmresearchgroup.org/Awards/Maussan_Jaime.htm. **Retrieved 2008-02-05.**

504

98. http://www.paradigmresearchgroup.org/stephenbassett.html Retrieved 2010-19- 05.

99. "Paradigm Research Group, Bassett bio".
http://www.paradigmresearchgroup.org/stephenbassett.html.

100. Clark, Jerome (1998). *The UFO Encyclopedia: The Phenomenon from the Beginning, Volume 2, A-K Detroit: Omnigraphics*. Detroit, MI: Omnigraphics. ISBN 0-7808-0097-4.

101. "Own site (Robert Dean): Beyond Zebra".
http://www.beyondzebra.com/bobdean.shtml. Retrieved 2008-01-01.

102. "Keyhole Publishing, Dolan bio".
http://keyholepublishing.com/richard_dolan-bio.html.

103. http://www.nicap.org/ray.htm

104. http://www.cufos.org/books.html

105. Greer, Steven (1999). "About Steven M. Greer, M.D.". *Extraterrestrial Contact: The Evidence and Implications*. USA: Cataloging-in-Publication (Quality Books, Inc.). pp. 525–526. ISBN 978-0-96-732380-0.
http://www.disclosureproject.org.

106. "Official site: Richard H. Hall". http://www.hallrichard.com/. Retrieved 2008-02-06.

107. "UFO EVIDENCE: UFO Researchers & People- James Harder, Ph.D.".
http://www.ufoevidence.org/Researchers/Detail77.htm. Retrieved 2008-02-06.

108. "Inruders Foundation: Budd Hopkins UFO Abduction Research Foundation". http://www.intrudersfoundation.org/. Retrieved 2008-02-10.

109. "(www.nexusmagazine.com) Bio: Linda Moulton Howe".
http://www.nexusmagazine.com/Conf%202005/Linda%20Moulton%20Howe.html. Retrieved 2008-02-10.

110. "The J. Allen Hynek for UFO Studies". http://www.cufos.org/org.html. Retrieved 2008-02-10.

111. Ronald Story, ed., *The Encyclopedia of Extraterrestrial Encounters*, (New York: New American Library, 2001), s.v. "Morris K. Jessup," pp. 276.

112. Jerome Clark, *The UFO Book: Encyclopedia of the Extraterrestrial*. Visible Ink, 1998. ISBN 1-57859-029-9.

113. http://www.thephoenixlights.net/ Retrieved 2010-05-23

114. "UFORC.com NEWS SERVICE: KLAS-TV Reporter - George Knapp to Give Keynote Banquet Speech at 4th UFO Crash Retrieval Conference".
http://www.uforc.com/news/KLAS-TV_Knapp_UFOCRC_111006.htm. Retrieved 2008-01-01.

115. "SCIFIPEDIA: APRO (the Aerial Phenomena Research Organization)".
http://scifipedia.scifi.com/index.php?title=APRO&printable=yes. Retrieved 2008-02-10.

116. "Dr. Bruce Maccabee Research Website". http://brumac.8k.com/. Retrieved 2008-08-24.

117. Berliner, Don. "Congressional Hearings on UFOs". ufoevidence.org. http://www.ufoevidence.org/documents/doc1981.htm.

118. "The John E Mack Institute (www.johnemackinstitute.org/)". http://www.johnemackinstitute.org/. Retrieved 2008-08-24

119. http://en.wikipedia.org/wiki/Bill_Moore_(ufologist) Retrieved 2010-05-28

120. *The Roswell Encyclopedia*, 2000, Collins Press, ISBN 0380798530

121. "Nick Redfern Official site (www.nickredfern.com/)". http://www.nickredfern.com/. Retrieved 2008-08-24.

122. Diana Palmer Hoyt, "UFOCRITIQUE: UFOs, Social Intelligence and the Condon Committee"; Master's Thesis, Virginia Polytechnic Institute, 2000

123. "MUFON 2007 Symposium, Brad Sparks bio". http://www.mufon.com/documents/2007MUFONSymposium-NewMJ12Revelations.pdf.

124. http://www.ufophotoarchives.org/ Retrieved 2010-05-23.

125. http://en.wikipedia.org/wiki/Leonard_H._Stringfield Retrieved 2010-05-26

126. "The CUFON Inteerview of Michael D. Swords, Ph.D". http://www.cufon.org/cufon/cir/swords.htm. Retrieved 2008-08-24.

127. Five Arguments Against the Extraterrestrial Origin of Unidentified Flying Objects - Jacques Vallée, Ph.D.

128. "Alfred Webre". Jerry Pippin. http://www.jerrypippin.com/UFO_Files_alfred_webre.htm.Retrieved 2008-01-01.

129. http://exopolitics.blogs.com/about.html Retrieved 2010-05-26

130. "The UFO Casebook". http://ufocasebook.com/011209.html/. Retrieved 2009-01-14.

131. "The Daily Grail". http://dailygrail.com/.

132. http://www.ufoevidence.org/researchers/detail59.htm Retrieved 2010-05-23.

133. http://translate.google.com/translate?hl=en&sl=pt&u=http://www.aforteanosla.com.ar/Colaboraciones/autores/brasil/ademar%2520gevaerd.htm&ei=zMT9S7i2EJKyNqH6sN4H&sa=X&oi=translate&ct=result&resnum=3&ved=0C Retrieved 2010-05-23.

134. Livingston, David. "Dr. Carol Rosin". The Space Show. http://www.thespaceshow.com/guest.asp?q=209. Retrieved 2008-01-01.

135. "UFO INFO: "Argentina: Ufologist to Visit Policeman in Disappearance Case." (Fabio Zerpa)". http://www.ufoinfo.com/news/argentinapoliceman3.shtml. Retrieved 2008-02-05

136. http://en.wikipedia.org/wiki/Silvia_Simondini Retrieved 2010-05-26

CHAPTER 34
AN HISTORICAL LOOK AT UFO INVESTIGATIONS BY GOVERNMENTS, MILITARIES AND INTELLIGENCE COMMUNITIES

http://www.huffingtonpost.com/alejandro-rojas/ufo-cia-fbi-investigation_b_1856461.html

http://www.collective-evolution.com/2013/06/29/wikileaks-cables-confirm-existence-of-extraterrestrial-life/

http://marilynkaydennis.wordpress.com/2012/04/25/a-regiment-disappears-inside-a-cloud/

http://beforeitsnews.com/paranormal/2013/02/266-british-soldiers-disappear-in-1915-without-a-trace-2447930.html

http://midimagic.sgc-hosting.com/howifoof.htm

http://en.wikipedia.org/wiki/Foo_fighter

http://www.dailymail.co.uk/sciencetech/article-1299994/Churchill-Eisenhower-agreed-cover-UFO-encounter-WWII.html

http://en.wikipedia.org/wiki/Foo_fighter

http://www.skepdic.com/saucers.html

Menzel, D. H.; Taves, E. H. (1977); The UFO enigma; Garden City (NY, USA); Doubleday

Sagan, Carl and Page, Thornton (1995); UFOs: A Scientific Debate; Barnes & Noble; pp. 310; ISBN 978076070916

McDonald, James. E.; (1968); Statement on Unidentified Flying Objects submitted to the House Committee on Science and Astronautics at July 29, 1968, Symposium on Unidentified Flying Objects; Rayburn Bldg., Washington, D.D.

COMETA Report http://www.ufoevidence.org/topics/Cometa.htm

Politicking and Paradigm Shifting: James E. McDonald and the UFO Case Study http://www.project1947.com/shg/mccarthy/shgintro.html

http://www.bibliotecapleyades.net/ciencia/ufo_briefingdocument/1946.htm

Jerome Clark, The UFO Book: Encyclopedia of the Extraterrestrial, 1998, Visible Ink Press, ISBN 1-57859-029-9

Timothy Good, Above Top Secret, 1988, William Morrow & Co., ISBN 0-688-09202-0

Timothy Good, Need to Know: UFOs, the Military, and Intelligence, 2007, Pegasus Books, ISBN 978-1-933648-38-5 http://en.wikipedia.org/wiki/Green_fireballs

Jerome Clark, The UFO Book: Encyclopedia of the Extraterrestrial, Visible Ink Press, 1998.

Edward J. Ruppelt, The Report on Unidentified Flying Objects, 1956, Chap. 4

Brad Steiger, Project Blue Book, Ballantine Books, 1976 (Contains letter from Dr. J. Allen Hynek of Dr. LaPaz expressing final opinion on green fireballs)

http://en.wikipedia.org/wiki/Unidentified_flying_object

http://www.openminds.tv/remarkable-pearl-ufo-reported-by-famous-chinese-medieval-scientist-shen-kuo-1079/22837

http://www.americanchronicle.com/articles/view/17732

http://en.wikipedia.org/wiki/Unidentified_flying_object

http://en.wikipedia.org/wiki/Australian_ufology#Government

http://en.wikipedia.org/wiki/Ghost_rockets#Greek_government_investigation

Timothy Good, Above Top Secret, 1988, William Morrow & Co., ISBN 0-688-09202-0

Donald Keyhoe, Aliens From Space, 1973, Doubleday & Co., ISBN 0-385-06751-8

Timothy Good, Need to Know: UFOs, the Military, and Intelligence, 2007, Pegasus Books, ISBN 978-1-933648-38-5

http://www.nickpope.net/official_history.htm

Ministry of Defense (December 2000). "Unidentified Aerial Phenomena in the UK Air Defence Region: Executive Summary" (PDF). pp. 4, 10, 11.

 http://www.mod.uk/NR/rdonlyres/7D2B11E0-EA9F-45EA-8883-A3C00546E752/0/uap_exec_summary_dec00.pdf."   Retrieved May 5, 2010.

http://www.ufoevidence.org/documents/doc682.htm

https://www.youtube.com/watch?v=1sjpKvxu0Bk

http://www.inexplicata.blogspot.com/

http://en.wikipedia.org/wiki/Barry_Goldwater#UFOs

http://en.wikipedia.org/wiki/Project_Blue_Book#USAF_current_official_statement_on_UFOs

CHAPTER 35
GOVERNMENT UFO STUDIES IN THE U.S.A.

http://en.wikipedia.org/wiki/Identification_studies_of_UFOs

Blum, Howard, *Out There: The Government's Secret Quest for Extraterrestrials*, Simon and Schuster, 1990

Jerome Clark, *The UFO* Book: Encyclopedia of the Extraterrestrial, Visible Ink, 1998; ISBN 1-57859-029-9

http://ufoinfo.com/ufobooks/projectbluebook.shtml

CLOSE ENCOUNTER WITH DR. J. ALLEN HYNEK By Dennis Stacy; An Interview With The Dean, 1985; Re-Edited for CUFON by Dale Goudie 1991

Edward J. Ruppelt, The Report on Unidentified Flying Objects, Doubleday & Co., 1956

http://en.wikipedia.org/wiki/Project_Sign

Air Intelligence Report AIR 100-203-79: ANALYSIS OF FLYING OBJECT INCIDENTS IN THE U.S. URL accessed February 23, 2007.

McCoy Memo - 1948;

http://www.project1947.com/fig/1948c.htm

http://en.wikipedia.org/wiki/Project_Sign

McCoy Memo - 1948;

Edward J. Ruppelt; The Report on Unidentified Flying Objects; 1956; Doubleday & Company

http://www.project1947.com/bg/ufogov.htm

J. Allen Hynek; The UFO Experience: A Scientific Inquiry; 1972; Henry Regnery Company

Kevin Randle; UFO Casebook; Warner Books; 1989; ISBN 0-446-35715-4

http://en.wikipedia.org/wiki/Estimate_of_the_Situation

Blum, Howard, Out There: The Government's Secret Quest for Extraterrestrials, Simon and Schuster, 1990

Edward J. Ruppelt, The Report on Unidentified Flying Objects

509

Swords, 98

http://en.wikipedia.org/wiki/Project_Grudge

Haines, Gerald K. (April 14, 2007). "CIA's Role in the Study of UFOs, 1947-90".

https://www.cia.gov/library/center-for-the-study-of-intelligence/csi-publications/csi-studies/studies/97unclass/ufo.html. Retrieved June 28, 2010.

http://en.wikipedia.org/wiki/Robertson_Panel

Report of Scientific Advisory Panel on Unidentified Flying Objects Convened by Office of Scientific Intelligence, CIA January 14 - 18, 1953

Clark, 1998

 Dr. Thornton Page's Review of the Condon Report

Jerome Clark, The UFO Book: Encyclopedia of the Extraterrestrial, ISBN 1-57859-029-9

http://en.wikipedia.org/wiki/Robertson_Panel

http://en.wikipedia.org/wiki/Robertson_Panel

Blum, Howard, Out There: The Government's Secret Quest for Extraterrestrials, Simon and Schuster, 1990.

David R. Saunders and R. Roger Harkins, UFOs? Yes! Where the Condon Committee Went Wrong, New York: Signet, 1968, p. 105

Blum, 1990,

http://en.wikipedia.org/wiki/Project_Blue_Book#USAF_current_official_statement_on_UFOs

24. "UNIDENTIFIED FLYING OBJECTS AND AIR FORCE PROJECT BLUE BOOK" URL accessed February 21, 2010
25. Dr. Michael D. Swords; "UFOs, the Military, and the Early Cold War Era", pages 82-121 in "UFOs and Abductions: Challenging the Borders of Knowledge" David M. Jacobs, editor; 2000, University Press of Kansas, ISBN 0-7006-1032-4; p103.
26. see Clark, 1998
27. Dr. Michael D. Swords; "UFOs, the Military, and the Early Cold War Era", p102
28. http://www.nicap.dabsol.co.uk/51-69.htm
29. Blum, Howard, Out There: The Government's Secret Quest for Extraterrestrials, Simon and Schuster, 1990

30. Jacques Vallee, Passport to Magonia: On UFOs, Folklore and Parallel Worlds (1969)

31. Jerome Clark, The UFO Book: Encyclopedia of the Extraterrestrial, 1998; Detroit: Visible Ink Press, ISBN 1-57859-029-9

32. Jerome Clark, The UFO Book: Encyclopedia of the Extraterrestrial, p468

33. Jerome Clark, The UFO Book: Encyclopedia of the Extraterrestrial, p592

34. Ann Druffel; Firestorm: Dr. James E. McDonald's Fight for UFO Science; 2003, Wild Flower Press; ISBN 0-926524-58-5, p63

35. see also Portage County UFO chase page for further details

36. Jerome Clark, The UFO Book: Encyclopedia of the Extraterrestrial, p477

37. Jerome Clark, The UFO Book: Encyclopedia of the Extraterrestrial, 1998; Detroit: Visible Ink Press, p478–479; emphasis as in original

38. http://www.nidsci.org/pdf/quintanilla.pdf | Quintanilla, H. (1974). UFOs, An Air Force Dilemma.

39. Jerome Clark, The UFO Book: Encyclopedia of the Extraterrestrial, p480

40. Proceedings of the Sign Historical Group UFO History Workshop

41. USAF Fact Sheet 95-03

42. Jenny Randles and Peter Houghe; The Complete Book of UFOs: An Investigation into Alien Contact and Encounters; Sterling Publishing Co, Inc, 1994; ISBN 0-8069-8132-6, p179

43. Jenny Randles and Peter Houghe; The Complete Book of UFOs: An Investigation into Alien Contact and Encounters; Sterling Publishing Co, Inc, 1994; ISBN 0-8069-8132-6, p202

44. J. Allen Hynek; The UFO Experience: A Scientific Inquiry; 1972; Henry Regenery Company, p180

45. J. Allen Hynek; The UFO Experience: A Scientific Inquiry; 1972; Henry Regenery Company, p175

46. J. Allen Hynek; The UFO Experience: A Scientific Inquiry; 1972; Henry Regenery Company, p187

47. J. Allen Hynek; The UFO Experience: A Scientific Inquiry; 1972; Henry Regenery Company, p103

- ABC's Peter Jennings reports on Blue Book and history of UFOs in America
- Project Blue Book Archive Online version of USAF Project Blue Book microfilm
- The Project Bluebook Story and Facts UFO.Whipnet.org
- *The Report on Unidentified Flying Objects* – Edward J. Ruppelt The book on-line
- Project Blue Book in FBI's FOIA reading room A summary of the project
- Footnote Project Blue Book Archive View and annotate original Project Blue Book report documents from the National Archives (NARA) microfilm collection.
- Official U.S. Air Force factsheet
- Informal Dept of Defense collection of references on the web

CHAPTER 36
UNIVERSITY OF COLORADO UFO PROJECT (CONDON REPORT)

https://www.youtube.com/watch?v=H8jlNttLUOQ

References

13. ^ Condon Report: Section II, Summary of the Study, Edward Condon
14. ^ *a b* Condon Report: Section I Conclusions and Recommendations, Edward Condon
15. ^ Review of the University of Colorado Report on Unidentified Flying Objects by a Panel of the National Academy of Sciences, 1969
16. ^ "An Analysis of the Condon Report on the Colorado UFO Project" by Peter A. Sturrock, 1987, J. Scientific Exploration, Vol. 1, No. 1, p. 75.
17. ^ *a b c d e f g h i j k l m n o p q r s* Clark, Jerome, *The UFO Book: Encyclopedia of the Extraterrestrial*, Visible Ink, 1998, pp. 593-604, ISBN 1-57859-029-9.
18. ^ *a b c d e f g h i j k l m* Hynek, J. Allen, *The UFO Experience: A Scientific Inquiry*, Henry Regnery Company, 1972, pp. 192-244.
19. ^ The UFO Report: Condon Study Falls Short, Robert L. M. Baker, Jr., Scientific Research, April 14, 1969, p. 41.
20. ^ UFO - An Appraisal of the Problem, 1968 Statement of the American Institute of Aeronautics and Astronautics (AIAA) Subcommittee on UFOs
21. ^ Condon Report, Photographic Case Studies: Cases 46 - 59
22. ^ Science in Default - Twenty-Two Years of Inadequate UFO Investigations - James E. McDonald -1969
23. ^ Dr. Thornton Page's Review of the Condon Report
24. ^ Review of Peter Sturrock's 'The UFO Enigma' - UFO Evidence

- "A Sledgehammer for Nuts"; *Nature*, Volume 221, March 8, 1969; pages 899-900
- C.D.B. Bryan; Close Encounters of the Fourth Kind: Alien Abduction, UFOs and the Conference at M.I.T.; Alfred A. Knopf, 1995; ISBN 0-679-42975-1
- Edward W. Condon, Director, and Daniel S Gillmor, Editor; Final Report of the Scientific Study of Unidentified Flying Objects; Bantam Books, 1968
- David Michael Jacobs; The UFO Controversy In America; Indiana University Press, 1975; ISBN 0-253-19006-1
- David R. Saunders and R. Roger Harkins; UFO's? Yes! Where the Condon Committee Went Wrong; World Publishing, 1969
- Peter A. Sturrock; The UFO Enigma: A New Review of the Physical Evidence; Warner Books, 1999; ISBN 0-446-52565-0

External links and Sources

- [Scientific Study of Unidentified Flying Objects, Dr. Edward U. Condon, Scientific Director, Daniel Gilmor, Editor](#)
- [UFO: An Appraisal of the Problem (from the AIAA)](#)
- [Robert M. L. Baker Jr., The UFO Report: Condon Study Falls Short](#)
- [Flying Saucer Fiasco, by John G. Fuller](#)
- [The Condon UFO Study: A Trick or a Conspiracy? by Philip J. Klass. 1986](#)
- [Science in Default: Twenty-Two Years of Inadequate UFO Investigations, by James E. McDonald](#)
- [UFOs And The Condon Report - A Scientist's Critique (James MacDonald Critique of the Condon Report)](#)
- [THE TRUTH ABOUT THE CONDON REPORT (from NICAP)](#)
- [Dr. Thornton Page's Review of The Condon Report](#)
- [An Analysis of the Condon Report on the Colorado UFO Project.](#) **Dr. Peter. A. Sturrock**
- [USAF-Sponsored Colorado Project for the Scientific Study of UFOs, by Michael D. Swords, Ph.D.](#)
- [Links to Related Online Documents regarding Scientific Study of Unidentified Flying Objects, Dr. Edward U. Condon, Scientific Director, Daniel Gilmor, Editor](#)

CHAPTER 37
GLOBAL CIVILIAN STUDIES OF THE UFO AND ETI PHENOMENON

The UFO Enigma: A New Review of the Physical Evidence; 2000; Edited by Peter A. Sturrock; Aspect Books; ISBN 0446677094

The UFO Enigma: A New Review of the Physical Evidence. 2000. Edited by Peter A. Sturrock. Aspect Books. ISBN 0446677094

ttp://www.ufoevidence.org/documents/doc604.htm

Kocher, George (November 1968). "UFOs: What to Do?" RAND Corporation.

http://www.theblackvault.com/documents/ufoswhattodo.pdf. Retrieved June 23, 2010.

http://en.wikipedia.org/wiki/Ufology

CNES (March 26, 2007). "GEIPAN UAP investigation unit opens its files". http://www.cnes.fr/web/CNES-en/5866-geipan-uap-investigation-unit-opens-its-files.php. Retrieved June 23, 2010.

GEIPAN, CNES (April 8, 2010). "GEIPAN statistics" (in French). http://www.cnes-geipan.fr/geipan/statistiques.html. Retrieved June 23, 2010.

La Dépêche du Midi (April 18, 2004). "'Yes, UFOs exist': Position statement by SEPRA head, Jean-Jacques Velasco". http://www.ufoevidence.org/documents/doc1627.htm. **Retrieved June 23, 2010.**

Photograph of United Nations meeting on UFOs, July 14, 1978 ufoevidence.org (Retrieved June 23, 2010)

A/DEC/32/424 UNBISnet- United Nations Bibliographic Information System, Dag Hammarskjöld Library (Retrieved June 23, 2010)

A/DEC/33/426, UNBISnet (Retrieved June 23 , 2010)

UN (December 18, 1978). "Recommendation to Establish UN Agency for UFO Research - UN General Assembly decision 33/426".

http://www.ufoevidence.org/documents/doc902.htm. Retrieved June 23, 2010.

COMETA Report, part 1 (July 1999). "UFOs and Defense: What Should We Prepare For?" (PDF). ufoevidence.org.

http://www.ufoevidence.org/newsite/files/COMETA_part1.pdf. **Retrieved June 23, 2010.**

"UFOs and Defense: What Should We Prepare For?" (PDF). ufoevidence.org. pp. 38.

http://www.ufoevidence.org/newsite/files/COMETA_part2.pdf. **Retrieved June 23, 2010.**"[…] almost certain physical reality of completely unknown flying objects […], apparently operated by intelligent [beings]." A single hypothesis sufficiently takes into account the facts […] . It is the hypothesis of extraterrestrial visitors.

COMETA Report, part 2 (July 1999). http://en.wikipedia.org/wiki/Ufology

Wired (5-10-2006). "It's Official: UFOs Are Just UAPs".

 http://www.wired.com/science/space/news/2006/05/70862. **Retrieved June 23, 2010.**

Ministry of Defense (December 2000). *"Unidentified Aerial Phenomena in the UK Air Defence Region: Executive Summary"* (PDF). pp. 4.

http://www.mod.uk/NR/rdonlyres/7D2B11E0-EA9F-45EA-8883-A3C00546E752/0/uap_exec_summary_dec00.pdf. Retrieved June 23, 2010. "That UAP exist is indisputable … [they] clearly can exhibit aerodynamic characteristics well beyond those of any known aircraft or missile – either manned or unmanned."

Telegraph (September 20, 2009). "Britain's X-Files: RAF suspected aliens of "tourist" visits to Earth".

http://www.telegraph.co.uk/news/newstopics/howaboutthat/ufo/6209684/Britains-X-Files-RAF-suspected-aliens-of-tourist-visits-to-Earth.html. Retrieved June 23, 2010.

514

"Unidentified Aerial Phenomena in the UK Air Defence Region: Executive Summary" (PDF). pp. 10. http://www.mod.uk/NR/rdonlyres/7D2B11E0-EA9F-45EA-8883-A3C00546E752/0/uap_exec_summary_dec00.pdf.

Ministry of Defense (December 2000). *"Unidentified Aerial Phenomena in the UK Air Defence Region: Executive Summary"* (PDF). pp. 11.

http://www.mod.uk/NR/rdonlyres/7D2B11E0-EA9F-45EA-8883-A3C00546E752/0/uap_exec_summary_dec00.pdf. Retrieved June 23, 2010.

Teodorani, Massimo (2004). "A Long-Term Scientific Survey of the Hessdalen Phenomenon". *Journal of Scientific Exploration* 18 (12): 222–224.

http://www.scientificexploration.org/journal/jse_18_2_teodorani.pdf.

Erling Strand. "Project Hessdalen 1984 - Final Technical Report".

http://www.hessdalen.org/reports/hpreport84.shtml. Retrieved June 23, 2010. "Beside the light measurements, it can be "measured" by radar and laser. Perhaps the measurements we did on the magnetograph and spectrum analyzer are due to this phenomenon as well. We have to do more measurements with these instruments, before we can be sure of that."

Erling Strand. "Project Hessdalen 1984 - Final Technical Report".

http://www.hessdalen.org/reports/hpreport84.shtml. Retrieved June 23, 2010. "We have not found out what this phenomenon is. That could hardly be expected either. But we know that the phenomenon, whatever it is, can be measured."

Teodorani, Massimo (2004). "A Long-Term Scientific Survey of the Hessdalen Phenomenon". Journal of Scientific Exploration 18 (12):

http://www.scientificexploration.org/journal/jse_18_2_teodorani.pdf. . "A self-consistent definitive theory of the phenomenon's nature and origin in all its aspects cannot be constructed yet quantitatively".

Massimo Teodorani, Gloria Nobili (2002). "EMBLA 2002 - An Optical and Ground Survey in Hessdalen" (PDF). pp. 16 http://www.hessdalen.org/reports/EMBLA_2002_2.pdf. Retrieved June 23, 2010. "Whatever these things are, if some "alien intelligence" is behind the Hessdalen phenomenon, that hypothetical intelligence has shown no interest in searching a direct, continuative and structurally evolved communication with mankind and went on behaving in such a way that the light phenomenon itself appears to be totally elusive."

Teodorani, Massimo (2004). "A Long-Term Scientific Survey of the Hessdalen Phenomenon". Journal of Scientific Exploration 18 (12): 217–251.

http://www.scientificexploration.org/journal/jse_18_2_teodorani.pdf.

"EMBLA 2002 - An Optical and Ground Survey in Hessdalen" (PDF). pp. 16 http://www.hessdalen.org/reports/EMBLA_2002_2.pdf. **Retrieved June 23, 2010**

Katelynn Raymer (May 10, 2001). "Group Calls for Disclosure of UFO Info". ABC News. http://abcnews.go.com/Technology/story?id=98572. Retrieved June 23, 2010.

Rob Watson (May 10, 2001). "UFO spotters slam 'US cover-up'". BBC News. http://news.bbc.co.uk/2/hi/americas/1322432.stm. Retrieved June 23, 2010.

Watson (2001)

Sharon Kehnemui (May 10, 2001). "Men in Suits See Aliens as Part of Solution, Not Problem". Fox News. http://www.foxnews.com/story/0,2933,24364,00.html. Retrieved June 23, 2010

Bonnie Malkin (November 14, 2007). "Pilots call for new UFO investigation". Telegraph.

http://www.telegraph.co.uk/news/worldnews/1569371/Pilots-call-for-new-UFO-investigation.html. Retrieved June 23, 2010.

"I touched a UFO: ex-air force pilot". The Sydney Morning Herald. November 13, 2007.

http://www.smh.com.au/news/unusual-tales/i-touched-a-ufo-exair-force-pilot/2007/11/13/1194766648633.html?page=fullpagecontentSwap1. Retrieved May 5, 2010. Retrieved June 23, 2010.

INDEX (VOLUME TWO)

M

M.I.C. black programs · 152
Mac Tonnies · 371
Maccabee, Dr. Bruce · 349
Madame Helena Petrovna Blavatsky · 90
Maj. Donald E. Keyhoe, · 303
Majestic 12" · 303, 424
Majestic Twelve · 425
Major Dewey J. Fournet · 446
Major Donald Keyhole · 276
Major General Charles P. Cabell · 443
Major Hector Quintanilla · 456, 462, 470
Major Robert J. Friend · 455
Mal-Var · 123, *See* Venusian
Mange · 158
Manhattan Project · 172
Mannequin · 180
ManTech Advanced Development Group · 113
Marcel Griaule · 279
Marian apparitions · 62, 373
Mariana UFO film footage · 446
Mark Rodeghier · 23, 194
Mars Global Surveyor · 318
Marshall Applewhite · 109, 113
Marshall Herff Applewhite · 28
Martin Anderson, · 248
Martin cannon · 182
Martin Cannon · 206, 238, 247, 250
Martin Jasek · 298
Marty Kottmeyer · 399
Mary Magdalene · 123
Matthew Williams · 179
Maurice Masse · 186
Mayo Clinic · 164
mental triggers · 230
Menzel · 40, 41, 42, 43, 45, 55
Merlin · 37, *See* King Arthur
Messenger of God · 85
metaphysical phenomena · *See* Bigfoot, Sasquatch, Chupacabra, Mothman, Loch Ness Monster, Earth spirit, spirit guides, and ghosts, *See* Bigfoot
MIC (Military Industrial Complex) · 150
Michael Brant Shermer · 48
Michael Cohen · 377
Michael D. Swords · 369, 448, 470
Michael Hall · 452
Michael Hesseman · 281
Michael Lindemann · 286
Michael Persinger · 44, 203
Michael Swords · 436
microchip mind control · 247
Midwest UFO Network · 16, 23
Mike Cataldo · 233, 234, *See* Natalie Chambers
Mike Rogers · 209, 218
Milab (Military Abductions) · 250
MiLabs - (Military Abductions) · 180
Military abductions (Milab), · 205

Military Industrial Complex · vi, 37, 150, 243, 275, 314
military industrial complex (MIC) · 174
Military Industrial Complex (MIC) · 37, 275
Millennial Generation · 363
Miracle at Fatima · 62, 373
missing time · 100, 182, 201, 219, 230, 238, 243, 244, 247
Mission Control Center · 45
MJ-12 · 42, 43, 304, 356, 424, 425
MKNAOMI, MKACTION, and MKSEARCH · 244, *See* *MKULTRA*
MKUltra · 240, *See* Artichoke, Bluebird, Pandora, MKDelta, MKSearch
MKULTRA · 238, 244, 247, 249
Mohammed · 115, 136
Monsignor Corrado Balducci · 284
Montauk Project · 179
Morris Ketchum Jessup · 340
mosquito-man · 159
Mothman · 78, 157
Mount Blanca Massif · 139
MUFON · 23, 24, 25, 79, 81, 297, 305, 319, 331, 339, 345, 347, 349, 350, 356, 365, 369, 370
MUFON (Mutual UFO Network) · 23, 321
MUFON (Mutual UFO Network), · 23
Mutual UFO Network · 370
Mutual UFO Network (MUFON) · 323
Mutual UFO Network (MUFON). · 379
Mutual UFO Network or MUFON · 16
mystical union · 201

N

Nancy Cartwright, · 138
Natalie Chambers · 233
National Aeronautics and Space Administration (NASA) · 239
National Aviation Reporting Center on Anomalous Phenomena (NARCAP). · 411
National Center for Space Studies (Centre National D'études Spatiales), or CNES · 487
National Center of Space Sciences (CNES) · 388
National Investigations Committee on Aerial Phenomena (NICAP) · 16, 330, 349, 448
National Investigations Committee On Aerial Phenomena (NICAP) · 345
National Press Club · 64, 199, 266, 319, 329, 490, 492
National Reconnaissance Office (NRO) · 239
National Security · 42, 148, 156, 171, 356, 387
National Security Agency (NSA) · 42, 239, 243
National Security Agency's (NSA) · 248
National Security Council · 444, 448
Neuro Linguistic · 175
New Paradigm Academy · 363
New Paradigm Project at Mikhail Gorbachev's State of the World Forum · 363
Nexia Biotechnologies · 161

523

T

U

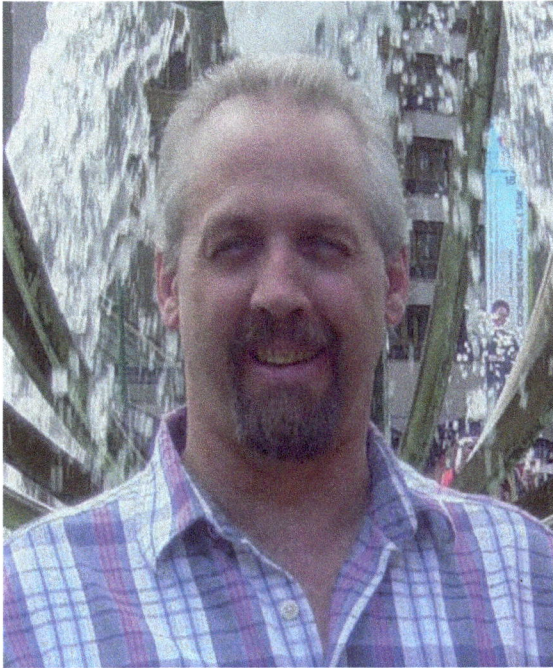

About the Author

This second book about UFOs and ETI is Terry Tibando's foray into the literary world, his background experience and understanding of this phenomenon spans 65 years of personal UFO sightings and ET contact that began at the age of five years. This childhood experience initiated a lifetime of many other-worldly sightings and encounters into a mysterious universe of Unidentified Flying Objects, Extraterrestrial Intelligence and the paranormal. As an experiencer, researcher, and investigator in Ufology he brings a unique and refreshing perspective on this subject based on a world view.

Terry began attending UFO lectures in the mid-sixties, meeting such people as Dr. Edward Edwards, a linguist from the University of Victoria and a fellow member of APRO (Aerial Phenomenon Research Organization and also, Daniel Fry from New Mexico, USA, well known contactee and UFO author.

He attended the University of Victoria in 1968 studying in the arts and sciences and majoring in astronomy and during those university years, there were many UFO sightings near his family's Victoria home leading Terry to theorized there that may be an undersea ET base off the coast of Vancouver Island which may account for the numerous UFO sightings seen over that Island.

He was a former UFO member of APRO and its Canadian sister organization CAPRO during the sixties. His investigative research culminated in the summer of 1996 when he met with Dr. Steven M. Greer during a one week "Ambassadors to the Universe" training seminar. They soon discovered that they shared similar UFO/ETI experiences during their early life.

Terry was a speaker at the Bellingham UFO Group (BUFOG) UFO seminar in 1996, and as a panel speaker along with Peter Davenport from NUFORC and Sharon Filip, alien abduction researcher.

He has been interviewed on such blog talk radio programs as Grimerica with Graham Dunlop in Calgary, Alberta; with Brian Ruhe on Bitchute.com blog radio Vancouver. B.C.; with David Twichell on WHFR.FM Michigan several times; with Jim Harold on The Paranormal Podcast show; Alan B. Smith Paranormal Now show on KGRA New York City; with Commander Cobra of Task Force Gryphon on KGRA Radio, NYC, and twice with Alfred L. Webre on Newsinsideout.com blog radio in Vancouver.

Terry has also appeared on the Discovery Channel during their "Alien Week" series in 1997 which had two ET spacecraft show up during the TV interview; he has appeared on BCTV News, and appeared briefly in Dr. Greer's successful documentary movie "Sirius" and was a major financial contributor to the current documentary "Unacknowledged"!

He was instrumental in coordinating, hosting and emceeing the first Disclosure Project event on UFOs and ETS in Canada as a part of Dr. Greer's Disclosure Witness Tour held at Simon Fraser University in Vancouver on September 9, 2001, which included guest speakers Dr. Steven Greer, Dr. Carol Rosin and Dr, Alfred Webre.

 For the last 25 years, Terry has been the field coordinator of CSETI Vancouver leading teams of people on field expeditions to successfully establish contact and communications with extraterrestrial intelligences visiting the Earth.

www.ingramcontent.com/pod-product-compliance
Lightning Source LLC
Chambersburg PA
CBHW080134240326

41458CB00128B/6459